APPLICATIONS OF CATEGORY THEORY TO FUZZY SUBSETS

THEORY AND DECISION LIBRARY

General Editors: W. Leinfellner (*Vienna*) and G. Eberlein (*Munich*)

Series A: Philosophy and Methodology of the Social Sciences

Series B: Mathematical and Statistical Methods

Series C: Game Theory, Mathematical Programming and Operations Research

Series D: System Theory, Knowledge Engineering and Problem Solving

SERIES B: MATHEMATICAL AND STATISTICAL METHODS

VOLUME 14

Scope: The series focuses on the application of methods and ideas of logic, mathematics and statistics to the social sciences. In particular, formal treatment of social phenomena, the analysis of decision making, information theory and problems of inference will be central themes of this part of the library. Besides theoretical results, empirical investigations and the testing of theoretical models of real world problems will be subjects of interest. In addition to emphasizing interdisciplinary communication, the series will seek to support the rapid dissemination of recent results.

The titles published in this series are listed at the end of this volume.

APPLICATIONS OF CATEGORY THEORY TO FUZZY SUBSETS

edited by

STEPHEN ERNEST RODABAUGH

Department of Mathematical and Computer Sciences,
Youngstown State University, Youngstown, Ohio, U.S.A.

ERICH PETER KLEMENT

Institut für Mathematik,
Johannes Kepler Universität, Linz, Austria

and

ULRICH HÖHLE

Fachbereich Mathematik,
Bergische Universität, Wuppertal, Germany

KLUWER ACADEMIC PUBLISHERS

DORDRECHT / BOSTON / LONDON

Library of Congress Cataloging-in-Publication Data

Applications of category theory to fuzzy subsets / edited by Stephen Ernest Rodabaugh, Erich Peter Klement, Ulrich Höhle.
p. cm. -- (Theory and decision library. Series B, Mathematical and statistical methods ; v. 14)
The material of this book has close relationship to the Eleventh International Seminar on Fuzzy Set Theory held in Linz, Austria, 1989.
Includes index.
ISBN 0-7923-1511-1
1. Fuzzy sets--Congresses. 2. Categories (Mathematics)--Congresses. I. Rodabaugh, Stephen Ernest. II. Klement, E. P. (Erich Peter) III. Höhle, Ulrich. IV. International Seminar on Fuzzy Set Theory (11th : 1989 : Linz, Austria) V. Series.
QA248.A76 1991
511.3'22--dc20 91-34781

ISBN 0-7923-1511-1

Published by Kluwer Academic Publishers,
P.O. Box 17, 3300 AA Dordrecht, The Netherlands.

Kluwer Academic Publishers incorporates
the publishing programmes of
D. Reidel, Martinus Nijhoff, Dr W. Junk and MTP Press.

Sold and distributed in the U.S.A. and Canada
by Kluwer Academic Publishers,
101 Philip Drive, Norwell, MA 02061, U.S.A.

In all other countries, sold and distributed
by Kluwer Academic Publishers Group,
P.O. Box 322, 3300 AH Dordrecht, The Netherlands.

Printed on acid-free paper

Printed in the Netherlands

TABLE OF CONTENTS

AUTHORS AND EDITORS

J. COULON
Institut des Sciences de la Matière
Institut de Mathématiques et Informatique
Université Lyon 1, 43 Bd. du 11 Novembere 1918
F-69622 Villeurbanne Cedex, France

J.-L COULON
Institut des Sciences de la Matière
Institut de Mathématiques et Informatique
Université Lyon 1, 43 Bd. du 11 Novembere 1918
F-69622 Villeurbanne Cedex, France

P. EKLUND
Department of Computer Science
Åbo Akademi University
Lemminkäinengatan 14-18
SF-20520 Åbo, Finland

W. GÄHLER
Akademie der Wissenschaften der DDR
Karl-Weierstraß-Institut für Mathematik
Mohrenstraße 39
D-1086 Berlin, Germany

G. GERLA
Dipartimento di Mat. ed Appl.
Renato Caccioppoli
Via Mezzocannone 8
I-80134 Napoli, Italy

H. HERRLICH
Universität Bremen
Fachbereich Mathematik und Informatik
Postfach 33 04 40
D-2800 Bremen, Germany

U. HÖHLE
Bergische Universität
Gesamthochschule Wuppertal
Fachbereich Mathematik 7
Gaußstraße 20, Postfach 100127
D-5600 Wuppertal 1, Germany

F. KLAWONN
Institut für Betriebssysteme und Rechnerverbund
Technische Universität Braunschweig
D-3300 Braunschweig, Germany

E. P. KLEMENT
Institut für Matematik
Johannes Kepler Universität
A-4040 Linz, Austria

W. KOTZÉ
Department of Mathematics (Pure and Applied)
Rhodes University
6140 Grahamstown, South Africa

T. KUBIAK
Department of Mathematics (Pure and Applied)
Rhodes University
6140 Grahamstown, South Africa

and

Institute of Mathematics
Adam Mickiewicz University
Matejki 48/49
60-769 Poznan, Poland

E. LOWEN
Department Wiskunde
Vrije Universiteit Brussel
Pleinlaan 2
B–1050 Brussels, Belgium

R. LOWEN
Dienst Wiskundige Analyse
Universiteit Antwerpen
Groenenborgerlaan 171
B-2020 Antwerp, Belgium

W. A. MACCAULL
Department of Mathematics and Computing Sciences
St. Francis Xavier University
Antigonish, Nova Scotia B2G-1C0, Canada

S. E. RODABAUGH
Department of Mathematical and Computer Sciences
Youngstown State University
Youngstown, Ohio 44555-3302, USA

L. N. STOUT
Department of Mathematics
Illinois Wesleyan University
Bloomington, Illinois 61702-2900, USA

M. W. WARNER
Department of Mathematics
City University
Northampton Square
London EC1 OHB, United Kingdom

PREFACE

This book has a fundamental relationship to the *International Seminar on Fuzzy Set Theory* held each September in Linz, Austria. First, this volume is an extended account of the eleventh Seminar of 1989. Second, and more importantly, it is the culmination of the tradition of the preceding ten Seminars.

The purpose of the Linz Seminar, since its inception, was and is to foster the development of the mathematical aspects of fuzzy sets. In the earlier years, this was accomplished by bringing together for a week small groups of mathematicians in various fields in an intimate, focused environment which promoted much informal, critical discussion in addition to formal presentations. Beginning with the tenth Seminar, the intimate setting was retained, but each Seminar narrowed in theme; and participation was broadened to include both younger scholars within, and established mathematicians outside, the mathematical mainstream of fuzzy sets theory.

Most of the material of this book was developed over the years in close association with the Seminar or influenced by what transpired at Linz. For much of the content, it played a crucial role in either stimulating this material or in providing feedback and the necessary screening of ideas. Thus we may fairly say that the book, and the eleventh Seminar to which it is directly related, are in many respects a culmination of the previous Seminars.

A key part of the distinctive atmosphere of the Seminar is due to its location, the *Bildungszentrum St. Magdalena*, and its most helpful staff. Further acknowledgements go to the Austrian *Bundesministerium für Wissenschaft und Forschung* in Vienna and the *Linzer Hochschulfonds* for their generous financial support for the Seminar over the years, and to the *Johannes Kepler Universität* (Linz) for continued logistical support.

Special commendation goes to the authors of this book, who after participating in the eleventh Seminar, agreed to rewrite, polish, and in many cases enlarge their presentations in response to the discussions and peer review of that Seminar.

We wish to express our deepest appreciation to Mrs. Karen DeMatteo for her indispensable, tireless, and professional efforts in typesetting the book in TEX. Finally, we acknowledge Kluwer Academic Publishers for their cooperation in bringing this project to completion, particularly to Ms. Marie M. Stratta for her capable assistance and patience, and to H. J. Skala for his supporting the inclusion of this book in the *Theory and Decision Library (Series B: Mathematical and Statistical Methods).*

The Editors
August 1991

INTRODUCTION

Since the introductory paper of L. A. Zadeh in 1965, significant strides have been made in the mathematics of fuzzy subsets. Much of this mathematical development has been increasingly categorical in nature, particularly in the areas of logic and topology, as well as in the overlap of these two areas.

A major stimulus for this categorical development has been the need to clarify the rather implicit definition of fuzzy sets originally given in [Zadeh 1965]. He states

> * "a **fuzzy set** (**class**) A in X [some given set] is characterized by a **membership** (**characteristic**) **function** $f_A(x)$ which associates with each point x in X a real number in the interval $[0,1]$, with the value of $f_A(x)$ at x representing the 'grade of membership' of x in A",

and concludes that

> * "essentially, fuzzy sets in X constitute a distributive lattice with a 0 and a 1".

In the first statement, the fuzzy set A is an undefined object **classified** by a lattice-valued map f_A, and in the second statement, the fuzzy set A is a **member** of a certain lattice. It is our view that these two statements represent fundamentally distinct concepts requiring distinct mathematical formulations and explanations — this requirement to delineate the mathematical meanings of these distinct concepts leads naturally to the use of categorical tools.

Categorically speaking, Zadeh's first statement, by not defining explicitly the fuzzy set A, forces the question: viewing the membership functions as generalized characteristic functions, in which sense are membership functions related to a certain kind of subobjects, or in a stricter sense, what kind of subobjects can be classified by membership functions? Answering this question leads to topos-like or model-theoretic approaches to fuzzy set theory, which often result in definitions of different categories for lattice-valued sets.

Categorically speaking, Zadeh's second statement, by viewing all fuzzy sets in X as a lattice $[0,1]^X$ replacing the usual powerset 2^X, has implicitly defined a new object function in the category **SET**, namely $X \longmapsto [0,1]^X$. This motivates yet other object functions in **SET** (or in even in other **ground categories** such as **SET** $\times$ **LOC**): if L is some sort of lattice, we can have $X \longmapsto L^X$, $X \longmapsto L^{L^X}$, etc. The choice of such an object function $F : |\mathbf{SET}| \longrightarrow |\mathbf{SET}|$ can have several consequences:

> * It could induce an algebraic theory in **SET** modifying the powerset theory, the double powerset theory, etc [cf. Manes 1976].

* It induces new notions of mathematical structures (e.g. new concepts of topological or measurable spaces): with each $X \in |\mathbf{SET}|$ may be associated some subset of $F(X)$.

* It could induce a natural transformation which lifts the usual adjoint operators between ordinary powersets of sets to adjoint operators between the images of these sets under F. This would provide morphisms (and a categorical framework) for the spaces included in the preceding statement, as well as a weakening of the adjunctions between **SET** and the category **CSLAT** (of complete join-semilattices) determined by the usual object function $X \longmapsto 2^X$.

In addition to serving as an approximate record of the *Eleventh International Seminar on Fuzzy Set Theory* (11-15 September 1989, Linz, Austria), this book demonstrates the influence of the issues developed in the preceding paragraphs on much of the current research in fuzzy sets.

* Part I, comprising the work of J. Coulon, J. L. Coulon, U. Höhle, and L. N. Stout, deals with variations of topos theory, non-unique classification of a certain type of subobjects, and local existence of entities in a monoidal setting.

* Part II, comprising the work of P. Eklund, W. Gähler, H. Herrlich, E. Lowen, R. Lowen, and S. E. Rodabaugh, uses categorical tools to study the relationship of fuzzy convergence theories to set functors and quasi-topoi, topological constructs applicable to certain categories of fuzzy topological spaces, the categorical significance of changing the underlying lattice of membership values, and the fuzzy topological aspects of certain types of lattices and functor categories.

* Part III, comprising the work of G. Gerla, F. Klawonn, W. Kotzé, T. Kubiak, W. A. MacCaull, and M. W. Warner, addresses the relationship between points and fuzzy spaces, model-theoretic approaches to fuzzy paths and fuzzy relational databases, the interplay between fuzzy sobriety and various Urysohn Lemmas, and the role Helly-like spaces play in the structure of the fuzzy unit intervals.

To summarize briefly: Part I presents the first comprehensive effort to mathematically clarify Zadeh's first statement, Part II develops the categorical tools to further explain Zadeh's second statement, and Part III studies questions related to Parts I and II.

We point out that the longer chapters of this book are intended as self-contained tutorials taking the reader to the frontier, while the shorter papers develop specific aspects of a topic. The book concludes with several appendices: transcripts of two informal Round Tables of the Linz Seminar, a shared and comprehensive bibliography, a common index of categories, and a common subject index.

We believe the material of this book is strategically located at the juncture of many mathematical disciplines: category theory, logic, topology, lattice theory, and theoretical computer science. It is therefore our hope that this book is useful for all those interested in mathematical foundations and their relations to the non-probabilistic treatment of uncertainty.

The Editors
August 1991

Part I:

Topos-like and Model-Theoretic Approaches

Chapter 1:

Classification of Extremal Subobjects of Algebras over **SM-SET**

S. E. Rodabaugh et al. (eds.), Applications of Category Theory to Fuzzy Subsets, 9–31.

CLASSIFICATION OF EXTREMAL SUBOBJECTS OF ALGEBRAS OVER SM-SET

J. Coulon
J.-L. Coulon
U. Höhle

§0. Introduction

The category **SM-SET** of strong M-valued sets can be motivated by a construction of a mathematical model satisfying Poincaré's Paradox (cf. [Höhle 1988b, 1989], [Menger 1966], [Poincaré 1902, 1904]). In this chapter we use **SM-SET** as a base category for two different algebraic theories. The first one means the addition of a restriction map on objects of **SM-SET** and leads to the category **RSM-SET**, while the second one is a generalization of the sheafification and generates the category **CSM-SET**. It is easy to see that **CSM-SET** is a full subcategory of **RSM-SET**.

The main purpose of this chapter is to investigate in **RSM-SET** or **CSM-SET** the fundamental problem of classifying extremal subobjects by characteristic morphisms, or more generally the problem of constructing strong extensions of extremal partial morphisms. The crucial step in this context lies in an appropriate generalization of the $\sim$-construction (cf. [Goldblatt 1979], [Kock and Wraith 1970], [Pénon 1977]).

In the first part of this chapter we demonstrate that a strong extension of extremal partial morphisms exists always, but unfortunately this extension is not unique. In **RSM-SET** the uniqueness of strong extensions depends essentially on the algebraic structure of the underlying monoid M: we show that every extremal partial morphism in **RSM-SET** has a unique strong extension if and only if M is determined by a complete Heyting algebra (i.e. the semigroup operation in M coincides with the lattice-theoretical meet operation). This situation cannot be improved, even if we replace **RSM-SET** by **CSM-SET**.

By the way, we deduce from the $\sim$-construction the pushout-theorem for **RSM-SET** (resp. **CSM-SET**).

In order to give a categorical understanding of the non-unique classification of extremal subobjects we introduce in the second part of this paper the weak subobject classifier axiom $(W\Omega)$. As its name indicates the $(W\Omega)$-axiom is a weakened version of the subobject classifier axiom due to [Lawvere 1970], but it is basically incomparable with the strong subobject classifier axiom which is satisfied by quasi-toposes (cf. [Pénon 1977], [Wyler 1976]). We show that various categories different from elementary toposes do satisfy the $(W\Omega)$-axiom — e.g. **CSM-SET** in the case of complete MV-algebras (cf. [Belluce 1986, 1989], [Chang 1958]) or collapsed toposes recently introduced by [Johnstone 1988].

PART I. STRONG M-VALUED SETS

§1. The Base Category SM-SET

Let $M = (L, \leq, *)$ be a **strong monoid** — i.e. M is an integral, commutative, completely lattice-ordered monoid with zero satisfying the additional axiom

$$\text{(S)} \quad \text{If } e * e = e \text{ and } e \leq \bigvee A, \text{ then } e \leq \bigvee\{\lambda * \lambda,\ \lambda \in A\}.$$

Standard examples of strong monoids are complete Heyting algebras ($* = \wedge$), Noetherian l-monoids, complete MV-algebras and the real unit interval equipped with left-continuous t-norms (cf. [Schweizer and Sklar 1983]).

A **strong M-valued set** is a pair (X, E) where X is a set and $E : X \times X \longrightarrow L$ is a map satisfying the following axioms:

$$\begin{array}{lll} \text{(E1)} & E(x,x) * E(x,y) * E(y,y) = E(x,y) & \textbf{(Strong Strictness)} \\ \text{(E2)} & E(x,y) = E(y,x) & \textbf{(Symmetry)} \\ \text{(E3)} & E(x,y) * E(y,z) \leq E(x,z) & \textbf{(Strong Transitivity)} \end{array}$$

We interpret $E(x,x)$ as the extent to which x exists and $E(x,y)$ as the degree to which x and y coincide. In particular X is also called the **support set** of (X, E).

A strong M-valued set (X, E) is said to be **separated** iff (X, E) satisfies the axiom

$$\begin{array}{lll} \text{(E4)} & E(x,x) \vee E(y,y) \leq E(x,y) & \textbf{(Separation)} \\ & \text{implies } x = y & \end{array}$$

It is interesting to note that the underlying lattice itself carries an intrinsic structure of a separated, strong M-valued set $(L, \leftrightarrow)$ where the bi-omplication $\leftrightarrow$ w.r.t. $*$ is determined by $\alpha \leftrightarrow \beta = \bigvee\{\lambda \in L \mid \alpha^*\lambda \leq \delta\beta,\ \beta^*\lambda \leq \alpha\}$.

The **data** of the category **SM-SET** are the following ones: **objects** are separated, strong M-valued sets, and **morphisms** are structure preserving maps — i.e. $\varphi : (X, E) \longrightarrow (Y, F)$ is a map from X to Y satisfying the following axioms

$$\begin{array}{lll} \text{(M1)} & F(\varphi(x), \varphi(x)) \leq E(x,x) & \textbf{(Strictness)} \\ \text{(M2)} & E(x_1, x_2) \leq F(\varphi(x_1), \varphi(x_2)) & \textbf{(Preservation of Equality)} \end{array}$$

The **composition** in **SM-SET** is the usual composition of maps and the **identity** of (X, E) is the identical map of X.

The empty set provided with the empty relation is the initital object 0 of **SM-SET**. The terminal object $1 = (I, E)$ of **SM-SET** is given as follows: I is the set of all idempotent elements $e \in L$ w.r.t. $*$ and E is the restriction of $*$ to $I \times I$. Further, **SM-SET** is finitely complete and cocomplete. In particular the product and coproduct of two strong, M-valued sets (X, E) and (Y, F) are determined by $(X, E) \times (Y, F) = (Z, E \times F)$, where

$$Z = \{(x, y) \in X \times Y \mid E(x, x) = F(y, y)\},$$

$$(E \times F)((x, y), (\bar{x}, \bar{y})) = E(x, \bar{x}) \wedge F(y, \bar{y})$$

$(X, E) \perp (Y, F) = (X \dot{\cup} Y, E \perp F)$, where $X \dot{\cup} Y$ is the disjoint union of X and Y, and $E \perp F$ is the unique extension of E and F to $X \dot{\cup} Y$ satisfying the condition $(E \perp F)(z_1, z_2) = 0$ whenever $(z_1, z_2) \notin (X \times X) \cup (Y \times Y)$.

Equalizers in **SM-SET** are constructed at the level of **SET**. In order to describe coequlizers in **SM-SET** we need the concept of M-valued strong equivalence relations on strong M-valued sets. A map $R : X \times Y \longrightarrow L$ is called a **strong M-valued equivalence relation** on (X, E) iff R is provided with the following properties (cf. [Höhle 1988a]):

(R01)	$E(x, x) * R(x, y) * E(y, y) = R(x, y)$	**(Strong Strictness)**
(R02)	$E(x, \bar{x}) * R(x, y) * E(y, \bar{y}) \leq R(\bar{x}, \bar{y})$	**(Strong Extensionality)**
(R1)	$E(x, x) \leq R(x, x)$	**(Reflexivity)**
(R2)	$R(x, y) = R(y, x)$	**(Symmetry)**
(R3)	$R(x, y) * R(y, z) \leq R(x, z)$	**(Strong Transitivity)**

PROPOSITION 1.1. **SM-SET** has coequalizers.

PROOF. Let (φ_1, φ_2) be a pair of **SM-SET**-morphisms from (X, E) to (Y, F). Then we define a map $S : Y \times Y \longrightarrow L$ by

$$S(y_1, y_2) = \bigvee\{F(\varphi_1(x), y_1) * F(\varphi_2(x), y_2) \mid x \in X\} \tag{1.1}$$

Since S is in general not reflexive, neither symmetric nor transitive, we choose the "smallest" strong M-valued equivalence relation R on (Y, F) "containing" S — i.e. if $\mathcal{E}$ is the set of all strong M-valued equivalence relations on (Y, F), then R is given by

$$R(y, z) = F(y, y) * F(z, z) * (\bigwedge\{W(y, z) \mid W \in \mathcal{E},\ S(u, v) \leq W(u, v),\ \forall\ u, v \in Y\})$$

Let Z be the set of all (ordinary) equivalence classes in Y w.r.t. the ordinary equivalence relation $y \approx z \Leftrightarrow F(y, y) = F(z, z) = R(y, z)$. Then $G : Z \times Z \longrightarrow L$ defined by $G([y_1], [y_2]) = R(y_1, y_2)$ is a separated, strong M-valued equality on Z, and the quotient map $\eta : Y \longrightarrow Z$ is a **SM-SET**-morphism from (Y, F) to (Z, G). Referring to (1.1) the relation $\eta \cdot \varphi_1 = \eta \cdot \varphi_2$ follows from the construction of $(\eta, (Z, G))$.

In order to verify the universal property of the coequalizer diagram let us consider a further strong M-valued set (U, H) and a **SM-SET**-morphism $\bar{\eta} : (Y, F) \longrightarrow (U, H)$ equipped with $\bar{\eta} \cdot \varphi_1 = \bar{\eta} \cdot \varphi_2$. From (1.1) and the construction of R we infer:

$$R(y_1, y_2) \leq H(\bar{\eta}(y_1),\ \bar{\eta}(y_2)) \text{ for all } y_1,\ y_2 \in Y \tag{1.2}$$

Since Y is separated, there exists a unique map $\Psi : Z \to U$ s.t. $\bar{\eta}(y) = \Psi(\eta(y))$ for all $y \in Y$. Using again (1.2) we see that Ψ is a **SM-SET**-morphism from (Z, G) to (U, H). □

In the following considerations we investigate the problem: in which sense **SM-SET** is an (E, M)-category.

An isomorphism $\varphi : (X, E) \longrightarrow (Y, F)$ in the sense of **SM-SET** is a surjective map $\varphi : X \to Y$ satisfying the following condition

$$\text{(EM)} \qquad E(x_1, x_2) = F(\varphi(x_1), \varphi(x_2)) \text{ for all } x_1, x_2 \in X.$$

A simple characterization of epimorphisms in **SM-SET** is given in

LEMMA 1.2. Let $\varphi_0 : (X, E) \longrightarrow (Y, F)$ be a **SM-SET**-morphism. Then the following assertions are equivalent:

(i) φ_0 is epic,
(ii) $\bigvee\{F(\varphi_0(x),\ Y) \mid x \in X\} = F(y, y)$ for all $y \in Y$.

PROOF. ((i) $\Rightarrow$ (ii)). Let us consider the following strong M-valued set $\Sigma(L) = (S(L), [\![\ ,\ ]\!])$, where $S(L) = \{(e, \lambda) \in L^2 \mid e * e = e,\ e * \lambda = \lambda\}$, $[\![(e, \alpha), (d, \beta)]\!] = e * d * \delta(\alpha \leftrightarrow \beta)$. Further we introduce two morphisms φ_1 and φ_2 from (Y, F) to $\Sigma(L)$ by

$$\varphi_1(y) = (F(y, y), F(y, y))$$
$$\varphi_2(y) = (F(y, y), \bigvee\{F(\varphi_0(x), y) \mid x \in X\})$$

Then (i) implies obviously (ii).

((ii) $\Rightarrow$ (i)). If Ψ_1 and Ψ_2 are two morphisms from (Y, F) to (Z, G) with $\Psi_1 \cdot \varphi_0 = \Psi_2 \cdot \varphi_0$, then we derive from assertion (ii) and axiom (S) the following chain of inequalities

$$\begin{aligned} f(y, y) &= \bigvee\{F(y, \varphi_0(x) * F(y, \varphi_0(x) \mid x \in X\} \\ &\leq \bigvee\{G(\Psi_1(y), \Psi_1(\varphi_0(x))) * G(\Psi_2(\varphi_0(x), \Psi_2(y)) \mid x \in X\} \\ &\leq \bigvee G(\Psi_1(y), \Psi_2(y)) \end{aligned}$$

Hence the assertion (i) follows from the separation axiom (E4). □

PROPOSITION 1.3. Let $\varphi : (X, E) \longrightarrow (Y, F)$ be a **SM-SET**-morphism. Then the following assertions are equivalent.

(i) φ is an extremal monomorphism.
(ii) φ satisfies the subsequent properties:
  1° (EM) $E(x_1, x_2) = F(\varphi(x_1), \varphi(x_2))$,
  2° $\qquad \varphi(X) = \{y \in Y \mid F(y, y) = \bigvee\{F(\varphi(x),\ y), x \in X\}\}$.

PROOF. The implication (i) $\Rightarrow$ (ii) follows immediately from Lemma 1.2. In order to verify (ii) $\Rightarrow$ (i) we observe first that because of (EM) the morphism φ is monic. Further let us consider the decomposition of φ,

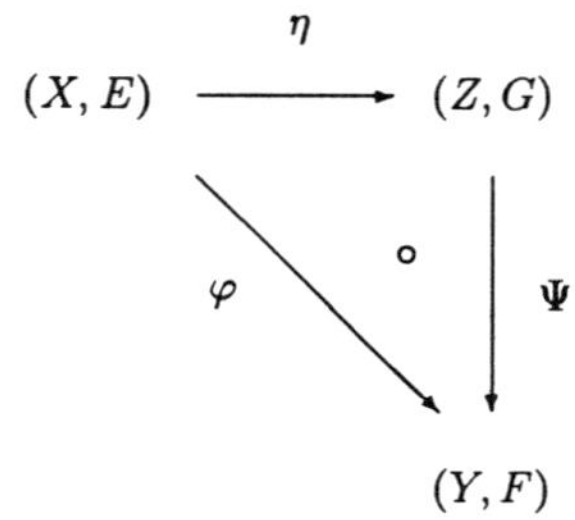

such that η is epic. We show that Ψ is an injective map. If $\Psi(z_1) = \Psi(z_2)$, then the relation

$$\begin{aligned} G(z_1, z_1) &= G(z_2, z_2) \\ &= \bigvee\{G(z_1, \eta(x)) * G(z_1, \eta(x)) * G(z_2, \eta(\bar{x})) * G(z_2, \eta(\bar{x})) \mid x, \bar{x} \in X\} \\ &\leq \bigvee\{G(z_1, \eta(x)) * F(\varphi(x), \varphi(\bar{x})) * G(\eta(\bar{x}), z_2) \mid x, \bar{x} \in X\} \\ &\leq G(z_1, z_2) \end{aligned}$$

follows from the axioms (S), (M2), (M2) and (EM); hence z_1 and z_2 coincide. Because of assertion (ii) and the injectivity of Ψ, the epimorphism η is an isomorphism. □

COROLLARY 1.4. **SM-SET** has the (epi, extremal mono)-diagonalization property.

PROOF. Let us consider a commutative diagram

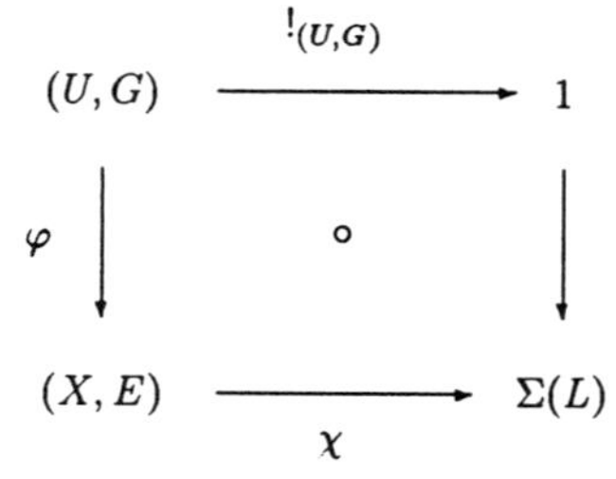

in which φ is epic and Ψ is extremally monic. Referring to Lemma 1.2 we obtain:

$$\begin{aligned} \hat{F}(\beta(\hat{x}), \beta(\hat{x})) = \hat{E}(\hat{x}, \hat{x}) &= \bigvee\{\hat{E}(\hat{x}, \varphi(x)) * F(\alpha(x), y) \mid x \in X, y \in Y\} \\ &\leq \bigvee\{\hat{F}(\beta(\hat{x}), \beta(\varphi(x))) * \hat{F}(\Psi(\alpha(x)), \Psi(y)) \mid x \in X, y \in Y\} \\ &\leq \bigvee\{\hat{F}(\beta(\hat{x}), \Psi(y)) \mid y \in Y\}. \end{aligned}$$

Since Ψ is extremally monic, we conclude from proposition 1.3 that there exists a unique arrow $k : (\hat{X}, \hat{E}) \to (Y, F)$ such that $\alpha = k \cdot \varphi$ and $\beta = \Psi \cdot k$ holds. □

It is easy to see that every **SM-SET**-morphism is (epi, extremal mono)-factorizable. Hence Corollary 1.4 together with 33.3 Theorem in [Herrlich and Strecker 1979] implies that **SM-SET** is an (epi, extremal mono)-category.

§2. Strong singletons of strong M-valued sets

Let (X, E) be a strong M-valued set and $s : X \longrightarrow L$ be a map. s is called a **strong singleton** of (X, E) iff s is provided with the properties

$$
\begin{array}{lll}
\text{(S0)} & 1^\circ \ s(x) * E(x,x) = s(x) & \\
 & 2^\circ \ s(x) * \mathbb{E}(s) = s(x) & \textbf{(Strong Strictness)} \\
 & \quad \text{where } \mathbb{E}(s) = \bigvee\{s(x) \mid x \in X\} & \\
\text{(S1)} & s(x) * E(x,y) \leq s(y) & \textbf{(Strong Extensionality)} \\
\text{(S2)} & s(x) * s(y) \leq E(x,y) & \textbf{(Strong Singleton Condition)}
\end{array}
$$

As an immediate consequence from (S02°) we obtain that $\mathbb{E}(s)$ is always idempotenmt w.r.t. $*$.

A simple example of a strong singleton of (X, E) can be given as follows: let $e \in L$ be a idempotent element and $x \in X$ be an arbitrary element; then $s_{(e,x)} : X \longrightarrow L$ determined by

$$s_{(e,x)}(z) = e * E(x,z), \ \forall \ z \in X \tag{2.1}$$

is a strong singleton of (X,E). If e coincides with the universal upper bound 1 (in L), then we introduce in this context a special notation: $s_{(1,x)}$ is also denoted by $\tilde{x}$.

Remark and Definition 2.1. Every strong singleton s of $\Sigma(L)$ (cf. Proof of Lemma 1.2) has the shape: $s = \widetilde{(\mathbb{E}(s), s(1,1))}$. We use this situation as a motivation for the following terminology: a separated strong M-valued set (X, E) is **complete** iff for each strong singleton s of (X, E) there exists an element $x \in X$ s.t. $s = \tilde{x}$. In the case of complete Heyting algebras this terminology goes back to D. Scott (cf. [Fourman and Scott 1979]).

Theorem 2.2. Let (X, E) be a strong M-valued set and $S(X, E)$ be the set of all strong singletons of (X, E). Then there exists a unique strong M-valued equality $\tilde{E}$ on $S(X, E)$ s.t.

(i) $\tilde{E}(s, \tilde{y}) = s(y)$ for all $y \in Y$,
(ii) $\tilde{E}(s, s) = \mathbb{E}(s)(= \bigvee\{s(x) \mid x \in X\})$.

In particular, $\tilde{E}(\tilde{x}, \tilde{y}) = E(x, y)$ — i.e. $\tilde{E}$ is an extension of E.

PROOF. Let F be a strong M-valued equality with (i) and (ii); then we obtain:

$$\begin{aligned}\bigvee\{s_1(y) * s_2(y) \mid y \in X\} &\leq F(s_1, s_2) \\ &= F(s_1, s_1) * F(s_1, s_1) * F(s_1, s_2) \\ &= \bigvee\{s_1(y) * s_1(y) * F(s_1, s_2) \mid y \in X\} \\ &\leq \bigvee\{s_1(y) * s_2(y) \mid y \in Y\}.\end{aligned}$$

On the other hand $\tilde{E}$ defined by $\tilde{E}(s_1, s_2) = \bigvee\{s_1(y) * s_2(y) | y \in X\}$ is a strong M-valued equality on $S(X, E)$ satisfying (i) and (ii). □

Motivated by property (ii) in Theorem 2.2 we interpret the quantity

$$\mathbb{E}(s) = \bigvee\{s(x) \mid x \in X\}$$

as the extent to which the strong singleton s exists. Strong singletons are irreducible in the following sense:

LEMMA 2.3. If $s_1, s_2 \in S(X, E)$ with $\mathbb{E}(s_1) = \mathbb{E}(s_2)$ and $s_1(x) \leq s_2(x)$ for all $x \in X$, then $s_1 = s_2$; i.e. comparable singletons with the same extent of existence coincide.

PROOF. Because of axiom (S) we obtain

$$\begin{aligned}s_2(x) &= s_2(x) * \mathbb{E}(s_1) * \mathbb{E}(s_1) \\ &= \bigvee\{s_2(x) * s_1(y) * s_1(y) \mid y \in X\} \\ &\leq \bigvee\{s_2(x) * s_2(y) * s_1(y) | y \in X\} \\ &\leq s_1(x) \square\end{aligned}$$

THEOREM 2.4. Let $\tilde{E}$ be the unique strong M-valued equality on $S(X, E)$ satisfying (i) and (ii) in Theorem 2.2. Then the following assertions are valid.

(a) $\Sigma(X, E) = (S(X, E), \tilde{E})$ is a separated, strong M-valued set.
(b) $\Sigma(X, E) = (S(X, E), \tilde{E})$ is a complete, strong M-valued set.

PROOF. The assertion (a) follows from Lemma 2.3 and the subsequent relation: $s_1(x) = s_1(x) * \mathbb{E}(s_1) = s_1(x) * \tilde{E}(s_1, s_2) \leq s_2(x)$. Further let σ be a strong singleton of $\Sigma(X, E)$. Because of

$$\begin{aligned}\mathbb{E}(\sigma) &= \bigvee\{\sigma(s) * \mathbb{E}(s) \mid s \in S(X, E)\} \\ &= \bigvee\{\sigma(s) * s(x) \mid s \in S(X, E), x \in X\} \\ &= \bigvee\{\sigma(\tilde{x}) \mid x \in X\}\end{aligned}$$

the restriction $\sigma_{|X}$ of σ to X is a strong singleton of (X, E). Then the assertion (b) follows from Lemma 2.3 and the subsequent relation

$$\begin{aligned}\tilde{E}(\sigma_{|X}, s) &= \bigvee\{s(x) * \sigma_{|X}(x) \mid x \in X\} \\ &= \bigvee\{s(x) * \sigma(\tilde{x}) \mid x \in X\} \\ &\leq \sigma(s) \ \square\end{aligned}$$

The construction of the singleton space $\Sigma(X, E)$ gives rise to an algebraic theory in clone form (here we are following the terminology of E. G. Manes (cf. [Manes 1976])): the object function is determined by Σ, the components of the natural transformation η are givem by $\eta_{(X,E)} : (X, E) \longrightarrow \Sigma(X, E)$, $\eta_{(X,E)}(x) = \tilde{x}$ for all $x \in X$; and we complete the picture, if we introduce the clone composition $\circ$ as follows

$$\varphi : (X, E) \longrightarrow \Sigma(Y, F), \ \Psi : (Y, F) \longrightarrow \Sigma(Z, G),$$

$$[(\Psi \circ \varphi)(x)](z) = \bigvee\{[\varphi(x)](y) * [\Psi(y)](z) \mid y \in Y\}.$$

THEOREM 2.5. $T = (\Sigma, \eta, \circ)$ is a degenerated algebraic theory (in clone form) in **SM-SET**.

PROOF. It is strightforward that T is an algebraic theory in clone form. The degeneracy (cf. [Schubert 1972]) of T follows from Theorem 2.4(b) and the fact that

$$\mu_{(X,E)} : \Sigma(\Sigma(X, E) \longrightarrow \Sigma(X, E)$$

is given by $\mu_{(X,E)}(\sigma) = \sigma_{|X}$. $\square$

Let $S_R(X, E)$ be the set of all strong singletons s of (X, E), which can be represented in the form of (2.1). Further let us denote by $\tilde{E}_R$ the restriction of $\tilde{E}$ to $S_R(X, E) \times S_R(X, E)$. Then $\Sigma_R(X, E) := (S_R(X, E), \tilde{E}_R)$ is a separated, strong M-valued set, and for each strong singleton $\sigma_{(e,s)}$ (where $s \in S_R(X, E)$) of $\Sigma_R(X, E)$, the restriction $\sigma_{(e,s)|X}$ of $\sigma_{(e,s)}$ to X is an element of $S_R(X, E)$.

PROPOSITION 2.6. $T_R := (\Sigma_R, \eta, \circ)$ is a subtheory of the algebraic theory $T = (\Sigma, \eta, \circ)$, which is also degenerated.

PROOF. Let us consider $\varphi : (X, E) \longrightarrow \Sigma_R(Y, F)$ and $\Psi : (Y, F) \longrightarrow \Sigma_R(Z, G)$. Because

$$\bigvee\{F(y_0, y) * [\Psi(y)](z) \mid y \in Y\} = [\Psi(y_0)](z)(\Psi \circ \varphi)(x)$$

is an element of $S_R(Z, G)$. Hence the assertion is an immediate consequence of those arguments used in 2.4 and 2.5. $\square$

An immediate consequence of the degeneracy of T (resp. T_R) is the following

COROLLARY 2.7. The Kleisli category $(\mathbf{SM\text{-}SET})^T$ (resp. $(\mathbf{SM\text{-}SET})^{T_R}$ is equivalent to the Eilenberg-Moore category $(\mathbf{SM\text{-}SET})^T$ (resp. $(\mathbf{SM\text{-}SET}^{T_R})$. In particular $(\mathbf{SM\text{-}SET})^T$ (resp. $(\mathbf{SM\text{-}SET})^{T_R}$) is a finitely complete and cocomplete category.

A strong M-valued set (X, E) is said to be **R-enriched** iff every strong singleton (of (X, E)) having the shape (2.1) is induced by an element of X — i.e. $s_{(e,x)} = \tilde{z}$.

In view of the following considerations we denote by **CSM-SET** (resp. **RSM-SET**) the full subcategory of **SM-SET** consisting of all complete (resp. R-enriched), strong M-valued sets. Because of 2.4 – 2.7 the category **CSM-SET** (resp. **RSM-SET**) is isomorphic to $(\mathbf{SM\text{-}SET})^T$ (resp. $(\mathbf{SM\text{-}SET})^{T_R}$). Since $\Sigma(L)$ is a complete, strong M-valued set (cf. 2.1), the concepts of epimorphisms and extremal monomorphisms in **CM-SET** (resp. **RSM-SET**) coincides with those of **SM-SET**.

Summing up, we obtain: **CSM-SET** and **RSM-SET** are finitely complete and cocomplete (epi, extremal mono)-categories.

REMARK 2.8. If the underlying monoid M is determined by a complete Heyting algebra Ω (i.e. $* = \wedge$), then **CSΩ-SET** (resp. **RSΩ-SET**) is the category of sheaves (resp. separated presheaves) over Ω, and $(\Omega\text{-}\mathbf{SET})^T$ (resp. $(\mathbf{S}\Omega\text{-}\mathbf{SET})^{T_R}$) is Higgs's topos of separated, Ω-valued sets (resp. Ponasse's category of separated fuzzy sets). In this case the content of Corollary 2.7 is well known (see e.g. [Fourman and Scott 1979], [Higgs 1973], [Ponasse 1988]).

§3. The $\sim$-Construction in RSM-SET (resp. CSM-SET)

The aim of this section is to study a generalized version of the $\sim$-construction (cf. [Kock and Wraith 1970]) and to explore its consequences for **RSM-SET** (resp. **CSM-SET**).

A **weak singleton** of a strong M-valued set (X, E) is a map $s : X \longrightarrow L$ satisfying the axioms (S0,1°), (S1), and (S2). Let us denote by $W(X, E)$ the set of all pairs (e, s) for which e is an idempotent element of L, s is a weak singleton of (X, E) and the following conditions hold:

(a) $e * s(x) = s(x)$ for all $x \in X$
(b) there exists $x_0 \in X$ s.t. $\bigvee\{\lambda \in L \mid \lambda * \lambda = \lambda,\ \lambda * (\bigvee_{x\in X} s(x)) = \lambda\} \leq s(x_0)$.

In the case of complete, strong M-valued sets the condition (b) is redundant. Further let us consider a strong M-valued equality $[\![\ ,\ ]\!]$ on $W(X, E)$ defined as follows:

$$[\![(e_1, s_1), (e_2, s_2)]\!] = e_1 * e_2 * (\bigwedge_{x\in X} (s_1(x) \leftrightarrow s_2(x))).$$

LEMMA 3.1. $\widetilde{(X, E)} := (W(X, E), [\![\ ,\ ]\!])$ is an R-enriched, strong M-valued set. In particular, if (X, E) is complete, then $\widetilde{(X, E)}$ is also complete.

PROOF. If $e \in L$ is idempotent, then the following relation holds

$$e * (\bigwedge_{i \in I} (\alpha_i \leftrightarrow \beta_i)) = e * (\bigwedge_{i \in I} ((e * \alpha_i) \leftrightarrow \beta_i)).$$

Hence, $(\widetilde{X, E})$ is R-enriched.

Further let σ be a strong singleton of $(\widetilde{X, E})$; then σ induces a weak singleton s_0 by $s_0(x) = \bigvee\{\sigma(e, s) * s(x) \mid (e, s) \in W(X, E)\}$. If (X, E) is complete, then $(\mathbb{E}(\sigma), s_0)$ is an element of $W(X, E)$. Because of $\sigma(e, s) \leq \bigwedge_{x \in X}(s_0(x) \leftrightarrow s(x))$ the completeness of $(\widetilde{X, E})$ follows from Lemma 2.3. □

LEMMA 3.2. Let (X, E) be a R-enriched, strong M-valued set. Then the map $\tilde{\eta}_{(X,E)}$: $X \longrightarrow W(X, E)$ defined by $\tilde{\eta}_{(X,E)}(x) = (E(x, x), x)$ is an extremal monomorphism from (X, E) to $(\widetilde{X, E})$ in the sense of **RSM-SET**.

PROOF. Obviously $\tilde{\eta}_{(X,E)}$ satisfies the condition (EM). In order to verify the condition 2° in Proposition 1.3 we proceed as follows. From $\bigvee\{[\![\eta_{(X,E)}(x), (e, s)]\!] \mid x \in X\} = e$, we infer $\bigvee\{s(x) \mid x \in X\} = e$. Because of (b) there exists an element $x_0 \in X$ with $s(x_0) = e$. Since (X, E) is R-enriched there exists an element $y \in X$ s.t. $s(z) = e * E(x_0, z) = E(y, z)$; hence the relation $\tilde{\eta}_{(X,E)}(y) = (e, s)$ holds. □

Now we need two further notions:

DEFINITION 3.3.

(a) A pair (j, φ) of morphisms with common domain is an **extremal partial morphism** iff j is an extremal monomorphism.
(b) Let (j, φ) be an extremal partial morphism in **RSM-SET** (resp. **CSM-SET**) such that (B, G) is the codomain of j and (X, E) the codomain of φ. A **RSM-SET**-morphism $\Psi : (B, G) \longrightarrow (\widetilde{X, E})$ is called a **strong extension** of (j, φ) iff the subsequent diagram is a pullback square:

$$\begin{array}{ccc} (B, G) & \xrightarrow{\quad \Psi \quad} & (\widetilde{X, E}) \\ j \Big\uparrow & \circ & \Big\uparrow \tilde{\eta}_{(X,E)} \\ (A, F) & \xrightarrow[\quad \varphi \quad]{} & (X, E) \end{array} \tag{D}$$

A characterization of strong extensions in **RSM-SET** (resp. **CSM-SET**) is given in

THEOREM 3.4. Let (j, φ) be an extremal partial morphism such that (B, G) is the codomain of j and $\widetilde{(X,E)}$ is the codomain of φ. If $\Psi : (B,G) \longrightarrow (X,E)$ is a **RSM-SET**-morphism, then the following assertions are equivalent:

(i) Ψ is a strong extension of (j,ϕ).

(ii) For every $b \in B$, the pair $\Psi(b) = (G(b,b), s_b)$ satisfies the subsequent conditions:

1° $\bigvee\{G(b,j(a)) * E(\varphi(a),x) \mid a \in A\} \leq s_b(x)$ for all $x \in X$.

2° If s_b is well-shaped — i.e. there exists a subset W of X s.t. $s_b(x) = \bigvee\{E(w,x), w \in W\}$ for all $x \in X$, then $s_b(x) \leq \bigvee\{G(b,j(a)) * E(\varphi(a),x) \mid a \in A\}$ for all $x \in X$.

PROOF. ((i) $\Rightarrow$ (ii)). Condition 1° follows from $s_b(x) \geq [\![\Psi(b), \Psi(j(a))]\!] * s_{j(a)}(x)$ and the commutativity of (D). Further let us fix $b \in B$, and let us assume that the weak singleton s_b is well-shaped — i.e. there exists a subset W of X s.t.

$$s_b(x) = \bigvee\{E(w,x), \ w \in W\} \tag{**}$$

Since (B,G) is R-enriched, there exists for every $w \in W$ an element $b_w \in B$ with

$$E(w,w) * G(b,z) = G(b_w,z);$$

Lemma 2.3 and the morphism axioms of Ψ imply: $\Psi(b_w) = (E(w,w), \tilde{w})$. Now we invoke the universal property of (D) and obtain for each $w \in W$ an element $a_w \in A$ *s.t.* $j(a_w) = b_w$ and $\varphi(a_w) = w$; hence condition 2° follows from (**).

((ii) $\Rightarrow$ (i)). The commutativity of (D) follows immediately from contiion 1° and Lemma 2.3. In order to verify the universal property of pullback squares we proceed as follows: Let us consider two morphism $\xi : (Z,G) \longrightarrow (B,G)$ and $\zeta : (Z,G) \longrightarrow (X,E)$ with $\Psi \cdot \xi = \tilde{\eta}_{(X,E)} \cdot \zeta$. From condition 2° we infer:

$$\begin{aligned} H(z,z) &= \bigvee\{E(\zeta(z), \ x), \ x \in X\} \\ &= \bigvee\{s_{\xi(z)}(x), \ x \in X\} \\ &= \bigvee\{G(\xi(z), \ j(a), \ a \in A\}. \end{aligned}$$

Since j is an extremal monomorphism there exists a unique $a_z \in A$ s.t. $\xi(z) = j(a_z)$. Using again 1° and 2° we obtain $\zeta(z) = \varphi(a_z)$. Hence there exists a unique morphism $\Upsilon : (Z,H) \longrightarrow (A,F)$ satisfying $j \cdot \Upsilon = \xi, \ \varphi \cdot \Upsilon = \zeta$. □

COROLLARY 3.5.

(a) Every extremal partial morphism (j,φ) in **RSM-SET** (resp. **CSM-SET**) has a strong extension.

(b) If the underlying monoid M is determined by a complete Heyting algebra (i.e. $* = \wedge$), then every extremal partial morphism in **RSM-SET** (resp. **CSM-SET**) has a **unique** strong extension.

PROOF. (a) We maintain the notations of the proof of Theorem 3.4. We fix $b \in B$ and put $e_b := \bigvee\{\lambda \in L \mid \lambda * \lambda = \lambda,\ \lambda * (\bigvee\{G(b, j(a)),\ a \in A\}) = \lambda\}$. Since (B, G) is R-enriched, and j is extremally monic, there exists an element $a_b \in A$ s.t.

$$e_b * G(b, z) = G(j(a_b), z) \text{ for all } z \in B \tag{**}$$

Further we associate with b a weak singleton s_b of (X, E) as follows:

$$s_b(x) = \bigvee\{G(b, j(a)) * E(\varphi(a), x) \mid a \in A\}$$

Because of $\bigvee\{s_b(x) \mid x \in X\} = \bigvee\{G(b, j(a) \mid a \in A\}$ we conclude from (**) that $(G(b,b), s_b)$ is an element of $W(X, E)$. Hence there exists a map $\Psi : B \longrightarrow W(X, E)$ defined by $\Psi(b) = (G(b,b), s_b)$. Obviously Ψ is a morphism from (B, G) to $(\widetilde{X, E})$ and by virtue of Theorem 3.4 a strong extension of (j, φ).

(b) If the semigroup operations $*$ coincides with $\wedge$, then every weak singleton is well-shaped. Hence the assertion (b) follows from Theorem 3.4.

COROLLARY 3.6 (**Pushout-Theorem**). The pushout of an extremal monomorphism in **RSM-SET** (resp. **CSM-SET**) by any morphism is again an extremal monomorphism, and the resulting diagram is a pullback also.

PROOF. The proof follows the usual lines (cf. [Kock and Wraith 1970]). In this context we have only to add an argument from the Proof of Proposition 1.3. □

Let 1 be the terminal object of **SM-SET**. We observe that $\tilde{1}$ is **SM-SET**-isomorphic to $\Sigma(L)$ (defined in the Proof of Lemma 1.2) and $\tilde{\eta}_1$ can be identified with $t : 1 \longrightarrow \Sigma(L)$ determined by $t(e) = (e, e)$ for each element e of the support set of 1. Since 1 is also the terminal object of **CSM-SET** (resp. **RSM-SET**), we deduce from 3.5 the following.

COROLLARY 3.7. Let $\varphi : (U, G) \longrightarrow (X, E)$ be an extremal monomorphism in **RSM-SET** (resp. **CSM-SET**). Then there exists a morphism $\chi : (X, E) \longrightarrow \Sigma(L)$ (so-called characteristic morphism) such that the diagram

$$\begin{array}{ccc} (X,E) & \xrightarrow{\ \varphi\ } & (\hat{X},\hat{E}) \\ {\scriptstyle\alpha}\big\downarrow & \circ & \big\downarrow{\scriptstyle\beta} \\ (Y,F) & \xrightarrow[\ \Psi\ ]{} & (\hat{Y},\hat{F}) \end{array} \tag{DD}$$

is a pullback square. In particular, if the underlying monoid M is determined by a complete Heyting algebra (i.e. $* = \wedge$), then there exists a unique characteristic morphism $\chi : (X, E) \longrightarrow \Sigma(L)$ such that (DD) is a pullback square.

REMARK 3.8. If $*$ is different from $\wedge$, then then classification of extremal subobjects (in **RSM-SET** or **CSM-SET**) as described in 3.7 is in general **not unique**. Let us consider the following counterexample:

$$\sqrt{\lambda} = \bigvee\{\mu \in L \mid \mu * \mu \leq \lambda\}, \ \chi_1(e,\lambda) = (e,\lambda), \ \chi_2(e,\lambda) = (e, e * \sqrt{\lambda})$$

Then χ_1 and χ_2 are different endormorphisms of $\Sigma(L)$. By virtue of the axiom (S) both endormorphisms classify the arrow t (= arrow true). Hence the classification of extremal subobjects in **RSM-SET** (resp. **CSM-SET**) by characteristic morphisms is unique iff the underlying monoid M is determined by a complete Heyting algebra.

DISCUSSION 3.9. In the case of complete Heyting algebras we conclude from corollary 3.5 that **RSΩ-SET** (= category of separated presheaves over Ω) is a quasi-topos. Of course, this is not a new result, rather it is a special case of a result due to [Johnstone 1979].

PART II. THE WEAK SUBOBJECT CLASSIFIER AXIOM.

The purpose of Part II is to introduce a categorical axiom dominating to a certain extent the classification of extremal subobjects in **CSM-SET**. In a topos (resp. quasi-topos) this axiom is given by the subobject classifier axiom (resp. strong subobject classifier axiom (cf. [Lawvere 1970], [Wyler 1976], [Pénon 1977]). In **CSM-SET** the situation is more complicated. In particular, if the underlying monoid M is not determined by a complete Heyting algebra (i.e. $* \neq \wedge$), then as we have already seen, there exist extremal subobjects in **CSM-SET** which can be classified by at least two different characteristic morphisms. In order to overcome these obstacles we select from the class of all characteristic morphisms appearing in the classification diagram (DD) a subclass consisting of t-tight morphisms.

As the reader will see below, tightness of morphisms requires a special concept of pseudo-complementation of extremal subobjects which we are now going to specify.

§4. Pseudo-complementation

Let $\mathcal{C}$ be a finitely complete and cocomplete (epi, extremal mono)-category and X be a $\mathcal{C}$-object. Then the set $\mathcal{ET}(x)$ of all extremal subobjects of X forms a lattice w.r.t. the usual partial ordering of subobjects. In particular we denote by $\wedge$ (resp. $\vee$) the meet (resp. join) operation in $\mathcal{ET}(X)$. Every extremal subobject $!_A : A \longrightarrow 1$ of 1 induces an extremal subobject $j_A : X_A \longrightarrow X$ of X according to the following pullback square

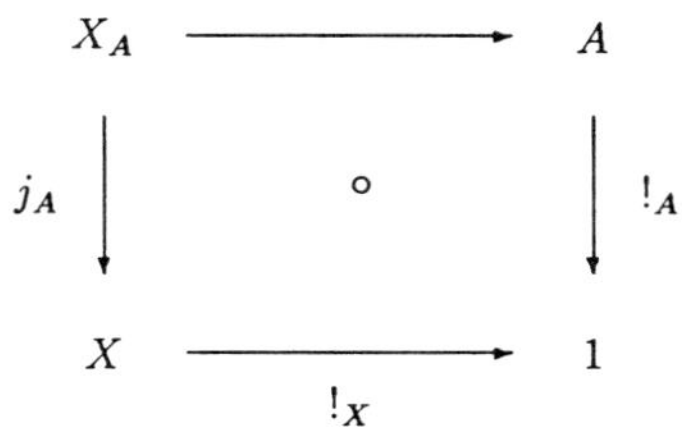

An **A-complement** of an extremal subobject $i : Y \longrightarrow X$ of X is an extremal subobject $n : U \longrightarrow X$ of X provided with the following properties:

(C1) The intersection $(Y, i) \wedge (U, n)$ is an extremal subobject of (X_A, j_A).

(C2) The diagram

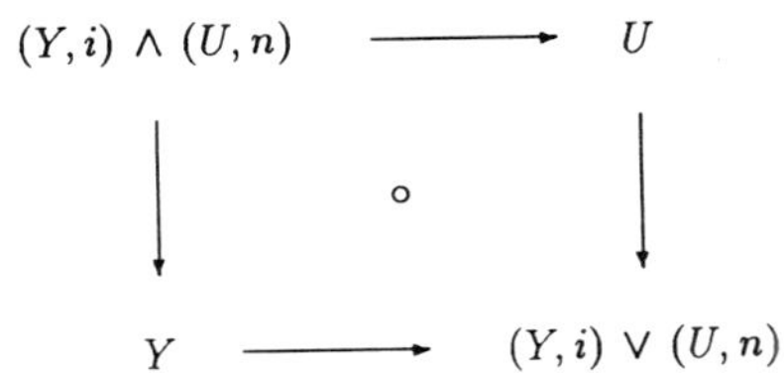

is a pulation square (i.e. pullback and pushout square).

(C3) For every extremal subobject $k : V \longrightarrow X$ of X satisfying (C1) and (C2) there exists a (unique) extremal monomorphisms $\ell : V \longrightarrow U$ s.t. $k = n \cdot \ell$.

An A-complement is unique up to an isomorphism. Further a finitely complete and cocomplete (epi, extremal mono)-category $\mathcal{C}$ has pseudo-complements iff for all $\mathcal{C}$-objects X, for all extremal subobjects (Y, i) of X, and for all extremal subobjects A of the terminal object 1, the A-complement exists.

PROPOSITION 4.1.

(a) Every (elementary) topos $\mathcal{C}$ has pseudo-complements. In particular, if $i : Y \longrightarrow X$ is an (extremal) subobject of X, then the pseudo-complement of (i, Y) is given by $(Y, i) \Rightarrow (X_A, j_A)$.

(b) Every finitely cocomplete, collapsed topos $\mathcal{C}$ ([Johnstone 1988]) has pseudo-complements. In particular, if $\mathcal{E}$ is the topos being reconstructed from $\mathcal{C}$ and if $\iota : \mathcal{E} \longrightarrow \mathcal{C}$ is the canonical coreflection corresponding to $\mathcal{C}$, then the A-complement of the (extremal) subobject $i : Y \longrightarrow X$ of the $\mathcal{C}$-object X is the coreflected A-complement of (Y, i) in $\mathcal{E}$ — i.e. the A-complement of (Y, i) in $\mathcal{C}$ is given by $\iota((Y, i) \Rightarrow (X, j_A))$.

PROOF. In every topos (resp. collapsed topos) subobjects and extremal subobjects coincide. Hence assertion (a) follows from the fact that the pulling-back functor has a right adjoint, and consequently preserves pushouts. In this context condition (C2) is redundant. In order to verify assertion (b) we proceed as follows. We consider the situation $\mathcal{C} = {}_{\mathcal{A}}\mathcal{E}$ — i.e. $\mathcal{E}$ is a topos, $\mathcal{A}$ is a subframe of $\mathrm{Sub}_{\mathcal{E}}(1)$, and $\mathcal{C}$ is the full subcategory of $\mathcal{E}$ of those $\mathcal{E}$-objects whose support $\sigma(X)$ (= image of the unique arrow $!_X : X \longrightarrow 1$) lies in $\mathcal{A}$. In particular the coreflection $\iota : \mathcal{E} \longrightarrow {}_{\mathcal{A}}\mathcal{E}$ is determined by the following pullback square

$$\begin{array}{ccc}
\iota(X) & \longrightarrow & X \\
\downarrow & \circ & \downarrow \\
\iota(\sigma(X)) & \longrightarrow & \sigma(X)
\end{array}$$

where $\iota(\sigma(X)$ is the largest subobject V of 1 in $\mathcal{A}$ s.t. $V \leq \sigma(X)$.

Further let $i : U \longrightarrow X$ be a subobject of X and $!_A : A \longrightarrow 1$ be a subobject of 1. Because of $(U,i) \wedge ((U,i) \Rightarrow (X_A, j_A)) \simeq (U,i) \wedge (X_A, j_A) \simeq U_A \simeq U \times A$ we obtain (cf. [Johnstone 1988])

$$\sigma((U,i) \wedge ((U,i) \Rightarrow (X_A, j_A))) = \sigma(U) \wedge A \tag{4.1}$$

Now let us assume that A, X, U are ${}_{\mathcal{A}}\mathcal{E}$-objects. Since ι preserves finite limits, we see immediately $\iota((U,i) \wedge \iota((U,i) \Rightarrow (X_A, j_A))) \simeq \iota((U,i) \wedge ((U,i) \Rightarrow (X_A, j_A)))$; hence we obtain from (4.1)

$$\begin{aligned}
\iota((U,i) \wedge \iota((U,i) \Rightarrow (X_A, j_A))) &\simeq (U,i) \wedge ((U,i) \Rightarrow (X_A, j_A)) \\
&\simeq (U,i) \wedge \iota((U,i) \Rightarrow (X_A, j_A))
\end{aligned}$$

where $\wedge$ denotes the intersection of subobjects in $\mathcal{E}$.

We show that $\iota((U,i) \Rightarrow (X_A, j_A))$ satisfies the conditions (C1)–(C3). In view of the preceeding considerations (C1) is obvious. Further let us consider the following pushout square in ${}_{\mathcal{A}}\mathcal{E}$:

$$\begin{array}{ccc}
(U,i) \wedge \iota((U,n) \Rightarrow (X_A, j_A)) & \longrightarrow & \iota((U,n) \Rightarrow (X_A, j_A)) \\
\downarrow & \circ & \downarrow \\
(U,i) & \longrightarrow & PO
\end{array}$$

Since the embedding ${}_{\mathcal{A}}\mathcal{E} \hookrightarrow \mathcal{E}$ preserves finite colimits, we obtain that PO is the join of (U, i) and $\iota((U, i) \Rightarrow (X_A, j_A))$ in $\mathcal{E}$. In particular the universal arrow from PO to X is monic (in $\mathcal{E}$). Hence PO is also the join of (U, i) and $\iota((U, i) \Rightarrow (X_A, j_A))$ in ${}_{\mathcal{A}}\mathcal{E}$. Now it is easy to see that $\iota((U, i) \Rightarrow (X_A, j_A))$ satisfies (C2). In order to verify (C3) we proceed as follows: First let us consider a pulation square

$$\begin{array}{ccc} (U,i) \mathring{\wedge} (V,k) & \longrightarrow & V \\ \downarrow & \circ & \downarrow \\ U & \longrightarrow & (U,i) \mathring{\vee} (V,k) \end{array}$$

in ${}_{\mathcal{A}}\mathcal{E}$, where $\mathring{\wedge}$ (resp. $\mathring{\vee}$) denotes the intersection (resp. join) in ${}_{\mathcal{A}}\mathcal{E}$. In particular the universal arrow $\Psi : (U, i) \mathring{\vee} (V, k) \longrightarrow X$ induced by i and k is a monomorphism in ${}_{\mathcal{A}}\mathcal{E}$. Since the embedding ${}_{\mathcal{A}}\mathcal{E} \hookrightarrow \mathcal{E}$ preserves finite colimits, the pushout-theorem in $\mathcal{E}$ implies that

$$\begin{array}{ccc} (U,i) \mathring{\wedge} (V,k) & \longrightarrow & V \\ \downarrow & \circ & \downarrow \\ U & \longrightarrow & (U,i) \mathring{\vee} (V,k) \end{array}$$

is a pullback square. Then $(U, i) \mathring{\wedge} (V, k) = (U, i) \wedge (V, k)$, because Ψ is monic. Further, if (V, k) fulfills (C1), we obtain $(V, k) \leq (U, i) \Rightarrow (X_A, j_A)$; hence (C3) follows from the idempotency of ι. $\square$

THEOREM 4.2.

(a) **CSM-SET** has pseudo-complements.
(b) If M satisfies the additional condition

$$(\hat{S})\begin{cases} \text{For each idempotent element } e_0 \text{ with } e_0 \leq \bigvee\{\lambda_i,\ i \in I\} \\ \text{there exists a family } \{e_i,\ i \in I\} \text{ of idempotent elements} \\ e_i \text{ with } e_i \leq \lambda_i \text{ for all } i \in I \text{ and } e_0 \leq \bigvee\{e_i,\ i \in I\}, \end{cases}$$

then **RSM-SET** has pseudo-complements.

The Proof of Theorem 4.2 requires the following

LEMMA 4.3. Let $m_1 : (U_1, F_1) \longrightarrow (X, E)$ and $m_2 : (U_2, F_2) \longrightarrow (X, E)$ be extremal monomorphisms in the sense of **SM-SET** and let j_1 and j_2 be morphisms determined by the subsequent pullback square

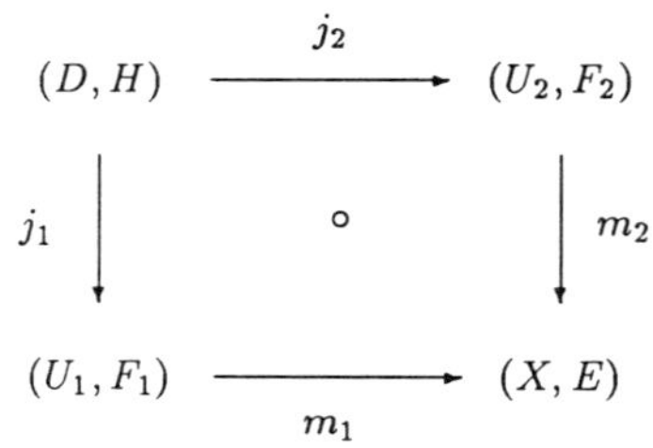

Further let $(PO(U_1, U_2), G)$ be the pushout of j_1 and j_2, and let $\Psi : (PO(U_1, U_2), G) \longrightarrow (X, E)$ be the universal arrow making the diagram

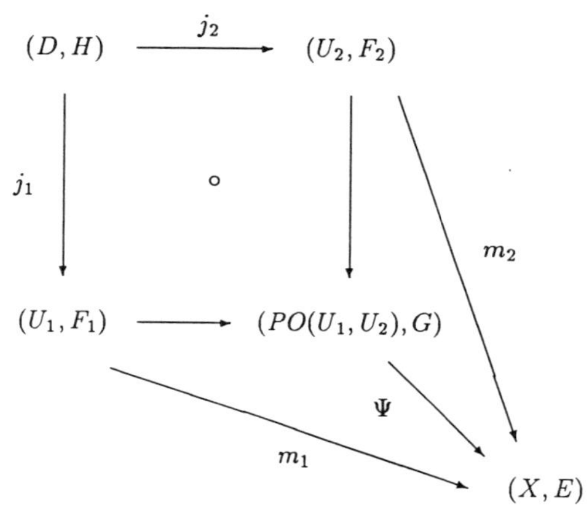

commutative. Then the following assertions are equivalent.

(i) Ψ satisfies condition (EM) (see Proposition 1.3).
(ii) $E(m_1(u_1), m_2(u_2)) = \bigvee_{z \in D} E(m_1(u_1), m_1 \cdot j_1(z)) \; * \; E(m_2 \cdot j_2(z), m_2(u_2))$ for all $(u_1, u_2) \in U_1 \times U_2$.

PROOF. Let V denote the support set $U_1 \times \{1\} \cup U_2 \times \{2\}$ of the coproduct

$$(U_1, F_1) \perp (U_2, f_2)$$

in **SM-SET**. Then the canonical embeddings

$$J_1 : (U_1, F_1) \longrightarrow (U_1, F_1) \perp (U_2, F_2), \; J_2 : (U_2, F_2) \longrightarrow (U_1, F_1) \perp (U_2, F_2)$$

PROOF. Let $m_1 : (U_1, F_1) \longrightarrow (X, E)$ and $m_2 : (U_2, F_2) \longrightarrow (X, E)$ be extremal subobjects of (X, E) in the sense of **RSM-SET** (resp. **CSM-SET**). Since **RSM-SET** (resp. **CSM-SET**) is an Eilenberg-Moore-category w.r.t. a degenerated algebraic theory T_R (resp. T) (cf. 2.5 and 2.6), the pushout of the diagram

$$\begin{array}{ccc} (U_1, F_1) \wedge (U_2, F_2) & \xrightarrow{\ j_2\ } & (U_2, F_2) \\ {\scriptstyle j_1}\big\downarrow & & \\ (U_1, F_1) & & \end{array} \tag{++}$$

is given by $T(PO(U_1, U_2), G))$, where $(PO(U_1, U_2), G)$ is the pushout of $(++)$ in the sense of **SM-SET**. By virtue of the special choice of the algebraic theory T_R (resp. T) the extension of **SM-SET**-morphisms according to the universal property of freely generated algebras preserves the condition (EM). Hence the assertion follows from Lemma 4.3. □

PROOF OF THEOREM 4.2. For each extremal subobject $!_A : A \longrightarrow 1$ there exists an idempotent element $e_A \in L$ such that $\{e \in L \mid e * e = e, e \le e_A\}$ is a support set of A. Further let $m : (U, F) \longrightarrow (X, E)$ be an extremal subobject of (X, E) (in the sense of **CSM-SET** or **RSM-SET**).

Verification of Assertion (a). First we put down the following notations:

$$\begin{aligned} U_A &:= \{x \in X \mid e_A * E(m(u), x) = E(m(u), x),\ \forall u \in U\}, \\ G &:= E_{|\,U_A \times U_A},\ n := \text{set inclusion of } U_A \text{ into } X \\ D &:= \{u \in U \mid F(u, u) \le e_A\},\ H := F_{|D \times D}, \\ j_1 &:= \text{set inclusion of } D \text{ into } U,\ j_2 := m_{|D}. \end{aligned}$$

Then (U_A, G) and (D, H) are complete, strong M-valued sets, and $n : (U_A, G) \longrightarrow (X, E)$ is an extremal subobject of (X, E). Further (D, H) is an extremal subobject of $(X, E)_A$, and the following diagram is a pullback square

$$\begin{array}{ccc} (D, H) & \xrightarrow{\ j_2\ } & (U_A, G) \\ {\scriptstyle j_1}\big\downarrow & \circ & \big\downarrow{\scriptstyle n} \\ (U, F) & \xrightarrow[\ m\ ]{} & (X, E) \end{array}$$

Since (U, F) is complete and m is extremally monic, there exists for each $u \in U$ and element

$u_A \in U$ s.t.

$$e_A * E(m(u), m(w)) = E(m(u_A),\ m(w)),\ \forall\, w \in U$$

Hence the relation

$$\begin{aligned} E(m(u), x) &= e_A * E(m(u),\ m(u)) * E(m(u), m(u)) * E(m(u), x)) \\ &\leq E(m(u),\ m(u_A)) * E(m(u_A), x) \\ &\leq \bigvee_{z \in D} E(m(u),\ m \cdot j_1(z)) * E(n \cdot j_2(z), n(x)) \end{aligned}$$

follows for all $(u, x) \in U \times U_A$. Now we invoke Corollary 4.4 and obtain that the pushout of the diagram

$$\begin{array}{ccc} (D, H) & \xrightarrow{\ j_2\ } & (U_A, G) \\ {\scriptstyle j_1}\big\downarrow & & \\ (U, F) & & \end{array}$$

is the join of $((U, F), m)$ and $((U_A, G), n)$. Because of the pushout-theorem (cf. Corollary 3.6) $((U_A, G), n)$ satisfies (C2). Further let $k : (V, Q) \longrightarrow (X, E)$ be an extremal subobject of (X, E) satisfying (C1) and (C2). We denote by P the set of all pairs $(u, v) \in U \times V$ with $m(u) = k(v)$. Then the following relations hold:

$$F(u, u) = Q(v, v) \leq e_A,\ \forall\, (u, v) \in P,$$

$$E(m(u), k(v) = \bigvee_{(u', v') \in P} E(m(u), m(u')) * E(k(v'), k(v))$$

Hence $k(V)$ is a subset of U_A — i.e. $((U_A, G), n)$ fulfills (C3). Summing up $((U_A, G), n)$ is a A-complement of $((U, F), m)$.

Verification of Assertion (b). Let us denote by $\mathcal{Z}$ the set of all extremal subobjects $((V, G), n)$ of (X, E) satisfying (C1) and (C2). In the framework of **RSM-SET** the conditions (C1) and (C2) are equivalent to:

1° $(u, v) \in U \times V \Rightarrow E(m(u), n(v) = e_A * E(m(u), n(v)),$
2° $(u, v) \in U \times V \Rightarrow E(m(u), n(v) = \bigvee_{z \in D} E(m(u), m \cdot j_1(z)) * E(n \cdot j_2(z), n(v))$ where D denotes a support set of $((U, F), m) \wedge ((V, G), n)$,
3° A support set of $((U, F), m) \vee ((V, G), n)$ is given by the set-theoretical union of $m(U)$ and $n(V)$.

We define an extremal monomorphism $\ell : (W, L) \longrightarrow (X, E)$ by

$$W := \{w \in X \mid E(w, w) =$$

$$\bigvee\{E(w, n(v)),\ v \in V \text{and}((V,G),\ n) \in \mathcal{Z}\}\},$$

$$L := E_{|W \times W},\ \ell := \text{set inclusion of } W \text{ into } X$$

In order to verify assertion (b) it is sufficient to show that $((W, L), \ell)$ is an element of $\mathcal{Z}$. From

$$E(m(u), w) = \bigvee\{E(m(u), n(v)) * E(n(v), w) \mid v \in V, ((V, G), n) \in \mathcal{Z}\} \tag{4.4}$$

we infer that $((W, L), \ell)$ satisfies 1°. Condition 2° follows immediately from (4.4) and the strong transitivity of E. In view of 3° let us assume

$$E(x, x) = (\bigvee\{E(x, m(u) \mid u \in U\}) \bigvee (\bigvee\{E(x, w) \mid w \in W\}) \tag{4.5}$$

We put down the following notations

$$e_U := \bigvee\{\alpha \in L \mid \alpha * \alpha = \alpha,\ \alpha\delta * (\bigvee\{E(x, m(u) \mid u \in U\}) = \alpha\}$$

$$e_V := \bigvee\{\alpha \in L \mid \alpha * \alpha = \alpha,\ \alpha * (\bigvee\{E(x, m(u) \bigvee E(x, n(v) \mid (u, v) \in U \times V\}) = \alpha\}$$

By virtue of $(\hat{S})$ we obtain $E(x, x) = \bigvee\{e_V \mid ((V, G), n) \in \mathcal{Z}\}$. Now we distinguish the following cases

Case 1: $e_U = E(x, x)$. Since $((U, F), m)$ is an extremal subobject (in the sense of **RSM-SET**), x is an element of $m(U)$.

Case 2: If $e_U \neq E(x, x)$, then $E(x, x) = \bigvee\{e_V \mid e_U \neq e_V, ((V, G), n) \in \mathcal{Z}\}$.

Referring to 3° we obtain for each $e_V \neq e_U$ an element $y_V \in V$ s.t. $e_V * E(x, n(v)) = E(n(y_V, n(v)), \forall\, v \in V$; hence $x \in W$ — i.e. $((W, L), \ell)$ satisfies also 3°. □

§5. The weak subobject classifier axiom

Let $\mathcal{C}$ be a category having pseudo-complements. Further let $\eta : X \longrightarrow Y$ be a $\mathcal{C}$-morphism and $\ell : Z \longrightarrow Y$ be an extermal subobject of Y. η is said to be **ℓ-tight** iff for all extremal subobjects $!_A : A \longrightarrow 1$ and for all A-complements (Z^A, ℓ^A) of (Z, ℓ) the following diagram is a pullback square

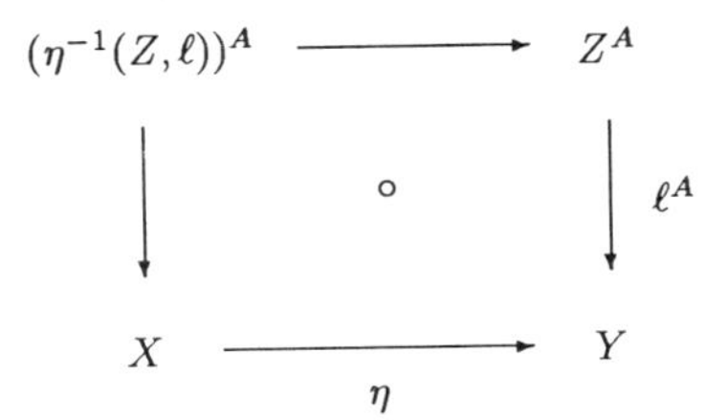

where $(\eta^{-1}(Z,\ell))^A$ is the A-complement of $\eta^{-1}(Z,\ell)$; i.e. η is ℓ-tight iff η preserves all A-complements of (Z,ℓ).

REMARK 5.1. If $\mathcal{C}$ is a topos (resp. collapsed topos), then every $\mathcal{C}$-morphism is tight.

Now we are ready to specify the weak subobject classifier axiom.

DEFINITION 5.2. Let $\mathcal{C}$ be a category having pseudo-complements, Ω be a $\mathcal{C}$-object and $t : 1 \longrightarrow \Omega$ be a $\mathcal{C}$-morphism. The pair (Ω, t) is called a **weak subobject classifier** iff (Ω, t) satisfies the following axioms

(WS1) For all $\mathcal{C}$-objects X and for all extremal subobjects (Y, φ), there exists a t-tight morphism $\chi : X \longrightarrow \Omega$ s.t. the diagram

$$\begin{array}{ccc} Y & \xrightarrow{\;!_Y\;} & 1 \\ {\scriptstyle\varphi}\downarrow & \circ & \downarrow{\scriptstyle t} \\ X & \xrightarrow[\;\chi\;]{} & \Omega \end{array}$$

is a pullback square.

(WS2) There exists a unique t-tight morphism $T : \Omega \longrightarrow \Omega$ s.t. the diagram

$$\begin{array}{ccc} 1 & \xrightarrow{\;!\;} & 1 \\ {\scriptstyle t}\downarrow & \circ & \downarrow{\scriptstyle t} \\ \Omega & \xrightarrow[\;T\;]{} & \Omega \end{array}$$

is a pullback square.

REMARK 5.2.

(a) A weak subobject classifier (Ω, t) is unique up to an isomorphism.

(b) Every topos (resp. collapsed topos (cf. [Johnstone 1988])) satisfies the weak subobject classifier axiom $(W\Omega) = (WS1) + (WS2)$.

(c) In general a quasi-topos does not satisfy the weak subobject classifier axiom $(W\Omega)$. Let us consider the following counterexample: Let $B = \{0, a, a^0, 1\}$ be the Boolean algebra consisting of four elements, and let $\mathcal{C}$ be the category of separated presheaves over B — i.e. $\mathcal{C}$ = **RSB-SET**. Obviously the classifier (Ω, t) is given by $\Omega = \Sigma(B) = (S(B), [\![\ ,\ ]\!])$, $t(e) = (e, e)$. Now let us consider the following non-extremal subobject $((X, E), i)$ of $\Sigma(B)$: $X = \{(1,1), (a,a), (a^0, a^0), (0,0), (a,0), (a^0, 0)\}$, $E = [\![\ \ ]\!]_{|X \times X}$, i = set inclusion of X into $S(B)$. Obviously t factors through $((X, E), i)$; in particular there exists an extremal monomorphism $c : 1 \longrightarrow (X, E)$ s.t. $i \cdot c = t$. It is easy to see that the unique classification χ_c of c coincides with i, but unfortunately i does not preserve the 0-complementation; hence χ_c is **not** t-tight.

THEOREM 5.3. Let M be a complete MV-algebra (see [Belluce 1986, 1989], [Höhle 1990a]). Then **CSM-SET** satisfies the weak subobject classifier axiom $(W\Omega) = (WS1) + (WS2)$.

PROOF. We show that $(\Sigma(L), t)$ (cf. Corollary 3.7) is a weak subobject classifier in **CSM-SET**. First let us consider an extremal monomorphism $\varphi : (Y, F) \longrightarrow (X, E)$ and a morphism $\chi : (X, E) \longrightarrow \Sigma(L)$ defined by $\chi(x) = (E(x,x), \mu(x))$ where $\mu(x) = \bigvee\{E(x, \varphi(y)) \mid y \in F\}$. Then we conclude from Corollary 3.5 and Corollary 3.7 that χ classifies φ . Referring to the construction of A-complements in **CSM-SET** (cf. part (a) of the Proof of Theorem 4.2) we obtain that χ is a t-tight classification; hence $(\Sigma(L), t)$ satisfies (WS1). The desired uniqueness of the t-tight classification of the arrow t follows from the fact that in M the "negation" (which sends α to $(\alpha \to 0)$) is an involution; hence $(\Sigma(L), t)$ satisfies also (WS2). □

REMARK 5.4. By addition of further axioms, the weak subobject classifier axiom can be enriched in such a way which leads to the concept of weak toposes (cf. [Höhle 1990b], [Höhle and Stout 1991]). In this context it is interesting to note that toposes and finitely cocomplete collapsed toposes are also weak toposes. Further we emphasize that weak toposes have an internal logic, which coincides in the case of **CMET1** (= category of complete metric spaces with non-expansive maps) with the $[0,1]$-valued Lukasiewicz logic (cf. [Höhle 1990b]).

Chapter 2:

M-valued Sets and Sheaves over Integral Commutative CL-Monoids

S. E. Rodabaugh et al. (eds.), Applications of Category Theory to Fuzzy Subsets, 33–72.

M-VALUED SETS AND SHEAVES OVER INTEGRAL COMMUTATIVE CL-MONOIDS

U. Höhle

§0. Introduction

The concept of Heyting algebra valued sets is intrinsically linked to intuitionistic models, sheaves and topos theory. Precursors of these ideas appear already in D. S. Scott's and R. Solovay's work on Boolean-valued models from the mid-sixties [Scott 1967]. Subsequently D. Higgs uses Boolean-valued sets in his unpublished, but widely circulated paper [Higgs 1973] and demonstrates that the category of sheaves over a complete Boolean algebra B is equivalent to the category of B-valued sets and maps in the original Scott-Solovay sense. The concept of Heyting algebra valued sets and its relationship to intuitionistic logic has been extensively studied by M. Fourman and D. S. Scott in their contributions to the *Research Symposium on Applications of Sheaf Theory*, Durham, NC, 1977 [Fourman and Scott 1979; Scott 1979]. We can summarize the situation as follows: let Ω be a complete Heyting algebra; then 1° Ω–valued sets are intuitionistic models of the formalized mathematical theory of identity with **existence** predicate, and 2° the category $sh(\Omega)$ of sheaves over Ω forms a topos. In this context we mention explicitly the fact that the unique classification of subobjects in $sh(\Omega)$ depends essentially on the existence of non-trivial objects with local support.

On the other hand Łukasiewicz logic is the internal logic of the category **MET1** of metric spaces and 1-Lipschitz continuous maps [Höhle 1990a]. Disregarding for a moment the initial object, **MET1** contains only objects with global support; hence the classification of extermal subobjects by (pseudo-)characteristic morphisms cannot be unique. In order to improve the classification of extremal subobjects in **MET1**, the situation in $sh(\Omega)$ suggests an introduction of "metric spaces with **local** support". This means syntactically an addition of an existence predicate to the formalized theory of identity or semantically an interpretation of an existence predicate with respect to Łukasiewicz logic.

If we step back for a moment we realize that all semantic interpretations of intuitionistic logic as well as of the Łukasiewicz-Wajsberg axioms by means of truth values are at least based on integral, commutative cl-monoids [Birkhoff 1973]. Hence with regard to non-classical interpretations of local existence we can also ask this question:

To which extent does sheaf theory depend on the cl-monoid structure of the underlying complete Heyting algebra Ω?

Beyond the logical aspects mentioned already above this question has a profound relationship to the problem of giving an appropriate, categorical foundation of fuzzy set theory initiated by L. A. Zadeh 1965. Looking into Zadeh's pioneering paper [Zadeh 1965] as well as into several papers of his successors, we find out immediately that fuzzy set theory is based on a fragment of Łukasiewicz logic. We describe this situation as follows: The set of truth values is the real unit interval $[0,1]$, but the set of logical connectives consists only of the negation $\neg$, conjunction $\wedge$ and disjunction $\vee$, where $\neg, \wedge, \vee$ are given by

$$\neg\alpha = 1 - \alpha \qquad \text{(Negation)}$$
$$\alpha \wedge \beta = Min(\alpha, \beta) \qquad \text{(Conjunction)}$$
$$\alpha \vee \beta = Max(\alpha, \beta) \qquad \text{(Disjunction)}$$

The preference of Łukasiewicz' negation, compared with the intuitionistic negation in $[0,1]$, is mainly due to the continuity of Łukasiewicz' negation with respect to the usual topology on $[0,1]$ which plays a crucial role in applications of fuzzy subsets to system theory. A glance at the set of connectives specified above shows that it does not determine uniquely a "logic" — e.g. we can add at least two different continuous implication operators $\rightarrow$ extending Łukasiewicz' negation (i.e. $1 - \alpha = \alpha \rightarrow 0$). We can use either Kleene-Dienes' implication $\overset{K}{\rightarrow}$ ($\alpha \overset{K}{\rightarrow} \beta = Max(1-\alpha, \beta)$) or Łukasiewicz implication $\overset{L}{\rightarrow}$ ($\alpha \overset{L}{\rightarrow} \beta = Min(1 - \alpha + \beta, 0)$). Here we take the freedom to add Łukasiewicz' implication $\overset{L}{\rightarrow}$ and observe that $\overset{L}{\rightarrow}$ is the residuation in the integral, commutative cl-monoid $([0,1], \leq, T_m)$, where T_m denotes the arithmetic conjunction [Frink 1938] and is defined by $T_m(\alpha, \beta) = Max(\alpha + \beta - 1, 0)$. This brings us back to the above question.

First we remark that a mere replacement of complete Heyting algebras by integral, commutative cl-monoids in the axioms of Ω-valued sets entails a lot of difficulties mainly caused by the lacking idempotency of the "arithmetic" conjunction. Hence a naive reinterpretation of the situation in $sh(\Omega)$ is impossible.

Our approach to a solution of the above problem consists now in the following steps: First we have to find in the framework of po-monoids a substitute for the missing law of idempotency — this is the axiom (L0) and leads to the concept of GL-monoids (cf. §1). Secondly we introduce for any given GL-monoid M the structure of a M-valued set in such a way that we re-obtain in the case of complete Heyting algebras the concept of Ω-valued sets. In particular the axioms of strictness and transitivity reflect in a distinctive way the absence of the law of idempotency.

In order to define sheaves over Gl-monoids we start from the fundamental observation that there exists a degenerated algebraic theory T in the category of Ω-valued sets and structure preserving maps such that the Eilenberg-Moore category of T-algebras is isomorphic to the category $sh(\Omega)$ of sheaves and the Kleisli category is isomorphic to Higgs' topos Ω-**SET**. In this context the equivalence between $sh(\Omega)$ and Ω-**SET** follows immediately from the degenercy of T [Schubert 1972] and has been already mentioned in the Boolean case.

By virtue of the axiom (L0) we are now in the position to reformulate this algebraic theory T for any given GL-monoid M. As a first result we quote that T is degenerated if and only if M is determined by a complete Heyting algebra — i.e. the semigroup operation $*$ and the meet operation coincide. Hence T reflects precisely the role of the law of idempotency. Motivated by the observation in the preceding paragraph we consider the Eilenberg-Moore category of T-algebras as the category $sh(M)$ of sheaves over M. In particular every "compatible" family has in any T-algebra a unique "join". Moreover we verify the following properties of $sh(M)$:

(1) $sh(M)$ is finitely complete.
(2) If M satisfies the additional axiom (L4) (which is simultaneously verified by complete Heyting algebras as well as by complete MV-algebras ([Belluce 1986, 1989], [Chang 1958])), then $sh(M)$ has coproducts.
(3) There exists a T-algebra $\bar{\Omega}$ such that strict, extensional, L-fuzzy subsets and T-homomorphisms from freely generated T-algebras to $\bar{\Omega}$ are the same things
(4) If Ω is the underlying lattice of M, then $sh(\Omega)$ is a full subcategory of $sh(M)$.

It is clear that this list of properties of $sh(M)$ is not exhaustive and is rather far from being complete. But we stop here and postpone this work to a future paper being very well aware that $sh(M)$ will open a fascinating, new area of fuzzy set theory and will play a role comparable to that of $sh(\Omega)$ at the beginning of topos theory.

§1. GL-monoids

The purpose of this section is to present a class of completely lattice-ordered, commutative monoids which covers simultaneously complete Heyting algebras as well as complete MV-algebras.

First we recall some basic definitions from the theory of lattice-ordered monoids. Referring to [Birkhoff 1973], a **commutative po-monoid** is an ordered triple $(L, \leq, *)$, where $(L, \leq)$ is a partially ordered set and $(L, *)$ is a commutative monoid satisfying the isotonicity condition

$$\alpha \leq \beta \quad \text{implies} \quad \alpha * \gamma \leq \beta * \gamma \tag{1.1}$$

A (commutative) po-monoid is **integral** iff the unit element $\mathbf{1}$ is also the universal upper bound in $(L, \leq)$ — i.e. $\alpha \leq \mathbf{1}$, $\forall \alpha \in L$. If an integral, commutative monoid has a zero element $\mathbf{0}$ (i.e. $\alpha * \mathbf{0} = \mathbf{0}$, $\forall \alpha \in L$), then $\mathbf{0}$ is necessarily the universal lower bound in $(L, \leq)$. An integral commutative po-monoid $M = (L, \leq, *)$ is the **dual** of a **divisibility** monoid iff M satisfies the axiom

$$\alpha \leq \beta \text{ implies the existence of } \gamma \in L \text{ s.t. } \alpha = \beta * \gamma \tag{L0}$$

A commutative po-monoid M is **residuated** iff there exists a further binary operation $\rightarrow$ on L satisfying the following condition

$$\beta * \alpha \leq \gamma \Leftrightarrow \alpha \leq (\beta \to \gamma), \ \forall \alpha, \beta, \gamma \in L \tag{L1}$$

i.e. for each (fixed) $\beta \in L$ the maps $\beta * -$ and $\beta \to -$ from L to L form a **Galois connection** between $(L, \leq)$ and the dual of the po-set $(L, \leq)$. A commutative po-monoid $M = (L, \leq, *)$ is **lattice-ordered** iff $(L, \leq)$ is a lattice and $*$ is distributive over joins — i.e. $\alpha * (\beta \vee \gamma) = (\alpha * \beta) \vee (\alpha * \gamma)$.

LEMMA 1.1. Let $(L, \leq)$ be a lattice with universal upper bound I and M be a commutative po-monoid. Then the following relations are valid:

(1) If M is integral, then $\alpha * \beta \leq \alpha \wedge \beta, \ \forall \alpha, \beta \in L$.
(2) If M is residuated, then
  1° $(\alpha \vee \beta) \to \gamma = (\alpha \to \gamma) \wedge (\beta \to \gamma)$
  $\alpha \to (\beta \wedge \gamma) = (\alpha \to \beta) \wedge (\alpha \to \gamma)$
  2° $(\alpha * \beta) \to \gamma = \alpha \to (\beta \to \gamma)$
  $\alpha \to \beta \leq (\alpha * \gamma) \to (\beta * \gamma)$.

A commutative po-monoid $M = (L, \leq, *)$ is **completely lattice-ordered** iff $(L, \leq)$ is a complete lattice and $*$ is distributive on arbitrary joins — i.e. M satisfies the **infinite distributive law**

$$\alpha * (\bigvee_{i \in I} \beta_i) = \bigvee_{i \in I} (\alpha * \beta_i). \tag{L2}$$

Every commutative, completely lattice-ordered po-monoid (= commutative cl-monoid) M with zero is residuated and the residuation in terms of implication is given by

$$\alpha \to \beta = \bigvee \{\lambda \in L \mid \alpha * \lambda \leq \beta\} \tag{1.2}$$

A **GL-monoid** is an integral, commutative cl-monoid M with zero which is the dual of a divisibility monoid.

LEMMA 1.2. Any Gl-monoid $(L, \leq, *)$ has the following properties:

(1) $(\bigvee_{i \in I} \alpha_i) \to \beta = \bigwedge_{i \in I} (\alpha_i \to \beta)$.
(2) $\alpha \to (\bigwedge_{i \in I} \beta_i) = \bigwedge_{i \in I} (\alpha \to \beta_i)$.
(3) $\alpha \wedge \beta = \alpha * (\alpha \to \beta)$.
(4) If $\alpha * \alpha = \alpha$, then $\alpha * \beta = \alpha \wedge \beta$.
(5) $(\alpha \to \beta) * ((\alpha \wedge \beta) \to \gamma) = \alpha \to (\beta \wedge \gamma)$.
(6) $\alpha \wedge (\bigvee_{i \in I} \beta_i) = \bigvee_{i \in I} (\alpha \wedge \beta_i)$ — i.e. the underlying lattice is a complete Heyting algebra.

PROOF. The relations (1) and (2) follow immediately from 1.1(2) 1° and (1.2). Obviously (L0) and (1.2) imply (3). If α is idempotent w.r.t. $*$, then we infer from (1.2) and 1.2(3):

$$\alpha \wedge \beta = \alpha * (\alpha \to \beta) = \alpha * \alpha * (\alpha \to \beta) = \alpha * (\alpha \wedge \beta) \leq \alpha * \beta;$$

hence (4) follows from 1.1(1). In view of 1.1(2), 1.2(1), 1.2(3), 1.1 and (L2) we obtain now:

$$(\alpha \to \beta) * ((\alpha \wedge \beta) \to \gamma) = (\alpha \to \beta) * ((\alpha * (\alpha \to \beta)) \to \gamma) =$$

$$(\alpha \to \beta) * ((\alpha \to \beta) \to (\alpha \to \gamma)) = (\alpha \to \beta) \wedge (\alpha \to \gamma) = \alpha \to (\beta \wedge \gamma),$$

$$\alpha \wedge (\bigvee_{i\in I} \beta_i) = (\bigvee_{i\in I} \beta_i) * ((\bigvee_{i\in I} \beta_i) \to \alpha) = (\bigvee_{i\in I} \beta_i) * (\bigwedge_{i\in I} (\beta_i \to \alpha)) \leq$$

$$\bigvee_{i\in I} (\beta_i * (\beta_i \to \alpha)) = \bigvee_{i\in I} (\alpha \wedge \beta_i)$$

Hence (5) and (6) are verified. □

Referring to 1.2(6) we can understand GL-monoids as complete Heyting algebras provided with an additional, monoidal structure. A Gl-monoid M is **strong** iff M satisfies the further axiom

$$\alpha * \beta \leq (\alpha * \alpha) \vee (\beta * \beta), \ \forall \alpha, \beta \in L \tag{L3}$$

In view of (L2), the condtion (L3) is equivalent to

$$(\bigvee A) * (\bigvee A) = \bigvee \{\alpha * \alpha \mid \alpha \in A\}, \ \forall A \subseteq L \tag{L3'}$$

LEMMA 1.3. Let M be a GL-monoid provided with the property

$$(\alpha \to \beta) \vee (\beta \to \alpha) \text{ is idempotent w.r.t. } * \text{ for all } \alpha, \beta \in L. \tag{L4}$$

Then M is a strong GL-monoid.

PROOF. Because of $\alpha * \beta \leq \alpha \vee \beta \leq (\alpha \to \beta) \vee (\beta \to \alpha)$, the idempotence of $(\alpha \to \beta) \vee \beta \to \alpha)$ implies (cf. 1.2(4)):

$$\alpha * \beta = (\alpha * \beta * (\alpha \to \beta)) \vee (\alpha * \beta * (\beta \to \alpha)) \leq (\alpha * \alpha) \vee (\beta * \beta)$$

A GL-monoid M is a **complete MV-algebra** ([Chang 1958]; [Belluce 1986, 1989]) iff the "complementation" based on $*$ is involutory — i.e. M fulfills the condition

$$(\alpha \to \mathbf{O}) \to \mathbf{O} = \alpha \text{ for all } \alpha \in L \, (\mathbf{O} \text{ is the universal lower bound in } (L, \leq)) \tag{L5}$$

LEMMA 1.4. Any complete MV-algebra M has the following properties:

(1) $(\bigwedge_{i\in I} \alpha_i) \to \mathbf{O} = \bigvee_{i\in I} (\alpha_i \to \mathbf{O})$.

(2) $\bigwedge_{i\in I} (\alpha \vee \beta_i) = \alpha \vee (\bigwedge_{i\in I} \beta_i)$ — i.e. the dual lattice of $(L, \leq)$ is also complete Heyting algebra.

(3) $\alpha \to \beta = (\alpha * (\beta \to \mathbf{O})) \to \mathbf{O}$.
(4) $\alpha \vee \beta = (\alpha \to \beta) \to \beta$.
(5) $\alpha \to (\alpha * \beta) = (\alpha \to \mathbf{O}) \vee \beta$.
(6) $\alpha * (\bigwedge_{i \in I} \beta_i) = \bigwedge_{i \in I} (\alpha * \beta_i)$.
(7) $(\alpha \to \beta) \vee (\beta \to \alpha) = \mathbf{1}$ (the universal upper bound in $(L, \leq)$).

PROOF. The relations (1) and (2) follow immediately from (L5), 1.2(1) and 1.2(6). Obviously (1.2) and (L5) imply (3). Because of 1.2(1), 1.2(3) and (L5) we infer from 1.4(3):

$$\alpha \vee \beta = [(\alpha \to \mathbf{O}) \wedge (\beta \to \mathbf{O})] \to \mathbf{O} =$$
$$[(\beta \to \mathbf{O}) * (((\beta \to \mathbf{O}) * \alpha) \to \mathbf{O})] \to \mathbf{O} =$$
$$(\alpha \to \beta) \to \beta,$$
$$(\alpha \to \mathbf{O}) \vee \beta = (\alpha \wedge (\beta \to \mathbf{O}) \to \mathbf{O} = (\alpha * (\alpha \to (\beta \to \mathbf{O}) \to \mathbf{O} =$$
$$[\alpha * ((\alpha * \beta) \to \mathbf{O})] \to \mathbf{O} = \alpha \to (\alpha * \beta)$$

In view of 1.2(2), 1.4(2) and 1.4(5) we obtain

$$\alpha \to (\alpha * (\bigwedge_{i \in I} \beta_i)) = \bigwedge_{i \in I} (\alpha \to (\alpha * \beta_i)) = \alpha \to (\bigwedge_{i \in I} (\alpha * \beta_i));$$

hence (6) follows from 1.1(1) and 1.2(3). In order to verify (7) we proceed as follows;

$$\begin{aligned}
\mathbf{1} &= [(\beta \to \alpha) * [((\beta \to \alpha) \to \beta * (((\beta \to \alpha) \to \beta) \to \alpha)]] \to \alpha \\
&= [(\beta * (\beta \to (\beta \to \alpha))) * (((\beta \to \alpha) \to \beta) \to \alpha)] \to \alpha \\
&= [(\beta \to (\beta \to \alpha)) * (((\beta \to \alpha) \to \beta) \to \alpha)] \to (\beta \to \alpha) = \\
&= (((\beta \to \alpha) \to \beta) \to \alpha) \to [(\beta \to (\beta \to \alpha)) \to (\beta \to \alpha)] \\
&= (((\beta \to \alpha) \to \beta) \to \alpha) \to [((\beta \to \alpha) \to \beta) \to \beta] \\
&= [((\beta \to \alpha) \to \beta) * (((\beta \to \alpha) \to \beta) \to \alpha)] \to \beta \\
&= [\alpha * (\alpha \to ((\beta \to \alpha) \to \beta))] \to \beta \\
&= ((\beta \to \alpha) \to (\alpha \to \beta)) \to (\alpha \to \beta) \\
&= (\beta \to \alpha) \vee (\alpha \to \beta). \quad \square
\end{aligned}$$

REMARK 1.5.

(a) 1.4(7) goes back to Theorem 3.3 in [Chang 1958]. In particular any complete MV-algebra is a strong GL-monoid (cf. 1.3).
(b) By virtue of 1.2(6) and 1.4(2) the Booleanization of any complete MV-algebra exists — i.e. there exists uniquely up to an isomorphism a complete Boolean algebra $\mathbb{B}$ satisfying the following property: There exists a lattice-theoretical embedding $i : L \longrightarrow \mathbb{B}$ s.t. i

preserves arbitrary meets and joins, and the Boolean algebra generated by the image $i(L)$ is a regular subalgebra of $\mathbb{B}$ (see [Funayama 1959], [Sikorski 1964], [Grätzer 1978]).

THEOREM 1.6. Let $(P, \leq, *, \rightarrow)$ be an integral, residuated, commutative po-monoid with zero satisfying (L0) and the additional axiom:

$$\text{if } \alpha \leq \beta^n \text{ (where } \beta^n \text{ is the nth power of } \beta \text{ } w.r.t. *) \text{ for all } n \in \mathbb{N}, \text{ then } \alpha = \alpha * \beta \quad \text{(L6)}$$

Then the MacNeille completion $(P^{\#}, \leq^{\#}, *^{\#})$ (cf. [MacNeille 1937]) is a GL-monoid satisfying also (L6).

PROOF. For each subset A of P let us denote by $U(A)$ (resp. $L(A)$) the set of all upper (resp. lower) bounds of A; then the MacNeille completion $(P^{\#}, \leq^{\#})$ of $(P, \leq)$ is determined by

$$P^{\#} = \{A \subseteq P \mid L(U(A)) = A\}$$
$$A \leq^{\#} B \Leftrightarrow A \subseteq B.$$

The semigroup operation $*$ induces on the power set $\mathcal{P}(P)$ of P the so-called Minkowski multiplication $A * B = \{a * b \mid a \in A, b \in B\}$.

(a) We show

$$U[L(U(A)) * B] = U(A * B) \tag{1.3}$$

The set inclusion $U[L(U(A)) * B] \subseteq U(A * B)$ is obvious. On the other hand we fix $c_0 \in L(U(A))$, $b_0 \in B$ and choose $\gamma \in U(A * B)$. Since the monoid is residuated, we infer from (L1): $b_0 \rightarrow \gamma \in U(A)$; hence $c_0 \leq b_0 \rightarrow \gamma$. Now we apply again (L1) and obtain $b_0 * c_0 \leq \gamma$ — i.e. $U(A * B)$ is a subset of $U[L(U(A)) * B]$.

(b) We introduce a binary operation $*^{\#}$ on $P^{\#}$ by $A *^{\#} B = L(U(A * B))$. Obviously

$$L(U(A * B)) = L(U[L(U(A)) * B]), \; A \subseteq P, \; B \subseteq P \tag{1.3'}$$

follows immediately from (1.3); hence $*^{\#}$ is associative, $L(U(\{\mathbf{1}\})) = L$ (resp. $L(U(\{\mathbf{0}\}) = \{\mathbf{0}\})$ acts as unit (resp. zero) element w.r.t. $*^{\#}$, and the embedding $i : P \rightarrowtail P^{\#}$ defined by $i(\alpha) = L(U(\{\alpha\}))$ is a semigroup homomorphism from $(P, *)$ to $(P^{\#}, *^{\#})$. In particular $(P^{\#}, \leq^{\#}, *^{\#})$ is an integral po-monoid with zero. Since $L(U(\bigcup_{i \in I} A_i))$ is the join of the family $\{A_i, i \in I\}$ in $(P^{\#}, \leq^{\#})$, we conclude from (1.3') that $(P^{\#}, \leq^{\#}, *^{\#})$ is also a completely lattice-ordered monoid.

(c) We show that $(P^{\#}, \leq^{\#}, *^{\#})$ satisfies (L0). For each pair $(A, B) \in P^{\#} \times P^{\#}$ with $A \subseteq B$ we define a subset C of P by

$$C := \{e \rightarrow a \mid e \in U(B), \; a \in A\} \tag{1.4}$$

The set inclusion $U(A) \subseteq U(C * B)$ follows from $(e \rightarrow a) * b \leq (e \rightarrow a) * e \leq a \leq \gamma$, where $e \in U(B)$, $b \in B$, $a \in A$, $\gamma \in U(A)$. In order to verify $U(C * B) \subseteq U(A)$ we first fix $\gamma \in U(C * B)$, $a \in A$ and demonstrate the implication

$$e \in U(B) \text{ implies } e * (a \rightarrow \gamma) \in U(B) \tag{1.5}$$

Because of $A \subseteq B$, we infer from (L0) (see also 1.2 (3)): $e * (e \to a) = a$ for all $e \in U(B)$; hence we obtain for all $b \in B$ and $e \in U(B), b = e * (e \to b) \le e * [(e * (e \to a)) \to ((e \to a) * b)] \le e * (a \to \gamma)$ and therewith (1.5) is established. A recursive application of (1.5) results in

$$(a \to \gamma)^n \in U(B),\ \forall\, n \in \mathbb{N} \tag{1.5'}$$

Since A is a subset of B, the relation (1.5') implies: $a \le (a \to \gamma)^n$, $\forall\, n \in \mathbb{N}$. Now we invoke (L6) and obtain: $a = a * (a \to \gamma) \le \gamma$; i.e. $U(C * B)$ is a subset of $U(A)$. Hence the equation $L(U(C)) *^{\#} B = A$ follows from $U(C * B) = U(A)$ and (1.3').

(d) Let us consider $A, B \in P^{\#}$ satisfying for all $n \in \mathbb{N}$ the condition $A \subseteq B^n$ where B^n is the nth power of B w.r.t. $*^{\#}$. In view of (1.3') this means

$$U(B \underbrace{* \dots *}_{n\ times} B) U(A),\ \forall\, n \in \mathbb{N} \tag{1.6}$$

In order to verify $A = A *^{\#} B$ it is sufficient to prove $U(A * B) \subseteq U(A)$. Therefore we choose $\gamma \in U(A * B)$ and $a \in A$; then $a \to \gamma$ is an upper bound of B and consequently $(a \to \gamma)^n \in U(B * \ \dots\ * B)$ for all $n \in \mathbb{N}$. Using (1.6) we obtain: $a \le (a \to \gamma)^n$ for all $n \in \mathbb{N}$. Now we invoke again (L6) and conclude that γ is an upper bound of A. □

The MacNeille completion preserves also the axioms (L3) and (L5). With respect to later applications we consider explicitly the case of MV-algebras.

COROLLARY 1.6'. Let $(P, \le, *, \to)$ be an **MV-algebra** — i.e. an integral, residuated, commutative po-monoid with zero satisfying (L0) and (L5). If $(P, \le, *, \to)$ fulfills (L6), then the MacNeille completion $(P^{\#}, \le^{\#}, *^{\#})$ is a complete MV-algebra.

PROOF. Referring to the proof of 1.6, the set $L(U(C))$ (where C is determined by (1.4)) is a subset of the residual of A by B in $(P^{\#}, \le^{\#}, *^{\#})$. Obviously $B \overset{\#}{\to} \mathbf{O}^{\#}$ coincides with $L(\{b \to \mathbf{O} \mid b \in B\}) =: D$. We show

$$\{d \to \mathbf{O} \mid d \in D\} = U(B) \tag{1.7}$$

If $d \in D$, then $d \le b \to \mathbf{O}$ for all $b \in B$; hence $d \to \mathbf{O}$ is an upper bound of B. On the other hand, if e is an upper bound of B, then $b*(e \to \mathbf{O}) = \mathbf{O}$ for all $b \in B$ — i.e. $e \to \mathbf{O} \in D$; hence (1.7) follows from (L5). By virtue of (1.7) we obtain: $B = L(U(B)) = (B \overset{\#}{\to} \mathbf{O}^{\#}) \overset{\#}{\to} \mathbf{O}^{\#}$. □

REMARK 1.7.

(a) Let $\mathcal{L}$ be the set of all well-formed formulas of the predicate calculus w.r.t. a generalized logic (e.g. intuitionistic logic, Łukasiewicz logic or a mixture of both). Then the MacNeille completion of the Lindenbaum-Tarski algebra (i.e. the algebra of all logically equivalent well-formed formulas) is a GL-monoid. In particular, under the hypothesis of (L5) the axiom (L6) is syntactically equivalent to an infinitary inference rule which forces the completeness of the predicate calculus w.r.t. the axioms of Łukasiewicz logic [Hay 1963].

(b) In view of 1.2(3) and 1.4(6) the axiom, (L6) is not only sufficient, but also necessary that the MacNeille completion of integral, commutative, residuated po-monoids with zero preserves the axioms (L0) and L5).

Let $(L, \leq, *)$ be an integral po-monoid. An element $\gamma \in L$ is called **square root** of a given element $\alpha \in L$ *w.r.t.* $*$ iff γ satisfies the following conditions

$$\gamma * \gamma = \alpha \tag{Q1}$$

$$\text{if } \beta * \beta \leq \alpha, \text{then } \beta \leq \gamma \tag{Q2}$$

If γ is a square root of α, then γ is uniquely determined by α; therefore γ is also denoted by $\sqrt{\alpha}$.

LEMMA 1.8. Let $(L, \leq, *, \rightarrow)$ be an integral, residuated, commutative po-monoid satisfying (L0).

(a) If the square roots of α and β exist, then the square root of $\alpha \rightarrow \beta$ exists and is given by $\sqrt{\alpha} \rightarrow \sqrt{\beta}$.
(b) If the square roots of α, β, and $\alpha * \beta$ exist, then the following relations are valid:
 1° $\sqrt{\alpha} * \sqrt{\beta} \leq \sqrt{\alpha * \beta}$.
 2° If $\alpha \rightarrow (\alpha * \beta) = \beta$, then $\sqrt{\alpha} * \sqrt{\beta} = \sqrt{\alpha * \beta}$.
(c) If the square roots of α and β exist, then
 3° $(\alpha \leq \beta \text{ and } \beta = \sqrt{\alpha} * \sqrt{\beta}) \Rightarrow \alpha = \beta$.

PROOF. $(\sqrt{\alpha} \rightarrow \sqrt{\beta}) * (\sqrt{\alpha} \rightarrow \sqrt{\beta}) \leq \alpha \rightarrow \beta$ follows from (Q1). If $\lambda * \lambda \leq \alpha \rightarrow \beta$, then $(\sqrt{\alpha} * \lambda) * (\sqrt{\alpha} * \lambda) \leq \beta$; hence (Q2) implies $\sqrt{\alpha} * \lambda \leq \sqrt{\beta}$ — i.e. $\lambda \leq \sqrt{\alpha} \rightarrow \sqrt{\beta}$; thus $\sqrt{\alpha} \rightarrow \sqrt{\beta}$ is the square root of $\alpha \rightarrow \beta$. The relation 1° is obvious. Further we apply (a) to $\alpha \rightarrow (\alpha * \beta) = \beta$ and obtain $\sqrt{\alpha} \rightarrow \sqrt{\alpha * \beta} = \sqrt{\beta}$; hence $\sqrt{\alpha} * \sqrt{\beta} = \sqrt{\alpha * \beta}$ follows from (L0). Finally the hypothesis of 3° implies: $\alpha = \sqrt{\alpha} * \sqrt{\beta} * (\sqrt{\beta} \rightarrow \sqrt{\alpha}) = \beta * (\sqrt{\beta} \rightarrow \sqrt{\alpha}) = \sqrt{\beta} * \sqrt{\alpha} = \beta$; hence 3° is verified. □

LEMMA 1.9. Let $M = (L, \leq, *)$ be a strong GL-monoid, and let every element $\alpha \in L$ be a square w.r.t. $*$ (i.e. there exists $\beta \in L$ *s.t.* $\beta * \beta = \alpha$). Then M has square roots — i.e. $\sqrt{\alpha}$ exists for all $\alpha \in L$.

PROOF. Motivated by (Q2) we put $\sqrt{\alpha} := \vee\{\lambda \in L \mid \lambda * \lambda \leq \alpha\}$. Then $\sqrt{\alpha} * \sqrt{\alpha} \leq \alpha$ follows from (L3'). Since α is a square, the equality $\sqrt{\alpha} * \sqrt{\alpha} = \alpha$ holds; hence $\sqrt{\alpha}$ is the square root of α. □

We finish this section with a list of examples of strong GL-monoids.

EXAMPLES 1.10.

(a) Let $(\Omega, \leq)$ be a complete Heyting algebra — i.e. $(\Omega, \leq)$ is a complete lattice satisfying the infinite distributive law

$$\alpha \wedge (\bigvee_{i \in I} \beta_i) = \vee_{i \in I} (\alpha \wedge \beta_i).$$

Then $M = (\Omega, \leq, \wedge)$ is a strong GL-monoid.

(b) As we have already mentioned in 1.5(a), complete MV-algebras are strong GL-monoids.

(c) Let $[0, 1]$ be the real unit interval and T be a continuous t-norm on $[0, 1]$, i.e. $T : [0,1] \times [0,1] \longrightarrow [0,1]$ is a continuous function provided with the following properties ([Schweizer and Sklar 1983]):

$T(x,1) = T(1,x) = x,\ T(0,x) = T(0,x) = 0$	**(Boundary Conditions)**	$(\mathcal{T}1)$
$T(x,y) = T(y,x)$	**(Symmetry)**	$(\mathcal{T}2)$
$T(x_1,y_1) \leq T(x_2,y_2)$ whenever $x_1 \leq x_2,\ y_1 \leq y_2$	**(Isotonicity)**	$(\mathcal{T}3)$
$T(x, T(y,z)) = T(T(x,y),z)$	**(Associativity)**	$(\mathcal{T}4)$

In particular continuous t-norms and continuous semigroup operations on $[0, 1]$ satisfying $(\mathcal{T}1)$ are the same things (cf. [Paalman-de Miranda 1964]). Then $M = ([0,1], \leq, T)$ is a GL-monoid.

REMARK 1.10'.

(a) The intersection of the class of complete Heyting algebras with the class of complete MV-algebras is precisely the class of complete Boolean algebras. The real unit interval equipped with the usual product (resp. with a strict Archimedean t-norm (cf. [Schweizer and Sklar 1983])) furnishes an example of a Gl-monoid which is neither a complete Heyting algebra (i.e. $* = \wedge$, see 1.10(a)) nor a complete MV-algebra.

(b) All examples in 1.10 fulfill (L4). This motivates the following **open question**: Does (L3) imply (L4)?

§2. The Category of M-Valued Sets

In this and the subsequent paragraphs $M = (L, \leq, *)$ denotes always a GL-monoid.

We introduce an M-**valued set** as an ordered pair (X, E), where X is a set (the so-called **carrier** of (X, E)) and $E : X \times X \longrightarrow L$ is a map subjected to the following axioms:

$E(x,y) \leq E(x,x) \wedge E(y,y)$	**(Strictness)**	(E1)
$E(x,y) = E(y,x)$	**(Symmetry)**	(E2)
$E(x,y) * (E(y,y) \rightarrow E(y,z)) \leq E(x,z)$	**(Transitivity)**	(E3)

Obviously every M-valued set can be viewed as a M-valued model of the formalized theory of identity with existence predicate [Höhle 1990b]. In particular $E(x,y)$ is the **degree** to which x and y **coincide**, and $E(x,x)$ is the **extent** to which x **exists**.

It is interesting to see that the Cartesian product of the underlying lattice $(L,\leq)$ carries itself an intrinsic structure of an M-valued set.

EXAMPLE 2.1. Let us consider a map $E_L : (L \times L) \times (L \times L) \longrightarrow L$ defined by

$$E_L((\alpha,\beta),(\bar{\alpha},\bar{\beta})) = (\alpha * (\beta \to \bar{\beta}) \wedge (\bar{\alpha} * (\bar{\beta} \to \beta))$$

Then $\mathcal{L} := (L \times L, E_L)$ is an M-valued set. The strictness and symmetry of E_L are evident. The transitivity of E_L follows from the subsequent estimations in which we use already the axiom (L0):

$$E_L((\alpha_1,\beta_1),(\alpha_2,\beta_2)) * (\alpha_2 \to E_L((\alpha_2,\beta_2),(\alpha_3,\beta_3))) \leq$$

$$\alpha_2 * (\alpha_2 \to (\alpha_1 * (\beta_1 \to \beta_2))) * (\alpha_2 \to (\alpha_2 * (\beta_2 \to \beta_3))) \leq$$

$$\alpha_1 * (\beta_1 \to \beta_2) * (\beta_2 \to \beta_3) \leq \alpha_1 * (\beta_1 \to \beta_3),$$

$$E_L((\alpha_1,\beta_1),(\alpha_2,\beta_2)) * (\alpha_2 \to E_L((\alpha_2,\beta_2),(\alpha_3,\beta_3))) \leq$$

$$\alpha_2 * (\beta_2 \to \beta_1) * (\alpha_2 \to (\alpha_3 * (\beta_3 \to \beta_2))) \leq$$

$$(\beta_2 \to \beta_1) * \alpha_3 * (\beta_3 \to \beta_2) \leq \alpha_3 * (\beta_3 \to \beta_1). \quad \square$$

A canonical construction of M-valued sets is explained in the following example:

EXAMPLE 2.2. Let X be a set, $d : X \longrightarrow L$, and $[\![\ ,\ ]\!] : X \times X \longrightarrow L$ are maps satisfying the following conditions:

$d(x) * [\![x,y]\!] \leq d(y)$	(**Extensionality**)	(EV0)
$[\![x,x]\!] = \mathbf{1}$	(**Global Existence**)	(EV1)
$[\![x,y]\!] = [\![y,x]\!]$	(**Symmetry**)	(EV2)
$[\![x,y]\!] * [\![y,z]\!] \leq [\![x,z]\!]$	(**Global Transitivity**)	(EV3)

$(X, [\![\ \ ]\!], d)$ induces an M-valued set $(X, E_{[\ ]})$ by

$$E_{[\ ]}(x,y) = d(x) * d(y) * [\![x,y]\!].$$

The strictness of $E_{[\ ]}$ follows immediately from $(EV0)$. The symmetry of $E_{[\ ]}$ is evident. Using again (EV0) we verify the transitivity of $E_{[\ ]}$ as follows

$$E_{[\ ]}(x,y) * (E_{[\ ]}(y,y) \to E_{[\ ]}(y,z)) =$$

$$(d(y) \to (d(x) * [\![x,y]\!])) * d(y) * d(y) * [(d(y) * d(y)) \to (d(y) * d(z) * [\![y,z]\!])]$$

$$= d(x) * d(z) * [\![x,y]\!] * [\![y,z]\!] \leq E_{[\ ]}(x,z). \quad \square$$

If M has square roots (see also 1.9), then every M-valued set (X, E) arises from a triple $(X, [\![\ ,\ ]\!], d)$ in the sense of 2.2. In particular $[\![\ ,\ ]\!]$ and d are determined by

$$[\![x, y]\!] = \sqrt{(E(x,x) \to E(x,y))} * \sqrt{(E(y,y) \to E(x,y))}, \ d(x) = \sqrt{E(x,x)}.$$

An M-valued set (X, E) is **separated** iff (X, E) satisfies the additional axiom

$E(x,x) \vee E(y,y) \le E(x,y)$ implies $x = y$ **(Separation)** (E4)

We can associate with every M-valued set (X, E) a separated M-valued set $(\hat{X}, \hat{E})$ in the following way: The axioms (E2) and (E3) guarantee that

$$x \approx y \Leftrightarrow E(x,x) \vee E(y,y) \le E(x,y)$$

is an (ordinary) equivalence relation on X; then $(\hat{X}, \hat{E})$ is given by:

$$\hat{X} = X / \approx, \ \hat{E}(\hat{x}, \hat{y}) = E(x', y') \text{ where } x' \in \hat{x}, \ y' \in \hat{y}$$

The category **M-SET** of M-valued sets consists of the following data: **Objects** are separated M-valued sets and **morphisms** are structure preserving maps — i.e. $\varphi : (X, E) \longrightarrow (Y, F)$ is a morphism iff $\varphi : X \to Y$ is a map satisfying the following axioms:

$F(\varphi(x), \varphi(x)) \le E(x,x)$ **(Strictness)** (M1)
$E(x_1, x_2) \le F(\varphi(x_1), \varphi(x_2))$ **(Preservation of Equality)** (M2)

The **composition** is the usual composition of maps and the **identity** of (X, E) is the **identity** map $id_X : X \to X$.

M-Set has an initial object $\mathbf{O} = (\emptyset, \emptyset)$ and a terminal object $\mathbf{1} = (L, \wedge)$, where L is the underlying lattice of the GL-monoid M and $\wedge$ is the meet operation in L.

PROPOSITION 2.3. **M-SET** has set-indexed products.

PROOF. Let $\mathcal{F} = \{(X_i, E_i) \mid i \in I\}$ be a family of separated M-valued sets. If $\mathcal{F}$ is void, then the terminal object **1** is the product of $\mathcal{F}$. Therefore we assume that $\mathcal{F}$ is non void, and we put

$$X_0 = \{(x_i)_{i\in I} \in \times_{i\in I} X_i \mid E_i(x_i, x_i) = E_j(x_j, x_j) \ \forall i, j \in I\}$$
$$(\times_{i\in I} E_i)((x_i)_{i\in I}, (\bar{x}_i)_{i\in I}) = \bigwedge_{i\in I} E_i(x_i, \bar{x}_i).$$

It is easy to see that $(X_0, x_{i\in I} E_i)$ is separated M-valued set. Further we denote by $\Pi_i : X_0 \longrightarrow X_i$ the restriction of the (set-theoretical) projection to X_0. Then $((X_0, x_{i\in I} E_i), (\Pi_i)_{i\in I})$ is obviously the product of $\mathcal{F}$. □

PROPOSITION 2.4. **M-SET** has multiple equalizers (which are constructed at the level of **SET**).

PROPOSITION 2.5. **M-SET** has inverse limits.

PROOF. Let $(I, \leq)$ be a downward-directed set and $((X_i, E_i), \varphi_{ji})$ be an inverse system in **M-SET** — i.e. for every pair $(i,j) \in I \times I$ with $j \leq i$, there exists a morphism $\varphi_{ji} : (X_i, E_i) \longrightarrow (X_j, E_j)$ such that the following two conditions are satisfied: $\varphi_{ii} = id_{(X_i,E_i)}, \varphi_{jk} = \varphi_{ji} \cdot \varphi_{ik}$ whenever $j \leq i \leq k$. Further let T_I be the subset of the carrier X_0 of $\times_{i\in I}(X_i, E_i)$ (cf. 2.3) defined by $T_I = \{(x_i)_{i\in I} \in X_0 \mid \varphi_{ji}(x_i) = x_j \text{ whenever } i \leq j\}$. We denote by E_I the restriction of $\times_{i\in I} E_i$ to T_I and by f_i the restriction of the canonical projection to T_I. Then it is not difficult to show that $((T_I, E_I), (f_i)_{i\in I})$ is the inverse limit of $((X_i, E_i), \varphi_{ji})$. □

PROPOSITION 2.6. **M-SET** has set-indexed coproducts.

PROOF. Let $\mathcal{F} = \{(X_i, E_i) | i \in I\}$ be a family of separated M-valued sets. If $\mathcal{F}$ is empty, then the initial object **O** is the coproduct of $\mathcal{F}$. Therefore we assume that $\mathcal{F}$ is non void and denote by $X_{00} = \dot{\cup}_{i\in I} X_i$ the disjoint union of $\{X_i \mid i \in I\}$. Further let $J_i : X_i \longrightarrow X_{00}$ be the inclusion map. We introduce on X_{00} an M-valued equality by

$$(+_{i\in I} E_i)(x,y) = \begin{cases} E_i(x,y) & : \text{ if } (x,y) \in X_i \times X_i \text{ for some } i \in I \\ \mathbf{0} & : \text{ if } (x,y) \notin X_i \times X_i \text{ for all } i \in I. \end{cases}$$

Then $((J_i)_{i\in I}, (X_{00}, +_{i\in I} E_i))$ is the coproduct of $\mathcal{F}$. □

In order to show that **M-SET** has multiple coequalizers we need the concept of M-valued equivalence relations.

DEFINITION 2.7. Let (X, E) be an M-valued set. A map $R : X \times X \longrightarrow L$ is called an **M-valued equivalence relation** on (X, E) iff R satisfies the following axioms

$R(x,y) \leq E(x,x) \wedge E(y,y)$	**(Strictness)**	(R01)
$(E(x,x) \to E(x,\bar{x})) * R(x,y) * (E(y,y) \to E(y,\bar{y}) \leq R(\bar{x},\bar{y})$	**(Extensionality)**	(R02)
$E(x,x) \leq R(x,x)$	**(Reflexivity)**	(R1)
$R(x,y) = R(y,x)$	**(Symmetry)**	(R2)
$R(x,y) * (R(y,y) \to R(y,z)) \leq R(x,z)$	**(Transitivity)**	(R3)

REMARK 2.8. Because of (R01), (R02), and (R1), every M-valued equivalence relation R on (X, E) fulfills the subsequent properties:

$$E(x,x) = R(x,x) \qquad \text{R1'}$$
$$E(x,y) \leq R(x,y) \qquad \text{R1''}$$

The relation (R1") motivates a partial ordering $\preceq$ on the set of $\mathcal{EV}(X, E)$ of all M-valued equivalence relations on (X, E) defined by

$$R_1 \preceq R_2 \Leftrightarrow R_1(x,y) \leq R_2(x,y), \text{ for all } x,y \in X$$

Then E is the universal lower bound in $(\mathcal{EV}(X,E),\preceq)$ — i.e. E is the finest M-valued equivalence relation on (X,E). The universal upper bound in $(\mathcal{EV}(X,E),\preceq$ (i.e. the coarsest M-valued equivalence relation on (X,E)) coincides with R_c, where R_c is defined by $R_c(x,y) = E(x,x)\wedge E(y,y)$. Moreover $(\mathcal{EV}(X,E),\preceq)$ is a complete lattice, and the meet of a family $\mathcal{F}=\{R_i \mid i\in I\}$ of M-valued equivalence relations R_i can be determined according to the following formula:

$$(\bigwedge \mathcal{F})(x,y) = \bigwedge_{i\in I} R_i(x,y), \text{ for all } x,y\in X$$

Proposition 2.9. **M-SET** has multiple coequalizers.

Proof. Let $\{\varphi_i | i\in I\}$ be a family of M-SET-morphisms $\varphi_i:(X,E)\longrightarrow(Y,F)$.

(a) We define a map $S: Y\times Y\longrightarrow L$ by

$$S(y_1,y_2) = \bigvee_{(i,j)\in I\times I} (\bigvee_{x\in X} F(\varphi_i(x),y_1) * (E(x,x)\rightarrow F(\varphi_j(x),y_2)))$$

Since S is in general not reflexive, neither symmetric, nor transitive, we consider the family $\mathcal{W}$ of all M-valued equivalence relations R on (Y,F) "containing" S — i.e. $S(y,z)\le R(y,z)$ for all $y,z\in Y$. Obviously $\mathcal{W}$ is non-void. We denote by R_0 the meet of $\mathcal{W}$ in $(\mathcal{EV}(Y,F),\preceq)$ and consider the separated, M-valued set $(Y/\approx,\hat{R}_0)$ associated with the M-valued set (Y,R_0). It is easy to see that the quotient map $\eta: Y\longrightarrow Y/\approx$ is a **M-SET**-morphism from (Y,F) to $(Y/\approx,\hat{R}_0)$. Because of the definition of S the morphism η satisfies the relation $\eta\cdot\varphi_i=\eta\cdot\varphi_j$ for all $i,j\in I$.

(b) We show that $(\eta,(Y/\approx,\hat{R}_0))$ satisfies the universal property of the coequalizer diagram. Therefore we consider a further morphism $\Psi:(Y,F)\longrightarrow(Z,G)$ provided with the property $\Psi\cdot\varphi_i=\Psi\cdot\varphi_j$ for all $i,j\in I$. The transitivity of G implies $S(y_1,y_2)\le G(\Psi(y_1),\Psi(y_2))$. Since (Z,G) is separated, there exists a unique map $\theta: Y/\approx\longrightarrow Z$ such that $\Psi(y)=\theta(\eta(y))$ for all $y\in Y$. Obviously θ is a M-SET-morphism. Hence the universal property is verified. □

Proposition 2.10. **M-SET** has direct limits.

Proof. Let $(I,\le)$ be an upward-directed set and $((X_i,E_i),\varphi_{ij})$ be a direct system — i.e. for every pair $(i,j)\in I\times I$ with $i\le j$ there exists a morphism $\varphi_{ji}:(X_i,E_i)\longrightarrow(X_j,E_j)$ such that the following two conditions are satisfied: $\varphi_{ii}: id_{(X_i,E_i)}$, $\varphi_{kj}=\varphi_{ji}\cdot\varphi_{ik}$ whenever $k\le i\le j$. Let us denote by X_{00} the carrier of the coproduct $((J_i)_{i\in I}, +_{i\in I}(X_i,E_i))$ (cf. 2.6). Then we define a map $S: X_{00}\times X_{00}\longrightarrow L$ by

$$S(u,v) = \bigvee_{i_0\le j_0} (\bigvee_{x_{i_0}\in X_{i_0}} (+_{i\in I} E_i)(u, J_{i_0}(x_{i_0}))$$

$$*(E_{i_0}(x_{i_0},x_{i_0})\rightarrow(+_{i\in I} E_i)(J_{j_0}\cdot\varphi_{j_0 i_0}(x_{i_0}),v))$$

and consider the "smallest" M-valued equivalence relation R_0 on $+_{i\in I}(X_i, E_i)$ "containing" S. Now we follow the trace of the proof of 2.9. First we denote by $(X_{00}/\approx, \hat{R}_0)$ the separated, M-valued set associated with (X_{00}, R_0) and by $\eta : +_{i\in I}(X_i, E_i) \longrightarrow (X_{00}/\approx, \hat{R}_0)$ the morphism induced by the quotient map $\eta : X_{00} \longrightarrow X_{00}/\alpha\approx$. Then $((\eta\cdot J_i)_{i\in I}, (X_{00}/\approx, \hat{R}_0))$ is the direct limit of $((X_i, E_i), \varphi_{ij})$. Indeed, if $\{\Psi_i \mid i \in I\}$ is another family of morphisms $\Psi_i : (X_i, E_i) \longrightarrow (Y, F)$ provided with the property $\Psi_i = \Psi_j \cdot \varphi_{ji}$ (whenever $i \leq j$), then the equivalence kernel of the universal arrow

$$+_{i\in I}\Psi_i : +_{i\in I}(X_i, E_i) \longrightarrow (Y, F) \text{ (i.e. } C(u,v) = F(+_{i\in I}\Psi_i(u), +_{i\in I}\Psi_i(v)))$$

contains S. Hence there exists an unique arrow $\theta : (X_{00}/\approx, \hat{R}_0) \longrightarrow (Y, F)$ satisfying the condition $\theta \cdot \eta \cdot J_i = \Psi_i$ for all $i \in I$. □

Summing up we obtain

FACT I. **M-SET** is a complete and cocomplete category.

In the following consideration we show that **M-SET** is an $(\mathcal{E}, \mathcal{M})$-category (for the definition of $(\mathcal{E}, \mathcal{M})$-categories see [Herrlich and Strecker 1979]). First we give in the case of strong GL-monoids a characterization of epimorphisms in **M-SET**.

PROPOSITION 2.11. Let M be a strong GL-monoid and (Y, F) be a separated, M-valued set such that for all $y \in Y$ the extent $F(y, y)$ of y is a square. Further let $\varphi : (X, E) \longrightarrow (Y, F)$ be an **M-SET**-morphism. Then the following assertions are equivalent:

(1) φ is an epimorphism

(2) $F(y,y) = \bigvee_{x\in X} (E(x,x) \rightarrow F(\varphi(x), y)) * F(\varphi(x), y)), \ \forall\, y \in Y \qquad (\mathcal{E})$

PROOF. ((1) $\Rightarrow$ (2)). Let $\hat{\mathcal{L}}$ be the separated M-valued set associated with $\mathcal{L} = (L \times L, E_L)$ (cf. 2.1). We define two maps μ_1 and μ_2 from Y to L by

$$\mu_1(y) = \bigvee\{\lambda \in L \mid \lambda * \lambda \leq \bigvee_{x\in X} (E(x,x) \rightarrow F(\varphi(x), y)) * F(\varphi(x), y)\},$$

$$\mu_2(y) = \bigvee\{\lambda \in L \mid \lambda * \lambda \leq F(y,y)\}$$

Now we invoke (L3') and obtain

$$\mu_1(y) * \mu_1(y) \leq \bigvee_{x\in X} (E(x,x) \rightarrow F(\varphi(x), y)) * F(\varphi(x), y)$$

Hence $\mu_1(y) * (F(y,y) \rightarrow F(y,\bar{y})) \leq \mu_1(\bar{y})$ follows from the transitivity of F; i.e. μ_i is an extensional L-fuzzy subset of (Y, F). Since F is strict, μ_2 is extensional too; and the hypothesis on the extent of existence in (Y, F) implies $\mu_2(y) * \mu_2(y) = F(y, y)$. By virtue of their extensionality both maps μ_1 and μ_2 determine respectively **M-SET**-morphisms χ_1 and χ_2 from (Y, F) to $\hat{L}$ as follows

$$\chi_i(y) = (F(y, y), \mu_i(y)), \ i = 1, 2.$$

Since $\chi_1 \cdot \varphi$ and $\chi_2 \cdot \varphi$ coincide, we conclude from assertion (1) that χ_1 and χ_2 also coincide — i.e. $F(y, y) = F(y, y) * (\mu_2(y) \to \mu_1(y))$ for all $y \in Y$; or equivalently $\mu_2(y) * \mu_2(y) = \mu_2(y) * \mu_1(y)$ for all $y \in Y$. Since $\mu_i(y)$ is the square root of $\mu_i(y) * \mu_i(y) (i = 1, 2)$, we apply 1.8(c) and obtain:

$$\begin{aligned} F(y, y) = \mu_2(y) * \mu_2(y) &= \mu_1(y) * \mu_1(y) \\ &\leq \vee_{x \in X}(E(x, x) \to F(\varphi(x), y)) * F(\varphi(x), y) \end{aligned}$$

Hence the assertion (2) is verified.

((2) $\Rightarrow$ (1)). Let us consider the following situation:

$$(X, E) \xrightarrow{\varphi} (Y, F) \underset{\Psi_2}{\overset{\Psi_1}{\rightrightarrows}} (Z, G)$$

with $\Psi_1 \cdot \varphi = \Psi_2 \cdot \varphi$. If we combine the assertion (2) with the transitivity axiom of M-valued equalities and the morphism axioms, then we obtain

$$\begin{aligned} G(\Psi_1(y), \Psi_2(y)) &\geq \vee_{x \in X} G(\Psi_1(y), \Psi_1 \cdot \varphi(x)) * (E(x, x) \to G(\Psi_2 \cdot \varphi(x), \Psi_2(y))) \\ &\geq \vee_{x \in X} F(y, \varphi(x) * (E(x, x) \to F(\varphi(x), y)) \\ &\geq F(y, y) \end{aligned}$$

Since (Z, G) is separated, Ψ_1 and Ψ_2 coincide. $\square$

A **M-SET**-morphism φ is monic iff φ is injective at the level of SET. A **M-SET**-morphism φ is an isomorphism iff φ is a bijective map satisfying

$$E(x_1, x_2) = F(\varphi(x_1), \varphi(x_2)) \text{ for all } x_1, x_2 \in X \tag{EM1}$$

In view of the following considerations we introduce two morphism classes in **M-SET**: $\mathcal{E}$ is the class of M-SET-morphisms satisfying the condition $\mathcal{E}$, and $\mathcal{M}$ is the class of M-SET-morphisms which fulfill (EM1) and the condition:

$$\varphi(X) = \{y \in Y \mid \bigvee_{x \in X} (E(x, x) \to F(\varphi(x), y)) * F(\varphi(x), y) = F(y, y)\}. \tag{EM2}$$

Obviously every element of $\mathcal{E}$ is an epimorphism (cf. part (b) of the proof of 2.11), and every element of $\mathcal{M}$ is a monomorphism. Moreover $\mathcal{E}$ and $\mathcal{M}$ are closed under composition. Now we are ready to verify

FACT II. **M-SET** is an $(\mathcal{E}, \mathcal{M})$-category.

LEMMA 2.12 (**Factorization**). Every **M-SET**-morphism $\varphi : (X,E) \longrightarrow (Y,F)$ can be factored as an $\mathcal{E}$-morphism $\eta : (X,E) \longrightarrow (Z,G)$ followed by a $\mathcal{M}$-morphism $\Psi : (Z,G) \longrightarrow (Y,F)$.

PROOF. Let Z be the subset of all $y \in Y$ satisfying

$$\bigvee_{x \in X} (E(x,x) \to F(\varphi(x),y)) * F(\varphi(x),y) = F(y,y)$$

Then G is the restriction of F to $Z \times Z$, Ψ is the inclusion map, and η is determined by φ — i.e. $\eta(x) = \varphi(x)$ for all $x \in X$. Obviously η is an $\mathcal{E}$-morphism, $\Psi \cdot \eta$ is a decomposition of φ, and Ψ fulfills (EM1). In order to complete the proof there is only one point left, namely the verification that Ψ satisfies (EM2). Therefore let us assume:

$$F(y_0,y_0) = \vee_{z \in Z}(G(z,z) \to F(\Psi(z),y_0) * F(\Psi(z),y_0)$$

Then we derive from the construction of $((Z,G),\Psi)$ the following relation

$$\begin{aligned} F(y_0,y_0) &= \bigvee_{z \in Z} (F(z,z) * (F(z,z) \to F(z,y_0)) * (F(z,z) \to F(z,y_0)) \\ &= \bigvee_{x \in X} (\bigvee_{z \in Z} E(x,x) * (E(x,x) \to F(\varphi(x),z)) * (E(x,x) \to F(\varphi(x),z)) \\ &\quad * (F(z,z) \to F(z,y_0)) * (F(z,z) \to F(z,y_0)) \\ &\leq \bigvee_{x \in X} E(x,x) * (E(x,x) \to F(\varphi(x),y_0)) * (E(x,x) \to F(\varphi(x),y_0)) \end{aligned}$$

Hence y_0 is an element of Z. □

LEMMA 2.13 (**Diagonalization property**). **M-SET** has the $(\mathcal{E},\mathcal{M})$-diagonalization property — i.e. for every commutative square in **M-SET**,

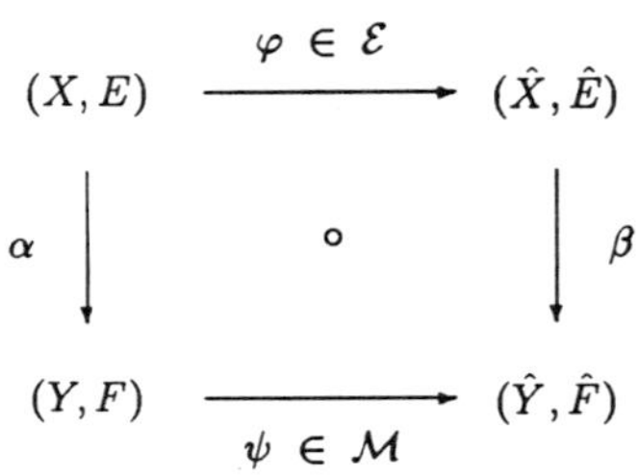

with $\varphi \in \mathcal{E}$ and $\Psi \in \mathcal{M}$ there exists a morphism $k : (\hat{X},\hat{E}) \longrightarrow (Y,F)$ such that $\alpha = k \cdot \varphi$ and $\beta = \Psi \cdot k$.

PROOF. By virtue of the transitivity of $\hat{F}$ the following relation holds:

$$\bigvee_{x \in X} (E(x,x) \to \hat{E}(\hat{x},\varphi(x))) * \hat{E}(\hat{x},\varphi(x)) * (F(y,y) \to F(y,\alpha(x))) * (E(x,x) \to F(y,\alpha(x)))$$

$$= \bigvee_{x \in X} (E(x,x) \to \hat{E}(\hat{x}, \varphi(x))) * \hat{E}(\hat{x}, \hat{x}) * (\hat{E}(\hat{x}, \hat{x}) \to \hat{E}(\hat{x}, \varphi(x))$$

$$*(F(y,y) \to F(y, \alpha(x))) * (E(x,x) \to F(y, \alpha(x)))$$

$$\leq \hat{E}(\hat{x}, \hat{x}) * [\bigvee_{x \in X} (E(x,x) \to \hat{F}(\beta(\hat{x}), \beta \cdot \varphi(x))) * (F(y,y) \to \hat{F}(\Psi(y), \beta \cdot \varphi(x)))$$

$$*(\hat{E}(\hat{x}, \hat{x}) \to \hat{F}(\beta(\hat{x}), \beta \cdot \varphi(x))) * (E(x,x) \to \hat{F}(\Psi(y), \beta \cdot \varphi(x)))]$$

$$\leq \hat{E}(\hat{x}, \hat{x}) * (F(y,y) \to \hat{F}(\beta(\hat{x}), \Psi(y))) * (\hat{E}(\hat{x}, \hat{x}) \to \hat{F}(\beta(\hat{x}), \Psi(y)))$$

Hence we obtain:

$$\bigvee_{y \in Y} (F(y,y) \to \hat{F}(\beta(\hat{x}), \Psi(y))) * \hat{F}(\beta(\hat{x}), \Psi(y))$$

$$\geq \bigvee_{x \in X} (\bigvee_{y \in Y} (E(x,x) \to \hat{E}(\hat{x}, \varphi(x))) * \hat{E}(\hat{x}, \varphi(x))$$

$$* (F(y,y) \to F(y, \alpha(x))) * (E(x,x) \to F(y, \alpha(x))))$$

$$= \bigvee_{x \in X} (E(x,x) \to \hat{E}(\hat{x}, \varphi(x))) * \hat{E}(\hat{x}, \varphi(x)) = \hat{E}(\hat{x}, \hat{x})$$

Now the condition (EM2) guarantees for each element $\hat{x} \in \hat{X}$ the existence of a unique element $y_{\hat{x}}$ with $\Psi(y_{\hat{x}}) = \beta(\hat{x})$; hence there exists a map $k : \hat{X} \longrightarrow Y$ satisfying the condition $\Psi \cdot k = \beta$. From (EM1) we conclude that k is a M-SET-morphism. Since Ψ is particularly a monomorphism, the relation $\Psi \cdot k \cdot \varphi = \beta \cdot \varphi = \Psi \cdot \alpha$ implies $k \cdot \varphi = \alpha$; therewith the assertion is verified. □

THEOREM 2.14.

(a) **M-SET** is an $(\mathcal{E}, \mathcal{M})$-category.
(b) If M has square roots, then **M-SET** is an (epi, extremal mono)-category.

PROOF. The assertion (a) follows from 2.12, 2.13 and Theorem 33.3 in [Herrlich and Strecker 1979]. The assertion (b) follows from 2.11, 2.12, 2.13 and from Theorem 33.3 and Corollary 33.7 in [Herrlich and Strecker 1979]. □

§3. An Algebraic Theory in M-SET

The intended, algebraic theory T is based on the concept of "**local points**" or singletons defined as follows: a **singleton** of a M-valued set (X, E) is a map $s : X \longrightarrow L$ provided with the properties

$$s(x) * (E(x,x) \to E(x,y)) \leq s(y) \qquad \textbf{(Extensionality)} \qquad \text{(S1)}$$

$s(x) * (\mathbb{E}(s) \to s(y)) \leq E(x,y)$ where $\mathbb{E}(s) = \bigvee_{x \in X} s(x)$ **(Singleton Condition)** (S2)

Referring to 1.2(3) the strictness of M-valued equalities implies also the strictness of singletons — i.e. every singleton satisfies the condition

$s(x) \leq E(x,x)$ for all $x \in X$ **(Strictness)** (S3)

A simple example of a singleton of (X, E) is generated by each pair $(x, \lambda) \in X \times L$ as follows:

$$s_{(x,\lambda)}(z) = (E(x,x) \to \lambda) * E(x,z) \; \forall z \in X \tag{3.1}$$

If $E(x,x) \leq \lambda$, then $s_{(x,\lambda)}$ is also denoted by $\tilde{x}$.

On the set $S(X,E)$ of all singletons of (X,E), there exists an intrinsic, M-valued equality $\tilde{E}$ defined by

$$\tilde{E}(s_1,s_2) = [\mathbb{E}(s_1) * \wedge_{x \in X}(s_1(x) \to s_2(x))] \wedge [\mathbb{E}(s_2) * \wedge_{x \in X}(s_2(x) \to s_1(x))]$$

The strictness and symmetry of $\tilde{E}$ is evident. In order to verify the transitivity we proceed as follows

$$\begin{aligned} &\tilde{E}(s_1,s_2) * (\mathbb{E}(s_2) \to \tilde{E}(s_2,s_3)) \\ &\leq [\mathbb{E}(s_2) \to (\mathbb{E}(s_1) * (\wedge_{x \in X}(s_1(x) \to s_2(x))))] \\ &* [\mathbb{E}(s_2) * [\mathbb{E}(s_2) \to (\mathbb{E}(s_2) * \wedge_{x \in X}(s_2(x) \to s_3(x)))]] \\ &\leq \mathbb{E}(s_1) * \wedge_{x \in X}(s_1(x) \to s_3(x)), \tilde{E}(s_1,s_2) * (\mathbb{E}(s_2) \to \tilde{E}(s_2,s_3)) \\ &\leq \mathbb{E}(s_2) * (\wedge_{x \in X}(s_2(x) \to s_1(x))) \\ &* (\mathbb{E}(s_2) \to (\mathbb{E}(s_3) * \wedge_{x \in X}(s_3(x) \to s_2(x)))) \\ &\leq \mathbb{E}(s_3) * (\wedge_{x \in X}(s_3(x) \to s_1(x))) \end{aligned}$$

i.e. $\tilde{E}(s_1,s_2) * (\tilde{E}(s_2,s_2) \to \tilde{E}(s_2,s_3)) \leq \tilde{E}(s_1,s_3)$. □

PROPOSITION 3.1.

(a) $\Sigma(X,E) := (S(X,E), \tilde{E})$ is a separated M-valued set. (b) $\tilde{E}(s,x) = s(x)$ for all $s \in S(X,E)$ and for all $x \in X$.

PROOF. (a) If $\mathbb{E}(s_1) = \mathbb{E}(s_1) * \wedge_{x \in X}(s_1(x) \to s_2(x))$, then $s_1(x) = \mathbb{E}(s_1) * (\mathbb{E}(s_1) \to s_1(x)) = s_1(x) * \wedge_{x \in X}(s_1(x) \to s_2(x)) \leq s_2(x)$. Interchanging now the role of s_1 and s_2 we can easily verify the separation axiom (E4).

(b) Let s be a singleton of (X,E); then we obtain from (S1) and (S2):

$$\bigwedge_{z \in X} (\tilde{x}(z) \to s(z)) = E(x,x) \to s(x), \quad \bigwedge_{z \in X} (s(z) \to \tilde{x}(z)) \geq \mathbb{E}(s) \to s(x)$$

hence $\tilde{E}(s,x) = s(x)$ follows from (S3). □

If (X, E) is a M-valued set, then $\eta_{(X,E)}$ denotes a map from X to $S(X, E)$ defined as follows

$$\eta_{(X,E)}(x) = \tilde{x} \text{ for all } x \in X \tag{3.2}$$

An immediate consequence of 3.1(b) is

COROLLARY 3.1'. $\eta_{(X,E)}$ is a **M-SET**-morphism from (X, E) to $\Sigma(X, E)$.

We show that (Σ, η) gives rise to an algebraic theory (in clone form) in **M-SET**. The following lemma starts with the definition of the clone composition.

LEMMA 3.2. If $\varphi : (X, E) \longrightarrow \Sigma(Y, F)$ and $\Psi : (Y, F) \longrightarrow \Sigma(Z, G)$ are **M-SET**-morphisms, then $\Psi \circ \varphi$ defined by

$$[(\Psi \circ \varphi)(x)](z) = \vee_{y \in Y}(F(y, y) \rightarrow [\varphi(x)](y) * [\Psi(y)](z) \tag{3.3}$$

is an **M-SET**-morphism from (X, E) to $\Sigma(Z, G)$.

PROOF. The extensionality of $(\Psi \circ \varphi)(x)$ is obvious. In order to verify the singleton condition (S2) we proceed as follows: $\mathbb{E}((\Psi \circ \varphi)(x)) = E(x, x)$ is an immediate consequence of the strictness of φ and Ψ. Since Ψ preserves the "equality" (cf. axiom (M2)), a repeated application of 1.2(3) leads to the following relation

$$(\mathbb{E}((\Psi \circ \varphi)(x)) \rightarrow [(\Psi \circ \varphi)(x)](z)) * [(\Psi \circ \varphi)(x)](\bar{z}) =$$

$$\bigvee_{y \in Y} (\mathbb{E}((\Psi \circ \varphi)(x)) \rightarrow [(\Psi \circ \varphi)(x)](z)) * [\varphi(x)](y) * (F(y, y) \rightarrow [\Psi(y)](\bar{z})) =$$

$$\bigvee_{y', y \in Y} [\varphi(x)](y') * (E(x, x) \rightarrow [\varphi(x)](y)) * (F(y', y') \rightarrow [\Psi(y')](z)) * (F(y, y) \rightarrow [\Psi(y)](\bar{z})) \leq$$

$$\bigvee_{y', y \in Y} F(y, y') * (F(y', y') \rightarrow [\Psi(y')](z)) * (F(y, y) \rightarrow [\Psi(y)](\bar{z})) \leq$$

$$\bigvee_{y', y \in Y} F(y', y') * (\wedge_{z' \in Z}([\Psi(y')](z') \rightarrow [\Psi(y)](z'))) * (F(y', y') \rightarrow [\Psi(y')](z)) *$$

$$(F(y, y) \rightarrow [\Psi(y)](\bar{z})) \leq$$

$$\bigvee_{y \in Y} [\Psi(y)](z) * (F(y, y) \rightarrow [\Psi(y)](\bar{z})) \leq G(z, \bar{z})$$

Hence the map $(\Psi \circ \varphi)(x)$ is a singleton of (Z, G) for each $x \in X$.

We show that $\Psi \circ \varphi$ is a **M-SET**-morphism. Obviously $\Psi \circ \varphi$ is strict. In order to verify (M2) we first observe that for any map $\varphi : X \longrightarrow S(Y, F)$, the axiom (M2) is equivalent to

$$E(x, \bar{x}) * (E(x, x) \rightarrow [\varphi(x)](y)) \leq [\varphi(\bar{x})](y), \ \forall\, x, \bar{x} \in X, \ \forall\, y \in Y$$

With regard to the preceding relation we can easily establish the following estimation

$$(E(x,x) \to E(x,\bar{x})) * [(\Psi \circ \varphi)(x)](z) =$$

$$\bigvee_{y \in Y} E(x,\bar{x}) * (E(x,x) \to [\varphi(x)](y)) * (F(y,y) \to [\Psi(y)](z)) \leq$$

$$\bigvee_{y \in Y} [\varphi(\bar{x})](y) * (F(y,y) \to [\Psi(y)](z)) = (\Psi \circ \varphi)(\bar{x})(z)$$

Hence $E(x,\bar{x}) \leq \tilde{G}((\Psi \circ \varphi)(x), (\Psi \circ \varphi)(\bar{x}))$ is verified for all $x, \bar{x} \in X$. □

THEOREM 3.3.

(a) Let Σ be the object function which associates with every separated, M-valued set (X,E) the singleton space $\Sigma(X,E)$ (cf. 3.1). Further η assigns to each separated, M-valued set (X,E) a **M-SET**-morphism $\eta_{(X,E)}$ determined by (3.2), and $\circ$ assigns to each ordered triple $((X,E),(Y,F),(Z,G))$ of separated, M-valued sets a "clone composition" function **M-SET** $((X,E),\Sigma(Y,F)) \times$ **M-SET**$((Y,F),\Sigma(Z,G)) \longrightarrow$ **M-SET** $((X,E),\Sigma(Z,G))$ defined in (3.3). Then the triple $T = (\Sigma, \circ, \eta)$ is an algebraic theory in clone form[1] in **M-SET**.

(b) T is degenerated[1] if and only if the underlying GL-monoid M is determined by a complete Heyting algebra — i.e. $* = \wedge$.

PROOF.

(a) It is easy to see that $\circ$ is associative. If $\varphi : (X,E) \longrightarrow \Sigma(Y,F)$ is a M-SET-morphism, then $\eta_{(Y,F)} \circ \varphi = \varphi$ follows from the extensionality of singletons. Further let $f : (X,E) \longrightarrow (Y,F)$ and $\Psi : (Y,F) \longrightarrow \Sigma(Z,G)$ be two **M-SET**-morphisms; then we obtain:

$$[\Psi(f(x))](z) \leq \bigvee_{y \in Y} F(f(x),y) * (F(y,y) \to [\Psi(y)](z)) \leq$$

$$\bigvee_{y \in Y} F(y,y) * ([\Psi(y)](z) \to [\Psi(f(x))](z)) * (F(y,y) \to [\Psi(y)](z)) \leq [\Psi(f(x))](z)$$

i.e. $(\Psi \circ (\eta_{(Y,F)} \cdot f)) = \Psi \cdot f$.

(b) First we note that T is degenerated iff for every separated, M-valued set (X,E) the **M-SET**-morphism $\mu_{(X,E)} := id_{\Sigma(X,E)} \circ id_{\Sigma(\Sigma(X,E))}$ is an **M-SET**-isomorphism. Because of 3.1(b), (3.2), and of $\mu_{(X,E)} \cdot \eta_{\Sigma(X,E)} = id_{\Sigma(X,E)}$, the degeneracy of $(\Sigma, \circ, \eta)$ is equivalent to the surjectivity of $\eta_{\Sigma(X,E)}$. Let us consider the separated M-valued set $(\{\cdot\}, \approx)$ whose carrier consists of a single element $\cdot$ with "global" extent of existence — i.e. $\cdot \approx \cdot = \mathbf{1}$. It is easy to see that $\Sigma(\{\cdot\}, \approx)$ is the terminal object $\mathbf{1}$ in **M-SET**; in particular $\eta_{\mathbf{1}} : \mathbf{1} \longrightarrow \Sigma(\mathbf{1})$ is determined by $[\eta_1(\alpha)](\beta) = \alpha \wedge \beta$. According to (3.1) we can associate with each $\alpha \in L$ a singleton $s_{(1,\alpha)}$ of $\mathbf{1}$. If η is surjective, then $[\eta_1(\alpha)](\beta) = \alpha * \beta$ holds for all $\alpha, \beta \in L$ —

[1]Here we follow the terminology presented in [Manes 1976] and [Schubert 1972].

i.e. $* = \wedge$, or in other words, M is determined by the underlying complete Heyting algebra L. On the other hand, if $*$ and $\wedge$ coincide, then $\tilde{E}$ and $\mu_{(X,E)}$ are given by

$$\tilde{E}(s_1, s_2) = \bigvee_{x \in X} (s_1(x) \wedge s_2(x))$$

$$[\mu_{(X,E)}(\sigma)](x) = \bigvee_{s \in S(X,E)} (\sigma(s) \wedge s(x)).$$

Then the surjectivity of $\eta_{\Sigma(X,E)}$ can be verified as follows:

$$\tilde{E}(\mu_{(X,E)}(\sigma), s') = \bigvee_{s \in S(X,E)} (\vee_{x \in X}(\sigma(s) \wedge s(x) \wedge s'(x))) =$$

$$\bigvee_{s \in S(X,E)} (\sigma(s) \wedge \tilde{E}(s, s')) = \bigvee_{s \in S(X,E)} (\sigma(s) \wedge (\mathbb{E}(s) \to \tilde{E}(s, s'))) = \sigma(s'). \quad \square$$

REMARK 3.4. Let M be a GL-Monoid determined by a complete Heyting algebra Ω — i.e. $* = \wedge$! Then the concepts of M-valued sets and Ω-valued sets [Fourman and Scott 1979] coincide. Moreover, the Eilenberg-Moore category $(\mathbf{M\text{-}SET})^T$ of T-algebras is isomorphic to the category $sh(\Omega)$ of sheaves over Ω, and the Kleisli category $(\mathbf{M\text{-}SET})_T$ is isomorphic to Higgs' topos Ω-**SET** (cf. [Higgs 1973]). In this context the equivalence between $sh(\Omega)$ and Ω-**SET** (see: [Forman and Scott 1979]) follows immediately from the degeneracy of the algebraic theory T.

§4. T-Algebras and Sheaves over GL-Monoids

Let $T = (\Sigma, \circ, \eta)$ be the algebraic theory (in clone form) specified in Section 3 above.

A **T-algebra** is a pair $((X, E), \xi)$ where (X, E) is a separated, M-valued set, and $\xi : \Sigma(X, E) \to (X, E)$ is a **M-SET**-morphism such that the following diagrams are commutative

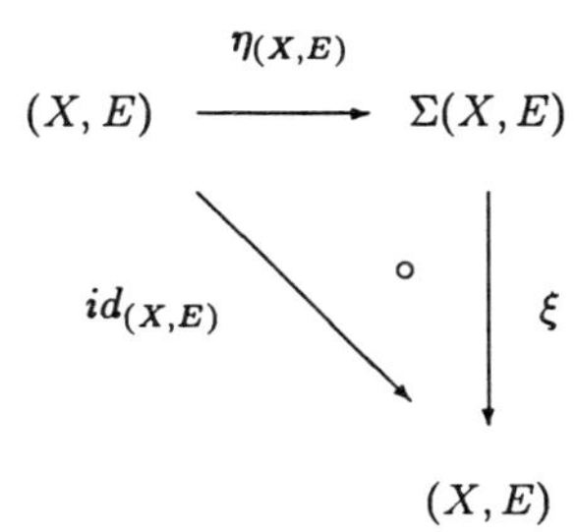

(4.1)

$$\begin{array}{ccc} \Sigma(\Sigma(X,E)) & \xrightarrow{\Sigma(\xi)} & \Sigma(X,E) \\ \mu_{(X,E)} \Big\downarrow & \circ & \Big\downarrow \xi \\ \Sigma(X,E) & \xrightarrow[\xi]{} & (X,E) \end{array} \tag{4.2}$$

where $\mu_{(X,E)} = id_{\Sigma(X,E)} \circ id_{\Sigma(\Sigma(X,E))}$ and $\Sigma(\xi) = (\eta_{(X,E)} \cdot \xi) \circ id_{\Sigma(\Sigma(X,E))}$.

A first example of a T-algebra is $(\Sigma(X,E), \mu_{(X,E)})$, the free T-algebra generated by (X,E). A further example is the following one: We put $R_{\mathcal{L}} := \{(\alpha,\beta) \in L \times L \mid \beta \leq \alpha\}$ and observe that the restriction $E_{\mathcal{L}}$ of E_L (see: 2.1) to $R_{\mathcal{L}} \times R_{\mathcal{L}}$ is a separated, M-valued equality. Further we denote by Ω the separated, M-valued set $(R_{\mathcal{L}}, E_{\mathcal{L}})$ and consider a **M-SET**-morphism $\xi_\Omega : \Sigma(\Omega) \longrightarrow \Omega$ defined as follows: $\xi_\Omega(s) = (\mathbb{E}(s), s(\mathbf{1},\mathbf{1}))$. Then the pair (Ω, ξ_Ω) is a T-algebra. In particular the conditions (4.1) and (4.2) follow from

$$\xi_\Omega(\eta_\Omega(\alpha,\beta)) = (\alpha,\, \alpha \wedge (\mathbf{1} * (\mathbf{1} \rightarrow \beta))) = (\alpha,\beta),$$

$$\xi_\Omega([\Sigma(\xi_\Omega)](\sigma)) = (\mathbb{E}(\sigma), \vee_{s \in S(R_{\mathcal{L}},E_{\mathcal{L}})} (\mathbb{E}(s) \rightarrow \sigma(s)) * E_{\mathcal{L}}(\xi_\Omega(s), (\mathbf{1},\mathbf{1})))) =$$

$$(\mathbb{E}(\sigma), \bigvee_{s \in S(R_{\mathcal{L}},E_{\mathcal{L}})} (\mathbb{E}(s) \rightarrow \sigma(s)) * (\mathbb{E}(s) \wedge s(\mathbf{1},\mathbf{1}))) =$$

$$(\mathbb{E}(\sigma), \bigvee_{s \in S(R_{\mathcal{L}},E_{\mathcal{L}})} (\mathbb{E}(s) \rightarrow \sigma(s)) * s(\mathbf{1},\mathbf{1}))) = \xi_\Omega(\mu_\Omega(\sigma)). \quad \square$$

Now we extend the concept of compatible subsets, joins of compatible subsets (see e.g. [Fourman and Scott 1979]) to the scope of M-valued sets.

DEFINITION 4.1.

Let (X,E) be a separated M-valued set.

(a) An element x_0 is called a **join** of a subset B of X iff $E(x_0,z) = \bigvee\{E(y,z) \mid y \in B\}$ for all $z \in X$.

(b) A subset B of X is said to be **compatible** iff

$$E(y_1,y_1) * ((\vee_{y' \in B} E(y',y')) \rightarrow E(y_2,y_2)) \leq E(y_1,y_2)$$

holds for all $y_1, y_2 \in B$.

Since we consider only separated, M-valued sets, joins (if they exist) are unique. A characterization of compatible subsets is given in

LEMMA 4.2. Let (X, E) be a separated, M-valued set and B be a subset of X. Then the following assertions are equivalent:

(1) B is a compatible subset of (X, E).
(2) The map $s : X \longrightarrow L$ defined by $s(z) = \bigvee\{E(y, z) \mid y \in B\}$ is a singleton of (X, E).

PROOF. The implication (2) $\Rightarrow$ (1) follows from the singleton condition (S2). In order to verify (1) $\Rightarrow$ (2) we first put $\kappa = \vee\{E(y, y) | y \in B\}$ and observe $\kappa = \mathbb{E}(s)$. The extensionality of s is obvious. The singleton condition follows from

$$s(z_1) * (\kappa \rightarrow s(z_2)) = \bigvee_{y \in B} E(y, z_1) * (\kappa \rightarrow s(z_2)) =$$

$$\bigvee_{y \in B} \kappa * (\kappa \rightarrow E(y, y)) * (E(y, y) \rightarrow E(y, z_1)) * (\kappa \rightarrow s(z_2)) =$$

$$\bigvee_{y, y' \in B} (\kappa \rightarrow E(y, y)) * (E(y, y) \rightarrow E(y, z_1)) * E(y', z_2) =$$

$$\bigvee_{y, y' \in B} E(y', y') * (\kappa \rightarrow E(y, y)) * (E(y, y) \rightarrow E(y, z_1)) * (E(y', y') \rightarrow E(y', z_2)) \leq$$

$$\vee_{y, y' \in B} E(y', y) * (E(y, y) \rightarrow E(y, z_1)) * (E(y', y') \rightarrow E(y', z_2)) \leq E(z_1, z_2). \quad \square$$

PROPOSITION 4.3. Let $((X, E), \xi)$ be a T-algebra and B be a compatible subset of X w.r.t. the M-valued equality E. Then B has a (unique) join.

PROOF. Referring to 4.2, each compatible subset B of X determines a singleton s by $s(z) = \bigvee_{y \in B} E(y, z)$. Then $s(z) \leq E(\xi(s), z)$ follows immediately from 3.1(b) and (4.1). On the other hand we obtain for $y \in B$:

$$E(y, y) * (\mathbb{E}(s) \rightarrow E(\xi(s), z)) = E(\xi(s), z) * (\mathbb{E}(s) \rightarrow E(y, y)) =$$

$$E(\xi(s), z) * (\mathbb{E}(s) \rightarrow E(\xi(s), y)) \leq E(y, z);$$

i.e.

$$E(\xi(s), z) = (\bigvee_{y \in B} E(y, y)) * (\mathbb{E}(s) \rightarrow E(\xi(s), z)) \leq \bigvee_{y \in B} E(y, z) = s(z)$$

Hence $\xi(s)$ is the join of B. $\square$

PROPOSITION 4.4. Every T-algebra $((X, E), \xi)$ induces a presheaf structure $(X, \downarrow, \mathbb{E})$ on X (over L) by $x \downarrow \lambda = \xi(s_{(x,\lambda)})$, where $s_{(x,\lambda)}$ is determined by (3.1), $\mathbb{E}(x) = E(x, x)$. In particular, $(X, \downarrow, \mathbb{E})$ is compatible with the underlying M-valued equality E in the following sense:

$$E(x \downarrow \alpha, x \downarrow \beta) = \mathbb{E}(x) \wedge \alpha \wedge \beta \tag{R}$$

for all $x \in X$ and $\alpha, \beta \in L$.

PROOF. The diagram (4.1) and the strictness of ξ imply: $\mathbb{E}(x \downarrow \lambda) = \mathbb{E}(x) \wedge \lambda$, $x \downarrow \mathbb{E}(x) = x$, $\forall\, x \in X$. For any given $\alpha, \beta \in L$, let us consider a singleton σ of $\Sigma(X, E)$ defined by

$$\sigma(s') = ((E(x,x) \wedge \alpha) \to \beta) * \tilde{E}(s_{(x,\alpha)}, s')\ \forall\ s' \in S(X,E).$$

Referring to 1.2(iii) and 3.1(b) we obtain:

$$\begin{aligned}
[\mu_{(X,E)}(\sigma)](z) &= \bigvee_{s' \in S(X,E)} ((E(x,x) \wedge \alpha) \to \beta) * \tilde{E}(s_{(x,\alpha)}, s') * (\mathbb{E}(s') * (\mathbb{E}(s') \to s'(z)) \\
&= ((E(x,x) \wedge \alpha) \to \beta) * (E(x,x) \to \alpha) * E(x,z) \\
&= (E(x,x) \wedge \alpha \wedge \beta) * (E(x,x) \to E(x,z) \\
&= (E(x,x) \to (\alpha \wedge \beta)) * E(x,z) \\
&= s_{(x,\alpha \wedge \beta)}(z), \\
([\sigma(\xi)](\sigma))(z) &= \vee_{s' \in S(X,E)}((E(x,x) \wedge \alpha) \to \beta) * \tilde{E}(s_{X,\alpha)}, s') * (\mathbb{E}(s') \to E(\xi(s'), z)) \\
&= ((E(x,x) \wedge \alpha) \to \beta) * E(\xi(s_{(x,\alpha)}), z) \\
&= ((E(x,x) \wedge \alpha) \to \beta) * E(x \downarrow \alpha) \downarrow \alpha, z)
\end{aligned}$$

Hence $x \downarrow (\alpha \wedge \beta) = \xi(\mu_{(X,E)}(\sigma)) = \xi * [\Sigma(\xi)](\sigma)) = (x \downarrow \alpha) \downarrow \beta$ follows from (4.2).

Because of $\tilde{E}(s_{(x,\alpha)}, s_{(x,\beta)}) = \alpha \wedge \beta \wedge E(x,x)$, the axiom (R) is an immediate consequence of the morphism axioms of ξ. □

Motivated by 3.4, 4.3 and 4.4 supra, it seems to be reasonable to call T-algebras (determined by $T = (\Sigma, \circ, \eta)$) also **sheaves over** M.

§5. The Eilenberg-Moore category (M-SET)$^{\mathbf{T}}$

Let $T = (\Sigma, \circ, \eta)$ be the algebraic theory (in clone form) specified in Section 3 (see also Section 4). The category $(\mathbf{M\text{-}SET})^T$ consists of the following data: **object** are T-algebras (cf. §4) and **morphisms** $\varphi : ((X,E), \xi) \longrightarrow ((Y,F), \zeta)$ are T-homomorphism — i.e. **M-SET**-morphisms $\varphi : (X,E) \to (Y,F)$ such that the following diagram is commutative

$$
\begin{array}{ccc}
\Sigma(X,E) & \xrightarrow{\Sigma(\varphi)} & \Sigma(Y,F) \\
\xi \downarrow & \circ & \downarrow \zeta \\
(X,E) & \xrightarrow[\varphi]{} & (Y,F)
\end{array}
\tag{5.1}
$$

where

$$([\Sigma(\varphi)](s))(y) = [(\eta_{(Y,F)} \cdot \varphi) \circ id_{\Sigma(X,E)}](y) = \bigvee_{x \in X} (E(x,x) \rightarrow s(x)) * F(\varphi(x), y)$$

The **identity** of $((X,E),\xi)$ is the identity of (X,E) and the **composition** is the composition in **M-SET**.

Any T-algebra $((X,E),\xi)$ itself furnishes a simple example of a T-homomorphism; indeed by virtue of (4.2) the **M-SET**-morphism ξ can be considered as a T-homomorphism from the freely generated T-algebra $(\Sigma(X,E),\mu_{(X,E)})$ to $((X,E),\xi)$; and this is also essentially the meaning of the axiom (4.2).

REMARK 5.1 (**Universal Property of** $(\Sigma(X < E), \mu_{(X,E)})$). Let (X,E) be a separated, M-valued set and $((Y,F),\zeta)$ be a T-algebra. Then for every **M-SET**-morphism $\varphi : (X,E) \longrightarrow (Y,F)$ there exists a unique T-homomorphism $\varphi^{\#} : (\Sigma(X,E),\mu_{(X,E)}) \longrightarrow ((Y,F),\zeta)$ making the following diagram commutative

$$
\begin{array}{ccc}
(X,E) & \xrightarrow{\eta_{(X,E)}} & \Sigma(X,E) \\
 & \overset{\varphi}{\searrow} \quad \circ & \downarrow \varphi^{\#} \\
 & & (Y,F)
\end{array}
\tag{5.2}
$$

In particular $\varphi^{\#}$ is determined by $\varphi^{\#}(s) = \zeta(\vee_{x\in X}(E(x,x) \rightarrow s(x)) * F(\varphi(x), -))$. In particular, if $((X,E),\xi)$ is any T-algebra, then we can view ξ itself as the extension $(id_{(X,E)})^{\#}$ of the identity of $id_{(X,E)})^{\#}$ of the identity of $id_{(X,E)}$. $\square$

A simple application of the universal property of free T-algebras permits an interpretation of strict, extensional, L-fuzzy subsets as T-homomorphisms.

REMARK 5.2. Let (X,E) be a separated, M-valued set. A map $v : X \longrightarrow L$ is called a **strict** and **extensional**, **L-fuzzy subset** of (X,E) iff v satisfies the conditions (S1) and (S3).

(a) Every strict and extensional, L-fuzzy subset v of (X,E) induces an **M-SET**-morphism $\chi_v : (X,E) \longrightarrow \Omega$ (cf. §4) by

$$\chi_v(x) = (E(x,x), v(x)), \forall x \in X \tag{5.3}$$

and vice-versa every **M-SET**-morphism $\chi : (X,E) \rightarrow \Omega$ can be represented in the form of (5.3) — i.e. there exists a unique strict and extensional, L-fuzzy subset v of (X,E) with $\chi(x) = (E(x,x), v(x))$ for all $x \in X$.

(b) Let (Ω, ξ_Ω) be the T-algebra defined at the beginning of Section 4. Referring to the universal property of $(\Sigma(X,E), \mu_{(X,E)})$ every T-homomorphism $\chi : (\Sigma(X,E), \mu_{(X,E)}) \longrightarrow (\Omega, \xi_\Omega)$ can be identified with a strict and extensional, L-fuzzy subset v of (X,E) and vice-versa. In particular $v^{\#}$ is determined by

$$v^{\#}(s) = \bigvee_{x \in X} (E(x,x) \rightarrow s(x)) * v(x) \text{ for all } s \in S(X,E) \tag{5.4}$$

LEMMA 5.3 (**Characterization of Isomorphisms in (M-SET^T)**). Let $\varphi : ((X,E),\xi) \longrightarrow ((Y,F),\xi)$ be a T-homomorphism. Then the following assertions are equivalent:

(i) φ is an isomorphism.
(ii) φ satisfies the following conditions:

$$F(y,y) = \bigvee_{x \in X} (E(x,x) \rightarrow F(\varphi(x),y) * F(\varphi(x),y) \text{ for all } y \in Y \tag{$\mathcal{E}$}$$

$$E(x_1,x_2) = F(\varphi(x_1),\varphi(x_2)) \text{ for all } x_1, x_2 \in X \tag{EM1}$$

PROOF. The implication (1) $\Rightarrow$ (2) is obvious. In order to prove the inverse direction (2) $\Rightarrow$ (1) it is sufficient to verify the surjectivity of φ. By virtue of ($\mathcal{E}$) and (EM1) every element $y \in Y$ induces a singleton s_y by $s_y(x) = F(y,\varphi(x))$ for all $x \in X$. In particular $\mathbb{E}(s_y)$ coincides with $F(y,y)$. Now we invoke again ($\mathcal{E}$) and obtain:

$$F(y,y) = \vee_{x \in X}(F(y,\varphi(x) * (E(x,x) \rightarrow s_y(x)))$$

$$\leq \vee_{x \in X}(F(y,\varphi(x) * (E(x,x) \rightarrow F(\varphi(x),\varphi(\xi(s_y)))))) \leq F(y,\varphi(\xi(s_y)))$$

Hence $y = \varphi(\xi(s_y))$ follows from the separation axiom of (Y,F). □

Since **M-SET** is especially finitely complete (cf. Fact I in §2), we obtain (cf. [Manes 1976]):

FACT III. The Eilenberg-Moore category $(\textbf{M-SET})^T$ is finitely complete.

If we slightly enrich the structure of the underlying GL-monoid, we can show that $(\mathbf{M\text{-}SET})^T$ has coproducts.

THEOREM 5.4. If the GL-monoid M satisfies (L4) (cf. 1.3), then $(\mathbf{M\text{-}SET})^T$ has coproducts.

The proof is based on several lemmas. First we introduce for any pair $((X,E),(Y,F))$ of M-valued sets the following notations:

$$X \oplus Y := \{(x,y) \in X \times Y \mid E(x,x) \wedge F(y,y) = \mathbf{O}\}$$
$$(E \oplus F)((x,y),(\bar{x},\bar{y})) = E(x,\bar{x}) \vee F(y,\bar{y})$$

LEMMA 5.5. If (X,E) and (Y,F) are separated, M-valued sets, then $(X,E) \oplus (Y,F) := (X \oplus Y, E \oplus F)$ is also a separated M-valued set.

PROOF. The strictness, symmetry and the separation axiom of $E \oplus F$ are obvious. The transitivity of $E \oplus F$ follows from

$$\begin{aligned}(E \oplus F)((x,y),(\bar{x},\bar{y})) * (E(\bar{x},\bar{x}) \vee F(\bar{y},\bar{y})) \to (E \oplus F)((\bar{x},\bar{y}),(\bar{\bar{x}},\bar{\bar{y}})) \leq \\ [E(x,\bar{x}) * [E(\bar{x},\bar{x}) \to ((E(\bar{x},\bar{\bar{x}}) \vee F(\bar{y},\bar{\bar{y}})) \wedge E(\bar{x},\bar{x}))]] \vee \\ [F(y,\bar{y}) * [F(\bar{y},\bar{y}) \to ((E(\bar{x},\bar{\bar{x}})) \vee F(\bar{y},\bar{\bar{y}})) \wedge F(\bar{y},\bar{y}))]] = \\ [E(x,\bar{x}) * (E(\bar{x},\bar{x}) \to E(\bar{x},\bar{\bar{x}}))] \vee [F(y,\bar{y}))] * (F(\bar{y},\bar{y}) \to F(\bar{y},\bar{\bar{y}}))] \leq E(x,\bar{\bar{x}}) \vee F(y,\bar{\bar{y}}).\end{aligned}$$
□

LEMMA 5.6. Let (X,E) and (Y,F) are separated, M-valued sets having an element with vanishing extent of existence (i.e. $\exists\, o \in X$ (resp. Y) s.t. $E(o,o) = \mathbf{O}$ (resp. $F(o,o) = \mathbf{O}$)). If M satisfies the additional axiom (L4), then there exists a $\mathbf{M\text{-}SET}$-isomorphism $\theta : \Sigma((X,E) \oplus (Y,F)) \longrightarrow \Sigma(X,E) \oplus \Sigma(Y,F)$ determined by

$$\theta(s) = (s_1, s_2), \text{ where } s_1(x) = s(x,o), s_2(y) = s(o,y), \ (x,y) \in X \times X \tag{5.5}$$

In particular for each $s \in S((X,E) \oplus (Y,F))$ there exists a unique pair (s_1,s_2) of $S(X,E) \times S(Y,F)$ satisfying the following conditions:

$$\mathbb{E}(s_1) \wedge \mathbb{E}(s_2) = \mathbf{O}, \ \mathbb{E}(s_1) \vee \mathbb{E}(s_2) = \mathbb{E}(s) \tag{5.5'}$$
$$s(x,y) = s_1(x) \vee s_2(y) \text{ whenever } (x,y) \in X \oplus Y \tag{5.5''}$$

PROOF. Because 1.2(iii) and (L4) we are in the position to derive from the strictness and extensionality of $s \in S((X,E) \oplus (Y,F))$ the following relation:

$$\begin{aligned} s(x,y) &= s(x,y) \wedge (E(x,x) \vee F(y,y)) \\ &= s(x,y) * [(E(x,x) \to F(y,y)) \vee (F(y,y) \to E(x,x))] \\ &= s(x,y) * [((E(x,x) \vee F(y,y)) \to F(y,y)) \vee ((E(x,x) \vee F(y,y)) \to E(x,x))] \\ &= s(x,y) * [((E(x,x) \vee F(y,y)) \to (E \oplus F)((x,y),(o,y))) \\ &\quad \vee ((E(x,x) \vee F(y,y)) \to (E \oplus F)((x,y),(x,o)))] \\ &\leq s(x,o) \vee s(o,y) \end{aligned}$$

On the other hand we obtain

$$s(x,o) \vee s(o,y) =$$
$$[s(x,o) * (E(x,x) \to (E \oplus F)((x,o),(x,y)))]$$
$$\vee [s(o,y) * (F(y,y) \to (E \oplus F)((o,y),(x,y)))] \leq s(x,y)$$

Hence we have verified (5.5"). The relation (5.5') follows immediately from

$$s(x,o) * ((\mathbb{E}(s_1) \vee \mathbb{E}(s_2)) \to s(o,y)) \leq (E \oplus F)((x,o),(o,y)) = \mathbf{0}$$

Further let $(s_1, s_2) \in S(X,E) \times S(Y,F)$ satisfying (5.5'). We show that $s_1 + s_2 : X \oplus Y \longrightarrow L$, defined by $(s_1+s_2)(x,y) = s_1(x) \vee s_2(y)$, is a singleton of $(X,E) \oplus (Y,F)$:

$$(s_1+s_2)(x,y) * [(E(x,x) \vee F(y,y)) \to (E \oplus F)((x,y),(\bar{x},\bar{y}))] =$$
$$[(((E(x,x) \vee F(y,y)) \to E(x,x)) * (E(x,x) \to s_1(x)))$$
$$\vee (((E(x,x) \vee F(y,y)) \to F(y,y)) * (F(y,y) \to s_2(y)))] * (E(x,\bar{x}) \vee F(y,\bar{y})) \leq$$
$$((E(x,x) \to s_1(x)) * (E(x,\bar{x})) \vee (F(y,\bar{y}) * (F(y,y) \to E(x,x)))$$
$$\vee (F(y,y) \to s_2(y)) * F(y,\bar{y})) \vee (E(x,\bar{x}) * (E(x,x) \to F(y,y))) \leq$$
$$s_1(\bar{x}) \vee s_2(\bar{y}) \vee (E(x,x) \wedge F(y,y)) = (s_1+s_2)(\bar{x},\bar{y});$$
$$(s_1+s_2)(x,y) * [(\mathbb{E}(s_1) \vee \mathbb{E}(s_2)) \to (s_1+s_2)(\bar{x},\bar{y})] =$$
$$[(((\mathbb{E}(s_1) \vee \mathbb{E}(s_2)) \to \mathbb{E}(s_1)) * (\mathbb{E}(s_1) \to s_1(x)))$$
$$\vee (((\mathbb{E}(s_1) \vee \mathbb{E}(s_2)) \to \mathbb{E}(s_2)) * (\mathbb{E}(s_2) \to s_2(y)))))] * (s_1(\bar{x}) \vee s_2(\bar{y})) \leq$$
$$((\mathbb{E}(s_1) \to s_1(x)) * s_1(\bar{x})) \vee (s_2(\bar{y}) * (\mathbb{E}(s_2) \to \mathbb{E}(s_1))) \vee$$
$$((\mathbb{E}(s_2) \to s_2(y)) * (s_2(\bar{y})) \vee (s_1(\bar{x}) * (\mathbb{E}(s_1) \to \mathbb{E}(s_2))) \leq$$
$$E(x,\bar{x}) \vee F(y,\bar{y}) \vee (\mathbb{E}(s_1) \wedge \mathbb{E}(s_2)) = (E \oplus F)((x,y),(\bar{x},\bar{y}))$$

Finally we show that θ and θ^{-1} are **M-SET**-morphisms. The strictness of θ and θ^{-1} is evident. Further we obtain:

$$(\widetilde{E \oplus F})(s,\bar{s}) \leq$$
$$[(\mathbb{E}(s_1) \vee \mathbb{E}(s_2)) * [(\wedge_{x \in X}(s(x,o) \to \bar{s}(x,o))) \wedge (\wedge_{y \in Y}(s(o,y) \to \bar{s}(o,y)))]] \wedge$$
$$[(\mathbb{E}(\bar{s}_1) \vee \mathbb{E}(\bar{s}_2)) * [(\wedge_{x \in X}(\bar{s}(x,o) \to s(x,o))) \wedge (\wedge_{y \in Y}(\bar{s}(o,y) \to s(o,y)))]] \leq$$
$$\tilde{E}(s_1,\bar{s}_1) \vee \tilde{F}(s_2,\bar{s}_2) \vee [(\mathbb{E}(s_1) * (\wedge_{x \in X}(s(x,o) \to \bar{s}(x,o)))) \wedge \mathbb{E}(\bar{s}_2)]$$
$$\vee [(\mathbb{E}(s_2) * (\wedge_{y \in Y}(s(o,y) \to \bar{s}(o,y))) \wedge \mathbb{E}(\bar{s}_1)] \leq$$
$$\tilde{E}(s_1,\bar{s}_1) \vee \tilde{F}(s_2,\bar{s}_2) \vee (\mathbb{E}(\bar{s}_1) \wedge \mathbb{E}(\bar{s}_2)) = (\tilde{E} \oplus \tilde{F})((s_1,s_2),(\bar{s}_1,\bar{s}_2)).$$

In order to demonstrate $\tilde{E}(s_1,\bar{s}_1) \vee \tilde{F}(s_2,\bar{s}_2) \le (\widetilde{E \oplus F})(s_1+s_2,\bar{s}_1+\bar{s}_2)$, we first observe

$$\begin{gathered}
[[((\mathbb{E}(s_1) \vee \mathbb{E}(s_2)) \to \mathbb{E}(s_1)) * (\mathbb{E}(s_1) \to \tilde{E}(s_1,\bar{s}_1))] \\
\vee\; [((\mathbb{E}(s_1) \vee \mathbb{E}(s_2)) \to \mathbb{E}(s_2)) * (\mathbb{E}(s_2) \to \tilde{F}(s_2,\bar{s}_2))]] * (s_1(\hat{x}) \vee s_2(\hat{y})) \le \\
([(\mathbb{E}(s_1) \to \tilde{E}(s_1,\bar{s}_1)) * \mathbb{E}(s_1) * (\mathbb{E}(s_1) \to s_1(\hat{x}))] \vee [s_2(\hat{y}) * (\mathbb{E}(s_2) \to \mathbb{E}(s_1))] \\
\vee\; [s_1(\hat{x}) * (\mathbb{E}(s_1) \to \mathbb{E}(s_2))] \vee [(\mathbb{E}(s_2) \to \tilde{F}(s_2,\bar{s}_2)) * \mathbb{E}(s_2) * (\mathbb{E}(s_2) \to s_2(\hat{y}))]) \le \\
([s_1(\hat{x}) * (\wedge_{x\in X}(s_1(x) \to \bar{s}_1(x)))] \vee [\mathbb{E}(s_1) \wedge \mathbb{E}(s_2)] \;\vee\; [s_2(\hat{y}) * (\wedge_{y\in Y}(s_2(y) \to \bar{s}_2(y))]) \\
\le \bar{s}_1(\hat{x}) \vee \bar{s}_2(\hat{y})
\end{gathered}$$

Hence the relation

$$(\tilde{E}(s_1,\bar{s}_1) \vee \tilde{F}(s_2,\bar{s}_2)) \le (\mathbb{E}(s_1) \vee \mathbb{E}(s_2)) * [\bigwedge_{(\hat{x},\hat{y})\in X\oplus Y} ((s_1(\hat{x}) \vee s_2(\hat{y})) \to (\bar{s}_1(\hat{x}) \vee \bar{s}_2(\hat{y})))]$$

follows. Analogously we verify

$$\begin{gathered}
\tilde{E}(s_1,\bar{s}_1) \vee \tilde{F}(s_2,\bar{s}_2) \le \\
(\mathbb{E}(\bar{s}_1) \vee \mathbb{E}(\bar{s}_2)) * [\bigwedge_{(\hat{x},\hat{y})\in X\oplus Y} ((\bar{s}_1(\hat{x}) \vee \bar{s}_2(\hat{y})) \to (s_1(\hat{x}) \vee s_2(\hat{y})))]
\end{gathered}$$

Summing up we obtain: $\tilde{E}(s_1,\bar{s}_1) \vee \tilde{F}(s_2,\bar{s}_2) = (\widetilde{E \oplus F})(s_1+s_2,\bar{s}_1+\bar{s}_2)$. □

In view of the subsequent considerations we stipulate the following **notations**: $\theta_{(X,E)(Y,F)} : \Sigma((X,E) \oplus (Y,F)) \longrightarrow \Sigma(X,E) \oplus \Sigma(Y,F)$ stands always for the **M-SET**-morphism specified in 5.6. Further for each pair (φ,Ψ) of M-SET-morphisms

$$\varphi : (X,E) \longrightarrow (\hat{X},\hat{E}),\;\; \Psi : (Y,F) \longrightarrow (\hat{Y},\hat{F})$$

the **M-SET**-morphism $\varphi \oplus \Psi : (X,E) \oplus (Y,F) \longrightarrow (\hat{X},\hat{E}) \oplus (\hat{Y},\hat{F})$ is determined by $(\varphi \oplus \Psi)(x,y) = (\varphi(x),\Psi(y))$ for all $(x,y) \in X \oplus Y$.

LEMMA 5.7. The following diagram is commutative

$$\begin{array}{ccc}
\Sigma((X,E) \oplus (Y,F)) & \xrightarrow{\;\Sigma(\varphi \oplus \Psi)\;} & \Sigma((\hat{X},\hat{E}) \oplus (\hat{Y},\hat{F})) \\
\Theta_{(X,E)(Y,F)} \Big\downarrow & \circ & \Big\downarrow \Theta_{(\hat{X},\hat{E})(\hat{Y},\hat{F})} \\
\Sigma(X,E) \oplus \Sigma(Y,F) & \xrightarrow[\;\Sigma(\varphi \oplus \Psi)\;]{} & \Sigma(\hat{X},\hat{E}) \oplus \Sigma(\hat{Y},\hat{F})
\end{array}$$

i.e. $\theta : \Sigma \cdot \oplus \to \oplus \cdot (\Sigma \times \Sigma)$ is a natural isomorphism.

Proof. Let s be a singleton of $S((X,E) \oplus (Y,F))$; we put $\theta_{(X,E)(Y,F)}(s) = (s_1, s_2)$. Then we obtain from (5.5"):

$$[(\Sigma(\varphi \oplus \Psi))(s)](\hat{x}, o) =$$

$$\bigvee_{(x,y) \in X \oplus Y} s(x,y) * ((E(x,x) \vee F(y,y)) \to \hat{E}(\varphi(x), \hat{x})) =$$

$$\bigvee_{(x,y) \in X \oplus Y} ([((E(x,x) \vee F(y,y)) \to E(x,x)) * (E(x,x) \to s_1(x)) * \hat{E}(\varphi(x), \hat{x})]$$

$$\bigvee [((E(x,x) \vee F(y,y)) \to F(y,y)) * (F(y,y) \to s_2(y)) * \hat{E}(\varphi(x), \hat{x})]) =$$

$$\bigvee_{(x,y) \in X \oplus Y} ([E(x,x) * (F(y,y) \to E(x,x)) * (E(x,x) \to s_1(x)) * (E(x,x) \to \hat{E}(\varphi(x), \hat{x}))]$$

$$\bigvee [\hat{E}(\varphi(x), \hat{x}) * (E(x,x) \to F(y,y)) * (F(y,y) \to s_2(y))]).$$

Now we invoke the axiom (L4) and the strictness of φ:

$$\begin{aligned} \hat{E}(\varphi(x), \hat{x}) * (E(x,x) \to F(y,y)) &\le E(x,x) \wedge F(y,y) = \mathbf{0}, \\ E(x,x) &= E(x,x) * [(E(x,x) \to F(y,y)) \vee (F(y,y) \to E(x,x))] \\ &= E(x,x) * (F(y,y) \to E(x,x)) \end{aligned}$$

Hence we can continue as follows:

$$[(\Sigma(\varphi \oplus \Psi))(s)](\hat{x}, o) =$$

$$\bigvee_{(x,y) \in X \oplus Y} E(x,x) * (E(x,x) \to s_1(x) * (E(x,x) \to \hat{E}(\varphi(x), \hat{x})) = [(\Sigma(\varphi))(s_1)](\hat{x}).$$

Analogously we verify: $[(\Sigma(\varphi \oplus \Psi))(s)](o, \hat{y}) = [(\Sigma(\Psi))(s_2)](\hat{y})$. Now we use (5.5) and obtain:

$$\begin{aligned} \theta_{(\hat{X},\hat{E})(\hat{Y},\hat{F})}((\Sigma(\varphi \oplus \Psi))(s)) &= ((\Sigma(\varphi))(s_1), (\Sigma(\Psi))(s_2)) = \\ &= [(\Sigma(\varphi) \oplus \Sigma(\Psi)) \cdot \theta_{(X,E)(Y,F)}](s). \quad \square \end{aligned}$$

MMA 5.8. The diagram

$$\begin{array}{ccc}
\Sigma(\Sigma((X,E)\oplus(Y,F))) & \xrightarrow{\Sigma(\Theta_{(X,E)(Y,F)})} & \Sigma(\Sigma(X,E)\oplus\Sigma(Y,F)) \\
\Big\downarrow & & \Big\downarrow \Theta_{\Sigma(X,E)\,\Sigma(Y,F)} \\
\mu_{(X,E)\oplus(Y,F)} & \circ & \Sigma(\Sigma(X,E))\oplus\Sigma(\Sigma(Y,F)) \\
\Big\downarrow & & \Big\downarrow \mu_{(X,E)\oplus(Y,F)} \\
\Sigma((X,E)\oplus(Y,F)) & \xrightarrow[\Theta_{(X,E)(Y,F)}]{} & \Sigma(X,E)\oplus\Sigma(Y,F)
\end{array}$$

ommutative.

OOF. From (5.5') and (5.5") we deduce the following relation

$$[\mu_{(X,E)\oplus(Y,F)}(\sigma)](x,o) = \bigvee_{s\in S((X,E)\oplus(Y,F))} (\mathbb{E}(s)\to\sigma(s)) * s(x,o) =$$

$$\bigvee_{(s_1,s_2)\in S(X,E)\oplus S(Y,F)} ((\mathbb{E}(s_1)\vee\mathbb{E}(s_2))\to\sigma(s_1,s_2)) * s_1(x) =$$

$$\bigvee_{\bar{s}_1\in S(X,E)} \Big(\bigvee_{(s_1,s_2)\in S(X,E)\oplus S(Y,F)} ((\mathbb{E}(s_1)\vee\mathbb{E}(s_2))\to\sigma(s_1,s_2)) *$$

$$\tilde{E}(s_1,\bar{s}_1) * (\mathbb{E}(\bar{s}_1)\to\bar{s}_1(x))\Big) =$$

$$[\mu_{(X,E)}([(\Sigma(\theta_{(X,E)(Y,F)}))(\sigma)](-,o))](x).$$

alogously we verify

$$[\mu_{(X,E)\oplus(Y,F)}(\sigma)](o,y) = [\mu_{(Y,F)}([(\Sigma(\theta_{(X,E)(Y,F)}))(\sigma)](o,-))](y)$$

nce the assertion follows from (5.5). □

MMA 5.9.

) For every separated, M-valued set (Z,G) there exists a **M-SET**-morphism $\Gamma_{(Z,G)} : (Z,G)\oplus(Z,G) \longrightarrow \Sigma(Z,G)$ determined by $[\Gamma_{(Z,G)}(z,\bar{z})](u) = G(z,u)\vee G(\bar{z},u)$.

(b) The following diagram is commutative

$$\begin{array}{ccc}
\Sigma((Z,G)\oplus(Z,G)) & \xrightarrow{\Theta_{(Z,G)(Z,G)}} & \Sigma(Z,G)\oplus\Sigma(Z,G) \\
 & & \Big\downarrow \Gamma_{\Sigma(Z,G)} \\
\Sigma(\Gamma_{(Z,G)}) \Big\downarrow & \circ & \Sigma(\Sigma(Z,G)) \\
 & & \Big\downarrow \mu_{(Z,G)} \\
\Sigma(\Sigma(Z,G)) & \xrightarrow[\mu_{(Z,G)}]{} & \Sigma(Z,G)
\end{array}$$

PROOF.

(a) The strictness of $\Gamma_{(Z,G)}$ is obvious. In order to verify the axiom (M2) we proceed as follows:

$$((G(z_1,z_1)\vee G(z_2,z_2))\to(G(z_1,\bar{z}_1)\vee G(z_2,\bar{z}_2)))*(G(z_1,u)\vee(G(z_2,u))=$$

$$(G(z_1,\bar{z}_1)\vee G(z_2,\bar{z}_2))*[[(G(z_1,z_1)\vee G(z_2,z_2))\to G(z_1,z_1))*(G(z_1,z_1)\to G(z_1,u))]\vee$$
$$[(G(z_1,z_1)\vee G(z_2,z_2))\to G(z_2,z_2))*(G(z_2,z_2)\to G(z_2,u))]].$$

We invoke (L4) and obtain in parallel to the proof of 5.7:

$$((G(z_1,z_1)\vee G(z_2,z_2))\to(G\oplus G)((z_1,z_2),(\bar{z}_1,\bar{z}_2)))*[\Gamma_{(Z,G)}(z_1,z_2)](u)=$$

$$[G(z_1,\bar{z}_1)*(G(z_1,z_1)\to G(z_1,u))]\vee$$
$$[G(z_2,\bar{z}_2)*(G(z_2,z_2)\to G(z_2,u)]\leq[\Gamma_{(Z,G)}(\bar{z}_1,\bar{z}_2)](u),\text{ i.e.}$$
$$(G\oplus G)((z_1,z_2),(\bar{z}_1,\bar{z}_2))\leq(G(z_1,z_1)\vee G(z_2,z_2))*$$
$$(\bigwedge_{u\in Z}([\Gamma_{(Z,G)}(z_1,z_2)](u)\to[\Gamma_{(Z,G)}(\bar{z}_1,\bar{z}_2)](u)))$$

Hence the assertion follows from the symmetry of M-valued equalities.

(b) Referring to (5.5") we obtain:

$$[\mu_{(Z,G)}((\Sigma(\Gamma_{(Z,G)}))(\hat{s})](u)$$

$$=\bigvee_{(z_1,z_2)\in Z\oplus Z}(\bigvee_{s\in S(Z,G)}((G(z_1,z_1)\vee G(z_2,z_2))\to\hat{s}(z_1,z_2))*$$

$$\tilde{G}(\Gamma_{(Z,G)}(z_1,z_2),s) * (\mathbb{E}(s) \to s(u)))$$
$$= \bigvee_{(z_1,z_2)\in Z\oplus Z} (((G(z_1,z_1) \vee G(z_2,z_2)) \to (\hat{s}(z_1,z_2)) * [\Gamma_{(Z,G)}(z_1,z_2)](u))$$
$$= \bigvee_{(z_1,z_2)\in Z\oplus Z} (((G(z_1,z_1) \vee G(z_2,z_2)) \to (G(z_1,u) \vee G(z_2,u))) * \hat{s}(z_1,z_2))$$
$$= \bigvee_{(z_1,z_2)\in Z\oplus Z} ([[((G(z_1,z_1) \vee G(z_2,z_2)) \to (G(z_1,z_1)) *$$
$$(G(z_1,z_1) \to \hat{s}_1(z_1))] \vee [((G(z_1,z_1) \vee G(z_2,z_2)) \to G(z_2,z_2)) *$$
$$(G(z_2,z_2) \to \hat{s}_2(z_2))]] * (G(z_1,u) \vee G(z_2,u))) =$$
$$= \bigvee_{(z_1,z_2)\in Z\oplus Z} ((\hat{s}_1(z_1) * (G(z_1,z_1) \to G(z_1,u))) \vee (\hat{s}_2(z_2) *$$
$$(G(z_2,z_2) \to G(z_2,u)))) = \hat{s}_1(u) \vee \hat{s}_2(u) =$$
$$[(\mu_{(Z,G)} \cdot \Gamma_{\Sigma(Z,G)} \cdot \theta_{(Z,G)(Z,G)})(\hat{s})](u). \quad \square$$

COROLLARY 5.10. Let $((X,E),\xi)$ and $((Y,F),\zeta)$ be two T-algebras, and let $\xi \perp \zeta : \Sigma((X,E)\oplus(Y,F)) \longrightarrow (X,E)\oplus(Y,F)$ be an **M-SET**-morphism determined by $\xi \perp \zeta = (\xi \oplus \zeta)\cdot\theta_{(X,E)(Y,F)}$. Then $((X,E)\oplus(Y,F), \xi \perp \zeta)$ is also a T-algebra.

PROOF.

$$\xi \perp \zeta \cdot \eta_{(X,E)\oplus(Y,F)} =$$
$$(\xi \oplus \zeta)\cdot(\theta_{(X,E)(Y,F)} \cdot \eta_{(X,E)\oplus(Y,F)}) = (\xi \oplus \zeta).(\eta_{(X,E)} \oplus \eta_{(Y,F)}) = id_{(X,E)\oplus(Y,F)}$$

Further we conclude from 5.7 and 5.8:

$$\xi \perp \zeta \cdot \mu_{(X,E)\oplus(Y,F)} = (\xi \oplus \zeta)\cdot(\theta_{(X,E)(Y,F)} \cdot \mu_{(X,E)\oplus(Y,F)}) =$$
$$(\xi \oplus \zeta)\cdot(\mu_{(X,E)} \oplus \mu_{(Y,F)})\cdot(\theta_{\Sigma(X,E)\Sigma(Y,F)} \cdot \Sigma(\theta_{(X,E)(Y,F)}) =$$
$$((\xi \cdot \mu_{(X,E)}) \oplus (\zeta \cdot \mu_{(Y,F)})\cdot \theta_{\Sigma(X,E)\Sigma(Y,F)} \cdot \Sigma(\theta_{(X,E)(Y,F)}) =$$
$$((\xi \cdot \Sigma(\xi)) \oplus (\zeta \cdot \Sigma(\zeta))\cdot \theta_{\Sigma(X,E)\Sigma(Y,F)} \cdot \Sigma(\theta_{(X,E)(Y,F)}) =$$
$$(\xi \oplus \zeta)\cdot(\Sigma(\xi) \oplus \Sigma(\zeta))\cdot \theta_{\Sigma(X,E)\Sigma(Y,F)} \cdot \Sigma(\theta_{(X,E)(Y,F)}) =$$
$$(\xi \oplus \zeta)\cdot\theta_{(X,E)(Y,F)} \cdot \Sigma(\xi \oplus \zeta)\cdot\Sigma(\theta_{(X,E)(Y,F)}) = (\xi \perp \zeta)\cdot\Sigma(\xi \perp \zeta). \quad \square$$

PROOF OF THEOREM 5.4.

Let $((X,E)\oplus(Y,F), \xi \perp \zeta)$ be the T-algebra given in 5.10. Further let us consider the canonical embeddings $j_{(X,E)} : (X,E) \longrightarrow (X,E)\oplus(Y,F)$ and $j_{(Y,F)} : (Y,F) \longrightarrow (X,E)\oplus(Y,F)$ defined by $j_{(X,E)}(x) = (x,o)$, $j_{(Y,F)} = (o,y)$ for all $x \in X$ and $y \in Y$. We note

$(j_{(X,E)} \cdot \xi)(s_1) = (\xi(s_1), o) = (\xi \oplus \zeta)(s_1, o) = [(\xi \oplus \zeta) \cdot \theta_{(X,E)(Y,F)} \cdot \Sigma(j_{(X,E)})](s_1) = [(\xi \perp \zeta) \cdot \Sigma(j_{(X,E)})](s_1)$; hence $j_{(X,E)}$ (resp. $j_{(Y,F)}$) is a T-homomorphism. Now we show that $((j_{(X,E)}, j_{(Y,F)}), ((X,E) \oplus (Y,F), \xi \perp \zeta))$ is the coproduct of $((X,E),\xi)$ and $((Y,F),\zeta)$ (in $(\mathbf{M\text{-}SET})^T$). Therefore let $((Z,G),\vartheta)$ be a further T-algebra and let us consider two T-homomorphisms $\varphi : ((X,E),\xi) \longrightarrow ((Z,G),\xi)$ and $\Psi : ((Y,F),\xi) \to ((Z,G),\vartheta)$. First we verify the following

Assertion. The following diagram commutes.

$$\begin{array}{ccc}
\Sigma(\Sigma(Z,G)) & \xrightarrow{\quad \Sigma(\vartheta) \quad} & \Sigma(Z,G) \\
\Gamma_{\Sigma(Z,G)} \Big\uparrow & \circ & \Big\uparrow \Gamma_{(Z,G)} \\
\Sigma(Z,G) \oplus \Sigma(Z,G) & \xrightarrow[\quad \vartheta \oplus \vartheta \quad]{} & (Z,G) \oplus (Z,G)
\end{array}$$

In fact: $[(\Sigma(\vartheta) \cdot \Gamma_{\Sigma(Z,G)})(s_1, s_2)](u) = \bigvee_{s \in S(Z,G)} (G(s_1, s) \vee G(s_2, s)) * (\mathbb{E}(s) \to G(\vartheta(s), u)) = G(\vartheta(s_1), u) \vee G(\vartheta(s_2), u) = [(\Gamma_{(Z,G)} \cdot (\vartheta \oplus \vartheta))(s_1, s_2)](u)$.

Now we continue the proof of 5.4 and introduce a $\mathbf{M\text{-}SET}$-morphism $\varphi \perp \Psi : (X,E) \oplus (Y,F) \longrightarrow (Z,G)$ as follows: $\varphi \perp \Psi = \vartheta \cdot \Gamma_{(Z,G)} \cdot (\varphi \oplus \Psi)$. If we combine 5.7 and 5.9 with the previous assertion, then we obtain:

$$\begin{aligned}
(\varphi \perp \Psi) \cdot (\xi \perp \zeta) &= \vartheta \cdot \Gamma_{(Z,G)} \cdot (\varphi \oplus \Psi) \cdot (\xi \oplus \zeta) \cdot \theta_{(X,E)(Y,F)} \\
&= \vartheta \cdot \Gamma_{(Z,G)} \cdot ((\vartheta \cdot \xi) \oplus (\Psi \cdot \zeta)) \cdot \theta_{(X,E)(Y,F)} \\
&= \vartheta \cdot \Gamma_{(Z,G)} \cdot (\vartheta \oplus \vartheta) \cdot (\Sigma(\varphi) \oplus \Sigma(\Psi)) \cdot \theta_{(X,E)(Y,F)} \\
&= \vartheta \cdot \Gamma_{(Z,G)} \cdot (\vartheta \oplus \vartheta) \cdot \theta_{(Z,G)(Z,G)} \cdot \Sigma(\varphi \oplus \Psi) \\
&= \vartheta \cdot \Sigma(\vartheta) \cdot \Gamma_{\Sigma(Z,G)} \cdot \theta_{(Z,G)(Z,G)} \cdot \Sigma(\varphi \oplus \Psi) \\
&= \vartheta \cdot \mu_{(Z,G)} \cdot \Gamma_{\Sigma(Z,G)} \cdot \theta_{(Z,G)(Z,G)} \cdot \Sigma(\varphi \oplus \Psi) \\
&= \vartheta \cdot \mu_{(Z,G)} \cdot \Sigma(\Gamma_{(Z,G)}) \cdot \Sigma(\vartheta \oplus \Psi) \\
&= \vartheta \cdot \Sigma(\vartheta) \cdot \Sigma(\Gamma_{(Z,G)}) \cdot \Sigma(\vartheta \oplus \Psi) \\
&= \vartheta \cdot \Sigma(\varphi \perp \Psi)
\end{aligned}$$

hence $\varphi \perp \Psi$ is a T-homomorphism with respect to the corresponding T-algebras. Further we observe:

$$((\varphi \perp \Psi) \cdot j_{(X,E)})(x) = (\vartheta \cdot \Gamma_{(Z,G)})(\varphi(x), 0) = \vartheta(\eta_{(Z,G)}(\varphi(x))) = \varphi(x)$$

i.e. the following diagram commutes.

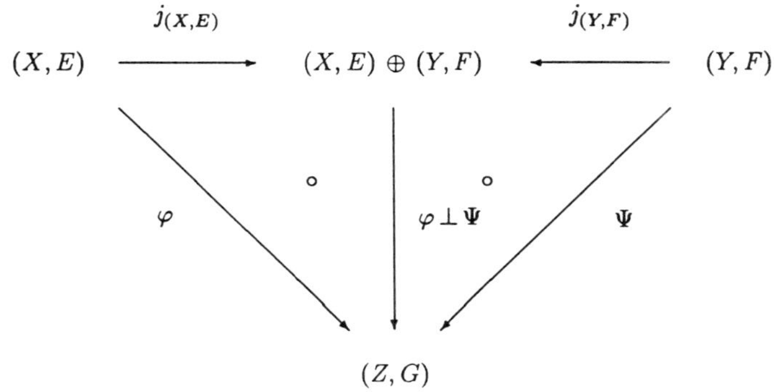

(**)

In order to verify the uniqueness of $\varphi \perp \Psi$ let us consider a further T-homomorphism

$$\Phi : ((X,E) \oplus (Y,F), \xi \perp \zeta) \longrightarrow ((Z,G),\vartheta)$$

making the diagram (**) commutative. Because of

$$(E \oplus F)((x,y),(x',y')) = (E \oplus F)((x,o),(x',y')) \vee (E \oplus F)((o,y),(x',y'))$$

we obtain

$$\begin{aligned}\Phi(x,y) =& [\Phi \cdot (\xi \perp \zeta) \cdot \eta_{(X,E)\oplus(Y,F)}](x,y)\\ =& [\vartheta \cdot \Sigma(\Phi) \cdot \eta_{(X,E)\oplus(Y,F)}](x,y)\\ =& [\vartheta \cdot \Gamma_{(Z,G)}](\phi(x,o),\phi(o,y)) = [\vartheta \cdot \Gamma_{(Z,G)}](\varphi(x),\Psi(y))\\ =& (\varphi \perp \Psi)(x,y). \ \square\end{aligned}$$

§6. A Comparison Between Sheaves over GL-monoids and Sheaves Over Complete Heyting Algebras.

In this section we distinguish between the given GL-monoid $M = (L, \leq, *)$ and underlying lattice L which is a complete Heyting algebra (cf. 1.2(6)). In particular $\rightarrow$ denotes always the implication based on $*$ (see: (1.2)).

Referring to the terminology stipulated e.g. in [Fourman and Scott 1979]), an L-valued set is a pair (X,E) such that X is a set and $E : X \times X \longrightarrow L$ is a map satisfying the following conditions:

$E(x,y) = E(y,x)$ **(Symmetry)** (H1)
$E(x,y) \wedge E(y,z) \leq E(x,z)$ **($\wedge$-Transitivity)** (H2)

Since (H2) implies (E1) the relation

$$E(x,y) * ((E(y,y) \to E(y,z)) \leq E(x,y) \wedge E(y,z)$$

follows from 1.2(3); hence every L-valued set is also a M-valued set. Moreover we can characterize compatibility of subsets in the sense of 4.1(b) as follows:

LEMMA 6.1. Let (X, E) be an L-valued set and B be a subset of X. Then the following assertions are equivalent:

(1) B is compatible in the sense of 4.1(b).
(2) $E(x,x) \wedge E(y,y) \leq E(x,y)$ for all $x, y \in B$.

PROOF. The implication (2) $\Rightarrow$ (1) is obvious. In order to verify (1) $\Rightarrow$ (2) we proceed as follows: first we put $\kappa = \bigvee\{E(y,y) \mid y \in B\}$ and observe

$$(\kappa \to (E(x,x) \wedge E(y,y)) * E(z,z) = (E(x,x) \wedge E(y,y)) * (\kappa \to E(z,z)) \leq E(x,z) \wedge E(x,y)$$

for all $x, y, z \in B$; now we invoke (H2) and obtain:

$$\bigvee_{z \in B} (\kappa \to (E(x,x) \wedge E(y,y))) * E(z,z) \leq E(z,y)$$

i.e. $E(x,x) \wedge E(y,y) \leq E(x,y)$. □

COROLLARY 6.2. Let (X, E) be an L-valued set and s be a singleton of (X, E) viewed as an M-valued set. Then the map $\bar{s} : X \longrightarrow L$ defined by

$$\bar{s}(z) = \bigvee_{x \in X} (s(x) \wedge E(x,z)) \tag{6.1}$$

satisfies the following conditions:

(1) $\bar{s}(z) \wedge E(z,y) \leq \bar{s}(y)$
(2) $\bar{s}(z) \wedge \bar{s}(y) \leq E(z,y)$;

i.e. $\bar{s}$ is a "singleton" in the sense of [Fourman and Scott 1979].

PROOF. The assertion (1) follows immediately from (6.1) and (H2). Repeating the proof of 6.1 verbatim we infer from (S2) that the inequality $s(x) \wedge s(y) \leq E(x,y)$ holds for all $x, y \in X$; hence the assertion (2) follows again from (H2). □

Let $T = (\Sigma, \circ, \eta)$ be the algebraic theory specified in paragraph 3. Then every complete L-valued set (X, E) (i.e. every sheaf over L (cf. [Fourman and Scott 1979]) can be viewed as a T-algebra $((X,E),\xi_X)$ in the following way: By virtue of 6.2 there exists a map $\xi_X : S(X,E) \longrightarrow X$ determined by

$$E(\xi_X(s),z) = \bar{s}(z) = \bigvee_{x\in X} (s(x) \wedge E(x,z)) \text{ for all } z \in X \tag{6.2}$$

From the relation

$$\mathbb{E}(s_1) \to \tilde{E}(s_1,s_2)) * \bar{s}_1(z) = (\bigwedge_{x\in X} (s_1(x) \to s_2(x)) * \bar{s}_1(z) \leq \bar{s}_2(z)$$

we conclude that $\xi_X : \Sigma(X,E) \longrightarrow (X,E)$ is an **M-SET**-morphism. Further the axiom (H2) implies:

$$\xi_X \cdot \eta_{(X,E)} = id_{(X,E)}, \xi_X \cdot \mu_{(X,E)} = \xi_X \cdot \Sigma(\xi_X),$$

$$E(\xi_X(s_{(x,\lambda)}),z) = E(x \downarrow \lambda,\ z) = \lambda \wedge E(x,z),\ \forall\ z \in X$$

i.e. $((X,E),\xi_X)$ is a T-algebra (in **M-SET**) and the restrictions in the sense of T-algebras (cf. (4.4)) and restrictions in the sense of the ordinary sheaf structure of (X,E) coincide.

Moreover **M-SET**-morphisms $\varphi : (X,E) \longrightarrow (Y,F)$ between complete L-valued sets (= sheaves over L) are also T-homomorphisms $\varphi : ((X,E),\xi_X) \longrightarrow ((Y,F),\xi_Y)$. In particular, it is not difficult to verify the following relation:

$$\overline{[\Sigma(\varphi)(s)]}(y) = \bigvee_{x\in X} (s(x) \wedge F(\varphi(x),y)) = \bigvee_{x\in X} (\bar{s})(x) \wedge F(\varphi(x),y))$$

Hence, if we identify complete, L-valued sets (X,E) with T-algebras $((X,E),\xi_X)$, then we obtain

FACT IV. $sh(L)$ is a full subcategory of $(\mathbf{M\text{-}SET})^T$.

§7. A Simple Application of $(\mathbf{M\text{-}SET})^T$ to Fuzzy Set Theory.

A major problem in fuzzy set theory is the question in which sense fuzzy subsets, viewed as generalized characteristic functions, describe subobjects within an appropriate, categorical framework. $(\mathbf{M\text{-}SET})^T$ provides us with a simple answer. As already mentioned above (cf. 5.3), strict and extensional, L-fuzzy subsets ν of a separated, M-valued set (X,E) can be identified with a T-homorphism $(\chi_\nu)^\# : (\Sigma(X,E),\mu_{(X,E)}) \longrightarrow (\Omega,\xi_\Omega)$. Further we observe that the terminal object $((L,\wedge),\xi_1)$ in $(\mathbf{M\text{-}SET})^T$ is freely generated by $(\{\cdot\},\approx)$ (see part (b) of the proof of 3.3). Hence there exists a unique T-homomorphism $t : ((L,\wedge),\xi_1) \longrightarrow (\Omega,\xi_\Omega)$ satisfying the condition $t(\mathbf{1}) = (\mathbf{1},\mathbf{1})$. Because of the special, semantic meaning of $\mathbf{1}$ the arrow t is also called the **arrow 'true'**.

Since $(\mathbf{M\text{-}SET})^T$ is finitely complete (cf. Fact III), every strict and extensional, L-fuzzy subset ν of (X,E) gives rise to a T-subalgebra $((Z_\nu,G_\nu),\xi_\nu)$ of the freely generated

T-algebra $(\Sigma(X,E),\mu_{(X,E)})$: $\Psi_\nu : ((Z_\nu, G_\nu), \xi_\nu) \longrightarrow (\Sigma(X,E),\mu_{(X,E)})$ is determined by the following pullback square

$$\begin{array}{ccc} ((Z_\upsilon, G_\upsilon), \xi_\upsilon) & \xrightarrow{\quad ! \quad} & (1, \xi_1) \\ \Psi_\upsilon \Big\downarrow & \circ & \Big\downarrow t \\ (\Sigma(X,E), \mu_{(X,E)}) & \xrightarrow[(\chi_\upsilon)^!]{} & (\Omega, \xi_\Omega) \end{array}$$

In particular

$$\begin{aligned} Z\nu &:= \{s \in S(X,E) \mid \bigvee_{x\in X} (s(x) * (E(x,x) \to \nu(x))) = \mathbb{E}(s)\}, \\ G_\nu &:= E_{|Z_\nu \times Z_\nu}, \\ [\xi_\nu(\sigma)](x) &:= \bigvee_{s\in Z_\nu} (\sigma(s) * (\mathbb{E}(s) \to s(x))), \ \forall \sigma \in S(\Sigma(Z_\nu, G_\nu), \\ \Psi_\nu &:= \text{ inclusion map.} \end{aligned}$$

In accordance with an interpretation of algebraic theories as non-deterministic theories (cf. [Manes 1976]) we can understand the T-subalgebra $((Z_\nu, G_\nu), \xi_\nu)$ associated with the L-fuzzy subset ν as follows: The elements of Z_ν are the **pure states** of ν, the singletons of $\Sigma(Z_\nu, G_\nu)$ are the **fuzzy states** of ν, and the **M-SET**-morphism $\xi_\mu : \Sigma(Z_\nu, G_\nu) \longrightarrow (Z_\nu, G_\nu)$ acts as a **decider** or **defuzzifier** of the **system** described by ν.

In general Z_ν is a subset of $U_\nu := \{s \in S(X,E) | s(x) \leq \nu(x) \ \forall x \in X\}$; but if M is determined by a complete Heyting algebra (i.e. $* = \wedge$!), then Z_ν and U_ν coincide, and ξ_ν is an isomorphism. Therefore in the case of complete Heyting algebras (i.e. $* = \wedge$) pure states and fuzzy states of a L-fuzzy subset coincide, and consequently defuzzification is trivial.

Chapter 3:

The Logic of Unbalanced Subobjects in a Category with Two Closed Structures

S. E. Rodabaugh et al. (eds.), Applications of Category Theory to Fuzzy Subsets, 73–105.

THE LOGIC OF UNBALANCED SUBOBJECTS IN A CATEGORY WITH TWO CLOSED STRUCTURES

L. N. Stout

§0. Prologue

Categories with two local closed structures and factorization of maps through canonical unbalanced (epi and mono) subobjects have an internal logic close to that used in fuzzy set theory. This chapter develops a first order type theory with appropriate axioms and inference rules which can be interpreted in such categories. The theory is shown to be sound for **SET(L)** and the category of tolerance spaces. Unbalanced powerobject representation gives a higher order type theory for **SET(L)**.

§1. Introduction

When we look for a categorical foundation for fuzzy set theory what we want is a category with all of the connectives normally used by fuzzy set theorists and with an elegant, systematic internal logic, preferably of higher order. A topos or quasitopos (as suggested in [Stout 1987] and in [Barr 1986]) has the elegant internal logic, but not the variety of connectives. Following work by [Pultr 1976a, 1976b], I suggested in [Stout 1987] that the appropriate setting for fuzzy set theory might be a category with two closed structures: a cartesian closed structure giving rise to the intuitionistic logic familiar in topoi and a second closed structure giving rise to the fuzzy (or Łukaciewcz) connectives. It turns out that the simplest definition of a category of fuzzy sets, that given in [Goguen 1974], has this kind of rich categorical structure. So do some other categories of interest, so this chapter develops the general theory of logic in such categories.

Categorical logic has concentrated in recent years on identifying topoi (and categories with less structure) with certain kinds of logical theories: local set theories, intuitionistic type theory, and various forms of typed lambda calculus (see for instance [Bell 1988, Makkai and Reyes 1977, Lambek and Scott 1986]). Categories with the structures we are positing in this chapter have a type theory with a richer collection of rules for creating new types than that used in local set theories. That type theory is developed in detail in section 4 of this chapter. In particular both kinds of products, both exponentials, and coproducts are used to produce derived types. If unbalanced powerobjects are representable, higher order statements will be interpretable analogously to the interpretation of local set theory.

The key observation in the theory developed here is that unbalanced subobjects are the correct ones to consider when looking for a foundation for fuzzy sets. In the Goguen category **SET(I)** the lattice of all subobjects of an object includes too much: it conflates the set theoretic character of subobjects which change underlying set with the fuzzy character of subobjects which change the degree of membership. Consideration of more restrictive kinds of subobjects separates these two aspects of monomorphisms in **SET(I)**: the strong monos which figure prominently in the quasitopos structure are entirely set-like in character; the unbalanced monos capture only the result of changing degrees of membership.

§2. Properties of the Categories Under Discussion

The type theory giving a logical setting for fuzzy set theory in section 4 will have natural interpretation in categories with a sufficiently rich structure. In this section I will discuss what structure we want our categories to have. The structure of a footing (the name I will use for a category with all of the structure I want) has several aspects: exactness properties, local closed structures, factorization properties for monomorphisms, and a hyperdoctrine relating slice categories to the lattices of unbalanced subobjects. These will be discussed in turn and then a summary definition will be given at the end of the section. Throughout this discussion I will highlight features a category **C** should have to be a footing.

2.1. EXACTNESS PROPERTIES. In order for **C** to have a satisfactory propositional logic we will ask that it have finite products and a terminal object. Since we will get the lattice of unbalanced subobjects $\mathcal{U}(A)$ as a subcategory of the slice category $\mathbf{C}/A$ (which has as objects the morphisms into A and as morphisms the morphisms of **C** making the triangles over A commute, see [MacLane 1971], we really want these structures in $\mathbf{C}/A$. Thus we will want pullbacks as well.

FOOTING FEATURE 1. **C** should have all finite limits.

As a consequence of the existence of finite limits if we are given a morphism $f : A \longrightarrow B$ we get a functor $\mathbf{f}^* : \mathbf{C}/B \longrightarrow \mathbf{C}/A$ obtained by pulling back. This functor always has a left adjoint Σ_f given by composition with f.

From the finite limits we will get the operation $\wedge$ in the lattice of unbalanced subobjects. To get $\vee$ we will also want coproducts. The top of the lattice comes from the terminal; the bottom will come from an initial object; construction of unions uses a pushout. In topos theory it is possible to deduce the existence of the colimits from the limits and the rest of the structure. This is not generally true in quasitopoi, so we will ask for these structures rather than trying to derive them.

FOOTING FEATURE 2. **C** should have finite colimits.

The operations posited so far give the intuitionistic connectives on subobjects characteristic of topos theory and quasitopos theory. To get a foundation which uses the fuzzy connectives we need a second product $\otimes$ on **C**. Since we will want the functor $-\otimes A$ to have a right adjoint it will follow that $\otimes$ distributes over $\times$ and has the terminal as a unit.

FOOTING FEATURE 3. Each slice **C**/X has a second monoidal structure given by the product $\otimes$ satisfying.

1. $A \otimes B \simeq B \otimes A$,
2. $(A \times B) \otimes C \simeq (A \otimes C) \times (B \otimes C)$,
3. $A \otimes (B \otimes C) \simeq (A \otimes B) \otimes C)$.

The structure given so far will produce the positive fragment of a propositional logic with two different notions of conjunction. We will need further structure to get implications, negations, and quantification.

2.2. LOCAL CLOSED STRUCTURES. In categorical logic quantification is defined along morphisms, rather than in terms of bound variables. Lawvere [Lawvere 1970] was the first to notice the connection between quantification and adjoints to the pulling back functor. In topoi one of the fundamental theorems is that the inverse image functor has both right and left adjoints. It is observed in [Penon 1977] that having a right adjoint is equivalent to having each slice category cartesian closed. This property is usually stated by saying that **C** is **locally cartesian closed**. The local cartesian closed structure will give intuitionistic implication and negation in the subobject lattices. To get the fuzzy implication we will also want a right adjoint to $\otimes$, a second local monoidal closed structure.

FOOTING FEATURE 4. **C** should be locally cartesian closed. This means that each slice **C**/A should be cartesian closed, giving an internal representation of the hom sets.

FOOTING FEATURE 5. **C** should have a second local monoidal closed structure. This means that the functor given by $\otimes$ with a given object has a right adjoint.

2.3. FACTORIZATION OF MORPHISMS. In order to make good use of the two local closed structures we will want a mechanism for transporting those structures from the slice categories to the lattices of unbalanced subobjects. This will be done by asking for a natural way to think of an unbalanced subobject as a morphism and an image factorization for morphisms.

DEFINITION 1. An **unbalanced monomorphism** is a morphism which is both monic and epic.

DEFINITION 2. An **unbalanced subobject** of an object A is an equivalence class of unbalanced monomorphisms where two unbalanced monos m and m' are equivalent if and only if there is an isomorphism i with $m = m'i$.

The unbalanced subobjects of an object A form a category using the order relation with $m \leq m'$ if and only if there is an unbalanced mono u with $m = m'u$. This category is designated by $\mathcal{U}(A)$. We will need a way to think of this as a subcategory of $\mathbf{C}/A$.

FOOTING FEATURE 6. $\mathbf{C}$ has canonical choice of representative for unbalanced subobjects in the form of a functor $\mathbf{i} : \mathcal{U}(A) \longrightarrow \mathbf{C}/A$ which is full and faithful.

We will want to be able to move from $\mathbf{C}/A$ to $\mathcal{U}(A)$ as well. One good way is to ask for the existence of an unbalanced image for any morphism $f : A \longrightarrow B$ in $\mathbf{C}$: a factorization of f as a morphism followed by an unbalanced mono $f = f_u \circ g$ such that any other such factorization $f = f' \circ g'$ gives rise to an unbalanced diagonal map d with $d \circ g = g'$ and $f' \circ d = f_u$. Such a factorization gives a left adjoint left inverse for $\mathbf{i}$.

FOOTING FEATURE 7. The functor $\mathbf{i} : \mathcal{U}(A) \longrightarrow \mathbf{C}/A$ has a left adjoint left inverse σ.

DEFINITION 3. Suppose we are given two categories $\mathbf{C}_1$ and $\mathbf{C}_2$ with reflective subcategories $\mathcal{U}_1$ and $\mathcal{U}_2$ with inclusions $\mathbf{i}_1$ and $\mathbf{i}_2$ with left adjoint left inverses σ_1 and σ_2 and we are also given a functor $\mathbf{f} : \mathcal{U}_1 \longrightarrow \mathcal{U}_2$. We say that a functor $\mathbf{F} : \mathbf{C}_1 \longrightarrow \mathbf{C}_2$ lifts $\mathbf{f}$ if $\mathbf{Fi} = \mathbf{if}$ and $\sigma\mathbf{F} = \mathbf{f}\sigma$.

If we know how to construct adjoints in the larger category, then the following proposition gives us a clue how to find them in the smaller category. It is proved by taking adjoints in the equations in the definition of a lifting.

PROPOSITION 1. If $\mathbf{F}$ lifts $\mathbf{f}$, then right adjoints (indicated here by a subscript r) and left adjoints (subscripted with l) will be related by the equations

$$\mathbf{f}_l\sigma_1 = \sigma_2\mathbf{F}_l \text{ and } \mathbf{i}_2\mathbf{f}_r = \mathbf{F}_r\mathrm{i}_1.$$

In the case of the reflective subcategories $\mathcal{U}(A)$ in $\mathbf{C}/A$, this will tell us that either a left or a right adjoint to a functor at the $\mathcal{U}(A)$ level which lifts can be found by transporting the situation up to the $\mathbf{C}/A$ level, taking the adjoint at that level, and then transporting back down using σ. In particular, the exponential adjoints in $\mathbf{C}/A$ can be transported down to give implications in the lattices of unbalanced subobjects.

The local cartesian closed property we have assumed gives us both right and left adjoints for the pulling back functor. These can be used to obtain quantification along morphisms of $\mathbf{C}$. This is summarized in the hyperdoctrinal diagram:

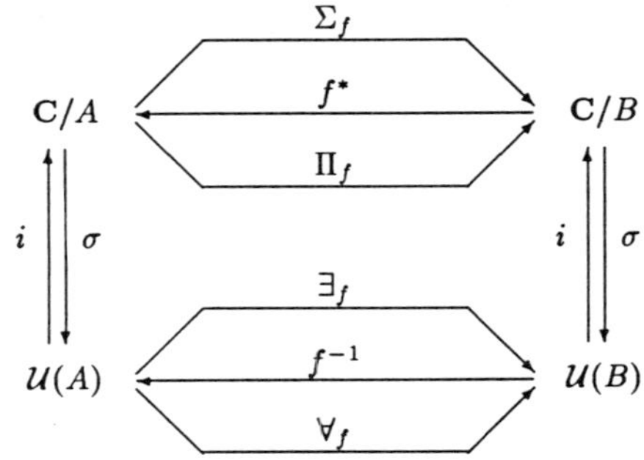

where

$$\Sigma_f \dashv f^* \dashv \Pi_f$$
$$\exists_f \dashv f^{-1} \dashv \forall_f$$

and the following equations hold:

$$\begin{aligned} \sigma\Sigma_f &= \exists_f\sigma \\ i\forall_f &= \Pi_f i \\ if^{-1} &= f^*i \\ \sigma f^* &= f^{-1}\sigma. \end{aligned}$$

2.4. Summary of the definition of a Footing. Let us summarize all of these properties in the form of a definition:

Definition 4. A **footing** is a category **C** satisfying the following conditions:

1. **C** has all finite limits and colimits.
2. Each slice **C**/X has a second monoidal structure given by the product $\otimes$ satisfying
   (a) $A \otimes B \simeq B \otimes A$,
   (b) $(A \times B) \otimes C \simeq (A \otimes C) \times (B \otimes C)$,
   (c) $A \otimes (B \otimes C) \simeq (A \otimes B) \otimes C$.
3. **C** is locally cartesian closed.
4. **C** has a second local monoidal closed structure.
5. **C** has canonical choice of representative for unbalanced subobjects in the form of a functor $\mathbf{i} : \mathcal{U}(A) \longrightarrow \mathbf{C}/A$ which is full and faithful.
6. The functor $\mathbf{i} : \mathcal{U}(A) \longrightarrow \mathbf{C}/A$ has a left adjoint left inverse σ.
7. The exponentials in $\mathbf{C}/A$ lift implications in $\mathcal{U}(A)$.
8. The functors Σ_f and Π_f lift quantifiers on $\mathcal{U}$.

The examples we will look at in the next section have these structures arise from quasitopos structures and initial category structures. To make those examples more transparent we include the next two definitions and a proposition.

DEFINITION 5. A **strong monomorphism** m in a category is a mono such that if e is an epimorphism and there are maps g and f with $ge = mf$ then there is a map t with $te = f$ and $mt = g$.

A typical example of a strong monomorphism is a subspace inclusion in the category of topological spaces.

DEFINITION 6 [Penon 1977]. A quasitopos is a category $\mathbf{Q}$ such that

1. $\mathbf{Q}$ has all finite limits and colimits,
2. $\mathbf{Q}$ is locally cartesian closed,
3. there is a representing object for strong subobjects.

Quasitopoi have a logical structure which is rich in much the same way as that of topoi, see for instance [Monro 1986a, 1986b]. They also occur much more often as topological categories than topoi do [Wyler 1976].

DEFINITION 7. A functor $\mathbf{F}: \mathbf{C}_1 \longrightarrow \mathbf{C}_2$ is an **initial structure functor** if given any family of morphisms $\{f_i : A \longrightarrow \mathbf{F}B_i\}$, there is a smallest B in $\mathbf{C}_1$ with morphisms $g_i : B \longrightarrow B_i$ such that $\mathbf{F}(B) = A$ and $\mathbf{F}\ (g_i) = f_i$.

PROPOSITION 2. If $\mathbf{C}$ is a quasitopos with an underlying set functor $\mathbf{U} : \mathbf{C} \longrightarrow \mathbf{SET}$ which is an initial structure functor which preserves and reflects unbalanced monos, then $\mathbf{C}$ satisfies parts 1, 3, 5, and 6 in the definition of a footing.

PROOF. Since quasitopoi have finite limits and colimits and are locally cartesian closed items 1 and 3 of the definition are satisfied.

The requirement that the underlying set functor be an initial structure functor entails the existence of right and left adjoints to the underlying set functor. In particular given any $f : B \longrightarrow A$ in $\mathbf{C}$, there is a smallest object $u : im_f \longrightarrow A$ in $\mathbf{C}/A$ with $\mathbf{U}im_f = A$, $\mathbf{U}u = id_a$, and $f = ug$ for some g with underlying function $\mathbf{U}f$. This is the image factorization which is induced by the final structure on $\mathbf{U}A$ making g a map in $\mathbf{C}$ over $\mathbf{U}f$.

An unbalanced mono in $\mathbf{C}$ has an isomorphism as its underlying set map. We can choose the canonical representative of an unbalanced subobject by requiring that the underlying set map be an identity. This makes the lattice of unbalanced subobjects of an object A isomorphic to the lattice of structures on the set $\mathbf{U}A$ which have morphisms in $\mathbf{C}$ into A with underlying map the identity. The image factorization gives the reflector. □

§3. Examples

There are three main examples of categories with the structure specified above. The motivating example is the category $\mathbf{SET(L)}$ where L is a complete lattice ordered semigroup with semigroup operation $*$. The category of tolerance spaces has a similar first order structure. The category of fuzzy sets with a non-transitive equality mentioned at the end of [Stout 1990] also has two local closed structures and an interesting logic of unbalanced subobjects.

3.1. **SET(L)**. Throughout this section we will assume that L is a complete lattice ordered semigroup. This means that L has a lattice operations $\wedge$ and $\vee$ and a semigroup operation $*$ which respects the order and thus distributes over $\wedge$. Completeness gives implications right adjoint to both products:

$$l_1 \to l_2 = \bigvee\{h \mid h * l_1 \leq l_2\}$$

and

$$l_1 \Rightarrow l_2 = \bigvee\{h \mid h \wedge l_1 \leq l_2\}.$$

The category **SET(L)** as defined in [Goguen 1974] has as its objects pairs (A, α), where $\alpha : A \longrightarrow L$ and as morphisms from (A, α) to (B, β) functions $f : A \longrightarrow B$ with $\beta(f(a)) \geq \alpha(a)$. In this category membership is fuzzy but equality is not, and functions are those which do not decrease the level of membership. It is a somewhat simpler category than the category of fuzzy sets on a locale discussed in [Bell 1988], p. 211. If the membership function is always 1 we will call the fuzzy set **crisp**. The full subcategory of crisp fuzzy sets is the category **SET**.

Goguen showed that completeness properties on L gave a cartesian closed structure on **SET(L)** and characterized categories of the form **SET(L)** in terms of the properties of the subobject lattices. [Pultr 1976a, 1976b] showed that the semigroup operation could be used to obtain a monoidal closed structure on **SET(L)** with a somewhat successful notion of powerobject.

3.1.1. Exactness and Initial Structures

Theorem 3. The underlying set functor $\mathbf{U} : \mathbf{SET(L)} \longrightarrow \mathbf{SET}$ is an initial structure functor with small fibers.

Proof. The small fibers part is obvious because each structure on A is a function to L, and there is only a set of such functions.

If we are given a family of functions $\{f_i : A \longrightarrow B_i\}$ and structures β_i on B_i for each i, we get the initial structure on A by taking $\alpha(a) = \bigwedge \beta_i(f_i(a))$. Similarly, if we are given a family of functions $\{g_i : C_i \longrightarrow A\}$ and structures γ_i on C_i, we get the final structure on A by taking $\alpha(a) = \bigvee_{\{c_i \mid g_i(c_i)=a\}} \gamma_i(c_i)$. □

Corollary 4. **SET(L)** has all limits and colimits.

Proof. Limits are constructed as in **SET** and then given the initial structure. Colimits are constructed as in **SET** and given the final structure. These constructions are standard and can be found in [Wischnewski 1972]. □

PROPOSITION 5. $\mathbf{SET(L)}/(A,\alpha)$ has a second monoidal structure for each (a,α) induced by the operation $*$ on L.

PROOF. The tensor is given by

$$((B,\beta) \to (A,\alpha)) \otimes ((C,\gamma) \to (A,\alpha)) =$$

$$((B \times_A C, \beta * \gamma) \to (A,\alpha))$$

The commutativity and associativity up to isomorphism of this operation follow from the commutativity and associativity up to isomorphism of pullback over A and the commutativity and associativity of $*$ in L. Distributivity over product follows from the distributivity of $*$ over $\wedge$. $\square$

3.1.2. LOCAL CLOSED STRUCTURES. Both of the products on $\mathbf{SET(L)}$ have right adjoints. The cartesian closed structure comes from the exponentiation:

$$(B,\beta)^{(A,\alpha)} = (\{f : A \longrightarrow B\}, \bigwedge_{a\in A} (\alpha(a) \Rightarrow \beta(f(a)))$$

which is right adjoint to the functor $- \times (A,\alpha)$. The closed structure using $\otimes$ uses the exponentiation

$$(B,\beta) \uparrow (A,\alpha) = (\{f : a \to B\}, \bigwedge_{a\in A} (\alpha(a) \to \beta(f(a)))$$

Both of these can be extended to local closed structures.

THEOREM 6. $\mathbf{SET(L)}$ is locally cartesian closed and has a second monoidal closed structure. This says that both products on $\mathbf{SET(L)}/A$ have right adjoints.

PROOF. The construction of the exponential in $\mathbf{SET(L)}/(A,\alpha)$ right adjoint to product is given by

$$(f : (B,\beta) \longrightarrow (A,\alpha))^{(g:(C,\gamma)\to(A,\alpha))} =$$

$$\pi_2 : (\{(h,a) \mid h : g^{-1}(a) \longrightarrow f^{-1}(a)\}, \zeta) \longrightarrow (A,\alpha)$$

where

$$\zeta(h,a) = \min\left(\alpha(a), \bigwedge_{c\in g^{-1}(a)} (\gamma(c) \Rightarrow \beta(h(c))\right)$$

The right adjoint to $-\otimes(g : (C,\gamma) \longrightarrow (A,\alpha))$ is given by

$$(f : (B\beta) \longrightarrow (A,\alpha)) \uparrow (g : (C,\gamma) \longrightarrow (A,\alpha)) =$$

$$\pi_2 : (\{(h,a) \mid h : g^{-1}(a) \longrightarrow f^{-1}(a)\}, \xi) \longrightarrow (A,\alpha)$$

where

$$\xi(h,a) = \min\left(\alpha(a), \bigwedge_{c\in g^{-1}(a)} (\gamma(c) \longrightarrow \beta(h(c))\right)$$

As Penon observed in [Penon 1977], the local cartesian closed structure is equivalent to the pullback functor

$$f^{-1} : \mathbf{SET(L)}/(B,\beta)\delta \longrightarrow \mathbf{SET(L)}/(A,\alpha)$$

having a right adjoint

$$\Pi_f : \mathbf{SET(L)}/(A,\alpha) \longrightarrow \mathbf{SET(L)}/(B,\beta)$$

(In any category with limits f^{-1} has a left adjoint Σ_f given by composition with f.) Again the construction uses the construction in **SET** as its carrier : the fiber over b is the product of the fibers over a taken over all of the set $f^{-1}(b)$. The extent of membership is then

$$\hat{\gamma}((c_a)_{a\in f^{-1}(b)}) = \bigwedge \gamma(c_a). \quad \square$$

3.1.3. SUBOBJECT REPRESENTATION. There are three kinds of subobjects of interest in the category $\mathbf{SET(L)}$, corresponding to the three kinds of monomorphisms. A subobject (with no further modifier) is an equivalence class of monomorphism $m : (A',\alpha') \longrightarrow (A,\alpha)$, where two monos are equivalent if there is an isomorphism i with $m_1 = im_2$. The lattice of subobjects was studied by Goguen. We will call it $\mathcal{P}(A,\alpha)$. For a quasitopos structure we need to identify the strong subobjects. In this case the strong subobjects are most easily characterized by characterizing the essential subobjects (those which appear as equalizers): a strong subobject is one of the form $(A',\alpha\mid_{A'}) \subseteq (A,\alpha)$. The lattice of strong subobjects $\mathcal{S}(A,\alpha)$ is the same as the lattice of subsets of A; we denote it by $\mathcal{S}(A,\alpha)$. An unbalanced subobject is one which is both epic and monic, so the underlying set map is an isomorphism. Unbalanced subobjects have canonical representatives of the form $m : (A,\alpha') \longrightarrow (A,\alpha)$. Note that the underlying set of the subobject is the same as that of (A,α). This comes closer to the usual meaning in fuzzy set theory than either strong subobjects or subobjects do.

Our description of the strong subobjects makes the following theorem easy:

THEOREM 7. The lattice of strong subobjects is representable by the usual characteristic morphism into the crisp two element set.

Corollary 8. $\mathbf{SET(L)}$ is a quasitopos.

PROOF. A quasitopos is a locally cartesian closed category which has finite colimits and has a strong subobject representor. $\square$

The category $\mathbf{SET(L)}$ has factorization of any morphism into a strong epi followed by a mono: $f : (B,\beta) \longrightarrow (A,\alpha)$ factors through the image of **U**f with the final structure induced by the structure β, $(f(B), f\beta)$. This factorization gives a left adjoint to the inclusion functor taking $\mathcal{P}(A)$ into $\mathbf{SET(L)}/(A,\alpha)$.

Any morphism can also be factored into an epi followed by a strong mono: f maps onto $(f(B), \alpha|_{f(B)})$. This gives a left adjoint for the inclusion of $\mathcal{S}(A,\alpha)$ into $\mathbf{SET(L)}/(A,\alpha)$.

Any monomorphism can be factored into a strong mono followed by an unbalanced mono: $m : (A', \alpha') \longrightarrow (A, \alpha)$ is factored as $s : (A', \alpha') \longrightarrow (A, \alpha')$ followed by $u : (A, \alpha') \longrightarrow (A, \alpha)$; here $\alpha'(a) = 0$ if $a \notin A'$. Combining this with the factorization of morphisms into strong epis followed by monos gives the left adjoint left inverse to the inclusion $\mathcal{U}(A,\alpha) \longrightarrow \mathbf{SET(L)}/(A,\alpha)$.

THEOREM 9. The inclusion $\mathbf{i} : \mathcal{U}(A,\alpha) \longrightarrow \mathbf{SET(L)}/(A,\alpha)$ has a left adjoint left inverse σ given by taking $f : (B,\beta) \longrightarrow (A,\alpha)$ to the set A with the final structure determined by the morphism f and the structure β.

Given $f : (A,\alpha) \longrightarrow (B,\beta)$ we get $f^{-1} : \mathcal{U}(B,\beta) \longrightarrow \mathcal{U}(A,\alpha)$ taking β' to

$$f^{-1}\beta' : a \longmapsto \min(\alpha(a), \beta'(f(a)))$$

and $\exists_f : \mathcal{U}(A,\alpha) \longrightarrow \mathcal{U}(B,\beta)$ taking α' to

$$\exists_f \alpha' : b \mapsto \begin{cases} 0, & \text{if no a maps to b} \\ \bigvee_{\{a|f(a)=b\}} \alpha'(a), & \text{otherwise} \end{cases}$$

and $\forall_f : \mathcal{U}(A,\alpha) \to \mathcal{U}(B,\beta)$ which takes α' to

$$\forall_f \alpha' : b \mapsto \begin{cases} \beta(b), & \text{if no a maps to b} \\ \bigwedge_{\{a|f(a)=b\}} \alpha'(a), & \text{otherwise} \end{cases}$$

These satisfy the usual adjointness conditions: if $\alpha'(a) \le \beta'(f(a))$, then

$$\begin{aligned} \exists_f \alpha' &= \bigvee_{\{a|f(a)=b\}} \alpha'(a) \\ &\le \bigvee_{\{a|f(a)=b\}} \beta'(f(a)) \\ &= \bigvee_{\{a|f(a)=b\}} \beta'(b) \\ &= \beta'(b). \end{aligned}$$

If $\beta'(f(a)) \le \alpha'(a)$, then

$$\begin{aligned} \beta'(b) &= \bigwedge_{\{a|f(a)=b\}} \beta'(b) \\ &\le \bigwedge_{\{a|f(a)=b\}} \alpha'(a) \\ &= \begin{cases} \forall_f \alpha'(b) & \text{if } \{a \mid f(a) = b\} \ne \emptyset \\ \beta'(b) & \text{otherwise} \end{cases} \\ &\le \forall_f \alpha'(b) \end{aligned}$$

These quantifiers lift so that the hyperdoctrine diagram is valid in **SET(L)**.

The products also lift and give implications in the external propositional logic. The implications are defined by

$$(\alpha' \Rightarrow \alpha'')(a) =$$

$$\bigvee\{h \mid h \leq \alpha(a) \text{ and } (h \wedge \alpha'(a)) \leq \alpha''(a)\}$$

and similarly for $\rightarrow$. Negations are then given by taking $\neg\alpha' = \alpha' \Rightarrow 0$ and $\sim \alpha' = \alpha' \rightarrow 0$. Thus $\sim \alpha'(a) = \min(\alpha(a), 1 - \alpha'(a))$.

Second order structures will be obtained using the representation of unbalanced powerobjects given by the next theorem:

THEOREM 10. Unbalanced subobjects are weakly representable in **SET(L)**. Either exponential can be used to extend this to a representation of the unbalanced powerobject. To avoid decisions about which implication to use, there is another natural choice for the unbalanced powerobject weak representor.

PROOF. Unbalanced subobjects can be weakly represented by morphisms to $(I, 1)$: an unbalanced subobject $(A, \alpha') \hookrightarrow (A, \alpha)$ has a natural characteristic map α' such that the pullback of $(I, \mathrm{id}) \hookrightarrow (I, 1)$ is the given unbalanced subobject. While natural, this choice of characteristic morphism is not unique if A has elements which have membership less than 1.

The representation of unbalanced powerobjects is most easily done by using the object

$$U(A, \alpha) = (\{f : A \longrightarrow I \mid \forall_{a \in A} f(a) \leq \alpha(a)\}, 1)$$

This becomes interesting when we give the unbalanced subobject

$$\epsilon_{(A,\alpha)} = (\{(f, a)\}, \epsilon(f, a) = f(a))$$

An unbalanced subobject $(B \times A, \phi)$ has characteristic morphism

$$\chi_\phi : (B, \beta) \longrightarrow U(A, \alpha) : b \mapsto \phi(b, -)$$

When you pull back $\epsilon_{(A,\alpha)}$ you get the original unbalanced subobject. This characteristic morphism need not be unique, however. If $\phi(a, b) = \beta(b) < \alpha(a)$ then b can be taken to a function taking a to any value less than or equal to $\alpha(a)$ and greater than or equal to $\beta(b)$. In some sense the characteristic map is the smallest available map which pulls ϵ back to the correct value. The structure given as powerobject representor is the strong subobject of crisp elements of $(I, 1)^{(A,\alpha)}$ with the induced membership. All of the characteristic functions of subobjects of $(B, \beta) \times (A, \alpha)$ factor through the strong subobject of crisp elements, so the unbalanced powerobject representor is a natural choice. □

3.2. TOLERANCE SPACES. Tolerance spaces were originally studied by [Zeeman and Buneman 1968] in an attempt to model visual discrimination. Two points were within tolerance if they could not be visually distinguished. Since a sequence of successively indistinguishable points can have points which can be distinguished as its endpoints, it is clear that the relation "within tolerance" need not be transitive. It should, however, be both reflexive and symmetric. The model example for tolerance spaces is a metric space (M, ρ) with the relation $a \sim b \Leftrightarrow \rho(a,b) \leq \tau$ for some fixed tolerance τ. This leads to the following definition.

DEFINITION 8. A **tolerance space** is a set A equipped with a symmetric, reflexive relation $\sim_A$. A function $f : A \longrightarrow B$ is said to be **tolerance preserving** if whenever $a \sim_A a'$ we get $f(a) \sim_B f(a')$. A function is said to reflect tolerance if whenever $f(a) \sim_B f(a')$ we get $a \sim_A a'$.

Observe that the composition of tolerance preserving functions is tolerance preserving and that identity functions are tolerance preserving. Hence we can talk about the category of tolerance spaces.

DEFINITION 9. The category of tolerance spaces and tolerance preserving functions is called **TOL**. Forgetting the tolerance relation gives the forgetful functor $\mathbf{U} : \mathbf{TOL} \longrightarrow \mathbf{SET}$.

The underlying set functor has both left and right adjoints, giving the smallest and largest tolerances possible on a set. The smallest tolerance is given by the identity relation (elements are only within tolerance of themselves). We will call such tolerance spaces **crisp**. The largest tolerance has everything within tolerance of everything else. This is the **complete** tolerance.

The category **TOL** includes discrete analogs of line segments of integral length and circles with integral circumference: an n-point line is the set $\{1, 2, \cdots, n\}$ with

$$i \sim j \text{ if and only if } |i - j| \leq 1$$

The n-point circle has $1 \sim n$ as well. Indeed much of topology has analogs in this category. For example there are two discrete forms of the Jordan Curve theorem (see [Stout 1988] and A. Rosenfeld's works on digital topology [Rosenfeld 1979, 1989]) and a fairly thoroughly developed homotopy theory.

The category **TOL** has both initial and final structures, so the underlying set functor has both right and left adjoints. In addition this tells us that **TOL** has both limits and colimits constructed in **SET** and then given the appropriate either initial or final tolerance structure.

THEOREM 11. The category **TOL** has both final and initial structures over **SET**.

PROOF. For **TOL** to have final structures means that for any family of functions $f_i : B_i \longrightarrow A$ and tolerances $\sim_i$ on B_i, there is a smallest tolerance on A making all of the f_i tolerance preserving. We obtain the tolerance on A by making $a \sim a'$ if either $a = a'$ or there is an f_i such that $a = f_i(b)$ and $a' = f_i(b')$ and $b \sim_i b'$. If f_i is to be tolerance preserving these two elements of A must be related, and the relation $\sim$ is reflexive and symmetric, as needed.

Similarly if $g_i : A \longrightarrow C_i$ is a family of functions and each C_i has a tolerance $\sim_i$, we get the initial tolerance on A by relating elements if every i has $f_i(a) \sim_i f_i(a')$. This gives the largest tolerance on A making the f_i all tolerance preserving. □

COROLLARY 12. **TOL** has arbitrary limits and colimits.

PROOF. To form the limit use the limit in **SET** equipped with the initial tolerance. For example the product $(A, \sim_A) \times (B, \sim_B) = (A \times B, \sim)$ where $(a, b) \sim (a', b')$ if and only if $a \sim_A a'$ and $b \sim_B b'$.

To form colimits take the colimit in **SET** and give it the final tolerance.

These constructions are standard in the theory of initial categories as in [Wischnewski 1972]. □

3.2.1. CLOSED STRUCTURES. The category **TOL** has two closed structures, arising from the two natural ways to define products. These are described in [Warner and Muir 1981] by giving two function space constructions. The cartesian closed structure gives a function space right adjoint to product. It uses the "usual tolerance" with $f \rho_\Pi g$ if and only if whenever $a \sim a'$, $f(a) \sim g(a')$. The coarse function space uses $f \rho_\otimes g$ if and only if $f(a) \sim g(a)$ for all $a \in A$. Both of these can be thought of as kinds of homotopy of length 2.

DEFINITION 10. The weak product of two tolerance spaces is given by $(A, \sim) \otimes (B, \sim) = (A \times B, \sim_\otimes)$, where $(a, b) \sim_\otimes (a', b')$ if and only if $a = a'$ and $b \sim b'$ or $a \sim a'$ and $b = b'$.

These definitions give rise to the following two theorems:

THEOREM 13. **TOL** is cartesian closed: the usual function space gives a right adjoint to $- \times (A, \sim)$.

PROOF. Suppose we are given a tolerance preserving map

$$h : (B, \sim_B) \times (A, \sim_A) \longrightarrow (C, \sim_C)$$

we get

$$\hat{h} : (B, \sim_B) \longrightarrow (C, \sim_C)^{(A, \sim_A)}$$

defined as in **SET** taking b to $\lambda a(h(b, a))$, which will be tolerance preserving. Elements of B which are within tolerance will be sent to functions which are within tolerance in the usual function space because if $a \sim_A a'$ and $b \sim_B b'$, then $h(a, b) \sim_C h(a', b')$.

Conversely, if we are given a tolerance preserving function g from $(B, \sim_B)$ to $(C, \sim_C)^{(A, \sim_A)}$, then induced map from $B \times A$ to C preserves the product tolerance. It takes (b, a) to $g(b)(a)$. If $(b, a) \sim_\Pi (b', a')$ then $b \sim_B b'$ and $a \sim_A a'$, so $g(b)(a) \sim_C g(b')(a')$ since $g(b) \rho_\Pi g(b')$. □

THEOREM 14. **TOL** has a second monoidal closed structure given by $\otimes$ and the coarse function space tolerance.

PROOF. The proof here looks like the one for the cartesian closed structure, with a little more care needed for certain parts. A function from $(B, \sim_B) \otimes (A, \sim_A)$ to $(C, \sim_C)$ induces a function from B to the set of tolerance preserving functions from A to C because the slices of $(B, \sim_B) \otimes (A, \sim_A)$ look just like $(A, \sim_A)$. The tolerance $\rho_\otimes$ only asks that values at the same a be within tolerance, so $b \sim_B b'$ and thus $h(b, a) \sim_C h(b', a)$ is sufficient to get $\hat{h}(b)\rho_\otimes\hat{h}(b')$ as needed.

Similarly the $\rho_\otimes$ tolerance is exactly what is needed to show that the induced map on the $\otimes$ product preserves tolerance. □

3.2.2. REPRESENTATION OF STRONG SUBOBJECTS. In order to talk about the quasitopos structure and the logic of unbalanced subobjects in **TOL** we will need to identify what the various kinds of subobjects are. This is the content of the following theorem:

THEOREM 15. A map in **TOL** is monic if the underlying map in **SET** is monic. It is a strong monic if it reflects tolerance as well as preserving it. It is an unbalanced mono if the underlying set does not change, only the tolerance.

PROOF. The only hard part is the part about strong monos. A strong mono m is a mono such that if e is epic and there are maps g and f with $ge = mf$, then there is a map t with $te = f$ and $mt = g$. Since all monos in **SET** are strong we can construct t as a map of sets. What we need to show is that t preserves tolerance if and only if m reflects tolerance. Let x and x' be within tolerance in the domain of t. Since e is epic, x and x' are in the image of e. Since g preserves tolerance, $g(x) \sim g(x')$. Now $ge = mf$ and m reflects tolerance so the elements of the domain of m which went to $g(x)$ and $g(x')$ must have been within tolerance. But we know that these elements are $t(x)$ and $t(x')$, so t preserves tolerance if m reflects tolerance.

If m does not reflect tolerance, then we can pick two elements of the image of m, call them y and y', which are within tolerance but do not come from elements within tolerance. Use the identity function from the discrete two point set to the two point line for e. Then let g map the end points to y and y'. This leads to an example where the induced t does not preserve tolerance. □

THEOREM 16. Any monomorphism can be factored into a strong mono followed by an unbalanced mono. This gives rise to a functor $\sigma : \mathbf{TOL}/(A, \sim_A) \longrightarrow \mathcal{U}(A, \sim_A)$ which is left adjoint and right inverse to the inclusion functor.

PROOF. A monomorphism is a morphism whose underlying function is monic. If $m : (A', \sim_{A'}) \to (A, \sim_A)$ is monic then we can form the unbalanced image of m by taking the tolerance $\sim_m$ on A which connects each element to itself and connects distinct elements if and only if they are both in A' and they are connected there. The construction of this tolerance guarantees that $m : (A', \sim_{A'}) \to (A, \sim_m)$ reflects tolerance, so it is a strong mono. Since $\sim_m$ has fewer pairs within tolerance than $\sim_A$, $id : (A, \sim_m) \hookrightarrow (A, \sim_A)$ is an unbalanced mono. It is clear that the construction gives the smallest unbalanced mono through which m can factor, hence giving the left adjoint to inclusion. That the unbalanced image is right inverse to inclusion follows from the fact that if m is unbalanced then $\sigma(m) = m$. □

The strong subobjects in **TOL** can be represented by characteristic morphisms to the complete tolerance space on two elements. This makes the category of tolerance spaces look just like the category of sets when we consider strong subobjects. In particular we have a strong partial map representor:

THEOREM 17. Given a tolerance space $(X, \sim)$, there is a tolerance space $(\tilde{X}, \sim)$ which has $(X, \sim)$ as a strong subobject such that tolerance preserving maps from $(T, \sim)$ to $(\tilde{X}, \sim)$ are in one to one correspondence, obtained using pull back, with the strong partial maps from $(T, \sim)$ to $(A, \sim)$.

PROOF. We obtain $(\tilde{X}, \sim)$ by adjoining a point to X which is within tolerance of all the old points. The underlying set is the partial map representor in sets. Any map induced by a tolerance preserving strong partial map will be tolerance preserving because the extra point is within tolerance of everything. □

COROLLARY 18. **TOL** is a quasitopos.

PROOF. All that is needed for a quasitopos is the existence of limits, colimits, cartesian closed structures, and partial map representation. □

In particular this is enough to show that **TOL** is locally cartesian closed, or equivalently that the pullback functor has both right and left adjoints. Since we will want to know what these adjoints look like when we develop quantification for unbalanced subobjects, we include the constructions in the next theorem:

THEOREM 19. A tolerance preserving map $f : (A, \sim_A) \longrightarrow (B, \sim_B)$ induces a functor "pullback along f"

$$\mathbf{f^{-1}} : \mathbf{TOL}/(B, \sim_B) \longrightarrow \mathbf{TOL}/(A, \sim_A)$$

with left adjoint "compose with f"

$$\Sigma_f : \mathbf{TOL}/(A, \sim_A) \longrightarrow \mathbf{TOL}/(B, \sim_B)$$

and right adjoint

$$\Pi_f : \mathbf{TOL}/(A, \sim_A) \longrightarrow \mathbf{TOL}/(B, \sim_B)$$

PROOF. The construction of $\mathbf{f^{-1}}$ and Σ_f are standard in any category with limits. The construction of Π_f mimics that of sets. If $f : (A, \sim_A) \to (B, \sim_B)$ and $h : (C, \sim_C) \to (A, \sim_A)$ are given, then $\Pi_f(h)$ has as its fiber over an element $b \in B$ the product of the fibers h over the elements in A which map to b under f. Thus an element of the fiber over b has the form

$(c_a)_{\{a|f(a)=b\}}$ where $h(c_a) = a$, unless $\{a \mid f(a) = b\} = \emptyset$ in which case the fiber has a single element $*$. The hard part is figuring out when two such tuples are within tolerance. Within the fiber over b we get $(c_a) \sim (c'_a)$ if and only if $c_a \sim c'_a$ for all a with $f(a) = b$. If our tuples are in different fibers, say $(c_a) \in f^{-1}(b)$ and $(c_{a'}) \in f^{-1}(b')$ over adjacent elements of B, then we ask that whenever $a \sim a'$ we get $c_a \sim c_{a'}$. If either of the fibers consists only of $*$ then we decide adjacency by the adjacency in B. □

The conjunctions in $\mathcal{U}(A, \sim_A)$ which lift to the two products are

$$a(\sim_1 \wedge \sim_2)a' \Leftrightarrow a \sim_1 a' \text{ and } a \sim_2 a'$$

and

$$a_1(\sim_1 * \sim_2)a_2 \Leftrightarrow a_1 = a_2$$

The triviality of this second product makes the implication $\rightarrow$ trivial as well (it gives $\sim_A$). If $f : (A, \sim_A) \rightarrow (B, \sim_B)$ and $\sim_1$ gives an unbalanced subobject of $(A, \sim_A)$ then

$$b(\exists_f \sim_1)b' \Leftrightarrow \text{ there are } a \text{ and } a' \text{ with } f(a) = b \text{ and } f(a') = b' \text{ and } a \sim_1 a'$$

and

$$b(\forall_f \sim_1)b' \Leftrightarrow \text{ for all } (a, a') \text{ with } f(a) = b \text{ and } f(a') = b' \; a \sim_1 a'$$

3.3. FUZZY SETS WITH NONTRANSITIVE EQUALITY. The two examples considered so far can be combined to give the category of fuzzy sets with nontransitive fuzzy equality discussed briefly at the end of [Stout 1987]. Again we get a quasitopos which is an initial category over **SET**. In this case we get a category enriched over **TOL**.

3.3.1. DEFINITIONS AND EXACTNESS PROPERTIES.

DEFINITION 11. Let **L** be a complete lattice ordered semigroup. The category **TOL(L)** has objects pairs (A, α) where $\alpha : A \times A \longrightarrow L$ has $\alpha(a, a') = \alpha(a'a)$ and $\alpha(a, a') \leq \alpha(a, a)$. Morphisms from (A, α) to (B, β) are functions $f : A \longrightarrow B$ such that $\beta(f(a), f(a')) \geq \alpha(a, a')$. Two morphisms f and g are adjacent if for all pairs $(a, a') \in A \times A$, $\alpha(a, a') \leq \beta(f(a), g(a'))$.

The conditions on the fuzzy tolerance relations are analogs of symmetry and a very weak form of the reflexive condition. **TOL** sits inside **TOL(L)** in much the same way that **SET** sits inside **SET(L)** : if we call a fuzzy tolerance space **crisp and reflexive** when the function α only takes on the values $\top$ and $\bot$ and always has $\alpha(a, a) = \top$, then the crisp and reflexive fuzzy tolerance spaces are just the ordinary tolerance spaces.

The adjacency relation on morphisms will make **TOL(L)** and enriched category with hom sets in **TOL**. The functor taking (A, α) to the fuzzy set $(A, \alpha\Delta)$, where $\Delta : A \rightarrow A \times A$ is the diagonal map, gives an underlying fuzzy set. The next theorem shows that **TOL(L)** is topological over **SET(L)** and thus also over **SET**.

THEOREM 20. The underlying fuzzy set functor $\mathbf{U} : \mathbf{TOL(L)} \longrightarrow \mathbf{SET(L)}$ has small fibers and is an initial structure functor.

PROOF. The part about small fibers is obvious since there are only a set of possible maps from $A \times A$ to L.

Now suppose that we are given a family $f_i : (A, \alpha) \longrightarrow (B_i, \beta_i\Delta)$ in $\mathbf{SET(L)}$. We want an initial fuzzy tolerance $\hat{\alpha}$ which makes all of the f_i tolerance preserving and which has $\alpha(a) = \hat{\alpha}(a, a)$. This is found by taking

$$\hat{\alpha}(a, a') = \alpha(a) \wedge \alpha(a') \wedge \bigwedge\{\beta_i(f_i(a), f_i(a'))\}. \quad \square$$

Initial structures based on the underlying set functor are a bit simpler: a family $\{f_i : A \longrightarrow B_i\}$ with structures β_i on B_i induces the initial structure

$$\alpha(a, a') = \bigwedge\{\beta_i(f_i(a), f_i(a'))\}$$

Final structures based on the underlying set functor will also prove useful. Given a family $\{g_i : B_i \rightarrow A\}$ and structures β_i on the B_i, we get a final structure $\tilde{\alpha}$ on A as

$$\begin{aligned} \tilde{\alpha}(a, a') &= \bot \text{ if either } a \text{ or } a' \text{ fails to be in the image of any } f_i \\ &= \bigvee_{\{i \mid \exists b, b' \in B_i \text{ with } f_i(b) = a \text{ and } f_i(b') = a'\}} \beta_i(b, b') \text{ otherwise} \end{aligned}$$

As usual, limits and colimits are constructed in $\mathbf{SET}$ and then given the appropriate initial or final structure:

COROLLARY 21. $\mathbf{TOL(L)}$ has products given by

$$(A, \alpha) \times (B, \beta) = (A \times B, \alpha \wedge \beta)$$

and coproducts given by

$$(A, \alpha) \coprod (B\beta) = (A \coprod B, \alpha \oplus \beta)$$

where

$$\begin{aligned} \alpha \oplus \beta(a, a') &= \alpha(a, a') \text{ if } a \text{ and } a' \text{ are both in } A \\ &= \beta(a, a') \text{ if } a \text{ and } a' \text{ are both in } B \\ &= \bot \text{ otherwise.} \end{aligned}$$

DEFINITION 12. The second monoidal structure on $\mathbf{TOL(L)}/(A, \alpha)$ is given by

$$((B, \beta) \longrightarrow (A, \alpha)) \otimes ((C, \gamma) \longrightarrow (A, \alpha)) = (B \times_A C, \beta * \gamma)$$

3.3.2. LOCAL CLOSED STRUCTURES. Both of the products on $\mathbf{TOL(L)}/(A,\alpha)$ have right adjoints. In the next subsection we will show that appropriate kinds of partial maps using strong subobjects are representable, giving a quasitopos structure.

THEOREM 22. $\mathbf{TOL(L)}$ is cartesian closed.

PROOF. The exponential is given by

$$(B,\beta)^{(A,\alpha)} = (\{f : A \to B \text{ in } \mathbf{SET}\}, \zeta)$$

where

$$\zeta(f,g) = \bigwedge_{(a,a')} (\alpha(a,a') \Rightarrow \beta(f(a), g(a')). \quad \Box$$

THEOREM 23. The functor $-\otimes (A,\alpha)$ has a right adjoint $-\uparrow (A,\alpha)$.

PROOF. The exponential is given by

$$(B,\beta) \uparrow (A,\alpha) = (\{f : A \longrightarrow B \text{ in } \mathbf{SET}\}, \xi)$$

where

$$\xi(f,g) = \bigwedge_{(a,a')} (\alpha(a,a') \to \beta(f(a), g(a')). \quad \Box$$

Both of these exponentials can be made local in similarly to the constructions of the local closed structures in $\mathbf{SET(L)}$.

THEOREM 24. $\mathbf{TOL(L)}$ is locally cartesian closed and has a second monoidal closed structure. This says that both products on $\mathbf{TOL(L)}/(A,\alpha)$ have right adjoints.

PROOF. The construction of the exponential in $\mathbf{TOL(L)}/(A,\alpha)$ right adjoint to product is given by

$$(f : (B,\beta) \longrightarrow (A,\alpha))^{(g:(C,\gamma)\to(A,\alpha))} =$$

$$\pi_2 : \left(\{(h,a) \mid h : g^{-1}(a) \longrightarrow f^{-1}(a)\}, \zeta\right) \longrightarrow (A,\alpha)$$

where

$$\zeta((h,a),(h',a')) =$$

$$\left(\alpha(a,a') \wedge \bigwedge_{c\in g^{-1}(a) \text{ and } c'\in g^{-1}(a')} (\gamma(c,c') \Rightarrow \beta(h(c), h'(c'))\right)$$

The right adjoint to $-\otimes (g : (C,\gamma) \to (A,\alpha))$ is given by

$$(f : (B\beta) \longrightarrow (A,\alpha)) \uparrow (g : (C,\gamma) \longrightarrow (A,\alpha)) =$$

$$\pi_2 : (\{(h,a) \mid h : g^{-1}(a) \longrightarrow f^{-1}(a)\}, \xi) \longrightarrow (A, \alpha)$$

where

$$\xi((h,a),(h',a')) =$$

$$\left(\alpha(a,a') \wedge \bigwedge_{c \in g^{-1}(a) \text{ and } c' \in g^{-1}(a')} (\gamma(c,c') \rightarrow \beta(h(c),h'(c'))) \right). \quad \square$$

3.3.3. SUBOBJECTS AND THE FACTORIZATION OF MONOS. As in our other two examples we can characterize three kinds of monomorphisms and the subobject lattices arising from them. A map $m : (A,\alpha) \longrightarrow (B,\beta)$ is monic if the underlying set map is monic. The notion of subobject this gives rise to is $(A',\alpha') \longrightarrow (A,\alpha)$ with $A' \subseteq A$ and $\alpha'(a,a') \leq \alpha(a,a')$. Strong monos are those which reflect tolerance as well as preserving it: $(A',\alpha') \hookrightarrow (A,\alpha)$ has $\alpha'(a,a') = \alpha(a,a')$. The strong subobjects then correspond to subsets. Unbalanced monos have underlying set map isomorphisms. The strong subobjects associated with strong monos have the form $(A,\alpha') \longrightarrow (A,\alpha)$ with the same underlying set.

THEOREM 25. The lattice of strong subobjects is representable by the usual characteristic morphism into the crisp two element set.

COROLLARY 26. **TOL(L)** is a quasitopos.

PROOF. A quasitopos is a locally cartesian closed category which has finite colimits and has a strong subobject representor. □

THEOREM 27. Given a fuzzy tolerance space (A,α), there is a fuzzy tolerance space $(\tilde{A},\alpha)$ which has (A,α) as a strong subobject such that tolerance preserving maps from (B,β) to $(\tilde{A},\alpha)$ are in one to one correspondence, obtained using pull back, with the strong partial maps from (B,β) to (A,α).

PROOF. We obtain $(\tilde{A},\alpha)$ by adjoining a point $*$ to A with $\alpha(*,a) = \alpha(a,a)$ and $\alpha(*,*) = \top$. The underlying set is the partial map representor in sets. Any map induced by a tolerance preserving strong partial map will be tolerance preserving because the extra point has maximal possible proximity to everything. □

The category **TOL(L)** has factorization of any morphism into a strong epi followed by a mono: $f : (B,\beta) \longrightarrow (A,\alpha)$ factors through the image of $\mathbf{U}f$ with the final structure induced by the structure β, $(f(B), f\beta)$. This factorization gives a left adjoint to the inclusion functor taking $\mathcal{P}(A)$ into $\mathbf{TOL(L)}/(A,\alpha)$.

Any morphism can also be factored into an epi followed by a strong mono: f maps onto $(f(B), \alpha|_{f(B)})$. This gives a left adjoint for the inclusion of $\mathcal{S}(A,\alpha)$ into $\mathbf{TOL(L)}/(A,\alpha)$.

Any monomorphism can be factored into a strong mono followed by an unbalanced mono: $m : (A',\alpha') \rightarrow (A,\alpha)$ is factored as $s : (A',\alpha') \rightarrow (A,\alpha')$ followed by $u : (A,\alpha') \rightarrow$

(A,α); here $\alpha'(a,a') = 0$ if $a \notin A'$ or $a' \notin A'$. Combining this with the factorization of morphisms into strong epis followed by monos gives the left adjoint left inverse to the inclusion $\mathcal{U}(A,\alpha) \longrightarrow \mathbf{TOL(L)}/(A,\alpha)$.

THEOREM 28. The inclusion $\mathbf{i} : \mathcal{U}(A,\alpha) \longrightarrow \mathbf{TOL(L)}/(A,\alpha)$ has a left adjoint left inverse σ given by taking $f : (B,\beta) \longrightarrow (A,\alpha)$ to the set A with the final structure determined by the morphism f and the structure β.

It remains to show that $\mathcal{U}(A,\alpha)$ has implications and quantifiers which lift. The constructions are very much like those in **SET(L)**. Suppose that we are given (A,α') and (A,α'') unbalanced subobjects of (A,α) and a fuzzy tolerance preserving function $f : (A,\alpha) \longrightarrow (B,\beta)$, then

$$(\alpha' \Rightarrow \alpha'')(a,a') = \alpha(a,a') \wedge (\alpha'(a,a') \Rightarrow \alpha'(a,a'))$$

$$(\alpha' \rightarrow \alpha'')(a,a') = \alpha(a,a') \wedge \delta(\alpha'(a,a') \rightarrow \alpha'(a,a'))$$

$$(\exists_f \alpha')(b,b') = \begin{cases} \bigvee_{f(a)=b \text{ and } f(a')=b'} \alpha'(a,a'), & \text{if such } a \text{ and } a' \text{ exist} \\ \bot, & \text{otherwise} \end{cases}$$

$$(\forall_f \alpha')(b,b') = \begin{cases} \bigwedge_{f(a)=b \text{ and } f(a')=b'} \alpha'(a,a'), & \text{if such } a \text{ and } a' \text{ exist} \\ \top, & \text{otherwise.} \end{cases}$$

give the adjoints to product, tensor, and pullback. The hyperdoctrinal diagram is valid.

§4. First Order Type Theory for Unbalanced Subobjects

There is a well developed connection between categories with various amounts of structure and logical theories. Cartesian closed categories give λ-calculi (see [Lambek and Scott 1986]), topoi give what Bell has called local set theories [Bell 1988]. In all of these theories we get a type theory with types given by the objects in a category with lots of structure. Here we want to use a category which is cartesian closed, has coproducts and has a second closed structure, so we will want all of those operations for type definition. We will want to have the morphisms between objects as function symbols, but we do not want to be too prescriptive, so that our logical theory has several possible categorical models. Following this tradition we define a type theory with two conjunctions and two associated λ-calculi by using the following definitions:

DEFINITION 13. Types are defined by giving

1. Basic types **A**, **B**, $\cdots$ including the types **1**, for the terminal object and **O** for the initial object.
2. If **A** and **B** are types then $\mathbf{A} \times \mathbf{B}$, $\mathbf{A} \otimes \mathbf{B}$, $\mathbf{A}^{\mathbf{B}}$, $\mathbf{A} \uparrow \mathbf{B}$, and $\mathbf{A} + \mathbf{B}$ are types.

We want to have function symbols that can be used to change the type of a term later on. These should be like morphisms of our category:

DEFINITION 14. For each pair $(\mathbf{A},\mathbf{B})$, we have a set of function symbols whose elements are written $f : \mathbf{A} \longrightarrow \mathbf{B}$. Choice of these sets is constrained by the following conditions:

1. For any $\mathbf{A}$ there are unique function symbols $? : 0 \to \mathbf{A}, ! : \mathbf{A} \to 1$, designated function symbols $id_A : \mathbf{A} \longrightarrow \mathbf{A}$ and $\Delta : \mathbf{A} \longrightarrow \mathbf{A} \times \mathbf{A}$, and
2. For each pair of types $(\mathbf{A},\mathbf{B})$ there are function symbols
   (a) $\pi_1 : \mathbf{A} \times \mathbf{B} \longrightarrow \mathbf{A}$
   (b) $\pi_2 : \mathbf{A} \times \mathbf{B} \longrightarrow \mathbf{B}$
   (c) $\iota_1 : \mathbf{A} \longrightarrow \mathbf{A} + \mathbf{B}$
   (d) $\iota_2 : \mathbf{B} \longrightarrow \mathbf{A} + \mathbf{B}$
   (e) $ev : (\mathbf{A}^{\mathbf{B}}) \times \mathbf{B} \longrightarrow \mathbf{A}$
   (f) $ev_{\otimes} : (\mathbf{A} \uparrow \mathbf{B}) \otimes \mathbf{B} \longrightarrow \mathbf{A}$
3. If $f : \mathbf{A} \longrightarrow \mathbf{B}$ and $g : \mathbf{B} \longrightarrow \mathbf{C}$ are function symbols so is $g \circ f : \mathbf{A} \longrightarrow \mathbf{C}$

We next use types and function symbols to define the terms of our theory:

DEFINITION 15. Terms are defined as follows:

1. There are no terms of type $\mathbf{O}$
2. There is a unique term $*$ of type $\mathbf{1}$. It is a constant.
3. For each other basic type $\mathbf{A}$ there is a set of variables x_i and a set of constants c_i which are terms of type $\mathbf{A}$
4. If $\mathbf{f} : \mathbf{A} \longrightarrow \mathbf{B}$ is a function symbol and t is a term of type $\mathbf{A}$ then $f(t)$ is a term of type $\mathbf{B}$.
5. If t_1 is a term of type $\mathbf{A}$ and t_2 is a term of type $\mathbf{B}$ then
   (a) (t_1, t_2) is a term of type $\mathbf{A} \times \mathbf{B}$
   (b) $t_1 \otimes t_2$ is a term of type $\mathbf{A} \otimes \mathbf{B}$
6. If $\mathbf{f} : \mathbf{A} \times \mathbf{B} \longrightarrow \mathbf{C}$ is a function symbol and t is a term of type $\mathbf{B}$ then $\lambda_a f(a,t)$ is a term of type $\mathbf{C}^A$ where a is a new bound variable of type $\mathbf{A}$.
7. If $\mathbf{f} : \mathbf{A} \otimes \mathbf{B} \longrightarrow \mathbf{C}$ is a function symbol and t is a term of type $\mathbf{B}$ then $\hat{\lambda}_a f(a \otimes t)$ is a term of type $\mathbf{C} \uparrow \mathbf{A}$ where a is a new bound variable of type $\mathbf{A}$.

To make our notation somewhat closer to ordinary mathematics we adopt the conventions:

1. If (t, a) is a term of type $\mathbf{C}^A \times \mathbf{A}$, then $t[a]$ is the term of type $\mathbf{C}$ given by $ev(t, a)$.
2. If $t \otimes a$ is a term of type $(\mathbf{C} \uparrow \mathbf{A}) \otimes \mathbf{A}$, then $t\widehat{[a]}$ is the term of type $\mathbf{C}$ given by $ev_{\otimes}(t \otimes a)$.

If we want these terms to reflect the structure of our category we need to impose an equality relation on terms.

DEFINITION 16. The equality relation on terms is the smallest equivalence relation such that:

1. Types and function symbols give a category:
   (a) $id \circ f = f$ and $f \circ id = f$
   (b) $f \circ (g \circ h) = (f \circ g) \circ h$
2. Products and the terminal work properly:
   (a) $\pi_1(t_1, t_2) = t_1$
   (b) $\pi_2(t_1, t_2) = t_2$
   (c) $!(t) = *$ for any term t
   (d) $\Delta(t) = (t, t)$
3. The two products have right adjoints:
   (a) $\lambda_x f(x, b)[a] = f(a, b)$
   (b) $\hat{\lambda}_x f(x \otimes b)\widehat{[a]} = f(a \otimes b)$

In general, terms of type $\mathbf{A}$ are variables with values in $\mathbf{A}$ which are constructed from simpler variables and constants. These give items we can ask questions about. These questions take the form of predicates which we can assign truth values to once all of the free variables have been assigned to constants.

DEFINITION 17. We define predicates as follows:

1. For each type there are predicate symbols indicated by lower case Greek letters called atomic predicates.
2. Two of these are designated: α takes the place of the predicate "true", and ω takes the place of "false".
3. If ϕ_1 and ϕ_2 are predicates of type $\mathbf{A}$, then so are
   (a) $\phi_1 \Rightarrow \phi_2$ and $\phi_1 \rightarrow \phi_2$,
   (b) $\phi_1 \wedge \phi_2$ and $\phi_1 * \phi_2$,
   (c) $\phi_1 \vee \phi_2$.
4. If ϕ is a predicate of type $\mathbf{B}$ and $\mathbf{f} : \mathbf{A} \rightarrow \mathbf{B}$ is a function symbol, then $\phi_f(x)$ is a predicate of type $\mathbf{A}$.
5. If ϕ is a predicate of type $\mathbf{A}$ and $\delta f : \mathbf{A} \rightarrow \mathbf{B}$ is a function symbol, then $\exists_f \phi$ and $\forall_f \phi$ are predicates of type $\mathbf{B}$.

In addition there are derived predicates: $\neg\phi = (\phi \Rightarrow \omega)$ and $\sim \phi = (\phi \rightarrow \omega)$. For each type $\mathbf{A}$ there is an equality predicate $=_{\mathbf{A}}$ given by $\exists_\Delta \alpha$.

DEFINITION 18. A formula of type $\mathbf{A}$ is an expression of the form $\phi(t)$ where ϕ is a predicate of type $\mathbf{A}$ and t is a term of type $\mathbf{A}$. It is a statement if it contains no unbound variables. With this definition, statements can be thought of as being *about* a particular term of type $\mathbf{A}$. If two terms of type $\mathbf{A}$ are being discussed, then what one is considering is a statement about a term of type $\mathbf{A} \times \mathbf{A}$.

4.1. AXIOMS AND INFERENCE. The logical system for unbalanced subobjects uses the notion of sequents. A sequent has the form $\phi_1(t) \vdash_{\mathbf{A}} \phi_2(t)$ where $\phi_1(t)$ and $\phi_2(t)$ are formulas of type $\mathbf{A}$ using the same term. The intuitive notion is that the truth values of

$\phi_1(t)$ is less than or equal to the truth value of $\phi_2(t)$. Thus the conclusion will be true to the extent that the premise is true.

The following logical axioms result from straightforward considerations of the meaning of $\vdash$ and the intended interpretation of equality, α and ω:

1. Tautology
$$S \vdash_{\mathbf{A}} S$$
2. Null
$$\omega(t) \vdash_{\mathbf{A}} \phi(t)$$
3. Truth
$$\phi(t) \vdash_{\mathbf{A}} \alpha(t)$$

In addition there are nonlogical axioms which result from the interaction of the two product structures:

1. Distributivity
$$\phi(t) * (\psi(t) \wedge \xi(t)) \vdash_{\mathbf{A}} (\phi(t) * \psi(t)) \wedge (\phi(t) * \xi(t))$$
$$(\phi(t) * \psi(t)) \wedge (\phi(t) * \xi(t)) \vdash_{\mathbf{A}} \phi(t) * (\psi(t) \wedge \xi(t))$$
2. Commutativity
$$\phi(t) * \psi(t) \vdash_{\mathbf{A}} \psi(t) * \phi(t)$$
3. $*$-weakening
$$\phi(t) * \psi(t) \vdash_{\mathbf{A}} \phi(t)$$

There are also rules of inference derived from standard logical principles and from the adjointness relationships between implications and conjunctions and between pulling back and quantification:

1. Cut
$$\frac{S \vdash_{\mathbf{A}} T \quad T \vdash_{\mathbf{A}} U}{S \vdash_{\mathbf{A}} U}$$
2. Transport along a function symbol $\mathbf{f} : \mathbf{A} \longrightarrow \mathbf{B}$
$$\frac{\phi(t) \vdash_{\mathbf{B}} \psi(t)}{\phi_f(t') \vdash_{\mathbf{A}} \psi_f(t')}$$
where t' is a term with $t = f(t')$.
3. Equal terms: If $t_1 = t_2$ then
$$\frac{\phi(t_1) \vdash_{\mathbf{A}} \psi(t_1)}{\phi(t_2) \vdash_{\mathbf{A}} \psi(t_2)}$$
4. Substituting a term: If t' is the term which results when all occurrences of a variable x in t are replaced with the term t'' then
$$\frac{\phi(t) \vdash_{\mathbf{A}} \psi(t)}{\phi(t') \vdash_{\mathbf{A}} \psi(t')}$$

5. $\rightarrow$-introduction

$$\frac{S * T \vdash_{\mathbf{A}} U}{S \vdash_{\mathbf{A}} T \rightarrow U}$$

6. $*$-introduction

$$\frac{S \vdash_{\mathbf{A}} T \rightarrow U}{S * T \vdash_{\mathbf{A}} U}$$

7. $\Rightarrow$-introduction

$$\frac{S \wedge T \vdash_{\mathbf{A}} U}{S \vdash_{\mathbf{A}} T \Rightarrow U}$$

8. $\wedge$-introduction

$$\frac{S \vdash_{\mathbf{A}} T \Rightarrow U}{S \wedge T \vdash_{\mathbf{A}} U}$$

9. Combined conclusions

$$\frac{S \vdash_{\mathbf{A}} T_1 \quad S \vdash_{\mathbf{A}} T_2}{S \vdash_{\mathbf{A}} T_1 \wedge T_2}$$

10. Combined premises

$$\frac{S_1 \vdash_{\mathbf{A}} T \quad S_2 \vdash_{\mathbf{A}} T}{S_1 \vee S_2 \vdash_{\mathbf{A}} T}$$

11. Weakening $\wedge$

$$\frac{r \vdash_{\mathbf{A}} p \wedge q}{r \vdash_{\mathbf{A}} p}$$

$$\frac{r \vdash_{\mathbf{A}} p \wedge q}{r \vdash_{\mathbf{A}} q}$$

12. $\forall_f$-introduction: If $\mathbf{f} : \mathbf{A} \longrightarrow \mathbf{B}$ and $f(t) = t'$ then

$$\frac{\phi_f(t) \vdash_{\mathbf{A}} \psi(t)}{\phi(t') \vdash_{\mathbf{B}} \forall_f \psi(t')}$$

13. $\forall_f$-elimination: If $\mathbf{f} : \mathbf{A} \longrightarrow \mathbf{B}$ and $f(t) = t'$ then

$$\frac{\phi(t') \vdash_{\mathbf{B}} \forall_f \psi(t')}{\phi_f(t) \vdash_{\mathbf{A}} \psi(t)}$$

14. $\exists_f$-introduction: If $\mathbf{f} : \mathbf{A} \longrightarrow \mathbf{B}$ and $f(t) = t'$ then

$$\frac{\psi(t) \vdash_{\mathbf{A}} \phi_f(t)}{\exists_f \psi(t') \vdash_{\mathbf{B}} \phi(t')}$$

15. $\exists_f$-elimination: If $\mathbf{f} : \mathbf{A} \longrightarrow \mathbf{B}$ and $f(t) = t'$ then

$$\frac{\exists_f \psi(t') \vdash_{\mathbf{B}} \phi(t')}{\psi(t) \vdash_{\mathbf{A}} \phi_f(t)}$$

The usual notion of a proof as a sequence of sequents in which each sequent is either an instance of an axiom or follows from previous sequents as a result of application of one of the rules of inference carries over to the current setting. In usual logic, however, there are axioms with no premises, so that a sequent of the form $\emptyset \vdash S$ tells us that S is a theorem. The logic of unbalanced subobjects only allows something to be a theorem to the maximum extent that it can be true. Thus we will say that a statement $\phi(t)$ is a theorem if we have a proof of the sequent $\alpha(t) \vdash_{\mathbf{A}} \phi(t)$. Since $\alpha(t)$ measures the extent to which it makes sense to talk about t, this is the best we can do.

§5. Interpretation in the Examples

5.1. SOUNDNESS FOR **SET(L)**. Let **SET(L)** be a category of fuzzy sets on a complete lattice with semigroup operation $*$ and thus implications $\rightarrow$ and $\Rightarrow$. The types will correspond to objects of **SET(L)**. The constants of type (A, α) and extent h are the elements a of A with $\alpha(a) \geq h$. They correspond to maps from the subobject $\mathbf{H} \hookrightarrow \mathbf{T}$ with $\alpha(*) = h$. Variables will be thought of as ranging over the elements of A. The predicates will be unbalanced subobjects of (A, α) with the designated predicates α given by (A, α) and ω given by $(A, 0)$.

DEFINITION 19. We define the truth value of a statement in the category **SET(L)**, $\|S\|$, as follows:

1. Atomic statements with constants of extent h give $\|\phi(a(*))\| = \phi(a) \wedge h$.
2. Logical connectives have the following effect:
   (a) $\|(\phi_1 \Rightarrow \phi_2)(t)\| = (\|\phi_1(t)\| \Rightarrow \|\phi_2(t)\|) \wedge \|\alpha(t)\|$;
   (b) $\|(\phi_1 \longrightarrow \phi_2)(t)\| = (\|\phi_1(t)\| \longrightarrow \|\phi_2(t)\|) \wedge \|\alpha(t)\|$;
   (c) $\|(\phi_1 \wedge \phi_2)(t)\| = \|\phi_1(t)\| \wedge \|\phi_2(t)\|$;
   (d) $\|(\phi_1 * \phi_2)(t)\| = \|\phi_1(t)\| * \|\phi_2(t)\|$;
   (e) $\|(\phi_1 \vee \phi_2)(t)\| = \|\phi_1(t)\| \vee \|\phi_2(t)\|$.
3. If $t = t'$, then $\|\phi(t)\| = \|\phi(t')\|$.
4. If $f(t') = t$ then $\|\phi_f(t')\| = \|\phi(t)\| \wedge \|\alpha(t')\|$
5. If $\mathbf{f} : \mathbf{A} \longrightarrow \mathbf{B}$ then

$$\begin{aligned} \|\exists_f \phi(t)\| &= \|\omega_B(t)\| && \text{if there is no } t' \text{ with } t = f(t') \\ &= \bigvee_{\{t' | f(t') = t\}} \|\phi(t')\| && \text{otherwise} \end{aligned}$$

6. If $\mathbf{f} : \mathbf{A} \longrightarrow \mathbf{B}$ then

$$\begin{aligned} \|\forall_f \phi(t)\| &= \|\alpha_B(t)\| && \text{if there is no } t' \text{ with } t = f(t') \\ &= \bigwedge_{\{t' | f(t') = t\}} \|\phi(t')\| && \text{otherwise} \end{aligned}$$

The truth value of a formula of type $\mathbf{A}$ which contains variables will be interpreted as a function $\|\phi(x)\| : (A, \alpha) \longrightarrow L$ taking a to $\|\phi(a)\|$.

Notice that with this definition the value of a statement of type $\mathbf{A}$ is an element of L. Statements with no explicit constants turn out to be statements of type $\mathbf{1}$, and thus also have valuation in L. However, if ϕ is a formula, x is a variable, and c is a constant all of type $\mathbf{A}$, the statements $\forall_x \phi(x)$ (interpreted as $\forall_\tau \phi(*)$ where $\tau : \mathbf{A} \to \mathbf{T}$ is the map to the terminal object) and $\phi(a)$ are of type $\mathbf{T}$ and $\mathbf{A}$, respectively. We can compare the truth values of these statements, but we cannot form the statement $\forall_x \phi(x) \Rightarrow \phi(a)$ because its constituents are not of the same type.

DEFINITION 20. A sequent $\phi(t) \vdash_{\mathbf{A}} \psi(t)$ is **satisfied** in $\mathbf{SET(L)}$ if $\|\phi(t)\| \leq \|\psi(t)\|$.

Once we have described a logical system and given a means for interpreting its statements in a category of fuzzy sets our first concern is with soundness of the system: are our axioms satisfied in $\mathbf{SET(L)}$ and do the rules of inference preserve satisfaction?

THEOREM 29. The axioms from section 2 are satisfied in $\mathbf{SET(L)}$.

PROOF. The satisfaction of the axiom of Tautology and the axiom of Equal Terms is clear, since the interpretation of the premise and conclusion is the same, hence they have the same truth values. The Null axiom becomes the statement that for any unbalanced subobject (A, α') and any element a in A, $\alpha'(a) \geq 0$. The Truth axiom says that $\alpha'(a) \leq \alpha(a)$, which is what it means for (A, α') to be unbalanced subobject of (A, α).

Satisfaction of the nonlogical axioms follows from properties of the operation $*$ in our lattice. Preservation of the order gives distributivity of $*$ over $\wedge$. The Commutativity axiom follows from the commutativity property for $*$. The axiom $*$-weakening follows from the identity axiom for the semigroup structure:

$$\begin{aligned}\|\phi(t) * \psi(t)\| &= \|\phi(t)\| * \|\psi(t)\| \\ &\leq \|\phi(t)\| * 1 \\ &= \|\phi(t)\|. \quad \square\end{aligned}$$

THEOREM 30. The rules of inference preserve satisfaction.

PROOF. We will consider each rule of inference in turn:

1. Cut

$$\frac{S \vdash_{\mathbf{A}} T \quad T \vdash_{\mathbf{A}} U}{S \vdash_{\mathbf{A}} U} \quad \text{becomes} \quad \frac{\|S\| \leq \|T\| \quad \|T\| \leq \|U\|}{\|S\| \leq \|U\|}$$

   which is just transitivity of $\leq$.
2. Transport along a function symbol $\mathbf{f} : \mathbf{A} \longrightarrow \mathbf{B}$

$$\frac{\phi(t) \vdash_{\delta B} \psi(t)}{\phi_f(t') \vdash_{\mathbf{A}} \psi_f(t')}$$

where t' is a term with $t = f(t')$ becomes

$$\frac{\|\phi(t)\| \leq \|\psi(t)\|}{\|\phi_f(t')\| \leq \|\psi_f(t')\|} \quad \text{or equivalently} \quad \frac{\|\phi(t)\| \leq \|\psi(t)\|}{\|\phi(t)\| \wedge \|\alpha(t')\| \leq \|\phi(t)\| \wedge \|\alpha(t')\|}$$

This last inference is valid because $\wedge$ preserves order.

3. Substituting a term: If t' is the term which results when all occurrences of a variable x in t are replaced with the term t'' then

$$\frac{\phi(t) \vdash_{\mathbf{A}} \psi(t)}{\phi(t') \vdash_{\mathbf{A}} \psi(t')} \text{ says } \frac{\|\phi(t)\| \leq \|\psi(t)\|}{\|\phi(t')\| \leq \|\psi(t')\|}$$

which just says that if two functions of several variables have a given order relation so do the functions which result from making substitutions for one of the variables.

4. $\rightarrow$-introduction

$$\frac{S * T \vdash_{\mathbf{A}} U}{S \vdash_{\mathbf{A}} T \rightarrow U} \text{ becomes } \frac{\|S\| * \|T\| \leq \|U\|}{\|S\| \leq \|T\| \rightarrow \|U\|}$$

This is the adjointness relationship between $*$ and $\rightarrow$.

5. $*$-introduction

$$\frac{S \vdash_{\mathbf{A}} T \rightarrow U}{S * T \vdash_{\mathbf{A}} U}$$

Soundness of this rule also follows from adjointness.

6. $\Rightarrow$-introduction

$$\frac{S \wedge T \vdash_{\mathbf{A}} U}{S \vdash_{\mathbf{A}} T \Rightarrow U}$$

and

7. $\wedge$-introduction

$$\frac{S \vdash_{\mathbf{A}} T \Rightarrow U}{S \wedge T \vdash_{\mathbf{A}} U}$$

follow similarly from the adjunction between $\wedge$ and $\Rightarrow$.

8. Combined conclusions

$$\frac{S \vdash_{\mathbf{A}} T_1 \quad S \vdash_{\mathbf{A}} T_2}{S \vdash_{\mathbf{A}} T_1 \wedge T_2} \text{ becomes } \frac{\|S\| \leq \|T_1\| \quad |S\| \leq \|T_2\|}{\|S\| \leq \|T_1\| \wedge \|T_2\|}$$

which follows because $\wedge$ is minimum.

9. Combined premises

$$\frac{S_1 \vdash_{\mathbf{A}} T \quad S_2 \vdash_{\mathbf{A}} T}{S_1 \vee S_2 \vdash_{\mathbf{A}} T} \text{ becomes } \frac{\|S_1\| \leq \|T\| \quad \|S_2\| \leq \|T\|}{\|S_1\| \vee \|S_2\|}$$

which follows from the fact that $\vee$ is maximum.

10. Weakening $\wedge$

$$\frac{r \vdash_{\mathbf{A}} p \wedge q}{r \vdash_{\mathbf{A}} p}$$

Soundness of this follows from the fact that $\|p\| \wedge \|q\| \leq \|p\|$.

11. $\forall_f$-introduction: If $\mathbf{f} : \mathbf{A} \longrightarrow \mathbf{B}$ and $f(t) = t'$ then

$$\frac{\phi_f(t) \vdash_{\mathbf{A}} \psi(t)}{\phi(t') \vdash_{\mathbf{B}} \forall_f \psi(t')}$$

and

12. $\forall_f$-elimination: If $\mathbf{f} : \mathbf{A} \longrightarrow \mathbf{B}$ and $f(t) = t'$ then

$$\frac{\phi(t') \vdash_{\mathbf{B}} \forall_f \psi(t')}{\phi_f(t) \vdash_{\mathbf{A}} \psi(t)}$$

have their soundness follow from the adjointness relation between pulling back and universal quantification. Similarly,

13. $\exists_f$-introduction: If $\mathbf{f} : \mathbf{A} \longrightarrow \mathbf{B}$ and $f(t) = t'$ then

$$\frac{\psi(t) \vdash_{\mathbf{A}} \phi_f(t)}{\exists_f \psi(t') \vdash_{\mathbf{B}} \phi(t')}$$

and

14. $\exists_f$-elimination: If $\mathbf{f} : \mathbf{A} \longrightarrow \mathbf{B}$ and $f(t) = t'$ then

$$\frac{\exists_f \psi(t') \vdash_{\mathbf{B}} \phi(t')}{\psi(t) \vdash_{\mathbf{A}} \phi_f(t)}$$

have soundness follow from the adjointness of pulling back and existential quantification. □

5.2. Interpretation in **TOL**. In **TOL** types will correspond to tolerance space and function symbols will be tolerance preserving maps. The interesting questions to ask about elements deal with how they are connected, not the extent to which they are members. Thus the constants should correspond to choices of pairs of elements of the tolerance space in question. Unbalanced subobjects give tolerances with fewer connections, we can ask whether a given pair is connected in an unbalanced subobject. This will be a two valued interpretation. Variables of type $(A, \sim_A)$ will be thought of a ranging over pairs of elements of A.

Proposition 31. The operations which give new terms from old (items 3 through 8 in the definition of terms) produce pairs of elements of the new types when given pairs of elements of the starting type.

It originally seemed to me that the constants and variables should be pairs of adjacent elements, but then $t_1 \otimes t_2$ and $\hat{\lambda}_a f(a \otimes t)$ might not give adjacent pairs.

Definition 21. We define the truth value of a statement in the category **TOL**, $||S||$, as follows:

1. $\|\alpha((a,b))\| = T$ if and only if $a \sim_A b$.
2. $\|\omega((a,b))\| = T$ if and only if $a = b$.
3. Atomic statements with constants give $\|\phi((a,b))\| = T$ if $a \sim_\phi b$. Otherwise the truth value is F.
4. Logical connectives have the following effect:
   (a) $\|(\phi_1 \Rightarrow \phi_2)(t)\| = \|\phi_1(t)\| \Rightarrow \|\phi_2(t)\|$;
   (b) $\|(\phi_1 \rightarrow \phi_2)(t)\| = T$:
   (c) $\|(\phi_1 \wedge \phi_2)(t)\| = \|\phi_1(t)\| \wedge \|\phi_2(t)\|$;
   (d) $\|(\phi_1 * \phi_2)(t)\| = F$;
   (e) $\|(\phi_1 \vee \phi_2)(t)\| = \|\phi_1(t)\| \vee \|\phi_2(t)\|$ where $\Rightarrow$ is the usual two valued implication.
5. If $t = t'$, then $\|\phi(t)\| = \|\phi(t')\|$
6. If $f(t') = t$, then $\|\phi_f(t')\| = \|\phi(t)\| \wedge \|\alpha(t')\|$
7. If $\mathbf{f} : \mathbf{A} \longrightarrow \mathbf{B}$, then

$$
\begin{aligned}
\|\exists_f \phi(t)\| &= \|\omega_B(t)\| && \text{if there is no } t' \text{ with } t = f(t') \\
&= \bigvee_{\{t' | f(t') = t\}} \|\phi(t')\| && \text{otherwise}
\end{aligned}
$$

8. If $\mathbf{f} : \mathbf{A} \longrightarrow \mathbf{B}$ then

$$
\begin{aligned}
\|\forall_f \phi(t)\| &= \|\alpha_B(t)\| && \text{if there is no } t' \text{ with } t = f(t') \\
&= \bigwedge_{\{t' | f(t') = t\}} \|\phi(t')\| && \text{otherwise}
\end{aligned}
$$

The definitions of the valuation for the $*$ and $\rightarrow$ connectives result from the fact that the tensor product on $\mathbf{TOL}/(A, \sim_A)$ becomes trivial for unbalanced monos. The $\otimes$ product can only have two elements connected if they lie in the fiber over the same element of A. The triviality of these connectives makes the logic reflect the second monoidal closed structure only in the construction of terms, not in the logic of statements.

The truth value of a formula of type $\mathbf{A}$ which contains variables will be interpreted as a function $\|\phi(x)\| : A \times A \rightarrow \{T, F\}$ taking (a, b) to $\|\phi((a,b))\|$.

DEFINITION 22. A sequent $\phi(t) \vdash_{\mathbf{A}} \psi(t)$ is **satisfied** in **TOL** if $\|\phi(t)\| \leq \|\psi(t)\|$, where we take $F \leq T$.

With these definitions and the reasoning for **SET(L)** as a guide, it is an easy exercise to prove the soundness of the interpretation in **TOL**.

5.3. INTERPRETATION IN **TOL(L)**. If we use **TOL(L)** for our category of types and function symbols, then the interesting question to ask about an unbalanced subobject is the extent to which it makes two elements of the underlying set equal. This will carry with it the extent of membership since an element is a member of an unbalanced subobject to the extent that it is equal to itself. Constants of type (A, α) will be pairs of elements of A. Notice that the term building operations can also be applied to pairs of elements to get pairs of elements. Variables will take pairs of elements of the same type as values. The predicates will be unbalanced subobjects of (A, α) with the designated predicates α given by (A, α) and ω given by $(A, 0)$.

DEFINITION 23. We define the truth value of a statement in the category **TOL(L)**, $\|S\|$, as follows:

1. Atomic statements with constants $\|\phi(a, a')\| = \phi(a, a')$.
2. Logical connectives have the following effect:
   (a) $\|(\phi_1 \Rightarrow \phi_2)(t)\| = (\|\phi_1(t)\| \Rightarrow \|\phi_2(t)\|) \wedge \|\alpha(t)\|$;
   (b) $\|(\phi_1 \rightarrow \phi_2)(t)\| = (\|\phi_1(t)\| \rightarrow \|\phi_2(t)\|) \wedge \|\alpha(t)\|$;
   (c) $\|(\phi_1 \ \wedge \ \phi_2)(t)\| = \|\phi_1(t)\| \wedge \|\phi_2(t)\|$;
   (d) $\|(\phi_1 \ * \ \phi_2)(t)\| = \|\phi_1(t)\| * \|\phi_2(t)\|$;
   (e) $\|(\phi_1 \ \vee \ \phi_2)(t)\| = \|\phi_1(t)\| \vee \|\phi_2(t)\|$.
3. If $t = t'$, then $\|\phi(t)\| = \|\phi(t')\|$.
4. If $f(t') = t$, then $\|\phi_f(t')\| = \|\phi(t)\| \wedge \|\alpha(t')\|$
5. If $\mathbf{f} : \mathbf{A} \longrightarrow \mathbf{B}$, then

$$\begin{aligned} \|\exists_f \phi(t)\| &= \|\omega_B(t)\| && \text{if there is no } t' \text{ with } t = f(t') \\ &= \bigvee_{\{t' | f(t') = t\}} \|\phi(t')\| && \text{otherwise} \end{aligned}$$

6. If $\mathbf{f} : \mathbf{A} \longrightarrow \mathbf{B}$, then

$$\begin{aligned} \|\forall_f \phi(t)\| &= \|\alpha_B(t)\| && \text{if there is no } t' \text{ with } t = f(t') \\ &= \bigwedge_{\{t' | f(t') = t\}} \|\phi(t')\| && \text{otherwise} \end{aligned}$$

The truth value of a formula of type **A** which contains variables will be interpreted as a function $\|\phi(x)\| : (A, \alpha) \rightarrow L$ taking a to $\|\phi(a)\|$.

Notice that with this definition the value of a statement of type **A** is an element of L, just like the interpretation of **SET(L)** did. Indeed the soundness of the system in this interpretation proceeds almost exactly the same as for **SET(L)**.

§6. Higher Order Types

There are two ways that we can introduce logic of higher order. If we have unbalanced subobject representation in some form, the way we do in **SET(L)**, we can form an internal second order logic. A more limited kind of higher order logic is given by a natural numbers object.

6.1. UNBALANCED POWEROBJECT REPRESENTATION. Categories of fuzzy sets share with topoi and quasitopoi the existence of a higher order logic resulting from representation of appropriate kinds of powerobject representation. Addition of this higher order structure will change the type theory developed in the preceding sections by adding a new type formation rule and a new designated predicate.

DEFINITION 24. The higher order type theory adds to the type formation rules the following:

If $\mathbf{A}$ is a type then so is $\mathbf{UA}$.

Terms are defined exactly as before. So is the equality relation on terms. With predicates, however, we replace the former definition with the following:

DEFINITION 25. A predicate in the higher order type theory is obtained by:

1. For each type $\mathbf{A}$ there is a predicate ϵ of type $\mathbf{UA} \times \mathbf{A}$.
2. If ϕ_1 and ϕ_2 are predicates of type $\mathbf{A}$, then so are
   (a) $\phi_1 \Rightarrow \phi_2$ and $\phi_1 \rightarrow \phi_2$,
   (b) $\phi_1 \wedge \phi_2$ and $\phi_1 * \phi_2$,
   (c) $\phi_1 \vee \phi_2$.
3. If ϕ is a predicate of type $\mathbf{B}$ and $\mathbf{f} : \mathbf{A} \longrightarrow \mathbf{B}$ is a function symbol, then $\phi_f(x)$ is a predicate of type $\mathbf{A}$.
4. If ϕ is a predicate of type $\mathbf{A}$ and $\mathbf{f} : \mathbf{A} \longrightarrow \mathbf{B}$ is a function symbol, then $\exists_f \phi$ and $\forall_f \phi$ are predicates of type $\mathbf{B}$.

The predicates of type $\mathbf{A}$ which were given by atomic predicates before now arise from function symbols $\ulcorner \chi_\phi \urcorner : 1 \rightarrow UA$ by taking

$$\phi = \epsilon_{\ulcorner \chi_\phi \urcorner \times id_A}$$

Since the characteristic map of an unbalanced subobject is not unique, this may result in several copies of the same predicate, corresponding to different forms of the characteristic function of ϕ.

6.2. NUMBERS. A natural numbers object as defined by Lawvere will not really work because of the existence of objects with full existence and incomplete membership. We extend it by taking a truncation: for any $h \hookrightarrow \top$ if we are given $a : h \longrightarrow A$ and $f : A \longrightarrow A$, then we get a unique arrow $\hat{f} : N \times h \longrightarrow A$ making the recursion diagram commute.

In $\mathbf{SET(L)}$ this gives the crisp natural numbers. The rationals are also crisp. The reals, however, can be defined in several different inequivalent ways. The following definition gives a version which seems particularly appropriate for fuzzy mathematics:

DEFINITION 26. The fuzzy Dedekind real numbers object is the unbalanced subobject of $UQ \times UQ$ with elements (L, U) such that

1. $U = \sim L$ (this is the negation associated with the second product),
2. $\forall_\pi (q_1 \in L * q_2 \leq q_1 \rightarrow q_2 \in L)$,

3. $\forall_\pi(q_1 \in U * q_2 \geq q_1 \rightarrow q_2 \in U)$,
4. $\forall_\pi(q_1 \in L \rightarrow \exists_\pi(q_2 > q_1 * q_2 \in L))$,
5. $\exists_\pi(q \in L)$ and $\exists_\pi(q_2 \in U)$.

In **SET(I)** there is a nice interpretation of the fuzzy Dedekind reals: since Q is the crisp rationals, an element L of UQ corresponds to an interval valued function on Q. The corresponding upper cut U is found by taking $U(q) = 1 - L(q)$. This pair of functions will be a fuzzy real to the extent given by the minimum of the numbers

1. $\sup(L(q))$, which gives the truth value of $\exists_\pi(q \in L)$,
2. $\sup(U(q))$, which gives the truth value of $\exists_\pi(q_2 \in U)$,
3. $1 - \sup_{\{q_1 < q_2\}}(L(q_2) - L(q_1))$, measuring the degree to which L is decreasing,
4. $1 - \sup_{q_1} \inf_{\{q_2 | q_2 > q_1\}}(L(q_2) - L(q_2))$, measuring the degree to which L is continuous from the right, the open cut requirement.

Thus a real number corresponds to an upper semicontinuous decreasing function which approaches 1 at large negative q and 0 at large positive q. Alternatively we can look at the upper cut and get real numbers as cumulative distribution functions. The advantage of our definition in the internal logic is that it also tells us the degree of membership in the reals of functions which look almost like cumulative distributions.

Part II:

Categorical Methods in Topology

Chapter 4:

Fuzzy Filter Functions and Convergence

S. E. Rodabaugh et al. (eds.), *Applications of Category Theory to Fuzzy Subsets*, 109–136.

FUZZY FILTER FUNCTORS AND CONVERGENCE

P. Eklund
W. Gähler

§1. Introduction

In this chapter we describe a fuzzy filter functor in a general framework of set functors. The general theory includes generalized Cauchy spaces, together with a construction for completions, and generalized pseudo-topologies, which in the case of the fuzzy filter functor results in a development of fuzzy convergence structures.

The outline of the chapter is as follows. In section 2, we describe functor structures as defined through covariant set functors. Examples include powerset, ideal and filter functors, and also the fuzzy filter functor. Section 3 deals with functors from the category **SET** of sets to the category **PROSET** of preordered sets. Here we also present Urysohn modifications of functors, which overcomes a lattice-theoretical barrier in connection with the completion construction. Section 4 presents details about the fuzzy filter functor. In section 5, we develop fuzzy convergence, which also includes fuzzy topologies. Incidently, concerning fuzzy topologies, the question about constants being open or not turns out to relate to the connectedness of the fuzzy filter functor. Section 6 contains the completion construction, and the chapter is concluded with section 7 on monads. The notion of a monad relates to the development of regularity and iteratedness, as well as to compactifications.

§2. Functor Structures

Many structures can be described by means of set functors, i,e. by functors the domain and codomain of which is the category **SET** of all sets. A necessary and sufficient condition for all structures of one and the same type to be describable by set functors has been given by [Kučera and Pultr 1972/1973] (see also [Adámek 1983]). To make this treatment simple we will here consider structures only on sets. However, we point out the fact that in the more general heterogeneous case, namely when the structures are e.g. on families of sets, a lot of interesting examples of structure types appear which are important in applications (see e.g. [Burmeister 1986, Gähler and Gähler 1989, Gähler 1984]). In the heterogeneous case, set–valued functors (i.e. functors with **SET** as codomain) has to be used instead of set functors.

Let φ be a covariant set functor. φ assigns to each set X and mapping $f : X \longrightarrow Y$ a set φX and mapping $\varphi f : \varphi X \longrightarrow \varphi Y$, respectively. By a **$\varphi$-structure** on a set X we mean any subset of φX. A set X equipped with a φ-structure on X is called a **φ-space.** A **φ-morphism** $f : (X,s) \longrightarrow (Y,t)$ between φ-spaces is a mapping $f : X \longrightarrow Y$ such that $\mathcal{M} \in s$ implies $\varphi f(\mathcal{M}) \in t$. Let **$\varphi$-SP** denote the category of all φ-spaces and φ-morphisms.

A **φ-convergence structure** on a set X is a relation between φX and X, that is, a subset of $\varphi X \times X$. Each pair consisting of a set and a φ-convergence structure on it, is a **φ-convergence space.** φ-convergence structures are the same as $\varphi \times id$-structures, where $\varphi \times id$ is the product of φ and the identity set functor id (which assigns each set and mapping to itself). Note that for covariant set functors $\varphi_i, i \in I$, with I any set, the **product** $\prod_{i\in I} \varphi_i$ of these functors is defined as the covariant set functor which assigns to each set X and mapping $f : X \longrightarrow Y$ the cartesian product $\prod_{i\in I} \varphi_i X$ and the mappings $(\mathcal{M}_i)_{i\in I} \mapsto (\varphi_i f(\mathcal{M}_i))_{i\in I}((M_i)_{i\in I} \in \prod_{i\in I} \varphi_i X)$, respectively. Instead of $\varphi \times id$-**SP** we also write φ-**CON**. The $\varphi \times id$-morphisms $f : (X,s) \longrightarrow (Y,t)$ are also called **φ-continuous** mappings. If s is a φ-convergence structure, then instead of $(\mathcal{M},x) \in s$ we also write $\mathcal{M} \xrightarrow[s]{} x$, and say that $\mathcal{M}$ **converges** to x (with respect to s). Then the condition that a mapping $f : (X,s) \longrightarrow (Y,t)$ between φ-convergence spaces is φ-continuous means that $\mathcal{M} \xrightarrow[s]{} x \Rightarrow \varphi f(\mathcal{M}) \xrightarrow[t]{} f(x)$.

EXAMPLES OF COVARIANT SET FUNCTORS.

(1) One of the standard examples is the (covariant) **powerset functor P**, which to each set X and mapping $f : X \longrightarrow Y$ assigns the powerset PX of X and the mappings $M \mapsto f[M]$ $(M \in PX)$, respectively. The **proper power set functor** P_0 is the subfunctor of P such that for each X, P_0X is the powerset of X without the empty set. Clearly, a covariant set functor φ is a **subfunctor** of a covariant set functor ϑ if for each set X, mapping $f : X \longrightarrow Y$ and $\mathcal{M} \in \psi X$ we have $\psi X \subseteq \varphi X$ and $\psi f(\mathcal{M})$ and $\varphi f(\mathcal{M})$.

(2) Another example is the **proper ideal functor** I_0. For each set X, I_0X consists of all proper ideals on X, i.e. all non-empty subsets $\mathcal{M}$ of P_0X such that $M \in \mathcal{M}$ and $N \subseteq M$ imply $N \in \mathcal{M}$ and from $M, N \in \mathcal{M}$ it follows $M \cup N \in \mathcal{M}$. For each mapping $f : X \longrightarrow Y$, $I_0 f$ is the mapping $\mathcal{M} \mapsto \{f[M] \mid M \in \mathcal{M}\}$ $(\mathcal{M} \in I_0X)$.

(3) Two further important examples are given by the **filter functor F** and the **proper filter functor** F_0. F assigns to each set X and mapping $f : X \longrightarrow Y$, respectively, the set of all filters on X and the mapping $\mathcal{M} \mapsto \{N \in PY \mid \exists M \in \mathcal{M}, f[M] \subseteq N\}$. F_0 is the subfunctor of F such that for each set X, F_0X consists of all proper filters on X, i.e. $F_0X = FX \setminus \{PX\}$.

Note that the composition $\psi \circ \varphi$ of two contravariant set functors is a covariant set functor. A typical example of a contravariant set functor is the **contravariant powerset functor** P^-, which assigns to each set X and mappings $f : X \longrightarrow Y$ the powerset PX the maps $M \mapsto f^{-1}[M]$ $(M \in PY)$ from PY to PX, respectively. We can easily see, that F, and hence also F_0, is a subfunctor of $P^- \circ P^-$, however, neither F nor F_0 is a subfunctor of $P \circ P$ (since for a mapping $f : X \longrightarrow Y$ and a filter $\mathcal{M}$ on X, $(P \circ P)f(\mathcal{M})$ need not be a filter on Y).

(4) There is the following generalization of the notion of proper filter functor. Let L be (at first) any set. The **contravariant hom-functor** related to L is the contravariant set function L^{id} which to each set X and mappings $f : X \longrightarrow Y$ assigns the set L^X of all mappings of X into L, and the mappings $g \mapsto g \circ f$ $(g \in L^Y)$, respectively. Then $L^{id} \circ L^{id}$ is a covariant set functor, which we also denote $L^{L^{id}}$. For any mapping $f : X \longrightarrow Y$ and $\mathcal{M} \in L^{L^X}$ we have $L^{L^{id}} f(\mathcal{M}) = \mathcal{M} \circ L^{id} f$, and hence,

$$L^{L^{id}} f(\mathcal{M})(g) = \mathcal{M}(g \circ f) \tag{2.1}$$

If a set X is fixed, then for any $\alpha \in L$, by $\bar{\alpha}$ we mean the constant mapping of X into L with value α.

Assume now that L is a meet semilattice. We may consider L as a partially ordered set such that each unordered pair $\{\alpha, \beta\}$ has the infirmum (meet) $\alpha \wedge \beta$. An element $\mathcal{M}$ of L^{L^X} will be called an $(L$-)**fuzzy filter (on** X**)**, if the following conditions are fulfilled:

(1) $\mathcal{M}(\bar{\alpha}) = \alpha$ for all $\alpha \in L$,
(2) $f, g \in L^X$ and $f \leq g$ imply $\mathcal{M}(f) \leq \mathcal{M}(g)$,
(3) for all $f, g \in L^X$ it holds $\mathcal{M}(f \wedge g) \geq \mathcal{M}(f) \wedge \mathcal{M}(g)$.

Clearly, $f \wedge g$ is defined componentwise. Note that from (2) and (3) it follows $\mathcal{M}(f \wedge g) = \mathcal{M}(f) \wedge \mathcal{M}(g)$ for all $f, g \in L^X$. By means of (2.1) we can easily show that for each mapping $f : X \longrightarrow Y$ and L-fuzzy filter $\mathcal{M}$ on X, $L^{L^{id}} f(\mathcal{M})$ is an L-fuzzy filter on Y. Hence, there is a subfunctor of $L^{L^{id}}$ assigning to each set X the set of all L-fuzzy filters on X. It will be called the L-**fuzzy filter functor** and will be written F_L.

In case of $(L = 2) = \{0, 1\})$ we have the following: Identifying each subset M of a set X with its characteristic function $\chi_M : X \longrightarrow 2$ leads to an identification of P^- with 2^{id} and hence also to an identification of $P^- \circ P^-$ with $2^{2^{id}}$. Changing in the case of $L = 2$ each $\mathcal{M} \in L^{L^{id}}$ to the related set system $\mathcal{F} \in P(PX)$ shows that the conditions (1), (2) and (3), respectively, mean

(1) $\emptyset \notin \mathcal{F}, \mathcal{F} \neq \emptyset$,
(2) $M \in \mathcal{F}$ and $M \subseteq N$ imply $N \in \mathcal{F}$,
(3) for all $M, N \in \mathcal{F}$ it holds $M \cap N \in \mathcal{F}$.

Hence, $\mathcal{M}$ is a 2-fuzzy filter on X if and only if $\mathcal{F}$ is a proper filter on X. Thus we also may identify the proper filter functor with the 2-fuzzy filter functor.

(5) A further interesting example of covariant set functors is for any set n, the $\mathbf{n^{th}}$ **power** id^n of the identity set functor. We have $id^n = \prod_{i \in n} \varphi_i$ with $\varphi_i = id$ for all $i \in n$. If, in particular, n is the set $\mathbb{N}$ of all non-negative integers, we get the **sequence functor** $seq(= id^{\mathbb{N}})$, which to each set X and mapping $f : X \longrightarrow Y$ resp. assigns the set of all sequences $(x_0, x_1, \ldots)$ on X and the mapping $(x_n)_{n \in \mathbb{N}} \mapsto (f(x_n))_{n \in \mathbb{N}}((x_n)_{n \in \mathbb{N}} \in seqX)$.

(6) The last example can be extended in a suitable way using the notions of constant set functor and of the set functor sum, which are defined as follows. If A is any set, by the constant set functor $A_{\mathbf{SET}}$ we mean the covariant set functor which assigns to each set X and mapping f the set A and the identity map 1_A, respectively. If for each element i of a set I, φ_i is a covariant set functor, then by the **sum** $\sum_{i\in I}\varphi_i$ of the functors φ_i we mean the covariant set functor which assigns to each set X and mappings $f : X \longrightarrow Y$ the disjoint union $\bigcup_{i\in I}(\{i\} \times \varphi_i X)$ of the set $\varphi_i X$ and the mapping $(i, \mathcal{M}) \mapsto (i, \varphi_i f(\mathcal{M}))((i,\mathcal{M}) \in (\sum_{i\in I}\varphi_i)X)$, respectively.

Let k be a cardinal number and $(\Omega_n)_{n\leq k}$ be a family of sets with index set $\{n \mid n \leq k\}$. Write $\Omega_n id^n$ instead of $(\Omega_n)_{\mathbf{SET}} \times id^n$. Note that the sum functor $\sum_{n\leq k}\Omega_n id^n$ assigns to each set X the set of all triples $(n, \omega, (x_i)_{i\in I})$ with $n \leq k$, $\omega \in \Omega_n$ and $(x_i)_{i\in I} \in X^n$.

By means of the examples of set functors given here, a lot of structure types can be described [Gähler 1981]. For instance, in case of $\varphi = F_0, F_L$, and seq, we have the proper filter convergence structures, the fuzzy filter convergence structures, and the sequence convergence structures, and in particular, the topologies, the fuzzy topologies, and the $\mathcal{L}$-structures (see e.g. [Fric, Koutnik 1979]).

Note that also algebraic structures can be considered as special φ-convergence structures. By an **operator domain** we mean a disjoint union $\Omega = \cup_{n\leq k}\{n\} \times \Omega_n$ of sets Ω_n where we take the sum over all cardinal numbers less than or equal to a cardinal number k. An **Ω-algebra** is a pair $(X, (s_{n\omega})_{n,\omega)\in\Omega})$ consisting of a set X and a family of mappings $s_{n\omega} : X^n \longrightarrow X$ (n-ary operations). Obviously, the Ω-algebras can be characterized as $\sum_{n\leq k}\Omega_n id^n$-convergence structures which are mappings $s : (\sum_{n\leq k}\Omega_n id^n)X \longrightarrow X$. These mappings and the family $(s_{n\omega})_{(n,\omega)\in\Omega}$ are in a one-to-one correspondance according to

$$s(n, \omega, (x_i)_{i\in n}) = s_{n\omega}((x_i)_{i\in n})$$

We will always identify s with its related family $(s_{n\omega})_{(n,\omega)\in\Omega}$. Then we get that the $\sum_{n\leq k}\Omega_n id^n$-morphisms between Ω-algebras are precisely the homomorphisms between these algebras. More general, we will identify in the same way $\sum_{n\leq k}\Omega_n id^n$-convergence structures which are partial mappings with partial Ω-algebraic structures. In this more general case we have analogously, that the $\sum_{n\leq k}\Omega_n id^n$-morphisms between partial Ω-algebras are the homomorphisms between these partial algebras.

Here we have reached one possibility to formulate axioms for φ-convergence structures. We mean both the axioms that a φ-convergence structure $s : \varphi X \longrightarrow X$, respectively, is a mapping of φX into X and a partial mapping between φX and X. The requirement for a φ-convergence structures to be a mapping is fulfilled very rarely in the "topological case" (see Proposition 4.1 in [Gähler 1988]). On the other hand, the axiom of a φ-convergence structure $s : \varphi X \longrightarrow X$ to be a partial mapping can be considered as the **Hausdorff axiom**:

$$\mathcal{M} \xrightarrow{s} x, y \Rightarrow x = y$$

Indeed, in case of a F_0-convergence structure it is the usual axiom T_2. Each φ-convergence structure s will be **separated** if s fulfills the Hausdorff axiom.

A further possibility to formulate an appropriate axiom is given if φ is **connected**, i.e. φE is a singleton provided E is a singleton (see [Wyler 1973]). Examples of connected covariant set functors are P_0, I_0, F_0 and seq (see Proposition 4.2 [Eklund and Gahler ****]). Each L-fuzzy filter functor F_L is connected, since for each singleton E, L^E consists only of the constant mappings $\bar{\alpha}$, and because of condition (1) for L-fuzzy filters, then $F_L E$ contains only the mapping $\bar{\alpha} \mapsto \alpha$ $(\alpha \in L)$. As is easily seen, also id^n is connected for every set n. P is not connected, since $PE = \{\emptyset, E\}$ for every singleton E. Moreover, $\sum_{n \leq k} \Omega_n id^n$ is not connected, provided Ω_n is non-empty for at least two n's.

Assume now that φ is connected. Then there is a unique natural transformation $\eta : id \longrightarrow \varphi$ (see [Eklund and Gähler ****]). For each singleton E, set X and $x \in X$,

$$\eta_X(x) = \varphi f(\mathcal{E})$$

where $\mathcal{E}$ is the only element of φE and $f : E \to X$ is the mapping with value x. A φ-convergence structure s of a set X is called **stable** provided $\eta_X(x) \underset{s}{\rightarrow} x$ holds for all $x \in X$. The requirement for a φ-convergence structure to be stable is an axiom which frequently appears in the case of "topological" structures. Note that stability can be defined also for any φ-structure on a set X, which means that $\eta_X(x)$ is an element of this structure for all $x \in X$. With φ, also $\varphi \times id$ is connected, and a φ-convergence structure is stable if and only if it is stable as a $\varphi \times id$–structure.

Each covariant set functor φ can be decomposed into its components (maximal connected set functors), and stability can then be meant in this general case of φ with respect to one of the components of φ. The components of any covariant set functor φ can be given easily in choosing a singleton E and in defining for each $a \in \varphi E$ a component φ_a of φ as follows: For each set X let

$$\varphi_a X = \{\mathcal{M} \in \varphi X \mid \varphi f(\mathcal{M}) = a \text{ with } f \text{ the unique mapping of } X \text{ into } E\}$$

The axiom of stability belongs to a larger class of axioms which can be formulated by means of natural transformations, and more general, by means of natural correspondences between covariant set functors ψ and χ related to φ. By a **natural correspondence** $\mu : \psi \circ\!\!\rightarrow \chi$ we mean a family $(\mu_X)_{X \in Ob\mathbf{SET}}$ of relations $\mu_X : \psi X \circ\!\!\rightarrow \chi X$ between ψX and χX such that $\mu_X \circ \psi f = \chi f \circ \mu_X$ for all mappings $f : X \longrightarrow Y$. Analogously as in [7] μ-**iteratedness** and μ-**regularity** can be defined with respect to any natural correspondence $\mu : \varphi \circ \varphi \circ\!\!\rightarrow \varphi$.

Some important axioms are formulated in using a preordering on each set φX. In the following sections we will turn to this problem.

§3. Functors from SET to PROSET

Most of the results in this section appear in [Eklund and Gähler ****].

PREORDERED SETS

Let $(M, \leq)$ be a preordered set shortly called a **proset**. By the **Urysohn modification** of $\leq$ we mean the preordering $\preceq$ of M given by

$$x \preceq y \Leftrightarrow \text{ for each } z \leq x \text{ there is a lower bound of } \{y, z\}$$

$(M, \preceq)$ will be called **Urysohn modification** of $(M, \leq)$. $x \leq y$ implies $x \preceq y$. If $\leq$ is not a partial ordering then also $\preceq$ is not. $\leq$ and $(M, \leq)$ will be said to be **Urysohn** provided $\leq$ and $\preceq$ coincide.

In the following we recall some propositions on Urysohn modifications.

PROPOSITION 3.1 ([Eklund and Gähler ****], 2.1). Each Urysohn modification is Urysohn.

PROPOSITION 3.2 ([Eklund and Gähler ****], 2.2). Let $x, y \in M$. $\{x, y\}$ has a lower bound in $(m, \leq)$ if and only if {x,y} has a lower bound in $(M, \preceq)$.

PROPOSITION 3.3 ([Eklund and Gähler ****], 2.3). Let $f : (M, \leq) \longrightarrow (N, \leq)$ be an isotone (i.e. a monotonely increasing) mapping between prosets which fulfills the following condition:

(E) For each $x \in M$ and $u \leq f(x)$, there is a $y \leq x$ with $f(y) \leq u$

Then $f : (M, \preceq) \longrightarrow (N, \preceq)$ is isotone and also fulfills (E) where $(M, \preceq)$ and $(N, \preceq)$ are the Urysohn modifications of $(M, \leq)$ and $(N, \leq)$, respectively.

PROPOSITION 3.4 ([Eklund and Gähler ****], 2.4). Let $(M, \leq)$ be a proset and let $x, y \in M$. Moreover let d be a **distinguished** upper bound of $\{x, y\}$ in $(M, \preceq)$, i.e. d is an upper bound such that for each $u \leq d$ there is a $v \leq u$ with $v \leq x$ or $v \leq y$. Then d is a supremum of $\{x, y\}$ in the Urysohn modification $(M, \preceq)$ of $(M, \leq)$. (Supremum means that $x \preceq d$ and $y \preceq d$ and from $x \preceq m$ and $y \preceq m$ it follows $d \preceq m$.) Moreover then d also is a distinguished upper bound of $\{x, y\}$ in $(M, \preceq)$.

In the sequel the prosets $(M, \leq)$ which fulfill the following condition will play an important role:

(S) For all $x, y \in M$ there exists a distinguished upper bound of $\{x, y\}$ in $(M, \leq)$

PROPOSITION 3.5 ([Eklund and Gähler ****], 2.4). Let $f : (M, \leq) \longrightarrow (HN, \leq)$ be an isotone mapping between prosets which filfills (E). For all $x, y \in M$, if d is a distinguished upper bound of $\{x, y\}$ in $(M, \leq)$, then $f(d)$ is a distinguished upper bound of $\{f(x), f(y)\}$ in $(N, \leq)$.

COROLLARY 3.6 ([Eklund and Gähler ****], 2.6). Let $(M, \leq)$ and $(N, \leq)$ be prosets fulfilling

(S) and $f : (M, \leq) \longrightarrow (N, \leq)$ an isotone mapping fulfilling (E). For all $x, y \in M$, if k is a supremum of $\{x, y\}$ in the Urysohn modification of $(M, \leq)$, then $f(k)($ is a supremum of $\{f(x), f(y)\}$ in the Urysohn modificaton of $(N, \leq)$.

A proset $(M, \leq)$ will be called a **(join) semiprelattice** provided for all $x, y \in M$, $\{x, y\}$ has a supremum. Join semiprelattices $(M, \leq)$ with $\leq$ a partial ordering may be identified with join semilattices. The main results in this section can be formulated as follows: for each proset $(M, \leq)$ fulfilling (S), the Urysohn modification $(M, \preceq)$ is a join semiprelattice. If $(M, \leq)$ and $(N, \leq)$ are presets fulfilling (S) and $f : (M, \leq) \longrightarrow (N, \leq)$ is an isotone mapping fulfilling (E), then $f : (M, \preceq) \longrightarrow (N, \preceq)$ is a join semiprelattice morphism, i.e. f preserves suprema of unordered pairs.

Covariant functors Φ : SET $\longrightarrow$ PROSET

Let **PROSET** denote the category of prosets (whose morphisms are the isotone mappings between prosets). Frequently a covariant functor Φ : **SET** $\longrightarrow$ **PROSET** will be written as a pair $(\varphi, \leq)$ where φ is the underlying set functor, that is, φ : **SET** $\longrightarrow$ **SET** is the composition of Φ with the forgetful functor of **PROSET**. For each set $X, \Phi X$ is a proset $(\varphi X, \leq)$. $\mathcal{M} \leq \mathcal{N}$ $(\mathcal{M}, \mathcal{N} \in \varphi X)$ may be read that $\mathcal{M}$ is **finer** than $\mathcal{N}$.

If $(\varphi, \leq)$ is a covariant functor from **SET** to **PROSET**, then also $(\varphi \circ \psi, \leq)$ is, where ψ is any covariant set functor and $((\varphi \circ \psi)X, \leq)$ of course means the proset $(\varphi Y, \leq)$ with $Y = \psi X$.

Suppose that $\Phi = (\varphi, \leq)$ is a covariant functor from **SET** to **PROSET** which fulfills the **extension condition**, that is, for each mapping $f : X \longrightarrow Y$, $\varphi f : (\varphi X, \leq) \longrightarrow (\varphi Y, \leq)$ fulfills condition (E). Because of Proposition then we may define as **Urysohn modification** of $(\varphi, \leq)$ the covariant functor $(\varphi, \preceq)$: **SET** $\longrightarrow$ **PROSET** which we get from $(\varphi, \leq)$ in replacing each preordered set $(\varphi X, \leq)$ by its Urysohn modification $(\varphi X, \preceq)$. $(\varphi, \preceq)$ also fulfills the extension condition. If $(\varphi, \leq)$ fulfills the extension condition, then also $(\varphi \circ \psi, \leq)$ fulfills this condition, where ψ is any covariant set functor.

The functor $(\varphi, \leq)$: **SET** $\longrightarrow$ **PROSET** will be called **Urysohn** provided it coincides with its Urysohn modification $(\varphi, \preceq)$. If $(\varphi, \leq)$ is Urysohn then also $(\varphi \circ \psi, \leq)$ is, where ψ is any covariant set functor.

Let **SEMIPLAT** denote the category of join semiprelattices whose morphisms are the mappings between join semiprelattices which preserve suprema of unordered pairs. We will mainly be interested in covariant functors $(\varphi, \leq)$: **SET** $\longrightarrow$ **PROSET** which are ranging in **SEMIPLAT**, i.e. for each set X and mapping $f : X \longrightarrow Y$, $(\varphi X, \leq)$ is an object and $\varphi f : (\varphi X, \leq) \longrightarrow (\varphi Y, \leq)$ a morphism of **SEMIPLAT**.

A covariant functor $(\varphi, \leq)$: **SET** $\longrightarrow$ **PROSET** will be said to fulfill the **supremum condition** provided each preordered set $(\varphi X, \leq)$ fulfills (S). If $(\varphi, \leq)$ fulfills the supremum condition, then also $(\varphi \circ \psi, \leq)$ fulfills this condition where ψ is any covariant set functor.

Proposition 3.7 ([Eklund and Gähler ****], 3.7). Let $(\varphi, \leq)$: **SET** $\longrightarrow$ **PROSET** be a covariant functor which fulfills the extension condition and the supremum condition. Then its Urysohn modification $(\varphi, \preceq)$ is ranging in **SEMIPLAT**.

If $(\varphi, \leq)$ is ranging in **SEMIPLAT** (resp. **SEMILAT**) then also $(\varphi \circ \psi, \leq)$ is, where ψ is any covariant set functor.

EXAMPLES.

(1) Let φ be the proper powerset functor P_0. In this case we define the preordering of each set φX by $M \leq N \Leftrightarrow M \subseteq N$ and then write $(P_0, \subseteq)$ for Φ.

(2) Another example is given as follows: Let φ be the proper ideal functor I_0. Here as preordering of each set I_0X will also be taken the inclusion and for Φ we write $(I_0, \subseteq)$.

(3) A standard example is the following: Let φ be the proper filter functor F_0. As preordering of each set φX is chosen the inversion of the inclusion: $\mathcal{M} \leq \mathcal{N} \Leftrightarrow \mathcal{M} \supseteq \mathcal{N}$. For Φ, here we write $(F_0, \supseteq)$.

(4) Example (3) can be generalized in the following way: Let φ be the L-fuzzy filter functor, where L is any meet semilattice. Here we take as preordering $\leq$ of each set φX the inversion $\geq^{\cdot}$ of the componentwise defined partial ordering $\leq^{\cdot}$: $\mathcal{M} \leq^{\cdot} \mathcal{N} \Leftrightarrow \mathcal{M}(f) \leq \mathcal{N}(f)$ for all $f \in L^X$. For Φ, we then write $(F_L, \geq^{\cdot})$. Clearly, using the identification of the proper filter functor with F_L for $L = 2$ given in section (2), it follows that $(F_0, \supseteq)$ coincides with $(F_2, \geq)$.

(5) Yet a further interesting example is as follows: Let φ be the sequence functor **seq**. The preordering of each set φX is defined by

$$(x_n)_{n\in\mathbb{N}} \leq (y_n)_{n\in\mathbb{N}} \Leftrightarrow (x_n)_{n\in\mathbb{N}} \text{ is a subsequence of } (y_n)_{n\in\mathbb{N}}$$

For Φ, here we write $(seq, \leq)$. On each set $seqX$, $\leq$ is indeed a preordering. called the **subsequence preordering**. Whereas in examples (1) to (4) all preorderings are partial orderings, for each set X which has at least two elements x and y, the subsequence preordering on $seqX$ is not a partial ordering. We namely have $(x, y, x, y, \ldots) \leq (y, x, y, x, \ldots) \leq (x, y, x, y, \ldots)$.

In the following part of this section we will investigate these examples.

PROPOSITION 3.8 (cf. [Eklund and Gähler ****], 3.2). $(P_0, \subseteq)$, I_0, $\subseteq)$ and $(seq, \leq)$ fulfill the extension condition. If L is a linearly ordered, completely distributive, complete lattice, then also $(F_L, \geq^{\cdot})$ satisfies the extension condition.

PROOF. We will show here only the last part and assume that L is a linearly ordered, completely distributive complete lattice. For each $\mathcal{M} \in F_LX$ and L-fuzzy filter $\mathcal{N}$ and Y with $\mathcal{N} \geq^{\cdot} F_Lf(\mathcal{M})$, define $\mathcal{L} \in L^{L^X}$ by $\mathcal{L}(h) = \mathcal{N}(h')$ for each $h \in L^X$, where $h' \in L^Y$ is the mapping given by

$$h(y) = \begin{cases} \bigvee_{x\in f^{-1}\{y\}} h(x), & \text{if } y \in f[X], \\ \bigvee_{x\in X} h(x), & \text{otherwise,} \end{cases}$$

for all $y \in Y$. It follows $\mathcal{L}(\bar{\alpha}) = \alpha$ for every $\alpha \in L$. Clearly, $\mathcal{L}$ is isotone. Since L is completely distributive, for all $h, k \in L^X$ and $y \in Y$ we have

$$(h' \wedge k')(y) = \bigvee_{x,u \in f^1\{y\}} (h(x) \wedge k(u))$$

Taking into account that L is linearly ordered, for all $x, u \in f^{-1}\{y\}$ we have $h(x) \leq h(u)$ or $h(x) \geq h(u)$, and consequently $h(x) \wedge k(u) \leq (h \wedge k)(u)$ or $h(x) \wedge k(u) \leq (h \wedge k)(x)$. Hence, it follows $(h \wedge k)' = h' \wedge k'$ and therefore $\mathcal{L}(h \wedge k) = \mathcal{L}(h) \wedge \mathcal{L}(k)$. Thus, $\mathcal{L}$ is an L-fuzzy filter on X. For each $h \in L^X$ we have $h' \circ f \geq h$, and because of $\mathcal{N} \geq^{\cdot} F_L f(\mathcal{M})$ therefore $\mathcal{L} \geq^{\cdot} \mathcal{M}$. On the other hand, since for each $g \in L^Y$ we have $(g \circ f)' \geq g$, it follows $F_L f(\mathcal{L}) \geq^{\cdot} \mathcal{N}$. Hence, $(F_L, \geq^{\cdot})$ satisfies the extension condition. □

REMARK. Clearly, as a special case, $(F_0, \supseteq)$ fulfills extension condition.

For each sequence $(x_n)_{n\in\mathbb{N}}$ on a set X, let $filt(x_n)_{n\in\mathbb{N}}$ denote the filter on X generated by $(x_n)_{n\in\mathbb{N}}$. It is well-known that $(x_n)_{n\in\mathbb{N}} \leq (y_n)_{n\in\mathbb{N}}$ implies $filt(x_n)_{n\in\mathbb{N}} \supseteq filt(y_n)_{n\in\mathbb{N}}$.

LEMMA 3.9 ([Eklund and Gähler ****], 3.3). Any two sequences $(x_n)_{n\in\mathbb{N}}$ and $(y_n)_{n\in\mathbb{N}}$ on a set X for which $filt(y_n)_{n\in\mathbb{N}} \supseteq filt(y_n)_{n\in\mathbb{N}}$, have a common subsequence.

PROPOSITION 3.10 ([Eklund and Gähler ****], 3.4). Let X be a set and $\leq$ the subsequences preordering on $seqX$. The Urysohn modification of $\leq$ is characterized by

$$(x_n)_{n\in\mathbb{N}} \preceq (y_n)_{n\in\mathbb{N}} \Leftrightarrow filt(x_n)_{n\in\mathbb{N}} \supseteq filt(y_n)_{n\in\mathbb{N}}$$

In the following part of this section we will no longer consider the fuzzy filter example, but we will treat it separately in a subsequent section.

PROPOSITION 3.11 (cf. [Eklund and Gähler ****], 3.5). $(P_0, \subseteq)$ and $(F_0, \supseteq)$ are Urysohn; however, $(I_0, \subseteq)$ and $(seq, \leq)$ are not.

REMARK. Note that for the proof of this proposition in case of $(F_0, \supseteq)$ in [Eklund and Gähler ****] , the notion of ultrafilters was used and also both the following properties of ultrafilters:

(1) For each proper filter $\mathcal{M}$ on a set X there exists an ultrafilters $\mathcal{U} \supseteq \mathcal{M}$ on X.
(2) Each proper filter $\mathcal{M}$ on a set X is the intersection of all ultrafilters $\mathcal{U} \supseteq \mathcal{M}$ on X.

Because of Proposition 3.11, $(I_0, \subseteq)$ and $(seq, \leq)$ differ from their Urysohn modifications $(I_0, \preceq)$ and $(seq, \preceq)$.

$(seq, \leq)$ is not ranging in **SEMIPLAT**, which was shown in [Eklund and Gähler ****].

PROPOSITION 3.12 (cf. [Eklund and Gähler ****], 3.6). $(P_0, \subseteq)$, $(I_0, \subseteq)$, $(F_0, \supseteq)$, and $(seq, \leq)$ fulfill the supremum condition.

Because of Propositions 3.7, 3.8 and 3.12, $(seq, \preceq)$ is ranging in **SEMIPLAT**. Obviously, all the functors $(P_0, \subseteq)$, $(I_0, \subseteq)$, $(I_0, \preceq)$, and $(F_0, \supseteq)$ are even ranging in the full subcategory **SEMILAT** of join semilattices.

Conditions for completion constructions

Let $\Phi = (\varphi, \leq)$ be a covariant functor of **SET** to **PROSET** such that φ is connected and let η denote the unique natural transformation from id to φ.

For completion constructions we will need some assumptions on Φ which we list in the following.

- (A1) For each set X and all $x, y \in X$ from $\eta_X(x) \leq \eta_X(y)$ it follows $x = y$.
- (A2) For $f : X \longrightarrow Y, \mathcal{M} \in \varphi X$ and $y \in Y$, if $\eta_Y(y) \leq \varphi f(\mathcal{M})$ then $y = f(x)$ for some $x \in X$.
- (A3) From $\mathcal{M} \leq \eta_X(x)$ it follows $\eta_X(x) \leq \mathcal{M}$, i.e. each $\eta_X(x)$ is minimal in $(\varphi X, \leq)$.
- (A4) For any set X and all $\mathcal{K}, L, M, N \in \varphi X$ if $\mathcal{K}$ is a supremum of $\{\mathcal{L}, \mathcal{M}\}$ such that $\mathcal{N} \leq \mathcal{K}$ holds, then $\{N, L\}$ or $\{\mathcal{N}, \mathcal{M}\}$ has a lower bound in $(\varphi X, \leq)$.
- (A5) For each set X, each $x \in X$ and all $\mathcal{M}, N \in \varphi X$ if there is a supremum $\mathcal{L}$ of $\{\mathcal{M}, \eta_X(x)\}$ such that $\mathcal{N} \leq \mathcal{L}$ then $\mathcal{N} \leq \mathcal{M}$ or $\eta_X(x) \leq \mathcal{N}$.

Note that condition (A4) is formulated in [Eklund and Gähler ****] in a different but equivalent way.

PROPOSITION 3.13. Assume that Φ is ranging in **SEMIPLAT** and fulfills (A4). Then Φ fulfills the supremum condition.

PROOF. Since Φ is ranging in **SEMIPLAT** for all $\mathcal{L}, M \in \varphi X$, a supremum $\mathcal{K}$ of $\{L, M\}$ exists. Because of (A4), $\mathcal{K}$ is a distinguished upper bound of $\{\mathcal{L}, M\}$. Hence, Φ fulfills the supremum condition. □

PROPOSITION 3.14 ([Eklund and Gähler ****], 5.1). Each of the functors $(P_0, \subseteq)$, $(I_0, \subseteq)$, $(F_0, \supseteq)$ and $(seq, \preceq)$ satisfies all conditions (A1) to (A5). If $(\psi, \leq)$ satisfies (A1) to (A5) and ψ is a connected covariant set functor such that the following conditions are fulfilled

- (B1) for each set X and all $x, y \in X$, from $\vartheta_X(x) = \vartheta_X(y)$ it follows $x = y$;
- (B2) for $f : X \longrightarrow Y$, $\mathcal{M} \in \psi X$, and $y \in Y$, if $\vartheta_Y(y) = \vartheta f(\mathcal{M})$, then $y = f(x)$ for some $x \in X$;

where ϑ is the unique natural transformation from id to ψ, then $(\varphi \circ \psi, \leq)$ also satisfies (A1) to (A5).

REMARK. (B1) and (B2) are e.g. fulfilled in the case of $\psi = id^n$ where n is any cardinal and, of course, if ψ is the underlying set functor of any covariant functor Ψ : **SET** $\longrightarrow$ **PROSET** which fulfills (A1) and (A2).

§4. The Case of Fuzzy Filter Functor

Let L be a meet semilattice.

PROPOSITION 4.1. $(F_L, \geq^{\cdot})$ is ranging in **SEMILAT**.

PROOF. Let X be a set and $\mathcal{M}, N \in F_L X$. Define $\mathcal{L} \in L^{L^X}$ by $\mathcal{L}(f) = \mathcal{M}(f) \wedge \mathcal{N}(f)$ for all $f \in L^X$. Obviously $\mathcal{L}$ fulfills the condition (i) and (ii) of a fuzzy filter. If $f, g \in L^X$ then $\mathcal{L}(f \wedge g) = \mathcal{M}(f \wedge g) \wedge \mathcal{N}(f \wedge g) = \mathcal{L}(f) \wedge \mathcal{L}(g)$. Hence also (iii) is fulfilled. Therefore L is an L-fuzzy filter. Clearly, then $\mathcal{L}$ is the supremum of $\{\mathcal{M}, \mathcal{N}\}$ with respect to $\geq^{\cdot}$, which will be denoted $\mathcal{M} \wedge^{\cdot} \mathcal{N}$.

Let $f : X \longrightarrow Y$ be a mapping. For all $\mathcal{M}, N \in F_L X$ and $g \in L^Y$ we have $F_L f(\mathcal{M} \wedge \mathcal{N})(g) = (\mathcal{M} \wedge^{\cdot} \mathcal{N})(g \circ f) = F_L f(\mathcal{M})(g) \wedge F_L f(\mathcal{N})(g)$. Hence, $F_l f$ preserves suprema of unordered pairs of fuzzy filters with respect to $\geq^{\cdot}$. □

This result may also be read as follows: For each set X, the infimum $\mathcal{M} \wedge^{\cdot} \mathcal{N}$ of every pair $\{\mathcal{M}, \mathcal{N}\}$ of L-fuzzy filters $\mathcal{M}$ and $\mathcal{N}$ with respect to $\leq^{\cdot}$ on X exists. For every mapping $f : X \longrightarrow Y$, $F_L f$ preserves infima of unordered pairs of fuzzy filters with respect to $\leq^{\cdot}$.

If L is a complete lattice we can show analogously the following: if X is any set and $f : X \longrightarrow Y$ is any mapping, then every subset K of $F_L X$ has the infimum $\dot{\bigwedge}_{\mathcal{M} \in K} \mathcal{M}$ in $(F_L X, \leq^{\cdot})$, where $(\dot{\bigwedge}_{\mathcal{M} \in K} \mathcal{M})(g) = \dot{\bigwedge}_{\mathcal{M} \in K} \mathcal{M})(g)$ for all $g \in L^X$ and, moreover, it holds

$$F_L f(\dot{\bigwedge_{\mathcal{M} \in K}} \mathcal{M}) = \dot{\bigwedge_{\mathcal{M} \in K}} F_L F(\mathcal{M})$$

In other words, if L is a complete lattice, then $(F_L, \geq^{\cdot})$ is ranging in the category **CSEMILAT** of complete join semilattices. **CSEMILAT** has as objects all complete lattices and as morphisms all mappings between complete lattices which preserve arbitrary suprema.

The following result gives a criterion on suprema of unordered pairs of fuzzy filters with respect to $\leq^{\cdot}$.

PROPOSITON 4.2. Assume that L is a complete distributive complete lattice. Let X be a set and $\mathcal{M}, N \in F_L X$. Then the following are equivalent:

(1) $\{\mathcal{M}, N\}$ has an upper bound with respect to $\leq^{\cdot}$.
(2) $\{\mathcal{M}, N\}$ has the supremum $\mathcal{M} \wedge^{\cdot} \mathcal{N}$ with respect to $\leq^{\cdot}$.
(3) For each $\alpha \in L$ and $f, g \in L^X$ with $f \wedge g = \bar{\alpha}$ we have $\mathcal{M}(f) \wedge \mathcal{N}(g) = \alpha$.

If the supremum $\mathcal{M} \vee^{\cdot} \mathcal{N}$ exists, then for every $h \in L^X$ we have

$$(\mathcal{M} \vee^{\cdot} \mathcal{N})(h) = \bigvee_{f \wedge g = h} \mathcal{M}(f) \wedge \mathcal{N}(g) \tag{4.3}$$

PROOF. Let $\mathcal{U}$ be an upper bound of $\{\mathcal{M}, \mathcal{N}\}$ with respect to $\leq^{\cdot}$. For each $\alpha \in L$ and

$f, g \in L^X$ with $f \wedge g = \bar{\alpha}$ we have $\bar{\alpha} \leq f, g$ and therefore $\alpha \leq \mathcal{M}(f) \wedge \mathcal{N}(g) \leq \mathcal{U}(f \wedge g) = \alpha$. Hence (1) implies (3).

Assume now that (3) is fulfilled. Taking into account that L is a complete lattice, we define $\mathcal{L} \in L^{L^X}$ by

$$\mathcal{L}(h) = \bigvee_{f \wedge g = h} \mathcal{M}(f) \wedge \mathcal{N}(g)$$

for all $h \in L^X$. Because of (3) for each $\alpha \in L$ we have

$$\mathcal{L}(\bar{\alpha}) = \bigvee_{f \wedge g = \bar{\alpha}} \mathcal{M}(f) \wedge \mathcal{N}(g) = \alpha$$

Fix now $h' \geq H$. Since L is distributive, $f \wedge g = h$ implies $h' = h \vee h' = (f \vee h') \wedge (g \vee h')$ and therefore

$$\mathcal{L}(h) \leq \bigvee_{f \wedge g = h} \mathcal{M}(f \vee h') \wedge \mathcal{N}(g \vee h') \leq \mathcal{L}(h')$$

Choose now $h, k \in L^X$. Taking into account that L is completely distributive we get

$$\mathcal{L}(h) \wedge \mathcal{L}(k) = \bigvee_{\substack{f \wedge g = h \\ u \wedge v = k}} \mathcal{M}(f) \wedge \mathcal{N}(g) \wedge \mathcal{M}(u) \wedge \mathcal{N}(v)$$

$$\leq \bigvee_{(f \wedge u) \wedge (g \wedge v) = h \wedge k} \mathcal{M}(f \wedge u) \wedge \mathcal{N}(g \wedge v) \leq \mathcal{L}(h \wedge K)$$

Thus $\mathcal{L}$ is a fuzzy filter. Taking $f = h$ and $g = \bar{1}$ as well as $f = \bar{1}$ and $g = h$ it follows $\mathcal{M}(h), \mathcal{N}(h) \leq \mathcal{L}(h)$ for all $h \in L^X$. For each upper bound $\mathcal{U} \geq^{\cdot} \mathcal{M}, \mathcal{N}$ we have $\mathcal{L}(h) \leq \bigwedge_{f \wedge g = h} \mathcal{U}(f) \wedge \mathcal{U}(g) = \mathcal{U}(h)$. Thus $\mathcal{L}$ is the supremum of $\{\mathcal{M}, \mathcal{N}\}$ with respect to $\leq^{\cdot}$ and therefore (3) implies (2). Clearly, (2) implies (1). Of course, the last part of the proposition is already shown.

For $\alpha \in L$ and A any subset of a set X, define the mapping $f_{\alpha,A} \in L^X$ by

$$f_{\alpha,A}(x) = \begin{cases} \alpha, & \text{if } x \in A, \\ 1, & \text{otherwise.} \end{cases}$$

If L is linearly ordered, then (3) is equivalent to the following condition as will be shown:

(3') For each $\alpha \in L$ and $A \subseteq X$ we have $\mathcal{M}(f_{\alpha,A}) \wedge \mathcal{N}(f_{\alpha,X \setminus A}) = \alpha$

Clearly, (3) implies (3'). On the other hand, if (3') is fulfilled and $f, g \in L^X$ are mappings such that $f \wedge g = \bar{\alpha}$ for some $\alpha \in L$, then we have $f_{\alpha,A} \geq f$, where $A = \{x \in X \mid f(x) = \alpha\}$. Moreover, then for each $x \in X \setminus A$ we have $f(x) > \alpha$, and since L is linearly ordered, therefore $g(x) = \alpha$. Hence, $f_{a,X \setminus A} \geq g$ and thus $\alpha \geq \mathcal{M}(f) \wedge \mathcal{N}(g)$. Because of $f, g \geq \bar{\alpha}$, then $\mathcal{M}(f) \wedge \mathcal{N}(g) = \alpha$. Therefore (3') implies (3). □

Note that for every fuzzy filter $\mathcal{M} \in F_L X$, each $A \subseteq X$ and $\alpha \in L$, because of $f_{\alpha,A} \geq \bar{\alpha}$ we have $\mathcal{M}(f_{\alpha,A}) \geq \alpha$. Clearly, $\mathcal{M}(f_{1,A}) = 1$ and $\mathcal{M}(f_{\alpha,A}) = \alpha$ always hold.

Call a fuzzy filter $\mathcal{M} \in F_L X$ **restricted** if for every subset A of X and all $\alpha, \beta \in L$ with $\beta < \alpha$ the following holds:

$$\mathcal{M}(f_{\beta,A}) = \beta \Rightarrow \mathcal{M}(f_{\alpha,A}) = \alpha$$

(or what is equivalent:

$$\mathcal{M}(f_{\alpha,A}) > \alpha \Rightarrow \mathcal{M}(f_{\beta,A}) > \beta)$$

A sufficient condition for $\mathcal{M} \in F_L X$ to be restricted is, that for every subset A of X and all $\alpha, \beta \in L$ with $\beta < \alpha$ we have

$$\mathcal{M}(f_{\alpha,A}) = \mathcal{M}(f_{\beta,A}) \vee \alpha. \tag{4.4}$$

Note that (4.4) can also be written in the form

$$\mathcal{M}(f_{\beta,A} \vee \bar{\alpha}) = \mathcal{M}(f_{\beta,A} \vee \mathcal{M}(\bar{\alpha})$$

Particularly, this condition is fulfilled in the case of $L = 2$. Hence all usual filters are restricted. For each set X, each $x \in X$, all $A \subseteq X$, and $\alpha \in L$, we have

$$\eta_X(x)(f_{\alpha,A}) = \begin{cases} \alpha, & \text{if } x \in A, \\ 1, & \text{otherwise} \end{cases}$$

Hence, every fuzzy filter $\eta_X(x)$ is restricted.

PROPOSITION 4.5. Let $\mathcal{M}$ and $\mathcal{N}$ be L-fuzzy filters on a set X. For each mapping $f : X \longrightarrow Y$, with $\mathcal{M}$ also $F_L f(\mathcal{M})$ is restricted. If both $\mathcal{M}$ and $\mathcal{N}$ are restricted, then $\mathcal{M} \wedge^{\cdot} \mathcal{N}$ also is. Moreover, if L is a linearly ordered, completely distributive, complete lattice, $\mathcal{M} \vee^{\cdot} \mathcal{N}$ exists, and $\mathcal{M}$ and $\mathcal{N}$ are restricted, then $\mathcal{M} \vee^{\cdot} \mathcal{N}$ also is restricted.

PROOF. Assume that $\mathcal{M}$ is restricted. Fix a mapping $f : X \longrightarrow Y$, a subset B of Y and $\alpha, \beta \in L$ with $\beta < \alpha$. Let $g_{\alpha,B} \in L^Y$ be defined analogously as $f_{\alpha,A} \in L^X$. Because of $g_{\alpha,A} \circ f = f_{\alpha,f^{-1}[B]}$, from $F_L f(\mathcal{M})(g_{\alpha,A}) > \alpha$ it follows $\mathcal{M}(f_{\alpha f^{-1}[B]}) > \alpha$ and therefore $F_L f(\mathcal{M})(g_{\beta,A}) = \mathcal{M}(f_{\alpha,f^{-1}[B]}) > \beta$. Thus, $F_L f(\mathcal{M})$ is restricted.

Assume now that $\mathcal{M}$ and $\mathcal{N}$ are restricted. Fix a subset A of X and $\alpha, \beta \in L$ with $\beta < \alpha$. From $(\mathcal{M} \wedge^{\cdot} \mathcal{N})(f_{\alpha,A}) > \alpha$ we get $\mathcal{M}(f_{\alpha,A}) > \alpha$ and $\mathcal{N}(f_{\alpha,A}) > \alpha$ and $\mathcal{N}(f_{\alpha,A}) > \alpha$, hence $\mathcal{M}(f_{\beta,A}) > \beta$ and $\mathcal{N}(f_{\beta,A}) > \beta$, and, since L is linearly ordered, therefore $(\mathcal{M} \wedge^{\cdot} \mathcal{N})(f_{\beta,A}) > \beta$. Thus $\mathcal{M} \wedge^{\cdot} \mathcal{N}$ is restricted.

If L is a linearly ordered, completely distributive complete lattice and $\mathcal{M} \vee^{\cdot} \mathcal{N}$ exists, then from (4.3) we get

$$(\mathcal{M} \vee^{\cdot} \mathcal{N})(f_{\alpha,A}) = \bigvee_{B \subseteq A} \mathcal{M}(f_{\alpha,B}) \wedge \mathcal{N}(f_{\alpha,A\setminus B})$$

Hence, then from $(\mathcal{M} \vee^{\cdot} \mathcal{N})(f_{\alpha,A}) > \alpha$ it follows that there exists a subset B_0 of A with $\mathcal{M}(f_{\alpha,B_0}) \wedge \mathcal{N}(f_{\alpha,A\setminus B_0}) > \alpha$ from which we get $\mathcal{M}(f_{\beta,B_0} \wedge \mathcal{N}(f_{\beta,A\setminus B_0}) > \beta$, and therefore $(\mathcal{M} \vee^{\cdot} \mathcal{N})(f_{\beta,A}) > \beta$. Hence, then also $\mathcal{M} \vee^{\cdot} \mathcal{N}$ is restricted. □

Because of the first assertion of this proposition, the **restricted fuzzy filter functor** $F_{L,rest}$ can be defined as a subfunctor of F_L assigning to each set X the set of all restricted fuzzy filters of X. When we equip each set $F_{L,rest}X$ with the inversion of the componentwise defined partial ordering, we have the functor $(F_{L,rest}, \dot{\geq}) :$ **SET** $\longrightarrow$ **PROSET**. Because of Proposition 4.1 and the second part of Proposition 4.5 we know that $F_{L,rest}, \dot{\geq})$ is ranging in **SEMILAT**. Finally, by means of the third part of Proposition 4.5 it follows that Proposition 4.2 analogously holds for $F_{L,rest}$, provided L is additionally assumed to be linearly ordered.

REMARK. As is easily seen, if L is a complete lattice and X is any set, then $\mathcal{M} : L^X \longrightarrow L$, defined by $\mathcal{M}(f) = \bigwedge_{x \in X} f(x)$ for all $f \in L^X$, is the least L-fuzzy filter in $(F_L X, \dot{\leq})$.

PROPOSITION 4.6. Assume that L is a completely distributive complete lattice. Let X be a set and K any non-empty subset of $F_L X$ which is directed (that is, for all $\mathcal{M}, \mathcal{N} \in F_L X$ there exists an upper bound of $\{\mathcal{M}, \mathcal{N}\}$ in $(F_L X, \dot{\leq})$. Then $\dot{\bigvee}_{\mathcal{M} \in K} \mathcal{M}$, defined by

$$(\dot{\bigvee_{\mathcal{M} \in K}} \mathcal{M})(f) = \bigvee_{\mathcal{M} \in K} \mathcal{M}(f)$$

for all $f \in L^X$, is the supremum of K in $(F_L X, \dot{\leq})$.

PROOF. Taking into account that K is non-empty, it is easily seen, that $\dot{\bigvee}_{\mathcal{M} \in K} \mathcal{M}$ fulfills the conditions (1) and (2) of an L-fuzzy filter. Fix now $f, g \in L^X$. We have

$$(\dot{\bigvee_{\mathcal{M} \in K}} \mathcal{M})(f \wedge g) \leq \bigvee_{\mathcal{M} \in K} (\mathcal{M}(f) \wedge \mathcal{M}(g))$$

Since K is directed, here even equality holds. Thus, since L is completely distributive, it follows

$$(\dot{\bigvee_{\mathcal{M} \in K}} \mathcal{M})(f \wedge g) = (\dot{\bigvee_{\mathcal{M} \in K}} \mathcal{M})(f) \wedge (\dot{\bigvee_{\mathcal{M} \in K}} \mathcal{M})(g)$$

Therefore $\dot{\bigvee}_{\mathcal{M} \in K} \mathcal{M}$ is an L-fuzzy filter. From the definition of $\dot{\bigvee}_{\mathcal{M} \in K} \mathcal{M}$ we immediately get that $(\dot{\bigvee}_{\mathcal{M} \in K} \mathcal{M})$ then is the supremum of K in $(F_L X, \dot{\leq})$. □

Because of this proposition and the remark given above, for each set X, all directed subsets of $(F_L X, \dot{\leq})$ have suprema. Hence, as a consequence of Zorn's lemma we get the following.

PROPOSITION 4.7. If L is a completely distributive complete lattice, then for each set X and each $\mathcal{M} \in F_L X$ there exists an **ultra fuzzy filter** $\mathcal{U}$ on X (i.e. a maximal element $\mathcal{U}$ in $(F_L X, \dot{\leq})$) such that $\mathcal{M} \dot{\leq} \mathcal{U}$ holds.

PROPOSITION 4.8. If L has the least and last element 0 and 1, respectively, such that $0 \neq 1$, then $(F_L, \dot{\geq})$ fulfills the conditions (A1) and (A2). Moreover, $(F_L, \dot{\geq})$ always fulfills the condition (A3).

PROOF. (Note that here as preordering $\leq$ on each set φX is used the inverse ordering of $\leq^{\cdot}$.) Suppose at first that L has the least and last element 0 and 1, respectively, and that $0 \neq 1$. Assume $\eta_X(x) \geq^{\cdot} \eta_X(y)$ for some $x, y \in X$. For each $f \in L^X$ we have $f(x) \geq f(y)$. Since in the case of $x \neq y$ there is a mapping $f \in L^X$ with $f(x) = 0$ and $f(y) = 1$, it follows $x = y$. Thus (A1) is proved.

Fix now $f : X \longrightarrow Y$, $\mathcal{M} \in F_L X$, and $y \in Y$ such that $\eta_X(y)(g) \geq F_L f(\mathcal{M})(g)$, i.e. $g(y) \geq \mathcal{M}(g \circ f)$ holds for all $g \in L^X$. Assume $y \neq f(x)$ for all $x \in X$. Define $g \in L^Y$ by $g(y) = 0$ and $g(z) = 1$ for all $x \neq y$. Then $0 \geq \mathcal{M}(\bar{1}) = 1$ which is a contradiction ($\bar{\alpha}$ always is the constant mapping with value α). Then there exists an $x \in X$ with $y = f(x)$. Hence (A2) is fulfilled.

Now let L be arbitrary. Fix $x \in X$ and $\mathcal{M} \in F_L X$ such that $\mathcal{M} \geq^{\cdot} \eta_X(x)$, i.e. $\mathcal{M}(g) \geq g(x)$ holds for all $g \in L^X$. Choose any $g \in L^X$ and define $h \in L^X$ by $h(x) = 1$ and $h(y) = \overline{g(x)}$ for all $y \neq x$. It follows $g \wedge h \leq \overline{g(x)}$ (constant mapping) and therefore $g(x) = \mathcal{M}(\overline{g(x)}) \geq \mathcal{M}(g \wedge h) = \mathcal{M}(g) \wedge \mathcal{M}(h)$. Since we have $\mathcal{M}(h) \geq h(x) = 1$, i.e. $\mathcal{M}(h) = 1$, we obtain $\mathcal{M}(g) = g(x)$. Thus $\mathcal{M} = \eta_X(x)$ and therefore (A3) is fulfilled. □

Because of Proposition 4.8, $(F_{L,rest}, \geq^{\cdot})$ also fulfills (A1), (A2) and (A3). We further have the following.

PROPOSITION 4.9. Assume that L is a linearly ordered, completely distributive lattice. Then $(F_{L,rest}, \geq^{\cdot})$ fulfills (A4).

PROOF. Let X be a set and $\mathcal{L}, \mathcal{M}, \mathcal{N} \in F_{L,rest}X$ and assume that $\{\mathcal{L} \wedge^{\cdot} \mathcal{M}, \mathcal{N}\}$ has an upper bound in $F_{L,rest}, \geq^{\cdot})$. Because of Proposition 4.2 (analogon) then for each $\alpha \in L$ and $f, g \in L^X$ with $f \wedge g = \bar{\alpha}$ we have

$$\mathcal{L}(f) \wedge \mathcal{M}(f) \wedge \mathcal{N}(g) = \alpha \tag{4.10}$$

Assume that neither $\{\mathcal{L}, \mathcal{N}\}$ nor $\{\mathcal{M}, \mathcal{N}\}$ has an upper bound in $(F_{L,rest}, \geq^{\cdot})$. Then there are $\alpha, \beta \in L$ and subsets A, B of X with $\mathcal{L}(f_{\alpha,A}) \wedge \mathcal{N}(f_{\alpha,X\setminus A}) > \alpha$ and $\mathcal{M}(f_{\beta,B}) \wedge \mathcal{N}(f_{\beta,X\setminus B}) > \beta$. We may assume $\beta \geq \alpha$ (linearity of the ordering). Since $\mathcal{M}$ and $\mathcal{N}$ are distinguished because of $\mathcal{M}(f_{\beta,B}) > \alpha$ and $\mathcal{N}(f_{\beta,X\setminus B}) > \beta$ er also have

$$\mathcal{L}(f_{\alpha,A\cap B}) \wedge \mathcal{M}(f_{\alpha,A\cap B}) \wedge \mathcal{N}(f_{\alpha,X\setminus(A\cap B)}) \geq \mathcal{L}(f_{\alpha,A}) \wedge (f_{\alpha,B}) \wedge \mathcal{N}(f_{\alpha,X\setminus(A\cap B)}) > \alpha$$

which contradicts (4.10). Thus $\{\mathcal{L}, \mathcal{N}\}$ or $\{\mathcal{M}, \mathcal{N}\}$ has an upper bound in $(F_{L,rest}, \geq^{\cdot})$, and therefore (A4) is fulfilled. □

REMARK. If L is a linearly ordered, completely distributive, complete lattice, then $(F_{L,rest}, \geq^{\cdot})$ fulfills the supremum condition. This follows by means of the Propositions 3.13 and 4.9 and taking into account that $(F_{L,rest}, \geq^{\cdot})$ is ranging in **SEMILAT**.

§5. Fuzzy Convergence

Let $(\varphi, \leq)$ be a covariant functor from **SET** to **PROSET** where the underlying set functor φ is connected. In the following we are interested in stable φ-convergence structures which fulfill some additional axioms given by the preordering of the prosets $(\varphi X, \leq)$. Let η be the unique natural transformation from id to φ.

Let X be a set. A φ-convergence structure t on X is called a **Φ-pseudo-topology** if the following conditions are satisfied:

(P1) $\eta_X(x) \underset{t}{\rightarrow} x$ for all $x \in X$ (stability).
(P2) $\mathcal{M} \rightarrow x$ and $\mathcal{N} \leq \mathcal{M}$ imply $\mathcal{N} \underset{t}{\rightarrow} x$.
(P3) If $\mathcal{M} \underset{t}{\rightarrow}$ and $\mathcal{N} \underset{t}{\rightarrow} x$, then there is an upper bound $\mathcal{U}$ of $\{\mathcal{M}, \mathcal{N}\}$ such that $\mathcal{U} \underset{t}{\rightarrow} x$.

X equipped with a Φ-pseudo-topology on X is called a **Φ-pseudo-topological space**.

We get an interesting special case if Φ is ranging in the category **POSET** of partially ordered sets and if for each $x \in X$ there exists among all elements of φX converging to x a largest one with respect to $\leq$. This largest element will be denoted $\mathcal{N}(x)$ and called the **neighbourhood object** of x. We then speak of a **Φ-pretopology space** instead of a Φ-pseudo-topology (respectively a Φ-pseudo-topological space). Each Φ-pretopology t on a set X is given by these largest elements $\mathcal{N}(x)$ where the only axiom is that $\eta_X(x) \leq \mathcal{N}(x)$ holds for all $x \in X$. In case of a Φ-pretopology t on a set X we have

$$\mathcal{M} \underset{t}{\rightarrow} x \Leftrightarrow \mathcal{M} \leq \mathcal{N}(x)$$

for all $x \in X$.

Let **Φ-PTOP** denote the category of all Φ-pseudo-topological spaces with all φ-continuous mappings between them as morphisms. Moreover, if Φ is ranging in **POSET**, let **Φ-PRETOP** denote the full subcategory of **Φ-PTOP** of all Φ-pretopological spaces.

PROPOSITION 5.1. If Φ is ranging in the category **CSEMILAT** of complete join semilattices, then **Φ-PRETOP** is a full bireflective subcategory of **Φ-PTOP**, where the reflection of each Φ-pseudo-topological space (X,t) is the identity mapping of onto $(X, pre\, t)$, with $pre\, t$ the Φ-pretopology given by

$$\mathcal{N} \underset{pre\, t}{\rightarrow} x \Leftrightarrow \mathcal{N} \leq \dot{\bigvee_{\mathcal{M} \underset{pre\, t}{\rightarrow} x}} \mathcal{M}$$

PROOF. Obvious. □

Here the neighbourhood objects with respect to $pre\, t$ will also be considered as neighbourhood objects with respect to t.

In the following we turn to the case of $\Phi = (F_L, \dot{\geq})$, where L is a meet semilattice. Recall that $(F_L, \dot{\geq})$ is ranging in **SEMILAT**, and if L is a complete lattice, even in **CSEMILAT**. Moreover, recall that also $(F_{L,rest}, \dot{\geq})$ is ranging in **SEMILAT**.

Because of (2.2), for the unique natural transformation $\eta : id \longrightarrow F_L$ and each set X, $x \in X$, and $f \in L^X$ we have

$$\eta_X(x)(f) = f(x) \tag{5.2}$$

Instead of an $(F_L, \geq^{\cdot})$-pseudo-topology and an $(F_L, \geq^{\cdot})$-pretopology we also speak of an (L-)**fuzzy pseudo-topology** and an (L-)**fuzzy pretopology**, respectively. Analogously, instead of an $(F_{L,rest}, \geq^{\cdot})$-pseudo-topology and an $(F_{L,rest}, \geq^{\cdot})$-pretopology we also speak of a **restricted** (L-)**fuzzy pseudo-topology** and a **restricted** (L-)**fuzzy pretopology**, respectively. For the related spaces we use the corresponding denotations (L-)**fuzzy pseudo-topological space** and (L-)**fuzzy pretopological space.**

Denote the category $(F_L, \geq^{\cdot})$-**PTOP** also by L-**FPTOP** and, if L is a complete lattice, $(F_L, \geq^{\cdot})$-**PRETOP** also by L-**FPRETOP**. Because of Proposition 5.1, if L is a complete lattice, L-**PRETOP** is a full bireflective subcategory of L-**FPTOP**.

In the following let (X, t) be an L-fuzzy pseudo-topological space. A fuzzy set $f \in L^X$ will be said to be **open** in (X, t) if $f(x) \leq \mathcal{M}(f)$ holds for all $x \in X$ and $\mathcal{M} \underset{t}{\rightarrow} x$. If L is a complete lattice, then the neighbourhood objects $\mathcal{N}(x)$ of the elements of X, which then are defined, will be called the **fuzzy neighbourhood filters**. When L is a complete lattice, openess of a fuzzy set $f \in L^X$ means that $f(x) \leq \mathcal{N}(f)(x)$ holds for all $x \in X$.

Assume that L is a complete lattice. By an (L-)**fuzzy topology** on a set X, in this chapter we mean any subset τ of L^X which is closed with respect to finite infima and arbitrary suprema, and which additionally fulfills the Lowen-condition that $\bar{\alpha} \in \tau$ for all $\alpha \in L$. X equipped with an L-fuzzy topology τ on X is an (L-)**fuzzy topological space.**

PROPOSITION 5.3. Assume that L is a complete lattice. Then for any L-fuzzy pseudo-topological space, the collection τ of all open fuzzy sets in (X, t) is an L-fuzzy topology on X.

PROOF. Follows easily from the fact that $\mathcal{M} \underset{t}{\rightarrow} x$ (respectively $\mathcal{N}$) are fuzzy filters. □

In the following assume that L is a completely distributive, complete lattice. Then each L-fuzzy topological space may be considered as special L-fuzzy pretopological spaces as will be shown in the following. This leads to a precise relationship to the classical theory of fuzzy topology.

Let (X, τ) be an L-fuzzy topological space. For each $x \in X$ define $\mathcal{N}(x) \in L^{L^X}$ through

$$\mathcal{N}(x)(f) = \bigvee_{\substack{g \leq f \\ g \in \tau}} g(x) \tag{5.4}$$

PROPOSITION 5.5. For each $x \in X, \mathcal{N}(x)$ is an L-fuzzy filter on X and it holds $\mathcal{N}(x) \leq \eta_X(x)$.

PROOF. For each $\alpha \in L$ we have

$$\mathcal{N}(x)(\bar{\alpha}) = \bigvee_{\substack{g \leq \bar{\alpha} \\ g \in \tau}} g(x) = \alpha$$

since $\bar{\alpha} \in \tau$. Clearly, $f \leq g$ implies $\mathcal{N}(x)(f) = \mathcal{N}(x)(g)$. Moreover, taking into account that L is completely distributive, for all $f, g \in L^X$ we get

$$\mathcal{N}(x)(f \wedge g) = \bigvee_{\substack{h \leq f \wedge g \\ h \in \tau}} h(x) \geq \bigvee_{\substack{h \leq f, k \leq g \\ h, k \in \tau}} (h(x) \wedge k(x)) = \mathcal{N}(x)(f) \wedge \mathcal{N}(x)(g)$$

Thus, $\mathcal{N}(x)$ is an L-fuzzy filter. For each $f \in L^X, \mathcal{N}(x)(f) \leq f(x)$, and because of (5.2), therefore $\mathcal{N}(x) \leq \eta_X(x)$ holds. □

$\mathcal{N}(x)$ defined by (5.4) is called the **fuzzy neighbourhood filter** of x in the fuzzy topological space (X, τ).

REMARK. Note that we do not have a notion of neighbourhoods. However, for each fuzzy set $f \in L^X$, $\mathcal{N}(x)(f)$ can be considered to measure the degree to which f is a neighbourhood of x.

Because of Proposition 5.5, the fuzzy neighbourhood filters in an L-fuzzy topoological space (X, τ) define an L-fuzzy pretopology t on X called the **associative L-fuzzy pretopology** of τ. A fuzzy set $f \in L^X$ is open in (X, t) if and only if

$$f(x) \leq \bigvee_{\substack{g \leq f \\ g \in \tau}} g(x)$$

holds for all $x \in X$, i.e.

$$f = \bigvee_{\substack{g \leq f \\ g \in \tau}} g$$

Hence, for an L-fuzzy topological space (X, τ), openess of a fuzzy set $f \in L^X$ with respect to the associated L-fuzzy pretopology, precisely means that f is an element of τ. This shows that we may identify each L-fuzzy topology with the associated L-fuzzy pretopology.

Recall that a mapping $f : (X, \tau) \longrightarrow (Y, \sigma)$ between L-fuzzy topologies is said to be **continuous** provided $g \circ f \in \tau$ whenever $g \in \sigma$.

PROPOSITION 5.6. A mapping $f : (X, \tau) \longrightarrow (Y, \sigma)$ between L-fuzzy topologies is continuous if and only if f is F_L-continuous with respect to the associated L-fuzzy pretopologies.

PROOF. 'If' part: assume that f is F_L-continuous. Let $\mathcal{N}(x), x \in X$ and $\mathcal{N}'(y), y \in Y$ denote the neighbourhood filters in (X, τ) and (Y, σ), respectively. Because of the F_L-continuity, and the fact that $\mathcal{N}(x) \to x$ for all $x \in X$, we have $F_L f(\mathcal{N}(x)) \geq \mathcal{N}'(f(x))$. Now if $g \in \sigma$, i.e. $\mathcal{N}'(y)(h) \geq g(y)$ for all $y \in Y$, then we have $\mathcal{N}(x)(g \circ f) = F_L f(\mathcal{N}(x))(g) \geq \mathcal{N}'(f(x))(g) \geq g(f(x)) = (g \circ f)(x)$ for all $x \in X$. Hence, $g \circ f \in \tau$ and then f is continuous with respect to τ and σ.

'Only if' part: assume that f is continuous with respect to the fuzzy topologies τ and σ. For any $x \in X$ and $g \in L^{id}Y$ we have

$$\begin{aligned} F_L f(\mathcal{N}(x))(g) &= \mathcal{N}_X(x)(g \circ f) \\ &= \bigvee_{\substack{h \leq g \circ f \\ h \in \tau}} h(x) \\ &\geq \bigvee_{\substack{k \circ f \leq g \circ f \\ k \in \sigma}} (k \circ f)(x) \\ &= \mathcal{N}'(f(x))(g) \end{aligned}$$

Hence, f is F_L-continuous. □

Let L-**FTOP** denote the category of all L-fuzzy topological spaces with all continuous mappings between these spaces as morphisms. Using the identification between L-fuzzy topological spaces and their associated L-fuzzy pretopological spaces, together with Proposition 5.6, shows that L-**FTOP** may be considered as a full subcategory of L-**FPRETOP**. Recall that we assume here that L is a completely distributive, complete lattice.

PROPOSITION 5.7. Let (X, t) be an L-fuzzy pretopological space and τ the L-fuzzy topology on X consisting of all open fuzzy sets in (X, t). Moreover, let t' be the L-fuzzy pretopology associated with τ. Then we have

$$\mathcal{N}'(x) \leq \mathcal{N}(x)$$

for the fuzzy neighbourhood filters of each $x \in X$ related to t and t', respectively.

PROOF. Taking into account that $g \in \tau$ implies $g(x) \leq \mathcal{N}(x)(g)$, it follows

$$\mathcal{N}'(x)(f) = \bigvee_{\substack{g \leq f \\ g \in \tau}} g(x) \leq \bigvee_{\substack{g \leq f \\ g \in \tau}} \mathcal{N}(x)(g) \leq \mathcal{N}(x)(f)$$

for all $f \in L^X$. □

Using the denotations of this proposition it is obvious that an L-fuzzy pretopological space (X, t) is an L-fuzzy topological space if and only if

$$\mathcal{N}(x)(f) = \bigvee_{\substack{g \leq f \\ g \in \tau}} g(x)$$

for all $x \in X$ and $f \in L^X$.

§6. Generalized Cauchy Spaces and Completions

Let $\Phi = (\varphi, \leq)$ be a covariant functor from **SET** to **PROSET** such that the underlying set functor is connected. Let η denote the unique natural transformation from id to φ.

A **Φ-Cauchy structure** on a set X is a subset s of φX satisfying

(C1) $\eta_X(x) \in s$ for all $x \in X$ (stability).
(C2) $\mathcal{M} \in s$ and $\mathcal{N} \leq \mathcal{M}$ imply $\mathcal{N} \in s$.
(C3) If $\mathcal{M}, \mathcal{N} \in s$ and $\{\mathcal{M}, \mathcal{N}\}$ has a lower bound in $(\varphi X, \leq)$ then there is an upper bound $\mathcal{U} \in$ of $\{\mathcal{M}, \mathcal{N}\}$.

The pair (X, s) is called a **Φ-Cauchy space**. Each φ-morphism between Φ-Cauchy spaces is also called a **Φ-Cauchy continuous mapping**.

Let (X, s) be a Φ-Cauchy space. The **associated Φ-pseudo-topology** t consists of all pairs $(\mathcal{M}, x) \in \varphi X \times X$ such that there is an upper bound $\mathcal{N}$ of $\{\mathcal{M}, \eta_X(x)\}$ in $(\varphi X, \leq)$ with $\mathcal{N} \in s$. t is indeed a Φ-pseudo topology on X as has been shown in [Eklund and Gähler ****]. s, and also (X, s), is said to be **separated** provided the associated Φ-pseudo-topology is. Further s is said to be **complete** if each $\mathcal{M} \in s$ converges, with respect to the associated Φ-pseudo-topology, to some element of X.

Let **Φ-CHY**, **Φ-CHY**$_{sep}$, and **Φ-CHY**$_{sep,cpl}$ denote the categories of all Φ-Cauchy spaces, separated Φ-Cauchy spaces, and both separated and complete Φ-Cauchy spaces, respectively, with all Φ-Cauchy continuous mappings between the corresponding spaces as morphisms.

PROPOSITION 6.1 ([Eklund and Gähler ****], 4.8 and 4.9). If Φ is ranging in **SEMIPLAT**, then in **Φ-CHY** all initial structures and final structures exist.

In the following assume that Φ is ranging in **SEMIPLAT**. In [Eklund and Gähler ****], a completion construction has been given where additionally Φ is assumed to fulfill the conditions (A1) to (A4). This construction is as follows.

Let $A = (X, s)$ be a separated Φ-Cauchy space. Define an equivalence relation $\cong$ on s by

$$\mathcal{M} \cong \mathcal{N} \Leftrightarrow \text{ there is an upper bound } \mathcal{K} \in s \text{ of } \{\mathcal{M}, \mathcal{N}\}$$

and write $\widehat{X} = \{\mathcal{M}^{\cong} \mid \mathcal{M} \in s\}$, where $\mathcal{M}^{\cong}$ is the equivalence class of $\mathcal{M}$ with respect to $\cong$. Moreover, define a mapping $\iota_A : X \to \widehat{X}$ by $\iota_A(x) = \eta_X(x)^{\cong}$ for all $x \in X$ and let

$$\widehat{s} = \{\mathcal{K} \in \varphi\widehat{X} \mid \text{there is an } \mathcal{M} \in s \text{ such that } \varphi\iota_A(\mathcal{M}) \leq \mathcal{L},$$

$$\text{and } \eta_X \sim (\mathcal{M}^{\sim}) \leq \mathcal{L} \text{ imply } \mathcal{K} \leq \mathcal{L})\}$$

As has been shown in [Eklund and Gähler ****], $\widehat{A} = (\widehat{X}, \widehat{s})$ is a separated, complete Φ-Cauchy space, and $\iota_A : A \to \widehat{A}$ is an epireflection of A with respect to **Φ-CHY**$_{sep,cpl}$. Hence, **Φ-CHY**$_{sep,cpl}$ is an epireflective subcategory of **Φ-CHY**$_{sep}$. For each separated Φ-Cauchy space A, the pair $(\iota_A, \widehat{A})$ may be considered as a completion of A. Assume in the following that L is a linearly ordered, completely distributive, complete lattice with different least and last element, i.e. $0 \neq 1$. Recall that $(F_{L,rest}, \geq^{\cdot})$ is ranging in **SEMILAT** and that, because of Propositions 4.8 and 4.9, $(F_{L,rest}, \geq^{\cdot})$ fulfills the conditions (A1) to (A4). Hence, separated $(F_{L,rest}, \geq^{\cdot})$-Cauchy spaces can be completed as above. It follows that $(F_{L,rest}, \geq^{\cdot})$-$\mathbf{CHY}_{sep,cpl}$ is an epireflective subcategory of $(F_{L,rest}, \geq^{\cdot})$-$\mathbf{CHY}_{sep}$.

Note that in [Eklund and Gähler ****] it has also been shown that for any covariant functor Φ : **SET** $\longrightarrow$ **PROSET** ranging in **SEMIPLAT**, which has a connected underlying set functor and which fulfills the conditions (A1) to (A5), Φ-$\mathbf{CHY}_{sep,cpl}$ is reflective in Φ-**CHY**. For further examples and the justified use of Urysohn modifications ranging in **SEMIPLAT**, we refer the reader to [Eklund and Gähler ****]. Compare also with [18]; in this paper a completion theory of generalized convergence groups is established.

§7. Monads

In his paper, [Kowalsky 1954] suggested a contraction mapping $\mu_X : FFX \longrightarrow FX$ given by

$$\mu_X(\mathcal{U}) = \bigcup_{\mathcal{R} \in \mathcal{U}} \bigvee_{\mathcal{M} \in \mathcal{R}} \mathcal{M}$$

Here $\bigwedge_{\mathcal{M} \in \mathcal{R}} \mathcal{M}$ means $\bigcap_{\mathcal{M} \in \mathcal{R}} \mathcal{M}$, if $\mathcal{R} \neq \emptyset$, and PX otherwise.

[Day 1975] showed that this construction, together with the natural transformation $\eta : id \longrightarrow F$ given by $\eta_X(x) = \{A \subseteq X \mid x \in X\}$, defines a monad, a concept defined in the following.

Let φ be a covariant set functor. Each natural transformation $\mu : \varphi \circ \varphi \longrightarrow \varphi$ will be called a **multiplication** on φ. A multiplication μ on φ is **associative** provided the following diagram

$$\begin{array}{ccc} \varphi \circ \varphi \circ \varphi & \xrightarrow{\varphi\mu} & \varphi \circ \varphi \\ \Big\downarrow{\scriptstyle \mu\varphi} & \circ & \Big\downarrow{\scriptstyle \mu} \\ \varphi \circ \varphi & \xrightarrow[\mu]{} & \varphi \end{array}$$

commutes. Let a multiplication μ on φ be fixed. A natural transformation $\eta : id \longrightarrow \varphi$ is a **left unit** (resp. **right unit**) with respect to μ provided the diagram

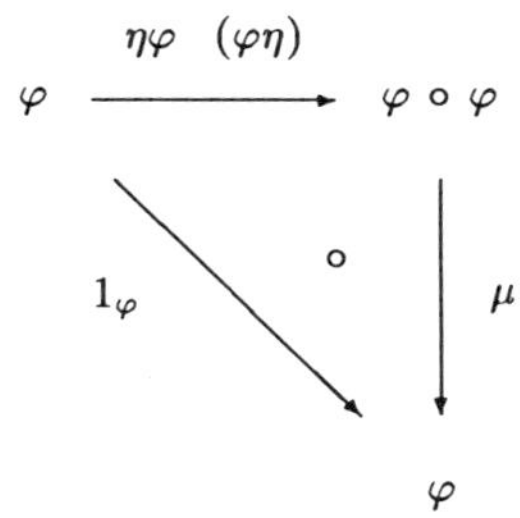

commutes. Here $\eta\varphi$ and $\varphi\eta$ are the natural transformations consisting of the mappings $\eta_{\varphi X} : \varphi X \longrightarrow \varphi\varphi X, X \in Ob(\mathbf{SET})$, respectively. Hence, for η to be a left unit (respectively a right unit) with respect to μ means that $\mu_X \circ \eta_{\varphi X} = 1_{\varphi X}$ (respectively $\mu_X \circ \varphi(\eta_X) = 1_{\varphi X}$) holds for each set X. A **unit** with respect to μ is a natural transformation that is a left and right unit with respect to μ. A **monad** (on **SET**) is a triple (φ, η, μ) consisting of a covariant set functor φ, an associative multiplication μ on φ, and a unit with respect to μ.

The notion of a monad does not only relate to adjoint functors and universal algebras (see e.g. [Manes 1976, Richter 1979]), but is indispensible also in topological analysis of results, i.e. concerning regularity and iteratedness (cf. [Gähler 1988]) as well as compactifications.

There are a lot of examples of monads. It is well-known that the proper filter functor F_0 becomes a monad (F_0, η, μ) in taking as η the unique natural transformation from id to φ and as μ the multiplication on φ given by

$$\mu_X(\mathcal{U}) = \bigcup_{R \in \mathcal{U}} \bigcap_{\mathcal{M} \in R} \mathcal{M}$$

for each set X and $\mathcal{U} \in F_0 F_0 X$.

For $\varphi = P_0 \circ id^2$ the situation is different. Since $\varphi = P_0 \circ id^2$ is connected, we know that there is an unique natural transformation $\eta : id \longrightarrow P_0 \circ id^2$, which consists of the injections $\eta_X : x \mapsto \{(x,x)\}$, $x \in X$, where X is any set. Now an assertion follows which implies that $P_0 \circ id^2$ cannot be made to a monad.

Proposition 7.1. The unique natural transformation $\eta : id \longrightarrow P_0 \circ id^2$ is not a unit with respect to a multiplication on $P_0 \circ id^2$.

Proof. Suppose there is a multiplication μ on $P_0 \circ id^2$ such that η is a unit with respect to μ. Let $X = \{1,2,3,4,\}, Y = \{1,2\}$, $M = \{(1,1),(2,2)\}$, and $N = \{(3,3),(4,4)\}$. Since η is a unit with respect to μ we have

$$\mu_Y\{(M,M)\} = M \tag{7.2}$$

and

$$\mu_Y\{(\{(1,1)\},\{(2,2)\})\} = \{(1,2)\} \tag{7.3}$$

Let $f : X \to Y$ be the mapping given by $f(1) = f(3) = 1$ and $f(2) = f(4) = 2$. Then $(f \times f)[M] = (f \times f)[N] = M$ and because of (7.2) and $(P_0 \circ id^2)f \circ \mu_X = \mu_Y \circ (P_0 \circ id^2)(P_0 \circ id^2)f$, therefore we have $(f \times f)[\mu_X\{(M,N)\}] = M$. Hence

$$\mu_X\{(M,N)\} \subseteq (\{1,3\} \times \{1,3\} \cup (\{2,4\} \times \{2,4\}$$

Using instead of f the mapping $g : X \longrightarrow Y$ given by $g(2) = g(3) = 1$ and $g(1) = g(4) = 2$ we analogously get

$$\mu_X\{(M,N)\} \subseteq (\{2,3\} \times \{2,3\} \cup (\{1,4\} \times \{1,4\}$$

Thus we have

$$\mu_X\{(M,N)\} \subseteq \{(1,1),(2,2),(3,3),(4,4)\} \tag{7.4}$$

Let $h : X \longrightarrow Y$ be the mapping given by $h(1) = h(2) = 1$ and $h(3) = h(4) = 2$. Then $(h \times h)[M] = \{(1,1)\}$ and $(h \times h)[N] = \{(2,2)\}$, and because of (7.3) therefore

$$(h \times h)[\mu_X\{(M,N)\}] = \{(1,2)\} \tag{7.5}$$

Hence, we have

$$\mu_X\{(M,N)\} \subseteq \{1,2\} \times \{3,4\}$$

which together with (7.4) implies that $\mu_X\{(M,N)\}$ is empty. This contradicts (7.5). □

Whereas $P_0 \circ id^2$ cannot be made to a monad, we at least have the following.

PROPOSITION 7.6. Let μ be the family consisting of the mappings

$$\mu_X : \mathcal{M} \mapsto \bigcup_{(M,N)\in\mathcal{M}} pr_1 M \times pr_2 N \quad (\mathcal{M} \subseteq P_0X^2 \times P_0X^2)$$

with X any set where $pr_1 M = \{x \in X \mid \exists y \in X,\ (x,y) \in M\}$ and $pr_2 N = \{y \in X \mid \exists x \in X, (x,y) \in N\}$. Then μ is an associative multiplication on $P_0 \circ id^2$ such that $\eta : id \longrightarrow P_0 \circ id^2$ is a right unit with respect to μ.

PROOF. If $f : X \longrightarrow Y$ is a mapping and M a subset of $P_0X^2 \times P_0X^2$, then $(f \times f)[\mu_X(\mathcal{M})] = \bigcup_{(M,N)\in\mathcal{M}} f[pr_1 M] \times f[pr_2 N] = \mu_Y((f \times f)[\mathcal{M}])$ because of $f[pr_1 M] = pr_1(f \times f)[M]$ and $f[pr_2 N] = pr_2(f \times f)[N]$. Hence μ is a natural transformation.

Let X be a set and $\varphi = P_0 \circ id^2$. If A is a subset of $P_0(P_0X^2 \times P_0X^2) \times P_0(P_0X^2 \times P_0X^2)$, then

$$\begin{aligned}
\mu_X(\mu_{\varphi X}(\mathcal{A})) &= \mu_X(\bigcup_{(\mathcal{M},\mathcal{N})\in\mathcal{A}} pr_1\mathcal{M} \times pr_2\mathcal{N}) \\
&= \bigcup_{(\mathcal{M},\mathcal{N})\in\mathcal{A}} \mu_X(pr_1\mathcal{M} \times pr_2\mathcal{N}) \\
&= \bigcup_{(\mathcal{M},\mathcal{N})\in\mathcal{A}} pr_1\mu_X(\mathcal{M}) \times pr_2\mu_X(\mathcal{N}) \\
&= \mu_X(\{(\mu_X(\mathcal{M}), \mu_X(\mathcal{M}) \mid (\mathcal{M},\mathcal{N}) \in \mathcal{A}\}) \\
&= \mu_X(\varphi\mu_X(\mathcal{A})).
\end{aligned}$$

Hence, μ is associative. If M is a subset of X^2, then

$$\mu_X((\eta_X \times \eta_X)[M]) = \mu_X\{(\{(x,x)\},\{(y,y)\} \mid (x,y) \in M\} = M.$$

Thus, η is a right unit with respect to μ. □

It is well-known that P can be made to a monad. Analogously, P_0 and $id^2 \circ F_0$ can be completed to monads. Since both these set functors are connected, the units of the related monads are uniquely determined. We can show that in both these cases the multiplications are uniquely given.

PROPOSITION 7.7. P_0 can be made only in one way to a monad (P_0,η,μ), namely in defining μ by

$$\mu_X(\mathcal{M}) = \bigcup_{M\in\mathcal{M}} M$$

for all sets X and $\mathcal{M} \in P_0P_0X$.

PROOF. Clearly, here $\eta_X(x) = \{x\}$ for each set X and $x \in X$. Let $\mathcal{M} \in P_0P_0X$, and choose a mapping $f : X \longrightarrow X$ which is an identity on $X \setminus \bigcup_{M\in\mathcal{M}} M$, and for which $f(\bigcup_{M\in\mathcal{M}}) = \{x\}$ for some $x \in X$. It follows $f(\mu_X(\mathcal{M})) = P_0f(\mu_X(\mathcal{M})) = \mu_X(P_0P_0f(\mathcal{M})) = \mu_X(\{\{x\}\})$. Now since $\mu_X \circ \eta_{P_0X} = 1_{P_0X}$, we have $f(\mu_X(\mathcal{M})) = \{x\}$, and therefore $\mu_X(\mathcal{M}) \subseteq \bigcup_{M\in\mathcal{M}} M$. Assume that $\mu_X(\mathcal{M}) \subset \bigcup_{M\in\mathcal{M}} M$ (strict inclusion) for some $\mathcal{M} \in \varphi X$. Then define $g : X \longrightarrow \{1,2\}$ through $g(\mu_X(\mathcal{M})) = \{1\}$ and $g(X \setminus \mu_X(\mathcal{M})) = \{2\}$. For $M \in \mathcal{M}$, $g(M)$ is either $\{1\},\{2\}$, or $\{1,2\}$. Hence $P_0P_0g(M)$ is either $\{\{1,2\}\},\{\{1\},\{2\}\},\{\{1\},\{2\},\{1,2\}\},\{\{1\},\{1,2\}\}$, or $\{\{2\},\{1,2\}\}$. In applying the permutation $1 \mapsto 2,\ 2 \mapsto 1$ of $Y = \{1,2\}$ on $\{1\} = \mu_Y(P_0P_0g(\mathcal{M}))$ it follows that

$$\{1\} = \mu_Y(\{1\}\{1,2\}) \text{ or } \{1\} = \mu_Y(\{2\},\{1,2\}). \tag{7.8}$$

Let $Z = \{1,2,3,4\}$ and $\mathcal{N} = \{\{1,2\},\{3,4\}\}$. We have $\mu_Z(\mathcal{N}) \subseteq Z$. Applying an arbitrary permutation of Z which leaves $\{1,2\}$ and $\{3,4\}$ fixed and maps $\{1,2\}$ onto $\{3,4\}$, and vice versa, we get that $\mu_Z(\mathcal{N}) = Z$. Applying on this set the mappings $1,2,3 \mapsto 1,\ 4 \mapsto 2$ and $1 \mapsto 1,2,3,\ 4 \mapsto 2$ of Z onto Y gives $\mu_Y(\{\{1\},\{1,2\}\}) = \mu_Y(\{\{2\},\{1,2\}\}) = \{1,2\}$, which contradicts (7.8). □

PROPOSITION 7.9. $id^2 \circ P_0$ can be made only in one way to a monad $(id^2 \circ P_0,\eta,\mu)$, namely in taking μ by

$$\mu_X(\mathcal{M},\mathcal{N}) = (\bigcup_{(A,B)\in\mathcal{M}} A, \bigcup_{(A,B)\in\mathcal{N}} B)$$

PROOF. Clearly, $\eta_X(x) = (\{x\},\{x\})$ for each set X and each $x \in X$. The proof is similar to the proof of Proposition 7.7, and therefore omitted. □

In the following we will show that for each set L, the functor $\varphi = L^{L^{id}}$ can be completed to a monad. For each set X, we define two mappings $\eta_X : X \longrightarrow \varphi X$ and $\mu_X : \varphi\varphi X \to \varphi X$ through

$$\eta_X(x)(f) = f(x) \tag{7.10}$$

(as in (5.1)) and

$$\mu_X(\mathcal{U}) = \mathcal{U} \circ \eta_{L^{id}X} \tag{7.11}$$

for each set X, $x \in X$, $f \in L^X$, and $\mathcal{U} \in \varphi\varphi X$. Note that φ is not connected if L contains at least two elements. For each mpapping $f : X \longrightarrow Y$ and each $g \in L^X$, because of (2.1), we have

$$(\varphi f \circ \eta_X)(x)(g) = \eta_X(x)(g \circ f) = (\eta_Y \circ f)(x)(g)$$

Hence $\eta = (\eta_X)_{X \in Ob SET} : id \longrightarrow \varphi$ is a natural transformation. In order to show that also $\mu = (\mu_X)_{X \in Ob\mathbf{SET}} : \varphi \circ \varphi \longrightarrow \varphi$ is a natural transformation, again take $f : X \longrightarrow Y$ and moreover, $\mathcal{U} \in \varphi\varphi X$ and $g \in L^X$. We must show that

$$\varphi f \circ \mu_X = \mu_Y \circ \varphi\varphi f$$

and because of (7.11), therefore that

$$\varphi f(\mathcal{U} \circ \eta_{L^{id}X}) = \varphi\varphi f(\mathcal{U}) \circ \eta_{L^{id}Y} \tag{7.12}$$

for all $\mathcal{U} \in \varphi\varphi X$.

From (2.1) it follows that for each $\mathcal{U} \in \varphi\varphi X$ and $g \in L^X$ we have

$$\varphi f(\mathcal{U} \circ \eta_{L^{id}X})(g) = (\mathcal{U}(\eta_{L^{id}X})(g \circ f))$$

Because of (7.10), $(\eta_{L^{id}X})(g \circ f)) : \varphi X \longrightarrow L$ is the mapping $(\mathcal{M} \mapsto \mathcal{M}(g \circ f))$. Hence,

$$\varphi f(\mathcal{U} \circ \eta_{L^{id}X})(g) = \mathcal{U}(\eta_{L^{id}Y}(g) \circ \varphi f)$$

and from which, because of (2.1), (7.12) follows. □

Further we have

PROPOSITION 7.13. $(L^{L^{id}}, \eta, \mu)$ is a monad.

PROOF. It remains to prove that μ is associative, and that η is a left and a right unit with respect to μ. Let again $\varphi = L^{L^{id}}$.

For the associativity of μ we have

$$\begin{aligned}(\mu_X \circ \varphi\mu_X(\mathcal{V})(g) &= \varphi\mu_X(\mathcal{V})(\eta_{L^{id}X}(g)) \\ &= \mathcal{V}(\eta_{L^{id}X}(g) \circ \mu_X) \\ &= \mathcal{V}(\mathcal{U} \mapsto \mathcal{U}(\eta_{L^{id}X}(g))) \\ &= \mathcal{V}(\eta_{\varphi X}(\eta_{L^{id}X}(g)))\end{aligned}$$

and

$$\begin{aligned}(\mu_X \circ \mu_{\varphi X}(\mathcal{V})(g) &= \mu_{\varphi X}(\mathcal{V})(\eta_{L^{id}X}(g)) \\ &= \mathcal{V}(\eta_{\varphi X}(\eta_{L^{id}X}(g)))\end{aligned}$$

The property of η to be a left unit follows from

$$\begin{aligned}(\mu_X \circ \eta_{\varphi X}(\mathcal{M})(g) &= \eta_{\varphi X}(\mathcal{M})(\eta_{L^{id}X}(g)) \\ &= \eta_{L^{id}X}(g)(\mathcal{M}) \\ &= \mathcal{M}(g)\end{aligned}$$

and the property to be right unit form

$$\begin{aligned}(\mu_X \circ \varphi\eta_X(\mathcal{M})(g) &= \varphi\eta_X(\mathcal{M})(\eta_{L^{id}X}(g)) \\ &= \mathcal{M}(\eta_{L^{id}X}(g) \circ \eta X) \\ &= \mathcal{M}(x \mapsto \eta_X(x)(g) = g(x)) \\ &= \mathcal{M}(g) \quad \square\end{aligned}$$

If (φ, η, μ) is any monad, then (φ', η', μ') will be called a **submonad** of (φ, η, μ) if

(i) φ' is a subfunctor of φ,
(ii) $\eta_X(x) = \eta'_X(x)$ for all $x \in X$,
(iii) $\mu'_X(\mathcal{U}) = \mu_X(\varphi e_X(\mathcal{U}))$,

where $e : \varphi' \longrightarrow \varphi$ is the natural transformation for which $e_x : \varphi' X \longrightarrow \varphi X$ are inclusion mappings. Note that $e\varphi \circ \varphi' e = \varphi e \circ e\varphi'$. Of course, a submonad is a monad.

PROPOSITION 7.14. Let L be a meet semilattice and $(L^{L^{id}}, \eta, \mu)$ be the monad as in 7.13. Let $e : F_L \longrightarrow L^{L^{id}}$ be the natural transformation with $e_X : F_L X \longrightarrow L^{L^X}$ inclusion mappings. Moreover, let η' denote the unique natural transformation from id to F_L. Then $(F_L, \eta', \mu\varphi e)$ is a submonad of $(L^{L^{id}}, \eta, \mu)$.

PROOF. It is straightforward to verify that $\eta_X(x) \in F_L X$, and so it remains to show that $\mu_X(\mathcal{U})$ is a fuzzy filter whenever $\mathcal{U}$ is.

First, $\mu_X(\mathcal{U})(\bar{\alpha}) = (\mathcal{U} \circ \eta_{L^{id}X})(\bar{\alpha}) = \mathcal{U}(\tilde{\alpha}) = \alpha$, since $\eta_{L^{id}X}(\bar{\alpha})(\mathcal{M}) = \mathcal{M}(\bar{\alpha}) = \alpha$, where $\tilde{\alpha}$ means the constant mappings on L^X with value α.

Second, suppose $g_1 \leq g_2$. Then $\eta_{L^{Id}X}(g_1)(\mathcal{M}) = \mathcal{M}(g_1) \leq \mathcal{M}(g_2) = \eta_{L^{id}X}(g_2)(\mathcal{M})$, which implies $\eta_{L^{id}X}(g_1) \leq \eta_{L^{id}X}(g_2)$. Hence, $\mu_X(\mathcal{U})(g_1) \leq \mu_X(\mathcal{U})(g_2)$.

Third, note that $\eta_{L^{id}X}(g_1 \wedge g_2)(\mathcal{M}) = \mathcal{M}(g_1 \wedge g_2) \geq \mathcal{M}(g_1) \wedge \mathcal{M}(g_2) = \eta_{L^{id}X}(g_1)(\mathcal{M}) \wedge \eta_{L^{id}X}(g_2)(\mathcal{M})$. Hence, $\eta_{L^{id}X}(g_1 \wedge g_2) \geq \eta_{L^{id}X}(g_1) \wedge \eta_{L^{id}X}(g_2)$, and consequently $\mu_X(\mathcal{U})(g_1 \wedge g_2) \geq \mu_X(\mathcal{U})(g_1) \wedge \mu_X(\mathcal{U})(g_2)$. $\square$

REMARK. For $L = 2$, the two-pointed lattice, we have the classical Kowalsky construction. In fact, suppose $N \in \bigcup_{\mathcal{R} \in \mathcal{U}} \bigcap_{\mathcal{M} \in R} \mathcal{M}$, or equivalently, there exists an $\mathcal{R} \in \mathcal{U}$ such that

$N \in \mathcal{M}$ for all $\mathcal{M} \in \mathcal{R}$. Then $\mathcal{U} \circ \eta_{2^{id}X}(N) = \mathcal{U}([N]) = 1$, since $N \in [N]$ (where [N] is the point filter of N). Conversely, if $\mathcal{U}([N]) = 1$, then $\mathcal{M}(N) = 1$ whenever $[N](\mathcal{M}) = 1$. Hence $N \in \bigcup_{\mathcal{R}\in\mathcal{U}} \bigcap_{\mathcal{M}\in\mathcal{R}} \mathcal{M}$.

REMARK. The universal algebra (Eilenberg-Moore) of the monad for the ordinary filter functor corresponds to a continuous lattice (see [Day 1975]). An analog relation was shown in [Gähler 1988] for the proper filter monad algebra.

Chapter 5:

Convenient Topological Constructs

S. E. Rodabaugh et al. (eds.), Applications of Category Theory to Fuzzy Subsets, 137–151.

CONVENIENT TOPOLOGICAL CONSTRUCTS

H. Herrlich

Dedicated to my friend R. Nakagawa on his 60th birthday

§0. Introduction

Many of the entities encountered in fuzzy topology are topological constructs. Thus the theory of topological constructs provides an appropriate framework for categorical fuzzy topology. The purpose of this chapter is to briefly delineate the main features of this theory in its present form. In particular the following convenience properties of topological constructs are investigated:

(1) Cartesian closedness (in particular concrete cartesian closedness on one hand and the existence of function spaces on the other hand).

(2) Representability of partial morphisms (in particular concrete representations).

(3) Quasitopoi (in particular universally topological constructs and topological universes).

Although the above concepts (at least (1) and (2)) seem rather unrelated, in the context of topological constructs each of them can be characterized by a condition stating that certain final sinks (e.g. all final sinks, resp., final epi-sinks, coproducts, quotients, discrete objects [= empty final sinks]) are preserved under pointwise pullbacks along certain morphisms (e.g., all morphisms, resp., projections, embeddings). Finally methods will be sketched which enable us to embedd 'reasonable' constructs into topological constructs with prescribed convenience features.

Many of the results presented here remain true in the more general context of concrete categories over (suitable) base categories **X**. Details, including complete proofs and further references, can be found in [Adàmek, Herrlich, and Strecker 1990].

§1. Topological Constructs

DEFINITION 1.1.

(1) A **construct** is a pair (**A**,**U**), consisting of a category **A** and a faithful functor $\mathbf{U} : \mathbf{A} \longrightarrow \mathbf{SET}$, called the **underlying** or **forgetful** functor of (**A**,**U**). To avoid uninteresting pathologies, we will assume throughout this paper that the fibre of the empty set has at most one member.

(2) A construct (**A**,**U**) is called **topological** provided every structured source $(X \xrightarrow{f_i} \mathbf{U}A_i)_{i\in I}$ has a unique initial lift $(A \xrightarrow{f_i} A_i)_{i\in I}$.

PROPERTIES OF TOPOLOGICAL CONSTRUCTS 1.2. For topological constructs (**A**,**U**) the following hold:

(1) [**Topological Duality Theorem**] Every structured sink $(\mathbf{U}A_i \xrightarrow{f_i} X)_{i\in I}$ has a unique initial lift $(A_i \xrightarrow{f_i} A)_{i\in I}$.
(2) For every set X the **fibre** of X (i.e., $U^{-1}(id_X)$) is a (possibly large) complete lattice. Its smallest member is called **discrete**, it largest member **indiscrete**.
(3) **A** is complete and cocomplete, and **U** lifts and preserves limits and colimits (including large ones).
(4) **U** is adjoint (its co-adjoint being the discrete functor) and co-adjoint (its adjoint being the indiscrete functor).
(5) In **A** every object A with $\mathbf{U}A \neq \emptyset$ is a separator; every indiscrete object A, such that $\mathbf{U}A$ has at least 2 members, is a coseparator.
(6) **U** preserves and reflects mono-sources and epi-sinks.
(7) For morphisms in **A** the following equivalences hold:
    (a) isomorphism ⇔ initial and bijective ⇔ final and bijective,
    (b) regular monomorphism ⇔ extremal monomorphism ⇔ **embedding** (i.e., initial and injective),
    (c) regular epimorphism ⇔ extremal epimorphism ⇔ **quotient map** (i.e., final and surjective).
(8) **A** has the following factorization structures for sources
    (a) (bimorphism, initial source),
    (b) (epimorphism, initial mono-source),
    (c) (quotient map, mono-source),
    and the following factorization structures for sinks:
    (d) (final sink, bimorphism),
    (e) (final epi-sink, monomorphism),
    (f) (epi-sink, embedding).
(9) **A** is extremally well powered and extremally co-wellpowered.
(10) Equivalent are:
    (a) **A** is wellpowered,
    (b) **A** is co-wellpowered,

(c) (**A**,**U**) is fibre-small.

EXAMPLES OF TOPOLOGICAL CONSTRUCTS 1.3.

(1) For every functor $\mathbf{T} : \mathbf{SET} \longrightarrow \mathbf{SET}$ the construction **SPA(T)**, whose objects are all **T-spaces** (i.e., pairs (X, α) with X a set and α a subset of $\mathbf{T}X$) and whose morphisms are all **T-maps** (i.e., maps $(X, \alpha) \xrightarrow{f} (Y, \beta)$ with $\mathbf{T}f[\alpha] \subset \beta$), is topological. Constructs of the form **SPA(T)** are called **functor-structured**.

(2) For every functor $\mathbf{T}$: $\mathbf{SET}^{op} \longrightarrow \mathbf{SET}$ the construct $\mathbf{SPA}^{op}(\mathbf{T})$, whose objects are all **T**-spaces and whose morphisms are all maps $(X, \alpha) \xrightarrow{f} (Y, \beta)$ with $Tf[\beta] \subset a$, is topological. Constructs of the form $\mathbf{SPA}^{op}(\mathbf{T})$ are called **functor-costructured**.

(3) For every small construct (**A**,**U**) define functors
   (a) $\mathbf{T}_{(\mathbf{A},\mathbf{U})} : \mathbf{SET} \longrightarrow \mathbf{SET}$, which sends each set X to the set $\mathbf{T}_{(\mathbf{A},\mathbf{U})}X$ of all pairs (A, a) consisting of an **A**-object A and a function $a : UA \longrightarrow X$, and each function $f : X \longrightarrow Y$ to the function $\mathbf{T}_{(\mathbf{A},\mathbf{U})}f : \mathbf{T}_{(\mathbf{A},\mathbf{U})}X \longrightarrow \mathbf{T}_{(\mathbf{A},\mathbf{U})}Y$, defined by $\mathbf{T}_{(\mathbf{A},\mathbf{U})}f(A, a) = (A, f \circ a)$.

   (b) $\mathbf{T}^{(\mathbf{A},\mathbf{U})} : \mathbf{SET}^{op} \longrightarrow \mathbf{SET}$, which sends each set X to the set $\mathbf{T}^{(\mathbf{A},\mathbf{U})}X$ of all pairs (a, A) consisting of an **A**-object A and a function $a : X \longrightarrow \mathbf{U}A$, and each function $f : X \longrightarrow Y$ to the function $\mathbf{T}^{(\mathbf{A},\mathbf{U})}f : \mathbf{T}^{(\mathbf{A},\mathbf{U})}Y \longrightarrow \mathbf{T}^{(\mathbf{A},\mathbf{U})}$, defined by $\mathbf{T}^{(\mathbf{A},\mathbf{U})}f(a, A) = (a \circ f, A)$.

   Then the constructions $\mathbf{REL(A,U)} = \mathbf{SPA}(\mathbf{T}_{(\mathbf{A},\mathbf{U})})$ and $\mathbf{COREL(A,U)} = \mathbf{SPA}^{op}(\mathbf{T}^{(\mathbf{A},\mathbf{U})})$ are topological. Constructs of the form **REL(A,U)**, resp. **COREL(A, U)**. are called **relational**, resp. **corelational**.

(4) For every algebra A of a certain type the construct **TOP**(A), whose objects are all A-**spaces** (i.e., pairs (X, α) with X a set and α a subalgebra of the product algebra A^X) and whose morphisms are all A-**continuous** maps (i.e., maps $(X, \alpha) \xrightarrow{f} (Y, \beta)$ such that $g \in \beta$ implies $(g \circ f) \in \alpha$), is topological. In particular if
   (a) A is a frame with precisely 2 elements, then **TOP**(A) is the construct **TOP** of topological spaces,
   (b) A is the frame $[0, 1]$, then **TOP**(A) is the construct **FTOP** of fuzzy topological spaces,
   (c) A is an arbitrary frame L, then **TOP**(A) is the construct **L-TOP** of L-fuzzy topological spaces.

(5) Each of the following familiar constructs is topological:
   TOP (topological spaces)
   CONV (convergence spaces)
   PRTOP (pretopological spaces)
   PSTOP (pseudotopological spaces)
   FTOP (fuzzy topological spaces)
   FLIM (fuzzy limit spaces)
   EPITOP (epitopological spaces)
   UNIF (uniform spaces)
   NEAR (nearness spaces)

MER (merotopic spaces)
CHY (Cauchy spaces)
DIST (distance spaces and non-expansive maps)
PMEL (pseudometric spaces and non-expansive maps)
AP (approach spaces)
CAP (convergence-approach spaces)
REL ((binary)relations)
RERE (reflexive relations)
ERERE (essentially reflexive relations, i.e., $x\rho y$ implies $x\rho x$ and $y\rho y$)
SYM (symmetric relations)
TRANS (transitive relations)
PROST (preordered sets, i.e., transitive, reflexive relations).

(6) None of the following familiar constructs is topological:
TOP$_0$ (T_0-spaces)
HUNIF (Hausdorff uniform spaces)
MET (metric spaces and non-expansive maps)
POS (partially ordered sets).

§2. Subconstructs and Topological Axioms

Topological constructs are often obtained in a 2-step process, namely as nicely embedded subconstructs of some (usually simpler) topological constructs. In fact (see below) topological constructs are (up to isomorphism) precisely those constructs, which can be defined as subconstructs of functor-structured constructs by means of topological axioms.

DEFINITION 2.1.

Let (**A**,**U**) be a topological construct.

(1) A **subconstruct** of (**A**,**U**) is a construct (**B**,**V**), where **B** is a full, isomorphism-closed subcategory of **A** and **V** is the restriction of **U** to **B**.
(2) A **topological axiom** in (**A**,**U**) (i.e. an **A**-morphism $A \xrightarrow{h} B$ with $\mathbf{U}h$ being an identity map) **holds** for an **A**-object C provided that $\mathbf{U}[\hom_{\mathbf{A}}(\mathrm{A},\mathrm{C})] = \mathbf{U}[\hom_{\mathbf{A}}(B,C)]$.
(3) A subconstruct (**B**,**V**) of (**A**,**U**) is said to be **definable by topological axioms**, provided that there exists a class $\mathcal{A}$ of topological axioms in (**A**,**U**) such that an object A belongs to **B** if and only if every topological axiom in $\mathcal{A}$ holds for A.

THEOREM 2.2.

For subconstructs (**B**,**V**) of topological constructs (**A**,**U**) the following conditions are equivalent:

(1) (**B**,**V**) is definable by topological axioms in (**A**,**U**),

(2) **B** is bireflective in **A**,
(3) **B** is epireflective in **A** and contains all indiscrete objects in (**A**,**U**),
(4) **B** is closed under the formation of initial sources in (**A**,**U**),
(5) **B** is closed under the formation of limits and indiscrete objects in (**A**,**U**).

In case, (**A**,**U**) is fibre-small, these conditions are equivalent to:

(6) **B** is closed under the formation of products, concrete subobjects and indiscrete objects in (**A**,**U**).

PROPOSITION 2.3. Every bireflective (and every bicoreflective) subconstruct of a topological construct is topological.

Vice versa: every topological subconstruct of a topological construct can be obtained in this way:

THEOREM 2.4. For subconstructs (**B**,**V**) of topological constructs (**A**,**U**) the following conditions are equivalent:

(1) (**B**,**V**) is topological,
(2) there exists a subconstruct (**C**,**W**) of (**A**,**U**) such that **B** is bireflective in **C** and **C** is bicoreflective in **A**,
(3) there exists a subconstruct (**C**,**W**) of (**A**,**U**) such that **B** is bicoreflective in **C**, and **C** is bireflective in **A**,
(4) there exists a **concrete functor** (**A**,**U**) $\xrightarrow{\mathbf{R}}$ (**B**,**V**) (i.e., a functor (**A** $\xrightarrow{\mathbf{R}}$ **B** with **U** = **V** ∘ **R**) that leaves each **A** object fixed.

TOPOLOGICAL STRUCTURE THEOREM 2.5. For fibre-small constructs (**A**,**U**) the following conditions are equivalent:

(1) (**A**,**U**) is topological,
(2) (**A**,**U**) is concretely isomorphic to a subconstruct of a functor-structured construct **SPA(T)** that is definable by topological axioms in **SPA(T)**,
(3) (**A**,**U**) is concretely isomorphic to a subconstruct of a functor-costructured construct $\mathbf{SPA}^{op}(\mathbf{T})$ that is definable by topological coaxioms in $\mathbf{SPA}^{op}(\mathbf{T})$.

EXAMPLES 2.6.

(1) **PROST** is definable by the following two topological axioms in **REL**. [Observe that **REL** is a relational, hence functor-structured construct]:

(A1) 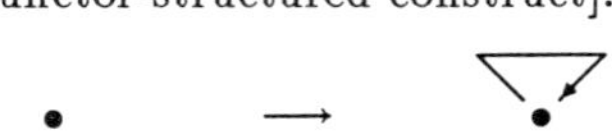

(A2) 

(2) **SYM** is definable in **REL** alternatively:
— by the single topological axiom:

$$\bullet \to \bullet \quad \longrightarrow \quad \bullet \leftrightarrow \bullet$$

— or by the single topological coaxiom:

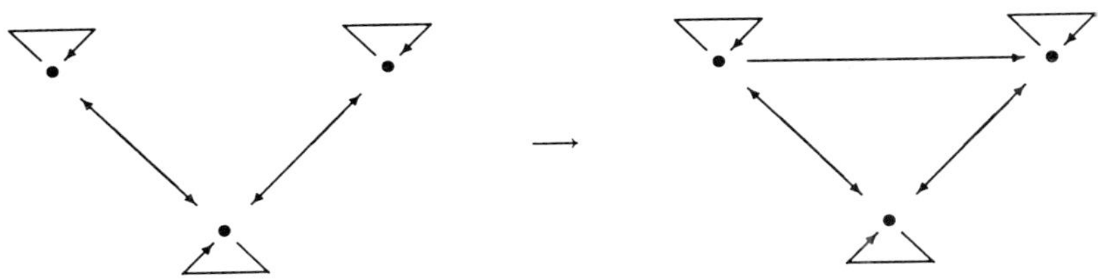

(3) **TOP** is definable by the following (proper) class of topological coaxioms in the functor-costructured construct $\mathbf{SPA}^{op}(\mathbf{P})$, where $\mathbf{P} : \mathbf{SET}^{op} \longrightarrow \mathbf{SET}$ is the contravariant power-set functor:

$$(\{0\}, \{\emptyset, \{0\}\}) \longrightarrow (\{0\}, \emptyset) \tag{C1}$$

$$(\{0,1,2,3\}, \{\ \{0\}, \{0,1\}, \{0,2\}\ \}) \longrightarrow (\{0,1,2,3\}, \{\ \{0,1\}, \{0,2\}\ \}) \tag{C2}$$

and for each set X and each subset $\mathcal{A}$ of $\mathbf{P}X$:

$$(\mathrm{C}_{(X,\mathcal{A})})\ (X, A \cup \{\bigcup \mathcal{A}\}) \to (X, \mathcal{A}).$$

Obviously a $\mathbf{SPA}^{op}(\mathbf{P})$-object $(X, \mathcal{T})$ satisfies:

$$(\mathrm{C1}) \text{ if and only if } \emptyset \in \mathcal{T} \text{and } X \in \mathcal{T},$$

$$(\mathrm{C2}) \text{ if and only if } \{A, B\} \subset \mathcal{T} \text{ implies } (A \cap B) \in \mathcal{T},$$

$$(\mathrm{C}_{(X,\mathcal{A})}) \text{ for all pairs } (X, \mathcal{A}) \text{ if and only if } \mathcal{L} \subset \mathcal{T} \text{ implies } (\bigcup \mathcal{L}) \in \mathcal{T}.$$

However, **TOP** is not definable by a set of topological (co)axioms in any functor-(co) structured construct. See [Adámek 1986b] and [Adámek and Reitermann 1986].

§3. Convenient Topological Constructs

As indicated in 1.2, topological constructs are in many respects very well behaved. However **SET** has several additional pleasant features, which a topological category needn't have (e.g., cartesian closedness). Surprisingly such additional convenient features can often be guaranteed by (in fact, among topological constructs, are equivalent to) the requirement that certain conglomerates of final sinks are stable under the formation of pointwise pullbacks along certain kinds of morphisms. This suggests the investigation of the following properties:

DEFINITION 3.1. In a topological construct $(\mathbf{A},\mathbf{U})$ consider the following classes of morphisms:

M1, the class of all $\mathbf{A}$-morphisms,
M2, the class of all projections $A \times B \xrightarrow{\pi_A} A$ in $\mathbf{A}$,
M3, the class of all embeddings in $(\mathbf{A},\mathbf{U})$,
and the following conglomerates of sinks:
S1, the conglomerate of all final sinks,
S2, the conglomerate of all final epi-sinks,
S3, the conglomerate of all colimits,
S4, the conglomerate of all cointersections,
S5, the conglomerate of all coproducts,
S6, the conglomerate of all singleton final epi-sinks (= quotient maps),
S7, the conglomerate of all empty final sinks (= discrete objects).

Then $(\mathbf{A},\mathbf{U})$ has the property (Mi,Sj) provided that S_j is stable under the formation of pointwise pullbacks along morphisms from M_i. A topological construct, satisfying (M1,S1), is called **universally topological**.

REMARKS 3.2.

(1) Pullbacks of f along embeddings are just suitable restrictions of f.
(2) Pullbacks of $C \xrightarrow{f} A$ along a projection $A \times B \xrightarrow{\pi_A} A$ have the form $C \times B \xrightarrow{f \times id_B} A \times B$.

RELATIONS BETWEEN THE PROPERTIES (Mi,Sj) 3.3. In topological constructs $(\mathbf{A},\mathbf{U})$ the following hold:

(1) For each j the following conditions are equivalent:
   (a) $(\mathbf{A},\mathbf{U})$ satisfies (M1,Sj),
   (b) $(\mathbf{A},\mathbf{U})$ satisfies (M2,Sj) and (M3,Sj).
(2) For each i the following conditions are equivalent:
   (a) $(\mathbf{A},\mathbf{U})$ satisfies (Mi,S1),
   (b) $(\mathbf{A},\mathbf{U})$ satisfies (Mi,S2) and (Mi,S7).
(3) For each i the following conditions are equivalent:
   (a) $(\mathbf{A},\mathbf{U})$ satisfies (Mi,S2),
   (b) $(\mathbf{A},\mathbf{U})$ satisfies (Mi,S3),
   (c) $(\mathbf{A},\mathbf{U})$ satisfies (Mi,S4), (Mi,S5), (Mi,S6).
   In case $(\mathbf{A},\mathbf{U})$ is fibre-small, these conditions are equivalent to:
   (d) $(\mathbf{A},\mathbf{U})$ satisfies (Mi,S5) and (Mi,S6).
(4) $(\mathbf{A},\mathbf{U})$ satisfies (M3,S7).

EXAMPLES 3.4.

(1) **REL** satisfies (M1,S1), hence all (Mi,Sj).
(2) **RERE** satisfies (M1,S2) (hence all (Mi,Sj) with $2 \leq j \leq 6$, and also (M3,S7), but not (M1,S1) (not even (M2,S7)).
(3) **PROST** satisfies (M2,S2), but not (M3,S2) (not even (M3,S6)).
(4) **PRTOP** satisfies (M3,S2) (even (M3,S1)), but not (M2,S2) (not even (M2,S6)).
(5) **TOP** satisfies (M2,S5), but not (M2,S6).
(6) **UNIF** satisfies (M2,S6), but not (M2,S5).

3.1. CARTESIAN CLOSEDNESS

DEFINITION 3.5. A category **A** is called **cartesian closed** provided it satisfies the following two conditions:

(1) **A** has finite products (in particular a terminal object),
(2) for each object A the functor $(A \times \_) : \mathbf{A} \longrightarrow \mathbf{A}$ is co-adjoint, i.e., for each object B there exist an object B^A (called **power object**) and a morphisms $ev : A \times B^A \longrightarrow B$ (called **evaluation morphism**) with the following property: for each object C and each morphism $f : A \times C \longrightarrow B$ there exists a unique morphism $f^* : C \longrightarrow B^A$ such that $f = ev \circ (id_A \times f^*)$.

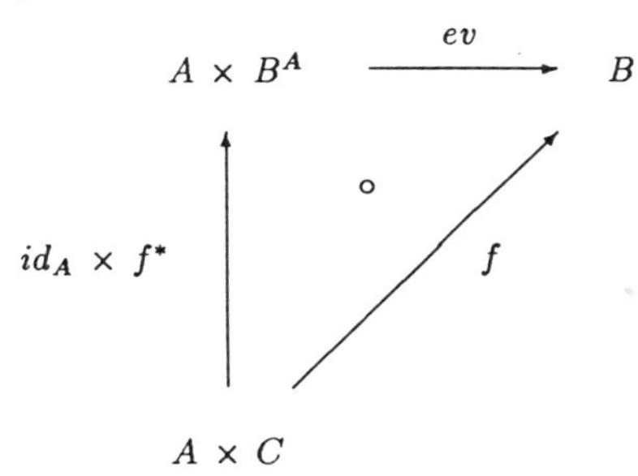

PROPERTIES OF CARTESIAN CLOSED TOPOLOGICAL CONSTRUCTS 3.6. In cartesian closed topological constructs **(A,U)** the following hold:

(1) **First Exponential Law**: $A^{B\times C} \cong (A^B)^C$.
(2) **Second Exponential Law**: $(\prod A_i)^B \cong \prod(A_i^B)$.
(3) **Third Exponential Law**: $A^{\coprod B_i} \cong \prod(A^{B_i})$.
(4) **Distributive Law**: $A \times \coprod B_i \cong \coprod(A \times B_i)$.
(5) Products with discrete factors A are coproducts, i.e.,
$$A \times B \cong \coprod_{x \in \mathbf{U}A} B = {}^{\mathbf{U}A}B.$$
(6) Power objects with discrete exponents A are products, i.e.,
$$B^A \cong \coprod_{x \in \mathbf{U}A} B = B^{\mathbf{U}A}.$$

(3) (**A**,**U**) is cartesian closed and fibres of one-element-sets contain each precisely one member.

3.2. Representability of Partial Morphisms

Definition 3.13. Let (**A**,**U**) be a construct.

(1) In (**A**,**U**) a **partial morphism** form A to B is a morphism from a concrete subobject of A to B.
(2) An embedding $B \overset{m}{\hookrightarrow} B^*$ **represents partial morphisms** into B provided for every partial morphism f form A into B there exists a unique morphism $f^* : A \longrightarrow B^*$ such that f is the corresponding restriction of f^*.

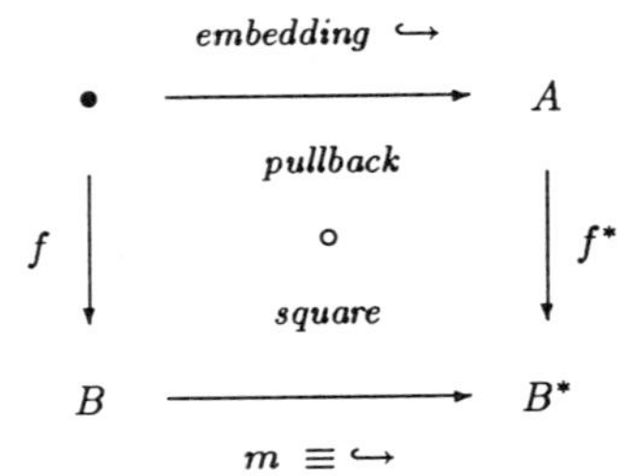

(3) (**A**,**U**) **has representable partial morphisms** provided that for every object B there exists an embedding $B \overset{m}{\hookrightarrow} B^*$ that represents partial morphisms into B.

Examples 3.14.

(1) The construct **SET** has representable partial morphisms. For each set B the one-point extension $B \hookrightarrow B \dot{\cup} \{\infty\}$ (where $\dot{\cup}$ denotes the disjoint union) represents partial morphisms into B.
(2) **REL** has representable partial morphisms. For each object $B = (X, \rho)$ the one-point-extension $B \hookrightarrow B^* = (X^*, \rho^*)$ with $X^* = X \dot{\cup} \{\infty\}$ and $\rho^* = \rho \cup (X^* \times \{\infty\}) \cup (\{\infty\} \times X^*)$ represents partial morphisms into B.
(3) **PRTOP**, **PSTOP**, **MER**, **SYM**, and **RERE** have representable partial morphisms.
(4) None of the following constructs has representable partial morphisms: **TOP**, **UNIF**, **NEAR**, **CHY**, **PMET**, **AP**, **TRANS**, and **PROST**.

Proposition 3.15. If a topological construct has representable partial morphisms, then the embeddings which represent partial morphisms into B are one-point extensions of B.

Characterization Theorem for Topological Constructs that have Representable Partial Morphisms 3.16. For topological constructs (**A**,**U**) the following conditions are equivalent:

(1) (**A**,**U**) has representable partial morphisms,

(2) (**A**,**U**) satisfies (M3,S1),
(3) (**A**,**U**) satisfies (M3,S2).

PROPOSITION 3.17. If (**A**,**U**) is a topological construct that has representable partial morphisms, then the following hold:

(1) An object A is injective if and only if there exist some embedding $B \hookrightarrow A$ that represents partial morphisms into A.
(2) Every object has an injective hull. If A is not injective, then an embedding $A \hookrightarrow B$ is an injective hull of A if and only if it represents partial morphisms into A.

3.3. Quasitopi and Topological Universes

DEFINITIONS 3.18. A topological construct (**A**,**U**) is called a **(concrete) topological quasitopos** provided it satisfies the following two conditions:

(1) (**A**,**U**) is (concretely) cartesian closed,
(2) (**A**,**U**) has representable partial morphisms.

A **well-fibred** (i.e., all fibres are small and fibres of one-element sets contain only one member) topological quasitopos is called a **topological universe.**

CHARACTERIZATION THEOREMS 3.19.

(1) A topological construct is a topological quasitopos if and only if it satisfies (M1,S2).
(2) A topological construct is a concrete topological quasitopos if and only if it satisfies (M1,S1), i.e., if and only if it is universally topological.
(3) A well-fibred topological construct is a topological universe if and only if it satisfies (M1,S2).

EXAMPLES 3.20.

(1) **REL**, **SYM**, and **ERERE** are universally topological constructs.
(2) **PSTOP**, **CONV**, **DIST**, **CAP**, and **RERE** are topological universes.
(3) None of the remaining constructs, mentioned in 1.3(5), is a quasitopos.

§4. Improvements of Constructs

Constructs may fail to have some of the properties, discussed in Sections 1 and 3. In such cases one might try to find better behaved subconstructs. Unfortunately, however, this approach often fails, as the following examples indicate:

EXAMPLES 4.0.

(1) $\mathbf{TOP}_0$ has only one topological subconstruct: it consists of all discrete spaces. Hence it is uninteresting.
(2) **MET** has infinitely many topological subconstructs. However each of these contains only metric spaces whose underlying topologies are discrete; thus none of them is very interesting.
(3) **TOP** has only two topological subconstructs, which have representable partial morphisms. They consist of all discrete spaces, respectively, of all indiscrete spaces. Hence they are uninteresting [Herrlich 1983a].

Surprisingly, however, under rather natural assumptions, any "not too big" construct can be concretely embedded into a topological quasitopos. Taking into account the following two rules:

(BB) **Bigger is Better**, i.e., stronger convenience stipulations require bigger extensions,
(SB) **Smaller is Better**, i.e. smaller extensions preserve more structure of the original construct,

it seems desirable to find, for a given construct (**A**,**U**) and a given convenience requirement φ, a smallest extension of (**A**,**U**) which satisfies φ. Such an extension will be called a φ-hull of (**A**,**U**). In the following we will briefly discuss

(1) topological hulls (= MacNeille completions [Herrlich 1976]),
(2) cartesian closed topological hulls ([Antoine 1966] and [Herrlich and Nel 1977]) and concretely cartesian closed topological hulls ([Strecker 1984]),
(3) extensional topological hulls ([Herrlich 1988]),
(4) topological universe hulls ([Wyler 1976]) and universally topological hulls ([Adámek 1986a]).

4.1. TOPOLOGICAL HULLS

DEFINITION 4.1. A full concrete embedding **E**: (**A**,**U**) $\longrightarrow$ (**B**,**V**) is called an **extension** of (**A**,**U**).

DEFINITION 4.2. An extension **E**: (**A**,**U**) $\longrightarrow$ (**B**,**V**) is called a **topological hull** of (**A**,**U**) provided that the following 3 conditions hold:

(1) (**A**,**U**) is a topological construct,
(2) **E**(Ob**A**) is finally dense in (**B**,**V**),
(3) **E**(Ob**A**) is initially dense in (**B**,**V**).

REMARKS 4.3.

(1) Topological hulls can be characterized by any of the following conditions:
   (a) as smallest topological extensions,

(b) as largest finally and initially dense extensions,
(c) as injective hulls in the quasicategory of constructs.

(2) Topological hulls preserve initial sources and final sinks.
(3) A topological hull of (**A**,**U**) exists, if and only if (**A**,**U**) has some topological extension. For other sufficient (and necessary) conditions see [Adámek, Herrlich, and Strecker 1979a]. If a topological hull exists, it is uniquely determined up to concrete isomorphisms.

EXAMPLES 4.4.

(1) **TOP** is a topological hull to $\mathbf{TOP}_0$.
(2) **UNIF** is a topological hull of **HUNIF**.
(3) **PMET** is a topological hull of **MET**.
(4) **PROST** is a topological hull of **POS**.

4.2. CARTESIAN CLOSED TOPOLOGICAL HULLS

We may start, for simplicity, with a topological construct. Depending on whether this is well-fibred or not, the following two concepts arise naturally:

DEFINITION 4.5. An extension **E**: (**A**,**U**) $\longrightarrow$ (**B**,**V**) of a well-fibred topological construct is called a **cartesian closed topological hull** of (**A**,**U**) provided that the following three conditions are satisfied:

(1) (**B**,**V**) is a well-fibred cartesian closed topological construct,
(2) **E**(Ob**A**) is finally epi-dense in (**B**,**V**),
(3) $\{\mathbf{E}A^{\mathbf{E}B} \mid A \in \mathrm{Ob}\mathbf{A} \text{ and } B \in \mathrm{Ob}\mathbf{A}\}$ is initially dense in (**B**,**V**).

DEFINITION 4.6. An extension **E**: (**A**,**U**) $\longrightarrow$ (**B**,**V**) of a topological construct is called a **concretely cartesian closed topological hull** of (**A**,**U**) provided that the following three conditions are satisfied:

(1) (**B**,**V**) is a concretely cartesian closed topological construct,
(2) **E**(Ob**A**) is finally dense in (**B**,**V**)
(3) $\{\mathbf{E}A^{\mathbf{E}B} \mid A \in \mathrm{Ob}\mathbf{A} \text{ and } B \in \mathrm{Ob}\mathbf{A}\}$ is initially dense in (**B**,**V**).

REMARKS 4.7. For (concretely) cartesian closed topological hulls, remarks similar to 4.3 apply. For details concerning cartesian closed topological hulls, see [Adámek, Reiterman, and Strecker 1985]. For details, concerning concretely cartesian closed topological hulls, see [Herrlich and Strecker 1986].

EXAMPLES 4.8.

(1) **EPITOP** is a cartesian closed topological hull of **TOP**.
PSTOP is a cartesian closed topological hull of **PRTOP**.

DIST is a cartesian closed topological hull of **PMET**.

(2) **REL** is a concretely cartesian closed topological hull of **TRANS**. (Observe that if $A = (\{0,1\},\{(0,1)\})$ and $B = (\{0,1\},\{(0,0),(0,1),(1,1)\})$, then A is finally dense in **REL** and $B^A = (X,\rho)$ with $X = \{c_0, c_1, id, s\}$ and $\rho = X^2 \setminus (\{c_1, s\} \times \{c_0, s\})$ is initially dense in **REL**).

(3) **PROST** is a cartesian closed (but not concretely cartesian closed) topological construct. **ERERE** is a concretely cartesian closed topological hull of **PROST**.

4.3. Extensional Topological Hulls

Definition 4.9. An extension **E: (A,U)** $\longrightarrow$ **(B,V)** of a topological construct is called an **extensional topological hull** of **(A,U)** provided the following three conditions are satisfied:

(1) **(B,V)** is topological and has representable partial morphisms,
(2) **E**(Ob**A**) is finally dense in **(B,V)**
(3) $\{A^* \mid A \in$ Ob**A** and **E**$A \longrightarrow A^*$ represents partial morphisms into **E**$A\}$ is initially dense in (**B**,V).

Remarks 4.10.

(1) If **E: (A,U)** $\longrightarrow$ **(B,V)** is a extensional topological hull of **(A,U)**, then **E**(Ob**A**) is finally epi-dense in **(B,V)**.
(2) Every fibre-small topological construct has a fibre-small extensional topological hull.

Examples 4.11.

(1) **REL** is an extensional topological hull of **TRANS**. (Observe that if $A = (\{0\},\emptyset)$, $B = (\{0,1\},\{(0,0),(0,1),(1,1)\})$, $A^* = (\{0,\infty\},\{(0,\infty),(\infty,0),(\infty,\infty)\})$, and $B^* = (X,\rho)$ with $X = \{0,1,\infty\}$ and $\rho = X^2 \setminus \{(1,0)\}$, then the inclusions $A \hookrightarrow A^*$ and $B \hookrightarrow B^*$ represent in **REL** partial morphisms into A resp. into B, and $\{A^*, B^*\}$ is initially dense in **REL**).
(2) **RERE** is an extensional topological hull of **PROST**.
(3) **PRTOP** is an extensional topological hull of **TOP**.

4.4. Universally Topological and Topological Universe Hulls

Definition 4.12. An extension **E: (A,U)** $\longrightarrow$ **(B,V)** of a well-fibred topological construct is called a **topological universe hull** of **(A,U)** provided that the following three conditions are satisfied:

(1) **(B,V)** is a topological universe,
(2) **E**(Ob**A**) is finally epi-dense in **(B,V)**,
(3) $\{A^{*B} \mid A \in$ Ob**A**, $B \in$ Ob**A**, and **E**$A \longrightarrow A^*$ represents partial morphisms into $EA\}$ is initially dense in **(B,V)**.

DEFINITION 4.13. An extension **E**: (**A**,**U**) $\longrightarrow$ (**B**,**V**) of a topological construct is called a **universally topological hull** of (**A**,**U**) provided that the following three conditions are satisfied:

(1) (**B**,**V**) is universally topological,
(2) **E**(Ob**A**) is finally dense in (**B**,**V**),
(3) $\{A^{*B} \mid A \in \text{Ob}\mathbf{A},\ B \in \text{Ob}\mathbf{A}$ and $\mathbf{E}A \longrightarrow A^*$ represents partial morphisms into $\mathbf{E}A\}$ is initially dense in (**B**,**V**).

REMARK 4.14. Topological universe hulls, provided they exist, can be constructed in two steps: first form an extensional topological hull **E**: (**A**,**U**) $\longrightarrow$ (**B**,**V**), then form a cartesian closed topological hull $\overline{\mathbf{E}}$: (**B**,**V**) $\longrightarrow$ (**C**,**W**). The composite $\overline{\mathbf{E}} \circ \mathbf{E}$: (**A**,**U**) $\longrightarrow$ (**C**,**W**) is a topological universe hull of (**A**,**U**). However, if one interchanges the order of the above two steps, the resulting category (**C**,**W**) may fail to be cartesian closed, and thus may fail to provide a topological universe hull of (**A**,**U**); see [Schwarz 1989]. A similar remark applies to universally topological hulls.

EXAMPLES 4.15.

(1) **PSTOP** is a topological universe hull of **TOP**.
(2) **DIST** is a topological universe hull of **PMET**.
(3) **RERE** is a topological universe hull of **PROST**.
(4) **ERERE** is a universally topological hull of **RERE** as well as of **PROST**.
(5) **REL** is a universally topological hull of **TRANS**.

§5. Problems

Determine for the construct **FTOP**:

(1) its cartesian closed topological hull (cf. [Min 1989]),
(2) its extensional topological hull,
(3) its topological universe hull.

Chapter 6:

A Topological Universe Extension of **FTS**

S. E. Rodabaugh et al. (eds.), Applications of Category Theory to Fuzzy Subsets, 153–176.

A TOPOLOGICAL UNIVERSE EXTENSION OF FTS

E. Lowen
R. Lowen

§0. Introduction

When building a nice theory of convergence suitable for applications in fuzzy set theory, one usually starts within the category **FTS** of fuzzy topological spaces or looks for subcategories with an even richer structure, such as the category of fuzzy neighborhood spaces or the category of fuzzy uniformizable spaces [R. Lowen 1981a, R. Lowen 1982] where convergence can easily be described.

For certain applications in Algebraic Topology or Functional Analysis these categories are not very convenient, the main reason being the fact that they are not cartesian closed. How important cartesian closedness is for applications is clearly demonstrated in the following papers: [Herrlich 1974, Dubuc 1979, Penon 1977].

In this paper, starting from convergence theory in fuzzy topological spaces as developed in [R. Lowen 1979] and [E. Lowen and R. Lowen 1985], we will slightly weaken the axioms of fuzzy topological convergence in order to obtain the nice cartesian closed topological category **FCS** of fuzzy convergence spaces containing **FTS** as fully embedded subcategory.

We will prove that the category of fuzzy convergence spaces is extremely nice to work in since it is even a topological quasi topos as introduced by Dubuc and Porta. Apart from being cartesian closed, in a topological quasi topos coproducts and quotients are hereditary. This property has proven to be very convenient for the development of a generalized connectedness theory [Salicrup 1988, Salicrup and Vasquez 1972].

Apart from the category **FTS** which is bireflectively embedded in **FCS**, the category of fuzzy convergence spaces contains other well-known categories as full subcategories. For example the category **PSCONV** of pseudotopological convergence spaces is bireflectively and bicoreflectively embedded. Whereas in a pseudotopological convergence space every filter has a limit set, i.e. the set of its convergence points, in a fuzzy convergence space every prefilter has a limit function, i.e. a function giving in each point the degree with which the given prefilter converges to the point.

Several examples of fuzzy convergence spaces arise naturally. We will show that a particular type of spaces studied in [Stout 1989] is fuzzy convergence spaces. For examples of fuzzy convergence in **FTS** we refer to [R. Lowen 1985].

§1. Notations

Recall that $I = [0,1]$ and that if X is a set, I^X denotes the set of all functions $X \longrightarrow I$, i.e. all fuzzy sets on X, equipped with the usual lattice structure. As such I^X is a complete and completely distributive lattice. If X is a set and $A \subset X$, then 1_A denotes the characteristic function of A, and if $A = \{x\}$, then we write 1_x for $1_{\{x\}}$. By $\alpha 1_x$ we mean the usual product of the constant function α with 1_x.

For filter theory in the lattice I^X we use the terminology as introduced by [Bourbaki 1965].

A prefilter (resp. prime prefilter) on X is a prefilter (resp. prime prefilter) in the lattice I^X. If $\mu \in I^X$ then we denote by $\dot{\mu}$ the principal prefilter $\{\nu \mid \nu \geq \mu\}$ and by $\hat{\mu}$ the prefilter $\{\nu \mid \forall x \in \mu^{-1}(]0,1]) : \nu(x) > \mu(x)\}$. When $\alpha \in I$, the prefilters $\alpha \dot{1}_x$ and $\alpha \hat{1}_x$ are prime. The set $\mathcal{P}(X)$ of all prefilters on X is ordered by inclusion, the important subcollection of $\mathcal{P}(X)$ consisting of all prime prefilters shall be denoted $\mathcal{P}_p(X)$. Given $\mathcal{F} \in \mathcal{P}(x)$, the set of all prime prefilters finer than $\mathcal{F}$ was shown to have minimal elements [R. Lowen 1979] and this latter collection is denoted $\mathcal{P}_m(\mathcal{F})$. This collection can be completely characterized by means of ultrafilters on X [R. Lowen 1979, E. Lowen and R. Lowen 1985].
When $\mathcal{F}$ is a prime prefilter we denote

$$c(\mathcal{F}) := inf\{\alpha \mid \alpha \in \mathcal{F}\}.$$

Here α is considered as both a number in I and a constant function in I^X. For $\mathcal{F} \in \mathcal{P}(X)$ we denote

$$c(\mathcal{F}) := \inf\{\alpha \mid \alpha \in \mathcal{F}\} = \sup_{\mathcal{G} \in \mathcal{P}_m(\mathcal{F})} c(\mathcal{G})$$

$$c^-(\mathcal{F}) = \inf_{\mathcal{G} \in \mathcal{P}_m(\mathcal{F})} c(\mathcal{G})$$

For any prefilter $\mathcal{F}$ we denote $\imath(\mathcal{F}) := \{\mu^{-1}(]0,1]) \mid \mu \in \mathcal{F}\}$. Given a collection $(\mathcal{F}_j)_{j \in \mathcal{J}}$ of prefilters on X and a filter $\mathcal{A}$ on $\mathcal{J}$ we denote the diagonal prefilter as in [E. Lowen and R. Lowen 1985],

$$\mathcal{D}((\mathcal{F}_j)_{j \in \mathcal{J}}, \mathcal{A}) := \bigvee_{A \in \mathcal{A}} \bigcap_{j \in A} \mathcal{F}_j$$

This concept is based on the analogous one in the theory of convergence spaces where it was introduced by Kowalsky in [Kowalsky 1954]. For more details we refer also to [E. Lowen and R. Lowen 1988].

FTS is the category of fuzzy topological spaces and fuzzy continuous maps. Fuzzy topological spaces can be defined by means of the open fuzzy sets, or the closed fuzzy sets or by means of a fuzzy closure operator. These different systems lead to isomorphic categories and we will not distinguish between them.

§2. Preliminaries

A concrete category C is called **topological** iff it satisfies the following conditions:

CAT TOP$_1$: EXISTENCE OF INITIAL STRUCTURES:

For any set X, any family $((X_i, \xi_i))_{i \in I}$ of C-objects indexed by a class I, and any family $(f_i : X \longrightarrow X_i)_{i \in I}$ of maps indexed by I, there exists a unique C-structure ξ on X which is initial with respect to the given source $(f_i : X \longrightarrow (X_i, \xi_i))_{i \in I}$ in the sense that for any C-object (Y, η), a map $g : (Y, \eta) \longrightarrow (X, \xi)$ is a C-morphism iff for every $i \in I$ the map $f_i \circ g : (Y, \eta) \longrightarrow (X_i, \xi_i)$ is a C-morphism.

CAT TOP$_2$: FIBRE-SMALLNESS:

For any set X, the C-fibre of X, i.e. the class of all C-structures on X, is a set.

CAT TOP$_3$: TERMINAL SEPARATOR PROPERTY:

For any set with cardinality one, there exists exactly one C–structure on X.

A topological category is complete and cocomplete, it has products, subspaces, coproducts, and quotients. It is well-powered and cowell-powered, the structures on a fixed set form a complete lattice, and constant maps are morphisms. For a survey on topological categories we refer to [Herrlich 1983b].

The categories **FTS**, **FNS**, or **FUS** (of fuzzy uniform spaces) in the sense of [R. Lowen 1974, 1982, 1981] are examples of topological categories with objects of a fuzzy nature.

A topological category C is **Cartesian closed topological** if for each pair (X, Y) of C-objects the set $Hom(X, Y)$ of all C-morphisms can be endowed with a C-structure such that the following are satisfied:

(1) The evaluation map $ev : X \times Hom(X, Y) \longrightarrow Y$ defined by $ev(x, f) = f(x)$ is a C-morphism.
(2) For each C-object Z and for every C-morphism $f : X \times Z \longrightarrow Y$, the map $f^* : Z \longrightarrow Hom(X, Y)$ defined by $f^*(z)(x) = f(x, z)$ is a C-morphism.

Cartesian closed topological categories have been studied in [Nel, 1976, Herrlich 1974]. One of their nice aspects is that they can also be characterized as topological categories in which coproducts and quotients are preserved under pullbacks along projections. We refer to [Herrlich 1984] for further details.

The categories **FTS** and **FNS** (resp. **FUS**) are not cartesian closed since they contain the non-cartesian closed categories **TOP** (resp. **UNIF**) as bicoreflective subcategories closed under the formation of finite products [Wuyts, Lowen and Lowen 1988; Lowen, Wuyts and Lowen 1989].

A topological category is a **topological quasi topos** if coproducts and quotients are universal, i.e. preserved under pullbacks along arbitrary morphisms. It means that apart from being cartesian closed, the topological category has hereditary coproducts and quotients. These supplementary properties are referred to as being **hereditary** (or **extensional** in the latest version), see e.g. [Herrlich 1988]. Quasi topoi were introduced in

[Dubuc and Porta 1971]. They were also studied in [Nel 1984], where they were referred to as **topological universes**, the terminology to which we shall adhere.

In our investigations the following results will be useful.

LEMMA 2.1. If $f : X \longrightarrow Y$ and $\mathcal{F} \in \mathcal{P}(X)$, then

$$\mathcal{P}_m(f(\mathcal{F})) \subset \{f(\mathcal{G}) \mid \mathcal{G} \in \mathcal{P}_m(\mathcal{F})\}.$$

PROOF. Let $\mathcal{H} \in \mathcal{P}_m(f(\mathcal{F}))$. Suppose that for all $\mathcal{G} \in \mathcal{P}_m(\mathcal{F})$ there exists $\mu_{\mathcal{G}} \in \mathcal{G}$ such that

$$f(\mu_{\mathcal{G}}) \notin \mathcal{H}$$

Then we can find a finite set $\mathcal{P}_0 \subset \mathcal{P}_m(\mathcal{F})$ such that

$$\sup_{\mathcal{G} \in \mathcal{P}_0} \mu_{\mathcal{G}} \in \mathcal{F}$$

and consequently

$$\sup_{\mathcal{G} \in \mathcal{P}_0} f(\mu_{\mathcal{G}}) \in f(\mathcal{F}) \subset \mathcal{H}$$

Since $\mathcal{H}$ is prime there exists $\mathcal{G} \in \mathcal{P}_0$ such that $f(\mu_{\mathcal{G}}) \in \mathcal{H}$, which is a contradiction.

Consequently there does not exist $\mathcal{G} \in \mathcal{P}_m(\mathcal{F})$ such that $f(\mathcal{G}) \subset \mathcal{H}$. Since $f(\mathcal{G})$ is prime and $f(\mathcal{F}) \subset f(\mathcal{G})$, it follows that $f(\mathcal{G}) = \mathcal{H}$ and we are done. □

LEMMA 2.2. If $f : X \longrightarrow Y$ and $\mathcal{F} \in \mathcal{P}_p(Y)$ has a trace on $f(X)$, then $\forall \mathcal{G} \in \mathcal{P}_m(f^{-1}(\mathcal{F}))$, $f(\mathcal{G}) = \mathcal{F}$.

PROOF. Clearly $f(\mathcal{G}) \supset \mathcal{F}$. To show the converse consider the ultrafilter $\mathcal{U} \supset \imath(f^{-1}(\mathcal{F})) = f^{-1}(\imath(\mathcal{F}))$ such that

$$\mathcal{G} = \omega(\mathcal{U}) \vee f^{-1}(\mathcal{F}),$$

and let $1_U \wedge f^{-1}(\mu) \in \mathcal{G}$ where $U \in \mathcal{U}$ and $\mu \in \mathcal{F}$. Then it follows that

$$f(1_U \wedge f^{-1}(\mu)) = 1_{f(U)} \wedge \mu$$

as can easily be verified. Now since $\mathcal{F}$ is prime we obtain

$$f(\mathcal{U}) \supset ff^{-1}(\imath(\mathcal{F})) \supset \imath(\mathcal{F})$$

and thus $f(\mathcal{U}) = \imath(\mathcal{F})$. This means that $1_{f(U)} \in \mathcal{F}$ and consequently too $1_{f(U)} \wedge \mu \in \mathcal{F}$ and we are done. □

LEMMA 2.3. If $(\mathcal{F}_i)_{i=1}^n$ are prefilters then

$$\mathcal{P}_m\left(\bigcap_{i=1}^n \mathcal{F}_i\right) \subset \bigcup_{i=1}^n \mathcal{P}_m(\mathcal{F}_i).$$

PROOF. If $\mathcal{G} \in \mathcal{P}_m(\bigcap_{i=1}^n \mathcal{F}_i)$ and for all $i = 1, \ldots, n$, we have $\mathcal{G} \not\supset \mathcal{F}_i$ then we can find $\nu_i \in \mathcal{F}_i \backslash \mathcal{G}, i = 1, \ldots, n$. Then however $\sup_{i=1}^n \nu_i \in \mathcal{G}$, which is a contradiction. Consequently there exists $i_0 \in \{1, \ldots, n\}$ such that $\mathcal{G} \supset \mathcal{F}_{i_0}$. That $\mathcal{G}$ is minimal is clear. □

LEMMA 2.4. Let $(\mathcal{F}_j)_{j\in J}$ be a family of prefilters on X and $\mathcal{A}$ a filter on J then the following properties hold:

1° $\mathcal{D}((\mathcal{F}_j)_{j\in J}, \mathcal{A}) = \bigcup_{A\in\mathcal{A}} \bigcap_{j\in A} \mathcal{F}_j$

2° If $(\mathcal{A}_1)_{1\in L}$ is a family of filters on J such that $\mathcal{A} = \bigcap_{1\in L} \mathcal{A}_1$, then

$$\mathcal{D}((\mathcal{F}_j)_{j\in J}, \mathcal{A}) = \bigcap_{1\in L} \mathcal{D}((\mathcal{F}_j)_{j\in J}, \mathcal{A}_1)$$

3° $\mathcal{D}((\mathcal{F}_j)_{j\in J}, \mathcal{A}) = \bigcap\{\mathcal{D}((\mathcal{G}_j)_{j\in J}, \mathcal{A}) \mid (\mathcal{G}_j)_{j\in J} \in \sqcap_{j\in J} \mathcal{P}_m(\mathcal{F}_j)\}$.

4° If each $\mathcal{F}_j, j \in J$ is prime and $\mathcal{A}$ is an ultrafilter, then $\mathcal{D}((\mathcal{F}_j)_{j\in J}, \mathcal{A})$ is prime.

PROOF. 1° If $\mu \in \mathcal{D}((\mathcal{F}_j)_{j\in J}, \mathcal{A})$, $A_1, \ldots, A_n \in \mathcal{A}$, and $\nu_i \bigcap_{j\in A_i} \mathcal{F}_j$ $i = 1, \ldots, n$, are such that $\nu_1 \wedge \ldots \wedge \nu_n \leq \mu$, then putting $A := A_1 \cap \ldots \cap A_n$, it follows that $\mu \in \bigcap_{j\in A} \mathcal{F}_j$.

2° One inclusion is trivial, to show the other one let us suppose

$$\nu \in \cap_{1\in L} \mathcal{D}((\mathcal{F}_j)_{j\in J}, \mathcal{A}_1)$$

Then making use of 1° it follows that for all $1 \in L$ there exists $A_1 \in \mathcal{A}_1$ such that $\nu \in \cap_{j\in A_1} \mathcal{F}_j$. If we let $A = \cup_{1\in L} A_1$ then $A \in \mathcal{A}$ and $\nu \in \cap_{1\in A} \mathcal{F}_1$.

3° Again one inclusion is quite clear, to show the other one let $\nu \notin \mathcal{D}((\mathcal{F}_j)_{j\in J}, \mathcal{A})$ and for each $j \in J$, if $\nu \notin \mathcal{F}_j$, choose $\mathcal{G}_j \in \mathcal{P}_m(\mathcal{F}_j)$ such that $\nu \notin \mathcal{G}_j$, whereas if $\nu \in \mathcal{F}_j$, choose $\mathcal{G}_j \in \mathcal{P}_m(\mathcal{F}_j)$ arbitrarily. Now by supposition and upon once again applying 1°, for each $A \in \mathcal{A}$ there is $j \in A$ such that $\nu \notin \mathcal{F}_j$, and thus $\nu \notin \mathcal{G}_j$. Consequently $\nu \notin \mathcal{D}((\mathcal{G}_j)_{j\in J}, \mathcal{A})$ and we are done.

4° Let $\mu \vee \nu \in \mathcal{D}((\mathcal{F}_j)_{j\in J}, \mathcal{A})$ and let $A \in \mathcal{A}$ be such that $\mu \vee \nu \in \bigcap_{j\in A} \mathcal{F}_j$. If we put

$$A_1 := \{j \in A \mid \mu \in \mathcal{F}_j\}$$

$$A_2 := \{j \in A \mid \nu \in \mathcal{F}_j\}$$

then f. i. $A_1 \in \mathcal{A}$, $\mu \in \bigcap_{j\in A_1} \mathcal{F}_j$, and we are done. □

§3. Fuzzy Convergence Structures

For applications in algebraic topology or functional analysis neither **FTS** nor any of its subcategories [R. Lowen 1982, Wuyts, R. Lowen, E. Lowen 1988] are very convenient, the main reason being the fact that they are not cartesian closed.

In the sequel, starting from the convergence theory in fuzzy topological spaces as developed in [E. Lowen and R. Lowen 1985, R. Lowen 1979], we will construct a nice cartesian closed topological category in which **FTS** is embedded as a full isomorphism-closed subcategory.

DEFINITION 3.1. A **fuzzy convergence space** is a pair $(X, \lim)$, where $X \in |\mathbf{SET}|$ and where

$$\lim : \mathcal{P}(X) \longrightarrow I^X : \mathcal{F} \longmapsto \lim \mathcal{F}$$

satisfies the conditions:

(PST) $\forall \mathcal{F} \in \mathcal{P}(X) : \lim \mathcal{F} = \inf_{\mathcal{G} \in \mathcal{P}_m(\mathcal{F})} \lim \mathcal{G}$;
(F1p) $\forall \mathcal{G} \in \mathcal{P}_p(X) : \lim \mathcal{G} \leq c(\mathcal{G})$;
(F2p) $\forall \mathcal{G}_1, \mathcal{G}_2 \in \mathcal{P}_p(X) : \mathcal{G}_1 \subset \mathcal{G}_2 \Rightarrow \lim \mathcal{G}_2 \leq \lim \mathcal{G}_1$;
(cl) $\forall x \in X,\ \forall \alpha \in I \backslash 0 : \lim \alpha \dot{1}_x \geq \alpha 1_x$;

Whenever necessary, to avoid confusion, we may write $\lim_X$ or similar instead of lim.

Whenever we define a fuzzy convergence space we will restrict ourselves to the definition of a limit of prime prefilters and it will be understood that the limit for arbitrary prefilters is derived by (PST).

DEFINITION 3.2. A function $f : (X, \lim_X) \longrightarrow (Y, \lim_Y)$ is called **continuous** if for every $\mathcal{F} \in \mathcal{P}(X)$ and $x \in Y$:

$$\lim_X \mathcal{F}(x) \leq \lim_Y f(\mathcal{F})(f(x))$$

PROPOSITION 3.3. A function $f : (X, \lim_X) \longrightarrow (Y, \lim_Y)$ is continuous if and only if for each $\mathcal{G} \in \mathcal{P}_p(X)$ and $x \in X : \lim_X \mathcal{G}(x) \leq \lim_Y f(\mathcal{G})(f(x))$.

PROOF. This follows immediately from the definitions. □

DEFINITION 3.4. We denote by **FCS** the concrete category with objects all fuzzy convergence spaces and morphisms all continuous maps.

THEOREM 3.5. **FCS** is a topological category.

PROOF. First we verify Cat Top$_1$. Let

$$(f_j : X \longrightarrow (X_j, \lim_j))_{j \in J}$$

be a source. We define a limit function on X in the following way. For $\mathcal{G} \in \mathcal{P}_p(X)$ we put

$$\lim \mathcal{G}(x) := \inf_{j \in J} \lim_j f_j(\mathcal{G})(f_j(x))$$

and for $\mathcal{F} \in \mathcal{P}(X)$ we put

$$\lim \mathcal{F}(x) := \inf_{\mathcal{G} \in \mathcal{P}_m(\mathcal{F})} \lim \mathcal{G}(x).$$

By definition lim satisfies (PST). If $\mathcal{G} \in \mathcal{P}_p(X)$ then

$$\lim \mathcal{G} \leq \inf_{j \in J} c(f_j(\mathcal{G})) = c(\mathcal{G})$$

and thus (F1p) is fulfilled. If $\mathcal{G}_1, \mathcal{G}_2 \in \mathcal{P}_p(X)$ are such that $\mathcal{G}_1 \subset \mathcal{G}_2$, then from the fact that for all $j \in J : f_j(\mathcal{G}_1) \subset f_j(\mathcal{G}_2)$, we immediately obtain (F2p). Finally from the fact that for all $j \in J : f_j(\alpha \dot{1}_x) = \alpha \dot{1}_{f_j(x)}$, we obtain (c1). To see that $(X, \lim)$ is initial for the given source, first notice that $f_j : (X, \lim) \to (X_j, \lim_j)$ is continuous for all $j \in J$. Next, if $(Y, \lim_Y) \in |FCS|$ and $g : Y \to X$ are such that $f_j \circ g$ is continuous for all $j \in J$, then for $\mathcal{G} \in \mathcal{P}_p(Y)$ we have

$$\begin{aligned} \lim_Y \mathcal{G}(y) &\leq \inf_{j \in J} \lim_j f_j(g(\mathcal{G}))(f_j(g(y))) \\ &= \lim_X g(\mathcal{G})(g(y)) \end{aligned}$$

for all $y \in Y$. Thus g is continuous. That lim is unique for the property of being initial follows from the fact that if $id_X : (X, \lim) \to (X, \lim')$ and $id_X : (X, \lim') \to (X, \lim)$ are both continuous then $\lim = \lim'$.

Cat Top$_2$ follows at once from the observation that for any set X with cardinality κ the number of limit functions on X is limited by $c^{2^{c^\kappa}}$.

To verify Cat Top$_3$, suppose $X = \{x\}$. The only prime prefilters on X then are $\alpha \dot{1}_x$ for $\alpha \in I_0$, and $\alpha \hat{1}_x$ for $\alpha \in I_1$. The structure defined by

$$\lim \alpha \dot{1}_x = \lim \alpha \hat{1}_x = \alpha 1_x$$

clearly is a fuzzy convergence on X, and moreover it follows from the successive application of (F1p), (F2p) and (cl) that it is the only one. □

§4. Stout's I–Tolerance Spaces

Tolerance spaces, introduced by Zeeman (see [Zeeman and Buneman 1968]) as a mathematical model of visual discrimination, have been generalized in [Stout 1989] to a finer and more discriminating model using fuzzy sets. An **I-tolerance space** (X, τ) is a pair where $X \in |\mathbf{SET}|$ and where

$$\tau : X \times X \longrightarrow I$$

satisfies (S) $\tau(x, y) = \tau(y, x)$ and (R) $\tau(x, y) \leq \tau(x, x)$ for all $x, y \in X$. For our purposes we must restrict ourselves to a stronger form of reflexivity, namely (R′) $\tau(x, x) = 1$ for all $x \in X$. A morphism $f : (X, \tau) \longrightarrow (X', \tau')$ is a function from X to X' such that $\tau(x, y) \leq \tau'(f(x), f(y))$ holds for all $x, y \in X$. We say that f is **tolerance–preserving. TOL′(I)** stands for the category with objects all tolerance spaces (satifying (S) and (R′) and morphisms all tolerance preserving functions.

THEOREM 4.1. **TOL′(I)** is isomorphic to a subcategory of **FCS**.

PROOF. Given $(X,\tau) \in |\mathbf{Tol'(I)}|$ we define the fuzzy limit $\lim_\tau$ by

$$\lim_{\tau} \mathcal{F}(x) := c(\mathcal{F}, \tau(\dot{x},\ \cdot))$$

for $x \in X$ and $\mathcal{F} \in \mathcal{P}_p(X)$, and where the notation on the right hand side has the following meaning. For any pair of prefilters $\mathcal{F}, \mathcal{H} \in \mathcal{P}(X)$

$$c(\mathcal{F},\mathcal{H}) := \begin{cases} c(\mathcal{F} \vee \mathcal{H}), & \exists\ \mathcal{F} \vee \mathcal{H} \\ 0, & \nexists\ \mathcal{F} \vee \mathcal{H} \end{cases}$$

That $\lim_\tau$ is indeed a fuzzy limit is easily seen. (PST) is fulfilled by definition. (F1p) follows from the fact that for any pair of prefilters $\mathcal{F},\mathcal{H} : c(\mathcal{F},\mathcal{H}) \leq c(\mathcal{F})$. (F2p) follows from the fact that $\mathcal{F} \subset \mathcal{H}$ implies $c(\mathcal{F}) \geq c(\mathcal{H})$. To prove (cl) let $x \in X, \alpha \in I$ then

$$\begin{aligned} \lim_{\tau}(\alpha \dot{1}_x)(x) &= c(\langle\{\alpha 1_x \wedge \tau(x,\cdot)\}\rangle) \\ &= \sup_{z \in X} \alpha 1_x(z) \wedge \tau(x,z) \\ &= \alpha \wedge \tau(x,x) = \alpha. \end{aligned}$$

If $(X,\tau),(X',\tau') \in |\mathbf{TOL'(I)}|$ and $f : X \to X'$ is tolerance preserving, then for any $x \in X$ and $\mathcal{F} \in \mathcal{P}_p(X)$ we have

$$\begin{aligned} \lim_{\tau'} f(\mathcal{F})(f(x)) &= c(f(\mathcal{F}) \vee \tau'(\dot{f(x)},\cdot)) \\ &= \inf_{\mu \in \mathcal{F}} \sup_{z \in X'} f(\mu)(z) \wedge \tau'(f(x),z) \\ &= \inf_{\mu \in \mathcal{F}} \sup_{y \in X} f(\mu)(f(y)) \wedge \tau'(f(x),f(y)) \\ &\geq \inf_{\mu \in \mathcal{F}} \sup_{y \in X} \mu(y) \wedge \tau(x,y) \\ &= c(\mathcal{F} \vee \tau(\dot{x},\cdot)) \\ &= \lim_{\tau} \mathcal{F}(x) \end{aligned}$$

Conversely if $f : (X, \lim_\tau) \longrightarrow (X', \lim_{\tau'})$ is continuous, then for any $x,y \in X$ we have

$$\begin{aligned} \tau'(f(x),f(y)) &= \sup_{z \in X'} 1_{f(x)}(z) \wedge \tau'(f(y),z) \\ &= c(1_{\dot{f(x)}} \vee \tau'(f(y),\cdot)) \\ &= \lim_{\tau'} 1_{\dot{f(x)}}(f(y)) \\ &\geq \lim_{\tau} 1_{\dot{x}}(y) \\ &= \tau(x,y). \ \square \end{aligned}$$

§5. Cartesian Closedness of FCS

We begin by defining a natural function space fuzzy convergence on the set $Hom(X,Y)$ of all continuous maps between two fuzzy convergence spaces $(X, \lim_X)$ and $(Y, \lim_Y)$. The function

$$ev : X \times Hom(X,Y) \longrightarrow Y$$

is defined by $ev(x,f) := f(x)$ and

$$pr_X : X \times Hom(X,Y) \longrightarrow X,\ pr_H : X \times Hom(X,Y) \longrightarrow Hom(X,Y)$$

are the projections. Given $\Psi \in \mathcal{P}_p(Hom(X,Y))$ and $f \in Hom(X,Y)$ we put

$$L(\Psi,f) := \{\alpha \in I \mid \forall\, x \in X, \forall\, \Theta \in \mathcal{P}_p(X \times Hom(X,Y)),\ pr_H\Theta \subset \Psi$$

$$\Rightarrow \lim_X pr_X\Theta(x) \wedge \alpha \leq \lim_Y ev\Theta(f(x))\}$$

Notice that $L(\Psi,f)$ is an interval of type $[0,l[$ or $[0,l[$ for some $l \in I$. We now define

$$\lim_H \Psi(f) := c(\Psi) \wedge \sup L(\Psi,f).$$

PROPOSITION 5.1. $(Hom(X,Y), \lim_H)$ is a fuzzy convergence space.

PROOF. (PST) and (F1p) are fulfilled by definition. If $\Psi_1, \Psi_2 \in \mathcal{P}_p(Hom(X,Y))$ such that $\Psi_1 \subset \Psi_2$, then $\lim_H \Psi_2 \leq \lim_H \Psi_1$ follows at once from the facts that $c(\Psi_2) \leq c(\Psi_1)$ and $(L(\Psi_2,f) \subset L(\Psi_1,f)$ for any $f \in Hom(X,Y)$. In order to verify (c1), let $\alpha \in I$ and $f \in Hom(X,Y)$. Since $c(\alpha\dot{1}_f) = \alpha$ we have

$$\lim_H \alpha\dot{1}_f(f) = \alpha \wedge \sup L(\alpha\dot{1}_f, f).$$

CLAIM: $1 \in L(\alpha\dot{1}_f, f)$. Let Θ be a prime prefilter on $X \times Hom(X,Y)$ such that $pr_H\Theta \subset \alpha\dot{1}_f$. From the fact that $pr_H\Theta$ is prime it follows that $1_f \in pr_H\Theta$. Put

$$\beta := \inf\{\mu(f) \mid \mu \in pr_H\Theta\}.$$

Then clearly either $pr_H\Theta = \beta\dot{1}_f$ or $pr_H\Theta = \beta\hat{1}_f$. In both cases we have:

$$\exists\, \nu_0 \in \Theta,\ \exists\, \gamma \geq \beta : pr_H(\nu_0) \leq \gamma 1_f,$$

and therefore also $pr_H(\nu_0)(g) = 0$ for all $g \neq f$. It follows that there exists $\nu_0 \in \Theta$ such that

$$\sup_{x \in X} \nu_0(x,g) = 0,\ \forall\, g \neq f \tag{*}$$

Now if $\nu \in \Theta$, then for any $y \in Y$:

$$\begin{aligned} f(pr_X(\nu \wedge \nu_0))(y) &= \sup_{x \in f^{-1}(y)} \sup_{h \in Hom(X,Y)} \nu \wedge \nu_0(x,h) \\ &= \sup_{x \in f^{-1}(y)} \nu \wedge \nu_0(x,f) \quad (\text{by } (*)) \\ &\leq \sup_{x \in f^{-1}(y)} \nu(x,f) \\ &\leq \sup_{(x,h) \in ev^{-1}(y)} \nu(x,h) \\ &= ev(\nu)(y) \end{aligned}$$

Now both $ev(\Theta)$ and $f(pr_X(\Theta))$ are prime prefilters on Y and hence, using the continuity of f we have for every $x \in X$:

$$\begin{aligned} \lim_Y ev(\Theta)(f(x)) &\geq \lim_Y f(pr_X(\Theta))(f(x)) \\ &\geq \lim_X pr_X(\Theta)(x) \end{aligned}$$

from which our claim follows.

Consequently $\alpha \leq \lim_H \alpha \dot{1}_f(f)$ and we are done. □

THEOREM 5.2. **FCS** is cartesian closed topological.

PROOF. Since we already proved that **FCS** is topological and since, for any pair of fuzzy convergence spaces $(X, \lim_X)$ and $(Y, \lim_Y)$, we have endowed $Hom(X,Y)$ with a canonical fuzzy convergence $\lim_H$, it suffices to show that $(Hom(X,Y), \lim_H)$ satisfies the following two conditions:

(1) Continuity of the evaluation. Let $x \in X, f \in Hom(X,Y)$ and $\Theta \in \mathcal{P}_p(X \times Hom(X,Y))$. Since $\lim_H pr_H(\Theta)(f) \in L(pr_H(\Theta), f)$ it follows that

$$\lim_X pr_X(\Theta)(x) \wedge \lim_H pr_H(\Theta)(f) \leq \lim_Y ev(\Theta)(f(x))$$

and thus also

$$(\lim_X \times \lim_H)(\Theta)(x,f) \leq \lim_Y ev(\Theta)(f(x)).$$

(2) Existence of a factorizing morphism. Let $(Z, \lim_Z)$ be a third fuzzy convergence space and let

$$(X, \lim_X) \times (Z, \lim_Z) \xrightarrow{f} (Y, \lim_Y)$$

be continuous. We define

$$(Z, \lim_X) \xrightarrow{f^*} (Hom(X,Y), \lim_H)$$

$$z \longrightarrow f(z) : X \longrightarrow Y$$

$$x \longmapsto f(x, z)$$

and we will show that f^* is continuous.

Let $z \in Z$ and let $\mathcal{G} \in \mathcal{P}_p(Z)$. First we show that $\lim_Z \mathcal{G}(z) \in L(f^*(\mathcal{G}), f^*(z))$. Therefore let $x \in X$ and let $\Theta \in \mathcal{P}_p(X \times Hom(X, Y))$ such that $pr_H \Theta \subset f^*(\mathcal{G})$.

CLAIM 1: *Θ lives on $X \times f^*(Z)$.* Indeed let $\nu \in \Theta$ and let $\mu \in \mathcal{G}$ be such that $f^*(\mu) \leq pr_H(\nu)$. Take $z \in Z$ such that $f^*(\mu)(f^*(z)) \neq 0$. Then it follows that

$$0 < \sup_{x \in X} \nu(x, f^*(z))$$

and thus there exists $x \in X$ such that $\nu(x, f^*(z)) \neq 0$, and Θ thus has a trace on $X \times f^*(Z)$. However, since Θ is a prime prefilter we can conclude that $1_{X \times f^*(Z)} \in \Theta$.

Now let $\mathcal{F} := (1_X \times f^*)^{-1}(\Theta)$.

CLAIM 2: *$\exists \mathcal{H} \in \mathcal{P}_m(\mathcal{F}) : pr_Z(\mathcal{H}) \subset \mathcal{G}$ and $pr_X(\mathcal{H}) = pr_X(\Theta)$.* For any $\mu \in \Theta$ and $z \in Z$ we have

$$\begin{aligned} pr_Z((1_X \times f^*)^{-1}(\mu))(z) &= pr_Z(\mu o(1_X \times f^*))(z) \\ &= \sup_{x \in X} \mu(x, f^*(z)) \\ &= \sup_{(x,h) \in pr_H^{-1}(f^*(z))} \mu(x, h) \\ &= pr_H(\mu)(f^*(z)) \\ &= f^{*-1}(pr_H(\mu))(z) \end{aligned}$$

which proves that $pr_Z(\mathcal{F}) = f^{*-1}(pr_H(\Theta))$. Consequently

$$pr_Z(\mathcal{F}) \subset \mathcal{G}.$$

By Lemma 2.1 there then exists $\mathcal{H} \in \mathcal{P}_m(\mathcal{F})$ such that

$$pr_Z(\mathcal{F}) \subset pr_Z(\mathcal{H}) \subset \mathcal{G}$$

which already proves the first part of our claim, and by Lemma 2.2 we then have

$$1_X \times f^*(\mathcal{H}) = \Theta$$

and hence also

$$pr_X(\mathcal{H}) = pr_X(1_X \times f^*)(\mathcal{H}) = pr_X(\Theta)$$

which proves the second part.

CLAIM 3: $ev(\Theta) \subset f(\mathcal{H})$. Indeed let $\mu \in \Theta$. Then $(1_X \times f^*)^{-1}(\mu) \in \mathcal{H}$, and moreover for any $y \in Y$,

$$\begin{aligned} f((1_X \times f^*)^{-1}(\mu))(y) &= \sup_{(x,z)\in f^{-1}(y)} (1_X \times f^*)^{-1}(\mu)(x,z) \\ &= \sup_{(x,z)\in f^{-1}(y)} \mu(x, f^*(z)) \\ &\leq \sup_{ev(x,h)=y} \mu(x,h) \end{aligned}$$

which proves our claim.

Finally, making use of all the foregoing and the continuity of f we can conclude that

$$\begin{aligned} \lim_Y ev(\Theta)(f^*(z)(x)) &\geq \lim_Y f(\mathcal{H})(f(x,z)) \\ &\geq \lim_{X\times Z} \mathcal{H}(x,z) \\ &= \lim_X pr_X(\mathcal{H})(x) \wedge \lim_Z pr_Z(\mathcal{H})(z) \\ &\geq \lim_X pr_X(\Theta)(x) \wedge \lim_Z \mathcal{G}(z) \end{aligned}$$

This shows that indeed $\lim_Z \mathcal{G}(z) \in L(f^*(\mathcal{G}), f^*(z))$. Since now also $\lim_Z \mathcal{G}(z) \leq c(\mathcal{G}) = c(f^*(\mathcal{G}))$, we finally have $\lim_Z \mathcal{G}(z) \leq \lim_H f^*(\mathcal{G})(f^*(z))$, which proves the continuity of f^* and we are done. □

§6. Extensionality of FCS

A topological category is called **extensional** [Herrlich 1988] provided final epi-sinks are hereditary, or equivalently if partial morphisms are representable (note that in the above mentioned paper "extensional" is called "hereditary").

If C is a topological category and $X, Y \in |C|$ then a **partial morphism** from X to Y is a morphism in $Hom_C(A, Y)$ where A is a subobject of X. Partial morphisms are said to be representable if each object $Y \in |C|$ can be embedded via the addition of a single point $\infty_Y \notin Y$ into an object $Y^\# \in |C|$ such that for every partial morphism $f : A \longrightarrow Y$ from X to Y the map

$$\begin{aligned} f^\# : X &\longrightarrow Y^\# \\ x &\longmapsto f(x) \qquad \text{if } x \in A \\ &\longmapsto \infty_Y \qquad \text{if } x \notin A \end{aligned}$$

is a morphism.

THEOREM 6.1. **FCS** is extensional.

PROOF. Let $(Y, \lim_Y) \in \mid \mathbf{FCS} \mid$ be given and put $Y^{\#} := Y \cup \{\infty_Y\}$, where $\infty_Y \notin Y$. We define a limit structure $\lim^{\#}$ on $Y^{\#}$ in the following way. For any prime prefilter $\mathcal{G}$ on $Y^{\#}$:

(1) if $\mathcal{G}$ has a trace on Y we put :
$\lim^{\#} \mathcal{G}(y) := \lim_Y(\mathcal{G}|_Y)(y)$ for $y \in Y$
$\lim^{\#} \mathcal{G}(\infty_Y) := c(\mathcal{G})$

(2) if $\mathcal{G}$ has no trace on Y, then necessarily $\mathcal{G} = \alpha \dot{1}_{\infty_Y}$ or $\mathcal{G} = \alpha \hat{1}_{\infty_Y}$ for some $\alpha \in I$, and then we put $\lim^{\#} \mathcal{G} \equiv \alpha$.

We define $\lim^{\#}$ for arbitrary prefilters by (PST), which therefore is fullfilled by definition.

(F1p) follows at once from the definition and from the fact that $c(\alpha \dot{1}_x) = c(\alpha \hat{1}_x) = \alpha$ for any α and any x. (C1) too follows at once from the definition.

In order to show (F2p) let $\mathcal{G}_1, \mathcal{G}_2 \in \mathcal{P}_p(X)$ be such that $\mathcal{G}_1 \subset \mathcal{G}_2$. If $\mathcal{G}_2$ has a trace on Y then so does $\mathcal{G}_1$ and then we have:

(1) $\lim^{\#} \mathcal{G}_2(\infty_Y) = c(\mathcal{G}_2) \leq c(\mathcal{G}_1) = \lim^{\#} \mathcal{G}_1(\infty_Y)$;
(2) $\lim^{\#} \mathcal{G}_2(y) = \lim_Y \mathcal{G}_2|_Y(y) \leq \lim_Y \mathcal{G}_1|_Y(y) = \lim^{\#} \mathcal{G}_1(y)$ for any $y \in Y$.

If $\mathcal{G}_2$ has no trace on Y then necessarily $\mathcal{G}_2 = \alpha \dot{1}_{\infty_Y}$ or $\mathcal{G}_2 = \alpha \hat{1}_{\infty_Y}$ for some $\alpha \in I$. Since $\mathcal{G}_1 \subset \mathcal{G}_2 \Rightarrow \imath(\mathcal{G}_1) = \imath(\mathcal{G}_2)$ it follows that also $\mathcal{G}_1 = \beta \dot{1}_{\infty_Y}$ or $\mathcal{G}_1 = \beta \hat{1}_{\infty_Y}$ for some $\beta \geq \alpha$ and again the conclusion follows at once. Now suppose $Z \subset X$ and suppose that

$$(Z, \lim_{X|Z}) \xrightarrow{f} (Y, \lim_Y)$$

is continuous. To show that $f^{\#}$ is continous let $\mathcal{H} \in \mathcal{P}_p(X)$.

CASE 1: *$\mathcal{H}$ has no trace on Z.* Then by the primeness of $\mathcal{H}$ it follows that $1_{X \setminus Z} \in \mathcal{H}$. Consequently

$$f^{\#}(\mathcal{H}) = \alpha \dot{1}_{\infty_Y} \text{ or } f^{\#}(\mathcal{H}) = \alpha \hat{1}_{\infty_Y}$$

for some $\alpha \in I$. For $z \in Z$ we then have

$$\begin{aligned} \overset{\#}{\lim}(\alpha \dot{1}_{\infty_Y})(f(z)) &= \alpha \\ &= c(f^{\#}(\mathcal{H})) \\ &= c(\mathcal{H}) \\ &\geq \lim_X \mathcal{H}(z) \end{aligned}$$

and for $x \in X \setminus Z$ we also have

$$\begin{aligned} \overset{\#}{\lim}(\alpha \dot{1}_{\infty_Y})(\infty_Y) &= \alpha = c(f^{\#}(\mathcal{H})) \\ &= c(\mathcal{H}) \geq \lim_X \mathcal{H}(x) \end{aligned}$$

The case of $\alpha \hat{1}_{\infty_Y}$ is perfectly the same.

CASE 2: *$\mathcal{H}$ has a trace on Z.* In this case it follows that also $f^{\#}(\mathcal{H})$ has a trace on Y and moreover that

$$f^{\#}(\mathcal{H})|_Y \subset f(\mathcal{H}|_Z)$$

as can easily be verified. Now if $z \in Z$, then

$$\begin{aligned}\overset{\#}{\lim}(f^{\#}(\mathcal{H}))(f(z)) &= \lim_Y(f^{\#}(\mathcal{H})|_Y)(f(z))\\ &\geq \lim_Y(f(\mathcal{H}|_Z))(f(z))\\ &\geq \lim_{X|Z}(\mathcal{H}|_Z))(z)\\ &= \lim_X(\mathcal{H}|_Z)(z)\\ &\geq \lim_X(\mathcal{H})(z),\end{aligned}$$

and if $z \in X \backslash Z$, then

$$\begin{aligned}\overset{\#}{\lim}(f^{\#}(\mathcal{H}))(\infty_Y) &= c(f^{\#}(\mathcal{H}))\\ &= c(\mathcal{H})\\ &\geq \lim_X \mathcal{H}(z)\end{aligned}$$

and we are done. □

THEOREM 6.2. **FCS** is a topological universe.

PROOF. This follows at once from the facts that **FCS** is both cartesian closed and extensional. □

§7. Embedding FTS in FCS

In [E. Lowen, R. Lowen 1985] the authors showed that fuzzy topologies can be described by means of fuzzy convergence of prefilters. The results there were formulated in terms of general prefilters whereas **FCS** is defined with strong emphasis on prime prefilters. Therefore we shall now give a description of fuzzy topologies mainly in terms of prime prefilters.

THEOREM 7.1. If (X, Δ) is a fuzzy topological space, then the map

$$\mathcal{P}(X) \longrightarrow I^X : \mathcal{F} \longmapsto \lim_{\Delta} \mathcal{F}$$

where $\lim_{\Delta} \mathcal{F}(x) := \inf_{\mathcal{G} \in \mathcal{P}_m(\mathcal{F})} \inf_{\mu \in \mathcal{G}} \bar{\mu}(x)$, satisfies the following properties:

(PST) $\forall\, \mathcal{F} \in \mathcal{P}(X) : \lim_\triangle \mathcal{F} = \inf_{\mathcal{G} \in \mathcal{P}_m(\mathcal{F})} \lim_\triangle \mathcal{G}$
(F1p) $\forall\, \mathcal{G} \in \mathcal{P}_p(X) : \lim_\triangle \mathcal{G} \leq c(\mathcal{G})$
(F2p) $\forall\, \mathcal{G}_1, \mathcal{G}_2 \in \mathcal{P}_p(X) : \mathcal{G}_1 \subset \mathcal{G}_2 \Rightarrow \lim_\triangle \mathcal{G}_2 \leq \lim_\triangle \mathcal{G}_1$
(C1) $\forall\, x \in X, \forall\, \alpha \in I\backslash 0 : \lim_\triangle \alpha\dot{1}_x \geq \alpha 1_x$
(C2p) $\forall\, \mathcal{F} \in \mathcal{P}(X),\ \forall \mathcal{G} \in \mathcal{P}_m(\mathcal{F}) : \lim_\triangle \mathcal{F} \leq \lim_\triangle \mathcal{G}$
(Ccp) $\forall\, J \in |\mathrm{SET}|,\ \forall\, (\mathcal{F}_j)_{j\in J} \subset \mathcal{P}_p(X),\ \forall\, \mathcal{A} \in \mathbf{F}(J) :\ \lim_\triangle \mathcal{D}((\mathcal{F}_j)_{j\in J}, \mathcal{A}) \geq \inf\ \lim_\triangle \mathcal{F}_j$
(Cdp) $\forall\, \mathcal{F} \in \mathcal{P}_p(X), \forall\, (\mathcal{S}(y))_{y\in X} \subset \mathcal{P}_p(X)$ such that $\rho_{\mathcal{S}} \in \mathcal{F}$: $\lim_\triangle \mathcal{D}((\mathcal{S}(y))_{y\in X}, \imath(\mathcal{F})) \geq \lim_\triangle \mathcal{F}$, where $\rho_{\mathcal{S}} := \lim \mathcal{S}(\cdot)(\cdot)$, and moreover the map

$$I^X \longrightarrow I^X : \mu \longmapsto \inf_{\mathcal{G} \in \mathcal{P}_m(\dot{\mu})} \lim_\triangle \mathcal{G}$$

coincides with the original fuzzy closure on X.

PROOF. (PST) This follows by definition and is actually implied by the remaining conditions (see following theorem).

(F1p) For $\mathcal{G} \in \mathcal{P}_p(X)$ we have $\lim_\triangle \mathcal{G} = \inf_{\mu\in\mathcal{G}}\ \bar{\mu} \leq \inf_{\mu\in\mathcal{G}} \sup_{x\in X} \mu(x) = c(\mathcal{G})$.

(F2p) For $\mathcal{G}_1, \mathcal{G}_2 \in \mathcal{P}_p(X)$, $\mathcal{G}_1 \subset \mathcal{G}_2$ implies $\lim_\triangle \mathcal{G}_1 = \inf_{\mu\in\mathcal{G}_1} \bar{\mu} \geq \inf_{\mu\in\mathcal{G}_2} \bar{\mu} = \lim_\triangle \mathcal{G}_2$.

(C1) For $x \in X$and $\alpha \in I\backslash 0$, since $\alpha\dot{1}_x$ is prime $\lim_\triangle \alpha\dot{1}_x = \overline{\alpha 1_x} \geq \alpha 1_x$.

(C2p) This follows by definition and actually is weaker than (PST).

(Ccp) Let $J, (\mathcal{F}_j)_{j\in J}$ and $\mathcal{A}$ be as given and put $\mathcal{D} := \mathcal{D}((\mathcal{F}_j)_{j\in J}, \mathcal{A})$. First of all observe that if $\mathcal{G} \in \mathcal{P}_m(\mathcal{D})$, i.e. there is some $U \in \mathbf{U}(\imath(\mathcal{D}))$ such that $\mathcal{G} = \omega(U) \vee \mathcal{D}$, and $\xi \in \mathcal{G}$, i.e. there are $\mathrm{U} \in U$ and $\mu \in \mathcal{D}$ such that $1_{\mathrm{U}} \wedge \mu \leq \xi$, then there must exist some $j \in J$ such that $\xi \in \mathcal{F}_j$. For if we suppose that for all $j \in J$ we have $1_{\mathrm{U}} \wedge \mu \notin \mathcal{F}_j$, then since there exists $A \in \mathcal{A}$ such that $\mu \in \mathcal{F}_j$ for all $j \in A$ and since $\mu = (1_{\mathrm{U}} \wedge \mu) \vee (1_{\mathrm{U}^c} \wedge \mu)$, it follows that $1_{\mathrm{U}^c} \wedge \mu \in \mathcal{F}_j$ for all $j \in A$ and thus that $1_{\mathrm{U}^c} \wedge \mu \in \mathcal{D}$. However, since $\mathcal{D} \subset \omega(U) \vee \mathcal{D}$, this is a contradiction. Making use of this observation we obtain

$$\begin{aligned}\lim_\triangle \mathcal{D} &= \inf_{\mathcal{G}\in\mathcal{P}_m(\mathcal{D})}\ \inf_{\mu\in\mathcal{G}} \bar{\mu}\\ &\geq \inf_{j\in J}\ \inf_{\mu\in\mathcal{F}_j} \bar{\mu}\\ &= \inf_{j\in J} \lim_\triangle \mathcal{F}_j\end{aligned}$$

(Cdp) Let $\mathcal{F}$ and $\mathcal{S}$ be as given and put

$$\mathcal{D} := \mathcal{D}((\mathcal{S}(y))_{y\in X}, \imath(\mathcal{F})).$$

From Lemma 2.4 we know that $\mathcal{D}$ is prime. Let $\mu \in \mathcal{D}$ then again there exists $F \in \imath(\mathcal{F})$ such that

$$\mu \in \bigcap_{y\in F} \mathcal{S}(y).$$

Let $\phi := 1_F \wedge \rho_{\mathcal{S}}$. Then $\phi \in \mathcal{F}$ and straightforward verification moreover shows that

$$\phi \leq \sup_{y\in\phi^{-1}(]0,1])} \lim_\triangle \mathcal{S}(y)$$

from which it follows that $\sup_{y\in\phi^{-1}(]0,1])} \lim_\triangle \mathcal{S}(y) \in \mathcal{F}$. Consequently then, since

$$\begin{aligned}\bar{\mu} &\geq \inf\{\bar{\xi} \mid \xi \in \cap_{y\in F} \mathcal{S}(y)\}\\ &\geq \inf\{\bar{\xi} \mid \xi \in \bigcap_{y\in\phi^{-1}(]0,1])} \mathcal{S}(y)\}\\ &\geq \sup_{y\in\phi^{-1}(]0,1])} \lim_\triangle \mathcal{S}(y)\end{aligned}$$

it follows that also $\bar{\mu} \in \mathcal{F}$. Consequently by idempotency of the fuzzy closure operator and the arbitrariness of $\mu \in \mathcal{D}$ it follows that $\lim_\triangle \mathcal{D} \geq \lim_\triangle \mathcal{F}$. That $\bar{\mu} = \sup_{\mathcal{G}\in\mathcal{P}_m(\dot{\mu})} \lim_\triangle \mathcal{G}$ holds for any $\mu \in I^X$ finally follows from [R. Lowen 1979], where it was shown that in a fuzzy topological space for any prefilter $\inf_{\mu\in\mathcal{F}} \bar{\mu} = \sup_{\mathcal{G}\in\mathcal{P}_m(\mathcal{F})} \lim_\triangle \mathcal{G}$. □

THEOREM 7.2. If

$$\mathcal{P}(X) \longrightarrow I^X : \mathcal{F} \longmapsto \lim \mathcal{F}$$

is a map which satisfies the properties (F1p), (F2p), (C1), (C2p), (Ccp) and (Cdp) then there exists a unique fuzzy topology $\triangle$ on X such that $\lim_\triangle$ coincides with lim and consequently lim also satisfies (PST).

PROOF. We define the map

$$I^X \longrightarrow I^X : \mu \longmapsto \bar{\mu} := \sup_{\mathcal{G}\in\mathcal{P}_m(\dot{\mu})} \lim \mathcal{G}$$

If α is constant, then by (F1p) we have

$$\bar{\alpha} = \sup_{U\in\mathbf{U}(X)} \lim \omega(U) \vee \dot{\alpha} \leq \alpha$$

If $\mu \in I^X$, then by (C1) we have

$$\begin{aligned}\bar{\mu} &= \sup_{\mathcal{U}\in\mathbf{U}(\imath(\dot{\mu})} \lim \omega(\mathcal{U}) \vee \dot{\mu}\\ &\geq \sup_{x\in\mu^{-1}(]0,1])} \lim \dot{\mu(x)1_x}\\ &\geq \sup_{x\in\mu^{-1}(]0,1])} \mu(x)1_x = \mu\end{aligned}$$

If $\mu, \nu \in I^X$ are such that $\mu \leq \nu$, then by (F2p) we have

$$\begin{aligned}\bar{\mu} &= \sup_{\mathcal{U}\in\mathbf{U}(\imath(\dot{\mu}))} \lim \omega(\mathcal{U}) \vee \dot{\mu}\\ &\leq \sup_{\mathcal{U}\in\mathbf{U}(\imath(\dot{\nu}))} \lim \omega(\mathcal{U}) \vee \dot{\nu}\\ &= \bar{\nu}\end{aligned}$$

From Lemma 2.3 we obtain that for $\mu, \nu \in I^X$,

$$\begin{aligned}\overline{\mu \vee \nu} &= \sup_{\mathcal{G} \in \mathcal{P}_m(\mu \dot\vee \nu)} \lim \mathcal{G} \\ &\leq \sup_{\mathcal{G} \in \mathcal{P}_m(\dot\mu)} \lim \mathcal{G} \vee \sup_{\mathcal{G} \in \mathcal{P}_m(\dot\nu)} \lim \mathcal{G}\end{aligned}$$

Thus we have that $\overline{\mu \vee \nu} = \bar{\mu} \vee \bar{\nu}$ holds for all $\mu, \nu \in I^X$. Now in order to show idempotency of $^-$ we first define $\lim' : \mathcal{P}(X) \to I^X$ by

$$\lim' \mathcal{F} = \begin{cases} \inf_{\mu \in \mathcal{F}} \bar{\mu}, & \mathcal{F} \text{ prime} \\ \inf_{\mathcal{G} \in \mathcal{P}_m(\mathcal{F})} \lim' \mathcal{G}, & \text{otherwise} \end{cases}$$

and show that $\lim'$ coincides with lim.

CLAIM 1: *lim and* $\lim'$ *coincide on prime prefilters.* If $\mu \in \mathcal{F}$, and $\mathcal{F}$ is prime, then since $\omega(\imath(\mathcal{F})) \vee \dot\mu \subset \mathcal{F}$ it follows from (F2p) that

$$\lim \mathcal{F} \leq \sup_{\mathcal{U} \in \mathbf{U}(\imath(\dot\mu))} \lim \omega(\mathcal{U}) \vee \dot\mu = \bar{\mu}$$

and thus $\lim \mathcal{F} \leq \lim' \mathcal{F}$. Conversely

$$\begin{aligned}\lim' \mathcal{F} &= \inf_{\mu \in \mathcal{F}} \sup_{\mathcal{U} \in \mathbf{U}(\imath(\dot\mu))} \lim \omega(\mathcal{U}) \vee \dot\mu \\ &= \sup_{\phi \in \sqcap_{\mu \in \mathcal{F}} \mathbf{U}(\imath(\dot\mu))} \inf_{\mu \in \mathcal{F}} \lim \omega(\phi(\mu)) \vee \dot\mu (_*)\end{aligned}$$

Now let us fix $\phi \in \sqcap_{\mu \in \mathcal{F}} \mathbf{U}(\imath(\dot\mu))$ and let us consider the family of prime prefilters

$$\{\omega(\phi(\mu)) \vee \dot\mu \mid \mu \in \mathcal{F}\}.$$

On the indexset $\mathcal{F}$ we taken the filter $\mathcal{A}$ generated by the basis $\{A_\mu \mid \mu \in \mathcal{F}\}$, where $A_\mu := \{\nu \in \mathcal{F} | \nu \leq \mu\}$. Straightforward verification now shows that

$$\mathcal{F} \subset \mathcal{D}((\omega(\phi(\mu)) \vee \dot\mu)_{\mu \in \mathcal{F}}, \mathcal{A})$$

By respectively (F2p) and (Ccp) we then obtain

$$\begin{aligned}\lim \mathcal{F} &\geq \lim \mathcal{D}((\omega(\phi(\mu)) \vee \dot\mu)_{\mu \in \mathcal{F}}, \mathcal{A}) \\ &\geq \inf_{\mu \in \mathcal{F}} \lim \omega(\phi(\mu)) \vee \dot\mu \qquad (**)\end{aligned}$$

By the arbitrariness of ϕ it follows from $(*)$ and $(**)$ that $\lim' \mathcal{F} \leq \lim \mathcal{F}$.

CLAIM 2: *lim* $=$ *lim'*. Let $\mathcal{F}$ be an arbitrary prefilter. If we apply (Ccp) to the family $\mathcal{P}_m(\mathcal{F})$ of prime prefilters with $\mathcal{A}$ the trivial filter then we already obtain

$$\lim \mathcal{F} \geq \inf_{\mathcal{G} \in \mathcal{P}_m(\mathcal{F})} \lim \mathcal{G}$$

From (C2p) it follows that

$$\begin{aligned}\lim \mathcal{F} &= \lim \omega(\imath(\mathcal{F})) \vee \mathcal{F} \\ &\leq \inf_{\mathcal{U} \in \mathbf{U}(\imath(\mathcal{F}))} \lim \omega(\mathcal{U}) \vee \mathcal{F} \\ &\leq \inf_{\mathcal{G} \in \mathcal{P}_m(\mathcal{F})} \lim \mathcal{G}\end{aligned}$$

Combining this with the first claim then shows that $\lim \mathcal{F} = \lim' \mathcal{F}$. Now to show that $^-$ is idempotent fix $\mu \in I^X$ and $\mathcal{G} \in \mathcal{P}_m(\dot{\bar{\mu}})$.

Claim 3: $\forall y \in X \; \exists \mathcal{S}(y) \in \mathcal{P}_m(\dot{\mu}) : lim\mathcal{S}(y)(y) = \bar{\mu}(y)$. Indeed, suppose on the contrary that for each $\mathcal{U} \in \mathbf{U}(\imath(\dot{\mu}))$ we have

$$\lim \omega(U) \vee \dot{\mu}(y) < \bar{\mu}(y)$$

Then for each $\mathcal{U} \in \mathbf{U}(\imath(\dot{\mu}))$ we can find $U_{\mathcal{U}} \in \mathcal{U}$ such that

$$\overline{1_{U_{\mathcal{U}}} \wedge \mu}(y) < \bar{\mu}(y)$$

Take $\mathcal{U}_1, \ldots, \mathcal{U}_n \in \mathbf{U}(\imath(\dot{\mu}))$ and $U_{\mathcal{U}_1}, \ldots, U_{\mathcal{U}_n}$ such that $\bigcup_{i=1}^n U_{\mathcal{U}_i} \in \imath(\dot{\mu})$ i.e. such that

$$\bigcup_{i=1}^{n} U_{\mathcal{U}_i} \supset \mu^{-1}(]0,1])$$

Then it follows that

$$\begin{aligned}\mu(y) &= \overline{1_{\bigcup_{i=1}^n U_{\mathcal{U}_i}} \wedge \mu}(y) \\ &= \sup_{i=1}^{n} \overline{1_{U_{\mathcal{U}_i}} \wedge \mu}(y) \\ &< \bar{\mu}(y)\end{aligned}$$

which is a contradiction, hence our claim holds.

Consider this selection $(\mathcal{S}(y))_{y \in X}$ as given in Claim 3. Then obviously $\rho_{\mathcal{S}} = \bar{\mu} \in \mathcal{G}$. Applying (Cdp) for this selection and for $\mathcal{G}$ given us,

$$\lim \mathcal{D} \geq \lim \mathcal{G}$$

where we have again put $\mathcal{D} := \mathcal{D}((\mathcal{S}(y))_{y \in X}, \imath(\mathcal{G}))$. Now since $\mu \in \mathcal{S}(y)$ for each $y \in X$, it follows that $\mu \in \mathcal{D}$. Consequently we obtain that

$$\bar{\mu} \geq \lim \mathcal{G}$$

which by the arbitrariness of $\mathcal{G} \in \mathcal{P}_m(\dot{\bar{\mu}})$ implies that $\bar{\mu} \geq \bar{\bar{\mu}}$. If $\triangle$ stands for the fuzzy topology associated with $^-$, then it follows that $\lim_{\triangle} = \lim' = \lim$. □

THEOREM 7.3. If $(X,\triangle),(X',\triangle') \in |FTS|$ and $f : X \longrightarrow X'$ then the following are equivalent:

(1) f is continuous;
(2) $\forall\, \mathcal{F} \in \mathcal{P}(X),\ \forall x \in X : \lim_{\triangle'} f(\mathcal{F})(f(x)) \geq \lim_{\triangle} \mathcal{F}(x)$;
(3) $\forall\, \mathcal{F} \in \mathcal{P}_p(X),\ \forall x \in X : \lim_{\triangle'} f(\mathcal{F})(f(x)) \geq \lim_{\triangle} \mathcal{F}(x)$.

PROOF. These results are well-known and for details we refer to [R. Lowen 1979; R. Lowen, P. Wuyts 1982]. □

The foregoing theorems prove that we can embed **FTS** and **FCS** and that we can identify **FTS** with the full subcategory of **FCS** the objects of which satisfy the supplementary conditions (Ccp) and (Cdp). If not necessary we shall not differentiate between **FTS** and its embedding in **FCS**. We shall now prove that this embedding actually is a very nice one.

THEOREM 7.4. **FTS** is a bireflective subcategory of **FCS**.

PROOF. Let $(X,\lim) \in |\mathbf{FCS}|$ and let $^-$ be the not necessarily idempotent fuzzy closure operator defined by

$$I^X \longrightarrow I^X : \mu \longmapsto \bar{\mu} := \sup_{\mathcal{G}\in\mathcal{P}_m(\dot{\mu})} \lim \mathcal{G}.$$

If $\mu \in I^X$, then

$$\bar{\mu} = \sup_{\mathcal{G}\in\mathcal{P}_m(\dot{\mu})} \lim \mathcal{G} \geq \sup_{x\in\mu^{-1}(]0,1])} \lim \mu(\dot{x})1_x \geq \mu.$$

If α is a constant fuzzy set, then

$$\bar{\alpha} = \sup_{\mathcal{G}\in\mathcal{P}_m(\dot{\alpha})} \lim \mathcal{G} \leq \sup_{\mathcal{G}\in\mathcal{P}_m(\dot{\alpha})} c(\mathcal{G}) \leq \alpha.$$

If $\mu,\nu \in I^X$ such that $\mu \leq \nu$, then from $\mathbf{U}(\imath(\dot{\mu})) \subset \mathbf{U}(\imath(\dot{\nu}))$ if follows that $\bar{\mu} \leq \bar{\nu}$, and if $\mu,\nu \in I^X$ are arbitrary, then

$$\begin{aligned}\overline{\mu \vee \nu} &= \sup_{\mathcal{G}\in\mathcal{P}_m(\mu\dot{\vee}\nu)} \lim \mathcal{G} \\ &\leq \sup_{\mathcal{G}\in\mathcal{P}_m(\dot{\mu})} \lim \mathcal{G} \vee \sup_{\mathcal{G}\in\mathcal{P}_m(\dot{\nu})} \lim \mathcal{G} \\ &= \bar{\mu} \vee \bar{\nu}.\end{aligned}$$

Thus we always have $\overline{\mu \vee \nu} = \bar{\mu} \vee \bar{\nu}$.

Now consider the collection

$$\{\mu \in I^X \mid \bar{\mu} = \mu\}$$

then it follows at once from the properties we have proved that this is precisely the collection of closed fuzzy sets for some fuzzy topology, say $\triangle$, on X. We shall prove that

$$(X,\lim) \stackrel{id_X}{\longrightarrow} (X,\lim_{\triangle})$$

is a bireflection. That id_X is continuous follows from the following observation. Let $\mathcal{G} \in \mathcal{P}_p(X)$ and let $\nu \in \mathcal{G}$ be closed w.r.t. Δ. Choose $\mathcal{H} \in \mathcal{P}_m(\dot{\nu})$ such that $\dot{\nu} \subset \mathcal{H} \subset \mathcal{G}$. Then

$$\lim \mathcal{G} \leq \lim \mathcal{H} \leq \bar{\nu} = \nu$$

and hence $\lim \mathcal{G} \leq \lim_\Delta \mathcal{G}$.
Now let (Y, Γ) be a fuzzy topological space and suppose

$$(X, \lim) \xrightarrow{f} (Y, \lim_\Gamma)$$

is continuous. Let ν be closed w.r.t. Γ, let $\mathcal{G} \in \mathcal{P}_m(f^{-1}\dot{(\nu)})$ and let $x \in X$. Choose $\mathcal{H} \in \mathcal{P}_m(\dot{\nu})$ such that $\mathcal{H} \subset f(\mathcal{G})$. Then it follows that

$$\begin{aligned}\lim \mathcal{G}(x) &\leq \lim_\Gamma f(\mathcal{G})(f(x)) \\ &\leq \lim_\Gamma \mathcal{H}(f(x)) \\ &\leq \bar{\nu}(f(x)) = \nu(f(x)) \\ &= f^{-1}(\nu)(x)\end{aligned}$$

Consequently

$$f^{-1}(\nu)(x) = \sup_{\mathcal{G} \in \mathcal{P}_m(f^{-1}\dot{(\nu)})} \lim \mathcal{G}(x) \leq f^{-1}(\nu)(x)$$

which by the arbitrariness of ν proves that also

$$(X, \lim_\Delta) \xrightarrow{f} Y, \lim_\Gamma)$$

is continuous, and we are done. □

§8. Embedding PSTOP in FCS

PSTOP stands for the category of pseudotopological convergence spaces (referred to shortly also as pseudotopological spaces). We recall the definition [G. Choquet 1948]. By a **pseudotopological convergence structure** on a set X we mean a map

$$q : X \longrightarrow 2^{\mathbf{F}(X)}$$

which satisfies the following axioms:

(PC1) $\dot{x} \in q(x)$;
(PC2) $\mathcal{F} \in q(x), \mathcal{G} \supset \mathcal{F} \Rightarrow \mathcal{G} \in q(x)$;
(PC3) $(\forall\ \mathcal{U} \in \mathbf{U}(\mathcal{F}) : \mathcal{U} \in q(x)) \Rightarrow \mathcal{F} \in q(x)$.

It turns out that **PSTOP** can be embedded in **FCS** in an extremely nice way. Given a pseudotopological convergence space (X, q) and $\mathcal{F} \in \mathbf{F}(X)$ we shall denote

$$L_q(\mathcal{F}) := \{x \in X | \mathcal{F} \in q(x)\}.$$

THEOREM 8.1. For $(X, q) \in |$ **PSTOP**$|$ the map

$$\lim_q : \mathcal{P}(X) \longrightarrow I^X : \mathcal{F} \longmapsto c^-(\mathcal{F}) \wedge 1_{L_q(\imath(\mathcal{F}))}$$

is a fuzzy convergence and the functor

$$\mathbf{PSTOP} \xrightarrow{\omega} \mathbf{FCS}$$

$$\begin{bmatrix} (X, q) \longrightarrow (X, \lim_q) \\ f \longrightarrow f \end{bmatrix}$$

is an embedding of **PSTOP** in **FCS**.

PROOF. Let $(X, q) \in |\mathbf{PSTOP}|$. First of all observe that $\lim_q$ is determined if we know its restriction to prime prefilters, more precisely that (PST) is fulfilled. Let $\mathcal{F} \in \mathcal{P}(X)$ then indeed

$$\begin{aligned} \lim_q \mathcal{F} &= c^-(\mathcal{F}) \wedge 1_{L_q(\imath(\mathcal{F}))} \\ &= (\inf_{\mathcal{G} \in \mathcal{P}_m(\mathcal{F})} c(\mathcal{G})) \wedge (\inf_{\mathcal{U} \in \mathbf{U}(\imath(\mathcal{F}))} 1_{L_q(\mathcal{U})}) \\ &= \inf_{\mathcal{G} \in \mathcal{P}_m(\mathcal{F})} (c(\mathcal{G}) \wedge 1_{L_q(\imath(\mathcal{G}))}) \\ &= \inf_{\mathcal{G} \in \mathcal{P}_m(\mathcal{F})} \lim_q \mathcal{G}, \end{aligned}$$

where we have of course used (PC2) and (PC3). (F1p), (F2p) and (C1) follow at once respectively, from the definition of $\lim_q$ and from (PC1). If $f : (X, q) \longrightarrow (X', q')$ is a morphism in **PSTOP** then it also follows at once from the definition that $f : (X, \lim_q) \longrightarrow (X', \lim_{q'})$ is a morphism in **FCS**. □

THEOREM 8.2. **PSTOP** is a simultaneously coreflective and bireflective subcategory of **FCS**.

PROOF.

COREFLECTIVITY: Given $(X, \lim) \in |\mathbf{FCS}|$, we define $q^{\lim} : X \longrightarrow 2^{\mathbf{F}(X)}$ by

(1) $\mathcal{U} \in \mathbf{U}(X) : \mathcal{U} \in q^{\lim}(x)$ iff $\forall \mathcal{F} \in \imath^{-1}(\mathcal{U}) : \lim \mathcal{F}(x) = c(\mathcal{F})$
(2) $\mathcal{F} \in \mathbf{F}(X) : \mathcal{F} \in q^{\lim}(x)$ iff $\mathbf{U}(\mathcal{F}) \subset q^{\lim}(x)$.

By (2), $q^{\lim}$ satisfies (PC2) and (PC3). So we only have to verify (PC1). Let $x \in X$ and let $\mathcal{F} \in \mathcal{P}_p(X)$ be such that $\imath(\mathcal{F}) = \dot{x}$. Then it follows at once that for some $\alpha \in I$ either $\mathcal{F} = \alpha\dot{1}_x$ or $\mathcal{F} = \alpha\hat{1}_x$. Consequently by (C1), $\lim \mathcal{F}(x) \geq \alpha = c(\mathcal{F})$ and thus $\dot{x} \in q^{\lim}(x)$. We claim that

$$(X, \lim_{q^{\lim}}) \xrightarrow{id_X} (X, \lim)$$

is a coreflection.

First we show that id_X is continuous. Put $\lim' := \lim_{q^{\lim}}$. If $\mathcal{F} \in \mathcal{P}_p(X)$ and $x \in X$ is such that $\lim' \mathcal{F}(x) > 0$ then

$$\lim{}' \mathcal{F}(x) = c(\mathcal{F}) \text{ and } \imath(\mathcal{F}) \in q^{\lim}(x).$$

Consequently $\lim \mathcal{F}(x) = c(\mathcal{F})$ which proves that $\lim' \leq \lim$.

Second, suppose $(Y, q) \in |\mathbf{PSTOP}|$ and suppose that

$$(Y, \lim_q) \xrightarrow{f} (X, \lim)$$

is continuous, i.e. if $y \in Y$ and $\mathcal{F} \in \mathcal{P}_p(Y)$, then

$$\imath(\mathcal{F}) \in q(y) \Rightarrow \lim f(\mathcal{F})(f(y)) \geq c(\mathcal{F}).$$

Let $y \in Y$ and let $\mathcal{U} \in q(y)$. Let $\mathcal{G} \in \mathcal{P}_p(X)$ be such that $\imath(\mathcal{G}) = f(\mathcal{U})$.

Claim: *$\exists \mathcal{F} \in \mathcal{P}_p(Y) : \imath(\mathcal{F}) = \mathcal{U}$ and $f(\mathcal{F}) = \mathcal{G}$.* Since $\imath(\mathcal{G}) = f(\mathcal{U})$ we have that $f^{-1}(\mathcal{G})$ exists. We always have $f(f^{-1}(\mathcal{G})) \supset \mathcal{G}$, but again from $\imath(\mathcal{G}) = f(\mathcal{U})$ we now also have that $1_{f(X)} \in \mathcal{G}$ and since for any $\mu \in \mathcal{G}$

$$f(f^{-1}(\mu)) \geq \mu \wedge 1_{f(X)}$$

we now also have $ff^{-1}(\mathcal{G}) = \mathcal{G}$. By Lemma 2.2 we then obtain that for any $\mathcal{F} \in \mathcal{P}_m(f^{-1}(\mathcal{G})) : f(\mathcal{F}) = \mathcal{G}$. Since

$$\mathcal{U} \supset f^{-1}(f(\mathcal{U})) = f^{-1}(\imath(\mathcal{G})) = \imath(f^{-1}(\mathcal{G}))$$

it follows that we can moreover find $\mathcal{F} \in \mathcal{P}_m(f^{-1}(\mathcal{G}))$ such that $\imath(\mathcal{F}) = \mathcal{U}$, which proves our claim.

Let $\mathcal{F}$ be such as in the claim. Then it follows that $\imath(\mathcal{F}) \in q(y)$ and thus

$$\begin{aligned} \lim \mathcal{G}(f(y)) &= \lim f(\mathcal{F})(f(y)) \\ &\geq c(\mathcal{F}) = c(f(\mathcal{F})) \\ &= c(\mathcal{G}), \end{aligned}$$

which by arbitrariness of $\mathcal{G}$ proves that $f(\mathcal{U}) \in q^{\lim}(f(y))$. This shows that also

$$(Y, \lim_q) \xrightarrow{f} (X, \lim_{q^{\lim}})$$

is continuous and we are done.

BIREFLECTIVITY: Given $(X, \lim) \in |\ \mathbf{FCS}\ |$ we define $q_{\lim} : X \longrightarrow 2^{\mathbf{F}(X)}$ by

(1) $\mathcal{U} \in \mathbf{U}(X) : \mathcal{U} \in q_{\lim}(x)$ iff $\lim \omega(\mathcal{U})(x) > 0$
(2) $\mathcal{F} \in \mathbf{F}(X) : \mathcal{F} \in q_{\lim}(x)$ iff $\mathbf{U}(\mathcal{F}) \subset q_{\lim}(x)$.

Again by (2), we only need to verify (PC1). If $x \in X$, then $\lim \omega(\dot{x})(x) = 1 > 0$ by (C1) and thus $\dot{x} \in q_{\lim}(x)$. We claim that

$$(X, \lim) \xrightarrow{id_X} (X, \lim_{q_{\lim}})$$

is a bireflection. Continuity of id_X follows at once from the fact that for any $\mathcal{F} \in \mathcal{P}_p(x)$ we have $\mathcal{F} \supset \omega(\imath(\mathcal{F}))$ and thus $\lim \mathcal{F} \leq \lim \omega(\imath(\mathcal{F}))$. Further let $(Y, q) \in |\mathbf{PSTOP}|$ and suppose that $(X, \lim) \xrightarrow{f} (Y, \lim_q)$ is continuous. Then if $\mathcal{U} \in q_{\lim}(x)$ we have:

$$\begin{aligned} \lim\ \omega(f(\mathcal{U}))(f(x)) &= \lim f(\omega(\mathcal{U}))(f(x)) \\ &\geq \lim \omega(\mathcal{U})(x) \\ &> 0 \end{aligned}$$

and thus $f(\mathcal{U}) \in q_{\lim_q} = q$. This proves that also $(X, \lim_{q_{\lim}}) \xrightarrow{f} (Y, \lim_q)$ is continuous and we are done. □

Chapter 7:

Categorical Frameworks for Stone Representation Theories

S. E. Rodabaugh et al. (eds.), *Applications of Category Theory to Fuzzy Subsets*, 177–231.

CATEGORICAL FRAMEWORKS FOR STONE REPRESENTATION THEORIES

S. E. Rodabaugh

§0. Introduction

The duality between topological spaces and lattices, first exploited by the famous Stone Representation Theorems [Stone 1936, 1937a, 1937b], is rightly regarded as one of the fundamental developments of twentieth century mathematics [Johnstone 1982]. The full expression of this duality in classical mathematics is seen in the relationship between topological spaces and frames/locales, culminating in the adjunction between topological spaces and locales [Papert-Papert 1958, Isbell 1972] and the resultant equivalence of sober topological spaces and spatial locales. For convenience, we dub this adjunction the **Stone adjunction**, and the resulting (sometimes dual) categorical equivalences the **Stone representation theory**. The most complete account of these matters, and related issues such as the Stone-Čech Compactification, may be found in [Johnstone 1982].

It is the purpose of this chapter to describe generalizations of the classical Stone adjunction and representations, and the categorical frameworks which make our generalizations possible. The consequences of our generalizations for the classical case include the following:

(1) The unit and counit^{op} of the Stone adjunction can be written, in our generalizations, in the explicit form of an **evaluation map**. (This **cannot** be done in the classical theory, as is seen later.) This consequence permits a streamlining and simplication of important parts of the classical theory.

(2) Two significant gaps of classical topology—non-spatial locales have no topological space representation, and localic products need not be product topologies—are filled in this chapter by generalizing the notion of topology.

(3) Generalizing the notion of topology also permits us to create space representations of functor categories which describe certain semilattice diagrams whose nodes are complete lattices.

This list of applications to classical topology and the classical Stone theory is augmented in later sections and at length in the author's four-part series on which this chapter is based [Rodabaugh ****]; it should be noted that the third application above is based on [Meßner-

Rodabaugh ****], the second part of that four-part series. That series also catalogues many applications to fuzzy topology including separation axioms and Stone-Čech compactification reflectors, and is based in turn on the preprint [Rodabaugh 1986–7]; lengthy abstracts have also appeared in [Rodabaugh 1986, 1991].

The crucial tools in all of this are the notions of fuzzy sets taken in their original **non-model-theoretic** sense [Zadeh 1965, Goguen 1967] and the corresponding notions of fuzzy topology [Chang 1968, Goguen 1973, Lowen 1976, Höhle 1978, Hutton 1980, and Rodabaugh 1983–7]. In keeping with [Rodabaugh 1991, ****], we call the non-model-theoretic set theory and topology used in this tutorial **point-set lattice-theoretic**, or **poslat** for short, to emphasize that we are dealing with sets and topological spaces grounded in notions of sets, functions, powersets, and powerset operators — the latter two being lattice-theoretic in nature and examined from a lattice-theoretic (and hence non-model-theoretic) point of view.

Thus the applications given above provide justifications of traditional or poslat (fuzzy) sets and topology, justifications additional to that already extant in the literature.

The adjunction diagrams used in this chapter are now given; we also use the duals of these diagrams (with $\mathcal{C}$ and $\mathcal{D}$ interchanged). Given categories $\mathcal{C}, \mathcal{D}$, a pair of functors $[\mathbf{F} : \mathcal{C} \longrightarrow \mathcal{D},\ \mathbf{G} : \mathcal{C} \longleftarrow \mathcal{D}]$ form an **adjunction**, denoted $\mathbf{F} \dashv \mathbf{G}$, iff the following two diagrams hold **with the indicated orderings of quantifiers and commutivities:**

Major Diagram

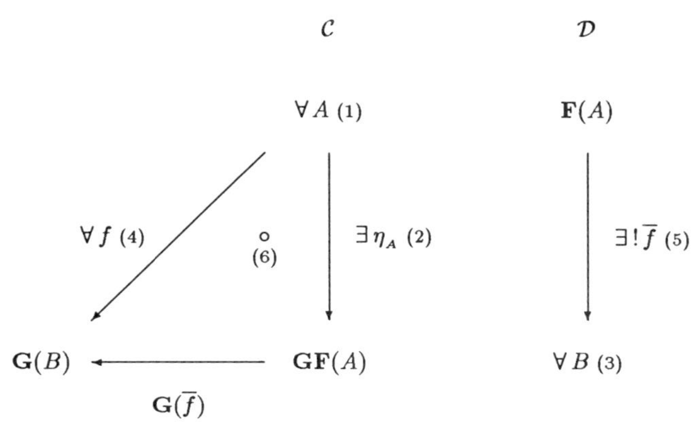

Minor Diagram (assuming the Major Diagram is satisfied)

$$\left[\begin{array}{ccccc} A_1 & \xrightarrow{\quad f \quad} & A_2 & \qquad & \mathbf{F}(A_1) \\ \eta_{A_1}\Big\downarrow & \circ & \Big\downarrow\eta_{A_2} & & \Big\downarrow !\,\overline{f \circ \eta_{A_2}} \\ \mathbf{GF}(A_1) & \xrightarrow[\mathbf{GF}(\overline{f \circ \eta_{A_2}})]{} & \mathbf{GF}(A_2) & & \mathbf{F}(A_2) \end{array}\right]$$

$$\Longrightarrow \mathbf{F}(f) = (or \equiv)\ \overline{f \circ \eta_{A_2}}$$

The option "or $\equiv$" in the Minor Diagram recognizes that the Major Diagram makes no use of the action of $\mathbf{F}$ on morphisms; so if $\mathbf{F}$ is only known on objects and the Major Diagram is satisfied, then the Minor Diagram constructs the **correct** action of $\mathbf{F}$ on morphisms. The arrow η in the above diagrams is called the **unit** of $\mathbf{F} \dashv \mathbf{G}$, and the analogous canonical arrow ε in the duals of the above diagrams (with $\mathcal{C}$ and $\mathcal{D}$ interchanged) is called the **counit** of $\mathbf{F} \dashv \mathbf{G}$.

Finally, an outline of this chapter by section is:

1. Ground Categories for Poslat Set Theories
2. General Frameworks for Poslat Topology
3. Lattice-Theoretic Frameworks
4. Classical Spatiality and Sobriety
5. Lattice-Dependent Spatiality and Sobriety
6. Lattice-"Independent" Spatiality and Sobriety
7. Final Remarks and Acknowledgements

§1. Ground Categories for Poslat Set Theories

In this section we shall have frequent recourse to the following categories: **SET**, the category of sets and functions; **CSLAT**, the category of complete join semi-lattices and functions preserving arbitrary $\bigvee$; **POSET**, the category of partially-ordered sets and order-preserving functions; and **CSLF**, the category of complete lattices and functions preserving arbitrary

$\bigvee$ and finite $\vee$. The name of the last category is an acronym on the phrase **C**(omplete join) **S**(emi-) **L**(attices with) **F**(ramelike morphisms), and is motivated by **FRM**, the category of complete lattices satisfying the first infinite distributive law and functions preserving arbitrary $\bigvee$ and finite $\wedge$. The category $\mathbf{LOC} = \mathbf{FRM}^{op}$ (see 1.2.1 below). Each lattice satisfies $0 \equiv \bigvee\{\ \} \neq \bigwedge\{\ \} \equiv 1$. Finally, recall that if L, M are pre-ordered sets, they may be regarded as pre-ordered categories; and if $f \in \mathbf{SET}(L, M)$, then f is a functor from L to $M \Leftrightarrow f$ is order-preserving.

Classical set theory is a poslat set theory in that functions between sets give rise to powerset operators between powersets, the latter two having important categorical and lattice-theoretic properties. More precisely, given $f : X \longrightarrow Y$, there is a unique lifting $f^{\rightarrow} : \mathcal{P}(X) \longrightarrow \mathcal{P}(Y)$ of f to the associated powersets which preserves arbitrary $\bigvee(= \cup)$; this unique lifting is the essence of the adjunction between **SET** and **CSLAT**. And $f^{\rightarrow}$ then uniquely generates the powerset operator $f^{\leftarrow} : \mathcal{P}(X) \longleftarrow \mathcal{P}(Y)$ via adjunction between the pre-ordered categories $\mathcal{P}(X)$ and $\mathcal{P}(Y)$, i.e. $f^{\rightarrow} \dashv f^{\leftarrow}$, using the pre-ordered version of the **Adjoint Functor Theorem**, both (dual) statements of which are now given (see [Herrlich-Strecker 1979] or [Johnstone 1982]).

1.1 Theorem (**Adjoint Functor Theorem (AFT)**). Let $f \in \mathbf{SET}(L, M)$ such that L has arbitrary $\bigvee$ and f preserves them. Then f is a functor from L to M; $\exists\ g$ a functor from M to L such that $f \dashv g$, where g is given by $g(b) = \bigvee\{a \in L : f(a) \leq b\}$; and g is the **unique** right adjoint if $L \in |\mathbf{POSET}|$.

1.2 Theorem (**Adjoint Functor Theorem (AFT)**). Let $g \in \mathbf{SET}(M, L)$ such that M has arbitrary $\bigwedge$ and g preserves them. Then g is a functor from M to L; $\exists\ f$ a functor from L to M such that $f \dashv g$, where f is given by $f(a) = \bigwedge\{b \in M : a \leq g(b)\}$; and f is the **unique** left adjoint if $M \in |\mathbf{POSET}|$.

1.3 Criteria. It would seem desirable in any poslat theory more general than classical set theory that the following properties be **captured** or at least **characterized**:

(1) The analogue of $f^{\rightarrow}[f^{\leftarrow}]$ be a unique lifting of $f^{\rightarrow}[f^{\leftarrow}]$, i.e. that the analogue of $f^{\rightarrow}[f^{\leftarrow}]$ be a **generalization** of $f^{\rightarrow}[f^{\leftarrow}]$ in a natural way.
(2) The analogue of $f^{\rightarrow}$ be left adjoint to the analogue of $f^{\leftarrow}$ i.e. the relationship between $f^{\rightarrow}$ and $f^{\leftarrow}$, in light of (1), be **generalized** in the new setting.

We find in the sequel that the requirements that these two properties be captured or characterized are met by all the poslat, non-model-theoretic, fuzzy set theories extant in the literature. And the meeting of these requirements guarantees the meaningfulness of fuzzy topology done in these settings.

§1.1 Poslat Fuzzy Set Theories Without Change of Basis

We now consider poslat set theories in which the **basis** or lattice L of membership values is held fixed, e.g. the set theories found in [Zadeh 1965], [Chang 1968], [Goguen 1967], 1973].

(The term "basis" has been suggested by U. Höhle.) Not all the restrictions places on L in these appraches will be required; we only assume $L \in |\mathbf{CSLF}|$. the choice of **CSLF** is defended in [Rodabaugh 1986, 1986–7, 1991, ****].

Let $X \in |\mathbf{SET}|$, $L \in |\mathbf{CSLF}|$, and endow L^X point-wise with $\leq, \bigvee, \bigwedge$ from L. Then $L^X \in |\mathbf{CSLF}|$, and is the lattice of **L-fuzzy subsets** of X because of:

$$\chi : \mathcal{P}(X) \longrightarrow \{0,1\}^X, \ \{0,1\}^X \hookrightarrow L^X$$

where $\chi(A) = \chi_A$ determines an isomorphism χ in **CSLF**, and $\hookrightarrow$ is a monomorphism (embedding) in **CSLF**. Let $\eta = \hookrightarrow \circ \chi$.

1.1.1 Criteria. Given $X, Y \in |\mathbf{SET}|$ and $L \in |\mathbf{CSLF}|$, there should exist $f_L^{\rightarrow}, f_L^{\leftarrow}$ such that the following restatements of Criteria 1.3(1,2) hold or are characterized:

(1) The following diagrams commute:

$$\begin{array}{ccc} X & \xrightarrow{\ f\ } & Y \\ \downarrow & \circ & \downarrow \\ \mathcal{P}(X) & \xrightarrow{\ !\, f^{\rightarrow}\ } & \mathcal{P}(Y) \\ \eta \downarrow & \circ & \downarrow \eta \\ L^X & \xrightarrow[\ ?\,!\, f_L^{\rightarrow}\ ]{} & L^Y \end{array} \qquad \begin{array}{ccc} \mathcal{P}(X) & \xrightarrow{\ !\, f^{\leftarrow}\ } & \mathcal{P}(Y) \\ \eta \downarrow & \circ & \downarrow \eta \\ L^X & \xrightarrow[\ ?\,!\, f_L^{\leftarrow}\ ]{} & L^Y \end{array}$$

We want $f_L^{\rightarrow}$, $f_L^{\leftarrow}$ to be the **unique** liftings of $f^{\rightarrow}$, $f^{\leftarrow}$ respectively, i.e. the **unique** arrows making the bottom squares commute.

(2) $f_L^{\rightarrow} \dashv f_L^{\leftarrow}$, viewing L^X and L^Y as pre-ordered categories.

1.1.2 Comment. These criteria are met in 1.1.4 and 1.1.6.1 below by determining $f_L^{\leftarrow}$ **first** and **then** determining $f_L^{\rightarrow}$, the obverse of the classical case. To do this we need monoids L of the form: $L \in |\mathbf{CSLF}|$, and $\exists$ an associative, commutative, binary operation $*$ on L having the upper [lower] bound 1 [0] as identity [zero]. We call such monoids **complete integral l-monoids** or **cil-monoids**. These should not be confused with the integral complete l-monoids or integral cl-monoids of [Birkoff 1967] since **no** distributivity of any sort of $*$ with respect to $\bigvee$ is assumed; cf. [Höhle 1978]. Note each object of **CSLF** is a cil-monoid if $*$ is chosen to be $\wedge$; and if $L = I$, we may choose $* = \cdot$ or a continuous t-norm T such as T_m — see [Klement 1981b, 1982a, 1982b] and [Höhle-Klement 1984].

Cil-monoids are needed for the following lemma (cf. [Negoita-Ralescu 1975]), which subsequently proves the theorem determining $f_L^{\leftarrow}$. *Sequens*, $\forall X, \forall L, \forall \alpha \in L$, $\underline{\alpha}$ denotes the **constant map** in L^X having value α.

1.1.3 LEMMA (**Decomposition Theorem**). Let $X \in |\mathbf{SET}|$ and L be a cil-monoid. Then each $b \in L^X$ is of the form: $b = \bigvee\{\underline{\alpha} * \chi_{[b \geq \alpha]} : \alpha \in L\}$.

PROOF. Let $x \in X$, set $\beta = $ b(x), and let $\alpha \in L$.∋. $x \in [b \geq \alpha]$. Then

$$b(x) \geq \alpha = (\underline{\alpha} * \underline{1})(x) = (\underline{\alpha} * \chi_{[b \geq \alpha]})(x),\ b(x) \geq (\bigvee\{\underline{\alpha} * \chi_{[b \geq \alpha]} : \alpha \in L\})(x)$$

On the other hand, $x \in [b \geq \beta]$, so that

$$b(x) \leq \beta = (\underline{\beta} * \underline{1})(x) = (\underline{\beta} * \chi_{[b \geq \beta]}(x),\ b(x) \leq (\bigvee\{\underline{\alpha} * \chi_{[b \geq \alpha]} : \alpha \in L\})(x)$$

which completes the proof. □

1.1.4 THEOREM (**Existence and Uniqueness of $\mathbf{f_L^{\leftarrow}}$**). Let L be a cil-monoid. Then ∃ **unique** $g \in \mathbf{CSLAT}(L^Y, L^X)$ satisfying:

(1) $\forall \alpha \in L,\ g(\underline{\alpha}) = \underline{\alpha}$;
(2) $\forall \alpha \in L,\ \forall B \subset Y,\ g(\underline{\alpha} * \chi_B) = \underline{\alpha}\ *\ g(\chi_B)$; and
(3) g lifts $f^{\leftarrow}$, i.e. $g \circ \eta = \eta \circ f^{\leftarrow}$ in 1.1.1(1).

Furthermore this g is given by: $\forall b \in L^Y,\ g(b) = b \circ f$. In particular, g is the **unique** choice of $f_L^{\leftarrow}$ with respect to (1-3) for each $L \in |\mathbf{CSLF}|$ for $* = \wedge$.

PROOF. We first show that

$$[g \in \mathbf{CSLAT}(L^Y, L^X) \text{ satisfying (1-3)}] \Rightarrow [\forall b \in L^Y,\ g(b) = b \circ f]$$

Letting $b \in L^Y$, we have from 1.1.3, $\forall b = \bigvee\{\underline{\alpha} * \chi_{[b \geq \alpha]} : \alpha \in L\}$, and so

$$\begin{aligned} g(b) &= g(\bigvee\{\underline{\alpha} * \chi_{[b \geq \alpha]} : \alpha \in L\} = \bigvee\{\underline{\alpha} * g(\chi_{[b \geq \alpha]}) : \alpha \in L\} \\ &= \bigvee\{\underline{\alpha} * g(\eta([b \geq \alpha])) : \alpha \in L\} = \bigvee\{\underline{\alpha} * \eta(f^{\leftarrow}([b \geq \alpha])) : \alpha \in L\} \\ &= \bigvee\{\underline{\alpha} * \chi_{f^{\leftarrow}([b \geq \alpha])} : \alpha \in L\} = \bigvee\{\underline{\alpha} * \chi_{[b \circ f \geq \alpha]} : \alpha \in L\} \\ &= b \circ f \end{aligned}$$

where in the next to last line, we used the fact that

$$\begin{aligned} f^{\leftarrow}[b \geq \alpha] &= \{x \in X : f(x) \in [b \geq \alpha]\} \\ &= \{x \in X : b(f(x) \geq \alpha\} \\ &= [b \circ f \geq \alpha] \end{aligned}$$

together with 1.1.3 applied to $b \circ f$. To complete the proof, we must show that

$$[\forall\, b \in L^X,\ g(b) = b \circ f] \Rightarrow [g \in \mathbf{CSLAT}(L^Y, L^X) \text{ satisfying (1-3)}]$$

Ad(1). Since $\underline{\alpha} \circ f = \underline{\alpha}$, g satisfies (1).

Ad(2). Let $\underline{\alpha} \in L$ and $B \subset Y$. Now $g(\underline{\alpha} * \chi_B) = (\underline{\alpha} * \chi_B) \circ f$ and $\alpha * g(\chi_B) = \underline{\alpha} * (\chi_B \circ f)$. To show these are equal, let $x \in X$. Then $(\underline{\alpha} * \chi_B)(f(x)) = \underline{\alpha}(f(x)) * (\chi_B(f(x)) = \alpha * \chi_B(f(x)) = \underline{\alpha}(x) * \chi_B(f(x))$.

Ad(3). Let $B \subset Y$ and $x \in X$. Then $[(g \circ \eta)(B)](x) = [g(\chi_B)](x) = [\chi_B \circ f](x) = \chi_B(f(x)) = \chi_{f^{\leftarrow}(B)}(x) = [\eta \circ f^{\leftarrow}(B)](x)$. Hence $g \circ \eta = \eta \circ f^{\leftarrow}$. □

1.1.5 REMARK. We are **compelled** to take $f_L^{\leftarrow}$ as the g in Theorem 1.1.4. This justifies the original definition, for $L = I$, given in [Zadeh 1965, p. 346]. Hence the **correct** definition of $f_L^{\leftarrow}(b)$, for each L and each b, is determined by completions of diagrams of the form:

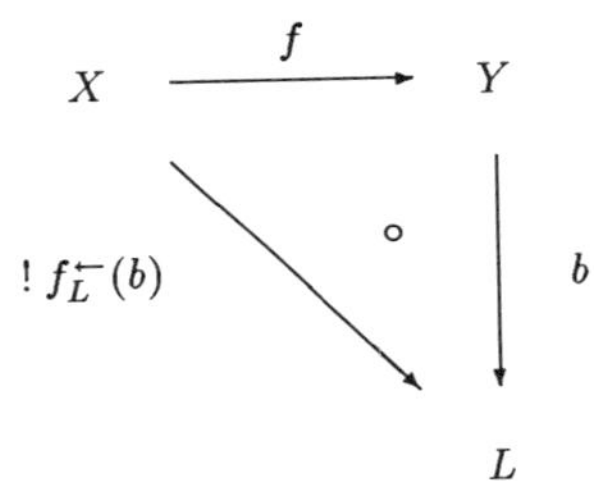

1.1.6 DETERMINATION OF $f_L^{\rightarrow}$. The determination of $f_L^{\leftarrow}$ effectively settles the choice of $f_L^{\rightarrow}$, as we now indicate.

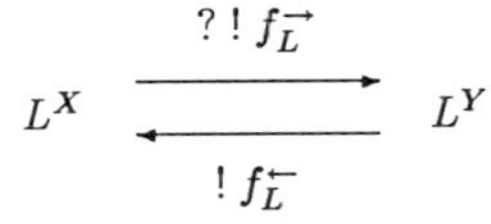

1.1.6.1 THEOREM. L^Y is in **POSET**, has arbitrary $\bigwedge$, and $f_L^{\leftarrow}$ preserves them. Hence $\exists$ (by AFT) a **unique** $h \in \mathbf{CSLAT}(L^X, L^Y)$ such that $h \dashv f_L^{\leftarrow}$, where h is given by

$$h(a)(y) = \bigvee\{a(x) : f(x) = y\} \tag{1.1.6.1(a)}$$

Furthermore, h lifts $f^{\rightarrow}$, i.e. $h \circ \eta = \eta \circ f^{\rightarrow}$ in 1.1.1(1); it is the **unique** such lifting left adjoint to $f_L^{\leftarrow}$. Thus h is the **correct** choice of $f_L^{\rightarrow}$.

PROOF. It is obvious that L^Y has arbitrary $\bigwedge$, and it is straightforward that $f_L^{\leftarrow}$ preserves arbitrary $\bigwedge$: $f_L^{\leftarrow}(\bigwedge_\gamma a_\gamma) = (\bigwedge_\gamma a_\gamma) \circ f = \bigwedge_\gamma (a_\gamma \circ f) = \bigwedge_\gamma f_L^{\leftarrow}(a_\gamma)$. Hence, by 1.2 (AFT), $\exists! h \in \mathbf{CSLAT}(L^X, L^Y)$ such that $h \dashv f_L^{\leftarrow}$. This h is given by (1.2):

$$h(a) = \bigwedge\{b \in L^Y : a \leq f_L^{\leftarrow}(b)\} \qquad (1.1.6.1(\text{b}))$$

It remains to show that 1.1.6.1(b) agrees with 1.1.6.1(a), and that h lifts the classical $f_L^{\leftarrow}$. For the former, note 1.1.6.1(b) can be rewritten

$$h(a) = \bigwedge\{b \in L^Y : a \leq b \circ f\}$$

Put $b^* = h(a)$ as defined in 1.1.6.1(a). It suffices to show

$$b^* = \bigwedge\{b \in L^Y : a \leq b \circ f\} \qquad (1.1.6.1(\text{c}))$$

Fix $y \in Y$. Since $b^*(y) = \bigwedge\{a(x) : f(x) = y\}$, $a(x) \leq b^*(y) = b^*(f(x))$, so "$\geq$" of 1.1.6.1(c) holds. On the other hand, if $a \leq b \circ f$, then

$$\begin{aligned} \bigvee\{a(x) : f(x) = y\} &\leq \bigwedge\{b(f(x)) : f(x) = y\} \\ &= \begin{cases} b(y), & f^{\leftarrow}\{y\} \neq \emptyset \\ 0, & \text{otherwise} \end{cases} \\ &\leq b(y) \end{aligned}$$

so that "$\leq$" holds in 1.1.6.1(c). Now for the lifting criterion, let $A \in \mathcal{P}(X)$ and $y \in Y$. Then

$$\begin{aligned} h(\eta(A))(y) = h(\chi_A)(y) = \bigvee\{\chi_A(y) : f(x) = y\} &= \begin{cases} 1, \exists\, x \in A \text{ such that } f(x) = y \\ 0, \text{ otherwise} \end{cases} \\ &= \chi_{f^{\rightarrow}(A)}(y) = \eta(f^{\rightarrow}(A))(y). \quad \Box \end{aligned}$$

1.1.6.2 META-COROLLARY. The criteria for a poslat set theory have been met for the fixed-basis case. In particular, the "Zadeh Extension Principle" of [Zadeh loc.cit.] is **correct**.

§1.2 POSLAT FUZZY SET THEORIES WITH CHANGE OF BASIS.

The original poslat fuzzy set theory of [Zadeh 1965] and [Chang 1968] used $L = I$ as the fixed basis. More general L were considered by [Goguen 1967, 1973], [Hutton 1975], [Höhle 1978–82], etc. In 1980, [Hutton 1980], followed by [Hutton-Reilly 1980] and [Erceg 1980], proposed the first categorical framework for change of basis. As demonstrated by [Rodabaugh 1983], Hutton's approach is a poslat set theory for singleton sets which allows for change of basis; i.e. Hutton's approach is purely a "lat" instead of "poslat" approach and incorporates **only** change of basis. A general poslat set theory allowing both change of set and change of basis was then proposed in [Rodabaugh 1983] and later reshaped and studied by [Rodabaugh 1986–1988a, 1991, ****], [Eklund 1984, 1986a, 1986b, ****], and [Meßner-Rodabaugh ****]. We now give this general poslat theory as outlined in [Rodabaugh 1986, 1986–7, 1991, ****].

1.2.1 DEFINITION. The category $\mathbb{L}$ is the dual of **CSLF**, i.e. is $\mathbf{CSLF}^{op}$. This means the objects of $\mathbb{L}$ are the same as those of **CSLF**, i.e. $|\mathbb{L}| = |\mathbf{CSLF}|$; and with regard to morphisms, both of the following hold:

(1) $\forall \phi \in \mathbb{L}\ (L, M),\ \exists!\, \psi \equiv \phi^{op} \in \mathbf{CSLF}(M, L)$;
(2) $\forall \psi \in \mathbf{CSLF}(M, L),\ \exists!\, \phi \equiv \psi^{op} \in \mathbb{L}(L, M)$.

Thus the morphisms of **CSLF** and $\mathbb{L}$ are in a bijection; the unique arrow in **CSLF** paired with a given ϕ in $\mathbb{L}$ is denoted ϕ^{op} (so above $\psi \equiv \phi^{op}$). The arrows of $\mathbb{L}$ need not be functions, i.e. need not be concrete. Note $\mathbb{L}$ has the same relationship to **CSLF** that **LOC** (category of **locales**) $\equiv \mathbf{FRM}^{op}$ has to **FRM**; one may say that **CSLF** and $\mathbb{L}$ are **FRM** and **LOC**, respectively, *sans* any distributivity.

1.2.2 LEMMA. $\forall \phi \in \mathbb{L}(A, B),\ \exists [\phi] \in \mathbf{POSET}(A, B)$, which is given by $[\phi](a) = \bigwedge\{b : a \leq \phi^{op}(b)\}$. It is also the case that $[\phi](0) = 0$.

PROOF. The first assertion follows since $\phi^{op} \in \mathbf{POSET}$, and the second follows since $\phi^{op}(0) = 0$. □

1.2.3 DEFINITION (**The Ground Category**). The category $\mathbf{SET} \times \mathbb{L}$ is defined as follows: **objects** are pairs (X, L), with $X \in |\mathbf{SET}|$ and $L \in |\mathbb{L}|$ **morphisms** are pairs (f, ϕ) with $f \in \mathbf{SET}$ and $\phi \in \mathbb{L}$; and **composition** is "slotwise", i.e. is that of **SET** [$\mathbb{L}$] in the 1st [2nd] slot. The phrase "ground category" means a poslat ground for topology and presages the next section; this phrase was taken from [Eklund 1986b]. From [Eklund 1984] comes the first description of a ground as a product category.

What sort of set theory must we in $\mathbf{SET} \times \mathbb{L}$ have? To answer that question we need two important arrows determined by ϕ and $[\phi]$. Let $X \in |\mathbf{SET}|$ and $\phi \in \mathbb{L}(L, M)$. Put $\langle[\phi]\rangle : L^X \longrightarrow M^X$, $\langle\phi^{op}\rangle : L^X \longleftarrow M^X$ by:

$$\langle[\phi]\rangle(a) = [\phi] \circ a, \quad \langle\phi^{op}\rangle(b) = \phi^{op} \circ b$$

1.2.4 LEMMA. $\langle[\phi]\rangle \in \mathbf{POSET}(L^X, M^X)$ and $\langle\phi^{op}\rangle \in \mathbf{CSLF}(M^X, L^X)$.

PROOF. It is straightforward to see that $\langle[\phi]\rangle$ and $\langle\phi^{op}\rangle$ inherit their claimed properties from $[\phi]$ and ϕ^{op}. □

1.2.5 DETERMINATION OF $(f, \phi)^{\rightarrow}$ AND $(f, \phi)^{\leftarrow}$ FROM (f, ϕ). Even though $(f, \phi)^{\leftarrow}$ is more amenable than $(f, \phi)^{\rightarrow}$, in the sequel we start discussing, but do not at that time resolve, $(f, \phi)^{\rightarrow}$; we next resolve $(f, \phi)^{\leftarrow}$ completely, and then return to resolve $(f, \phi)^{\rightarrow}$. Crucial use is made of $f_L^{\rightarrow}, f_L^{\leftarrow}, f_M^{\rightarrow}, f_M^{\leftarrow}$ uniquely determined in §1.1. Thus, the goal of the following material is to show that the conditions of 1.3(1,2) are either captured or characterized in the ground category **SET** $\times$ $\mathbb{L}$: 1.3(1) is partly addressed in 1.2.5.1-1.2.5.5, and both of 1.3(1,2) are finally resolved in 1.2.5.7-1.2.5.11, especially 1.2.5.7. We begin by analyzing each of two diagrams in connection with 1.3(1).

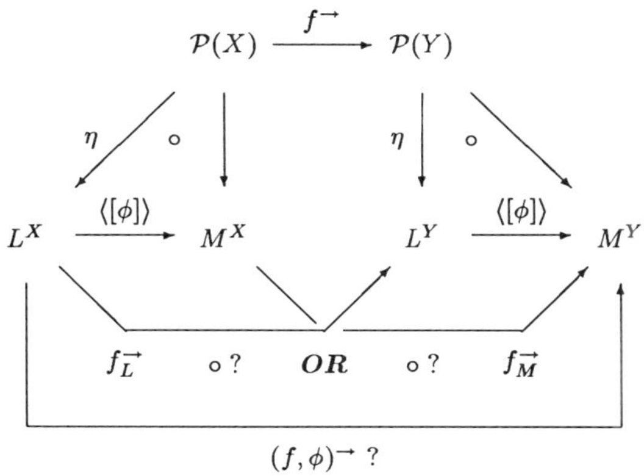

1.2.5.1 LEMMA. If $\langle[\phi]\rangle(\underline{1}) = \underline{1}$, then in both the upper two triangles of the above diagram, $\langle[\phi]\rangle \circ \eta = \eta$ (mapping to M^X in the left triangle and to M^Y in the right triangle).

PROOF. We do the left upper triangle case, the other case being similar. Let $A \subset X$ and $x \in X$. Then $(\langle[\phi]\rangle(\eta(A)))(x) = ([\phi] \circ \chi_A)(x) = [\phi](\chi_A(x))$. The latter is 0 if $x \notin A$ by 1.2.2, and the latter is 1 if $x \in A$ by the hypothesis — for if $x \in A$, we have $[\phi](\chi_A(x)) = [\phi](\underline{1}(1(x)) = (\langle[\phi]\rangle(\underline{1}))(x) = \underline{1}(x) = 1$. $\square$

1.2.5.2 QUESTION. Which commutivity, i.e. lifting, of the above diagram should define $(f,\phi)^{\rightarrow}$ (note the two possible definitions do not always coincide)? See 1.2.5.7, 1.2.5.11 below.

1.2.5.3 LEMMA. Both possible definitions of $(f,\phi)^{\rightarrow}$ yield arrows in **POSET**. The definition imposed by the left [right] bottom triangle of the above diagram is equivalent to putting $(f,\phi)^{\rightarrow}(a) \equiv \bigwedge\{b : f_L^{\rightarrow}(a) \leq \langle\phi^{op}\rangle(b)\}$ $[\bigwedge\{b : f_M^{\rightarrow}(\langle[\phi]\rangle(a)) \leq b\}$, respectively$]$.

PROOF. We do the left bottom triangle case, the other case being similar. Now the definition actually imposed on $(f,\phi)^{\rightarrow}$ by the left bottom triangle is $(f,\phi)^{\rightarrow} \equiv \langle[\phi]\rangle \circ f_L^{\rightarrow}$, which definition evaluated at an $a \in L^X$ and $y \in Y$ yields:

$$(\langle[\phi]\rangle(f_L^{\rightarrow}(a))(y) = ([\phi] \circ f_L^{\rightarrow}(a))(y) = [\phi](f_L^{\rightarrow}(a)(y)) = \bigwedge\{\beta : f_L^{\rightarrow}(a)(y) \leq \phi^{op}(\beta)\}$$

We must verify that $\bigwedge\{\beta : f_L^{\rightarrow}(a)(y) \leq \phi^{op}(\beta)\} = (\bigwedge\{b : f_L^{\rightarrow}(a) \leq \langle\phi^{op}\rangle(\beta)\})(y)$, i.e. that $\bigwedge\{\beta : f_L^{\rightarrow}(a)(y) \leq \phi^{op}(\beta)\} = \bigwedge\{b(y) : f_L^{\rightarrow}(a) \leq \langle\phi^{op}\rangle(b)\}$. It suffices to show

$$\{\beta : f_L^{\rightarrow}(a)(y) \leq \phi^{op}(\beta)\} = \{b(y) : f_L^{\rightarrow}(a) \leq \phi^{op} \circ b\}$$

Given β from the left set, define $b \in M^Y$ by

$$b(z) = \begin{cases} \beta, & z = y \\ 1, & z \neq y \end{cases}$$

Since $f_L^{\rightarrow}(a)(y) \leq \phi^{op}(\beta) = \phi^{op}(b(y))$, it follows $f_L^{\rightarrow}(a) \leq \phi^{op} \circ b$; and so $\beta = b(y)$ is in the right set. Now given $b(y)$ in the right set, put $\beta = b(y)$; it is easy to show $b(y) = \beta$ is in the left set—this is left to the reader. $\square$

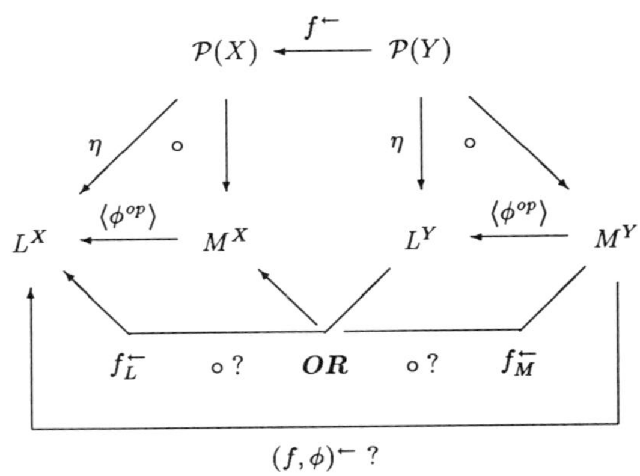

1.2.5.4 QUESTION. Which commutativity, i.e. lifting of the second diagram above, should define $(f, \phi)^{\leftarrow}$?

1.2.5.5 LEMMA. Both commutativities, of the second diagram above, give the **same definition** for $(f, \phi)^{\leftarrow}$, and the resulting arrow is in **CSLF**. Thus $(f, \phi)^{\leftarrow}$ is **doubly uniquely** determined and completes **uniquely** all squares of the following form:

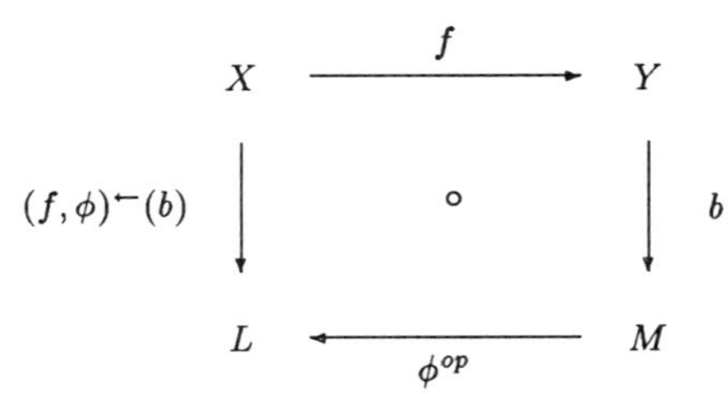

PROOF. We show that the two definitions coincide with each other; in so doing, their coincidence with the squares will also emerge. The other claims are left to the reader. Let $b \in M^Y$. Then we have:

$$\begin{aligned}(\langle\phi^{op}\rangle \circ f_M^{\leftarrow})(b) &= \langle\phi^{op}\rangle(f_M^{\leftarrow}(b)) = \phi^{op} \circ f_M^{\leftarrow}(b) = \phi^{op} \circ (b \circ f) \\ &= (\phi^{op} \circ b) \circ f = f_L^{\leftarrow}(\phi^{op} \circ b) = f_L^{\leftarrow}(\langle\phi^{op}\rangle(b)) \\ &= (f_L^{\leftarrow} \circ \langle\phi^{op}\rangle)(b). \quad \square\end{aligned}$$

1.2.5.6 DEFINITION. A statement in the literals X, Y, f holds **universally** iff it holds $\forall X \in |\mathbf{SET}|$, $\forall Y \in |\mathbf{SET}|$, and $\forall f \in \mathbf{SET}(X, Y)$. This terminology is needed in the following theorem.

1.2.5.7 FUNDAMENTAL THEOREM (**General Poslat Set Theory With Unique Liftings**). Let $\phi \in \mathbb{L}(L, M)$. Then the following are equivalent:

(1) ϕ^{op} preserves arbitrary $\bigwedge$;
(2) $[\phi]$ preserves arbitrary $\bigvee$;
(3) $[\phi] \dashv \phi^{op}$;
(4) $(f,\phi)^{\rightarrow} \dashv (f,\phi)^{\leftarrow}$ universally for at least one definition of $(f,\phi)^{\rightarrow}$;
(5) $(f,\phi)^{\rightarrow} \dashv (f,\phi)^{\leftarrow}$ universally for both definitions of $(f,\phi)^{\rightarrow}$;
(6) both definitions of $(f,\phi)^{\rightarrow}$ coincide universally;
(7) at least one definition of $(f,\phi)^{\rightarrow}$ preserves arbitrary $\bigvee$ universally;
(8) both definitions of $(f,\phi)^{\rightarrow}$ preserve arbitrary $\bigvee$ universally;
(9) $(f,\phi)^{\leftarrow}$ preserves arbitrary $\bigwedge$ universally.

OUTLINE OF PROOF. Full details are given in [Rodabaugh ****]. The proof hinges almost entirely around the "universal" character of the statements of the theorem and around the AFT (1.1/1.2). (1) $\Leftrightarrow$ (2) $\Leftrightarrow$ (3) is an immediate consequence of the AFT. Now (5) $\Rightarrow$ (4) is trivial, and the AFT assures that (4) $\Rightarrow$ (1,2,3) $\Rightarrow$ (5). By the AFT, adjoints are unique, so (5) $\Rightarrow$ (6), but (6) $\Rightarrow$ (1,2,3) requires the universality of (6). Also by the AFT, (4) $\Leftrightarrow$ (7) and (5) $\Leftrightarrow$ (8). But it can be shown independently of the AFT that (2) $\Leftrightarrow$ (7) and (2) $\Leftrightarrow$ (8), so that (7) $\Leftrightarrow$ (8); and it is also independent of the AFT that (1) $\Leftrightarrow$ (9). □

1.2.5.8 COROLLARY. (1-9) of 1.2.5.7 hold under the following conditions:

(1) M has an order-reversing involution and ϕ^{op} preserves all involutes;
(2) ϕ is an isomorphism in $\mathbb{L}$.

The definitions and lifting properties of $f_L^{\rightarrow}$ and $f_L^{\leftarrow}$ are recovered by setting $M = L$ and $\phi = id_L$.

1.2.5.9 DEFINITION (**The Category DMRG**). Corollary 2.3.5.8(1) justifies pinpointing those $L \in |\mathbf{CSLF}|$ which have an order-reversing involution and those morphisms of **CSLF** which preserve the involution. Because the deMorgan laws hold in each such L for arbitrary $\bigvee$ and arbitrary $\bigwedge$, it shall be called a **deMorgan algebra**—this is a non-standard definition since no distributivity is assumed; but our axioms for a deMorgan algebra are the minimum needed for the deMorgan laws to hold for arbitrary $\bigvee$ and arbitrary $\bigwedge$. Let **DMRG** denote the subcategory of such objects and morphisms. Note each arrow in **DMRG** preserves arbitrary $\wedge$ as a consequence of the deMorgan laws.

1.2.5.10 REMARK. The factorization of $(f,\phi)^{\leftarrow}$ has been studied at length in [Eklund 1986, ****] under the restriction of 1.2.5.8(1).

1.2.5.11 DEFINITION OF $(f,\phi)^{\rightarrow}$). If (1–9) of 2.3.5.7 fail to hold, we choose the "left bottom" definition of $(f,\phi)^{\rightarrow}$, namely $(f,\phi)^{\rightarrow} = \langle[\phi]\rangle \circ f_L^{\rightarrow}$ (or the equivalent given in 1.2.5.3), as our general definition. Thus $(f,\phi)^{\rightarrow}$ may be viewed as completing **uniquely** all lower triangles of squares of the following form:

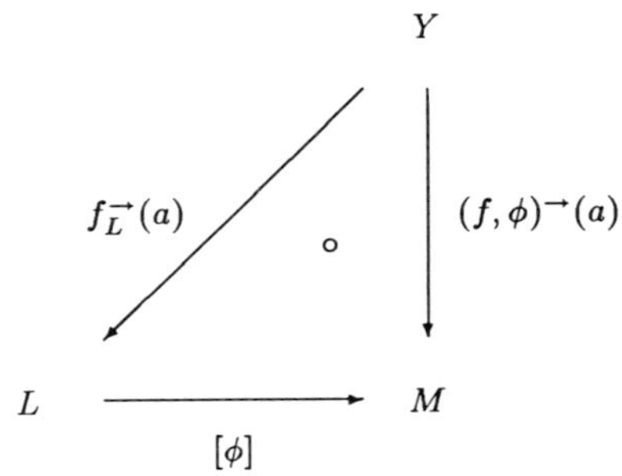

§2. General Frameworks for Poslat Topology

In this section are developed the topological frameworks pertinent to Stone Representations and Compactifications. These frameworks are based on the ground categories presented in §1. In counterdistinction to §1, we begin with the "change-of-basis" topological frameworks of [Rodabaugh 1983–1991, ****] and [Meßner-Rodabaugh ****]: the approach of Rodabaugh yields a category $\mathbb{T}$ based on the ground **SET** $\times$ $\mathbb{L}$, and that of Meßner-Rodabaugh yields a category $\mathbb{T}_D$ also based on the ground **SET** $\times$ $\mathbb{L}$. In the development of these topological frameworks, what is taken from the ground category is explicitly distinguished from what is intrinsically topological.

We also discuss the change-of-basis framework of [Hutton 1980], embedded as a subcategory of $\mathbb{T}$, and give the relationship between $\mathbb{T}$ and $\mathbb{T}_D$ in detail. Finally, we shall consider "fixed-basis" and "fixed-change-of-basis" topological frameworks: these include the categories of [Chang 1968], [Goguen 1973], and [Lowen 1976] embedded as important and sometimes crucial subcategories of the more general frameworks $\mathbb{T}$ and $\mathbb{T}_D$.

2.1 DEFINITION (**The Category** $\mathbb{T}$). The data for $\mathbb{T}$ are as follows.

(1) **Objects**. Objects are ordered triples (X, L, τ), called (**fuzzy**) **topological spaces**, satisfying the following axioms:
   (i) **Ground Axiom**. $(X, L) \in |\mathbf{SET} \times \mathbb{L}|$.
   (ii) **Topological Axiom**. $\tau \subset L^X$ such that $\hookrightarrow \in$ **CSLF**, i.e. τ is closed under arbitrary $\bigvee$ and finite $\wedge$; τ is called a **topology**, and its members are called **open sets**.

(2) **Morphisms**. Morphisms are ordered pairs $(f, \phi) : (X_1, L_1, \tau_1) \longrightarrow (X_2, L_2, \tau_2)$, called (**fuzzy**) **continuous**, satisfying the following axioms:

   (i) **Ground Axiom**. $(f,\phi) \in [\mathbf{SET} \times \mathbb{L}]\ [(X_1, L_1), (X_2, L_2)]$.
   (ii) **Continuity Axiom**. $(f,\phi)^{\leftarrow}|_{\tau_2} \in \mathbf{CSLF}(\tau_2, \tau_1)$, i.e. $\forall\, v \in \tau_2,\ (f,\phi)^{\leftarrow}(v) \in \tau_1$.
(3) **Composition**. As in $\mathbf{SET} \times \mathbb{L}$.

Subcategories of $\mathbb{T}$, defined later, demonstrate that $\mathbb{T}$ enriches the approaches of Chang, Goguen, Lowen, and Hutton mentioned above. The category $\mathbb{T}$ is in turn enriched by $\mathbb{T}_D$ defined below.

2.2 DEFINITION (**The Category $\mathbb{T}_D$**). The data for $\mathbb{T}_D$ are as follows:

(1) **Objects**. Objects are ordered quadruples (X, L, τ, D), called (**fuzzy**) **topological spaces**, satisfying the following axioms:
   (i) **Ground Axiom**. $(X, L) \in |\mathbf{SET} \times \mathbb{L}|$.
   (ii) **Topological Axiom**. $\tau \subset L^X$ such that $\hookrightarrow \in \mathbf{CSLF}$, i.e. τ is closed under arbitrary $\vee$ and finite $\wedge$; so $(X, L, \tau) \in |\ \mathbb{T}\ |$.
   (iii) **Special Topological Axiom**. $D \subset L$ such that $\hookrightarrow \in \mathbf{CSLF}$, i.e. D is closed under arbitrary $\vee$ and finite $\wedge$.
   (iv) **Stratification Axiom**. $\forall\, \alpha \in D$, $\underline{\alpha}$ from L^X is in τ. We put $\delta : D \longrightarrow \tau$ by by $\delta(\alpha) = \underline{\alpha}$. It follows $\delta \in \mathbf{CSLF}(D, \tau)$.
(2) **Morphisms**. Morphisms are ordered pairs
$$(f,\phi) : (X_1, L_1, \tau_1, D_1) \longrightarrow (X_2, L_2, \tau_2, D_2),$$
called (**fuzzy**) **continuous**, satisfying the following axioms:
   (i) **Ground Axiom**. $(f,\phi) \in [\mathbf{SET} \times \mathbb{L}]\ [(X_1, L_1), (X_2, L_2)]$.
   (ii) **Continuity Axiom**. $(f,\phi)^{\leftarrow}|_{\tau_2} \in \mathbf{CSLF}(\tau_2, \tau_1)$, i.e. $\forall\, v \in \tau_2$, $(f,\phi)^{\leftarrow}(v) \in \tau_1$.
   (iii) **Special Continuity Axiom**. $\phi^{op}|_{D_2} \in \mathbf{CSLF}(D_2, D_1)$, i.e. ϕ^{op} maps D_2 into D_1.
(3) **Composition**. As in $\mathbf{SET} \times \mathbb{L}$.

2.3 REMARK. Given $(X, L, \tau, D) \in |\mathbb{T}_D|$, D is the **categorically recognized degree of stratification**. Special continuity insures the comparability of recognized degrees of stratification. The relationship of $\mathbb{T}_D$ to other categories is given below; however, we point out that with various choices of D, we capture various types of fuzzy topological spaces, as we now indicate:

$D = \{0, 1\}$: (X, L, τ, D) is recognized as a **Chang-Goguen** space

$D = L$: (X, L, τ, D) is recognized as a **Lowen** space

$D = D(\tau) \equiv \{\alpha \in L : \underline{\alpha} \in \tau\}$: (X, L, τ, D) is recognized as having $(X, L, \tau)'s$ **existing degree of stratification**

2.4 MNEMONIC FOR $\mathbb{T}$ AND $\mathbb{T}_D$. If $(f,\phi) \in \mathbb{T}_D$, the subdiagrams of the following diagram in **CSLF** commute. Hence the following diagram captures the lattice-theoretic data determined by morphisms of $\mathbb{T}_D$; it implicitly captures the data determined by morphisms of $\mathbb{T}$ as well (middle and right vertical arrows).

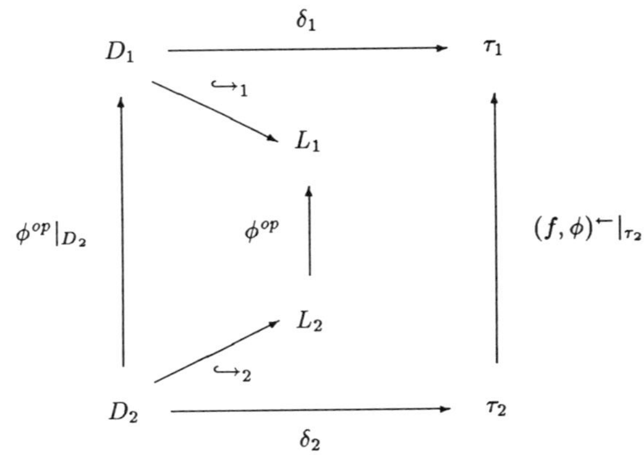

2.5 DISCUSSION. How different are the two categories $\mathbb{T}$ and $\mathbb{T}_D$? Let (X, L, τ) be an object in $\mathbb{T}$, and (as above) put

$$D(\tau) = \{\alpha \in L : \underline{\alpha} \in \tau\}$$

Then it can be easily checked that $D(\tau) \hookrightarrow L$ is in **CSLF**, i.e. $(X, L, \tau, D(\tau))$ is an object in $\mathbb{T}_D$. It can also be checked that if $f : (X_1, L_1, \tau_1) \longrightarrow (X_2, L_2, \tau_2)$ is in $\mathbb{T}$, then $\forall\, \beta \in L_2$,

$$(f,\phi)^{\leftarrow}(\underline{\beta}) = \underline{\phi^{op}(\beta)} \text{ as members of } L_1^{X_1}$$

whence it follows that $\forall\, \beta \in D(\tau_2)$, we have $\underline{\beta} \in \tau_2$, $(f,\phi)^{\leftarrow}(\underline{\beta}) \in \tau_1$, $\underline{\phi^{op}(\beta)} \in \tau_1$, and $\phi^{op}(\beta) \in D(\tau_1)$. So $(f,\phi) : (X_1, L_1, \tau_1, D(\tau_1)) \longrightarrow (X_2, L_2, \tau_2, D(\tau_2))$ is a morphism in $\mathbb{T}_D$. To summarize, let

$$\mathbf{S}_D : (X, L, \tau) = (X, L, \tau, D(\tau)),\ \mathbf{S}_D(f,\phi) = (f,\phi).$$

Clearly $\mathbf{S}_D$ preserves composition and identities, and so $\mathbf{S}_D : \mathbb{T} \longrightarrow \mathbb{T}_D$ is a functor, namely the (**degree of**) **stratification recognition** functor. Now let $\mathbf{S}_n : \mathbb{T} \longrightarrow \mathbb{T}_D$ be defined by

$$\mathbf{S}_n(X, L, \tau) = (X, L, \tau, \{0,1\}),\ \mathbf{S}_n(f,\phi) = (f,\phi)$$

Then it is easy to check that $\mathbf{S}_n$ is a functor; we call $\mathbf{S}_n$ the **non-stratification recognition functor**. Finally, let $\mathbf{V}$: $\mathbb{T}_D \longrightarrow \mathbb{T}$ be the **forgetful functor**, i.e.,

$$\mathbf{V}(X, L, \tau, D) = (X, L, \tau),\ \mathbf{V}(f,\phi) = (f,\phi)$$

We now are in a position to completely answer the question posed at the beginning of this discussion.

2.6 THEOREM. $\mathbf{S}_D \dashv \mathbf{V} \dashv \mathbf{S}_n$, and these adjunctions guarantee that $\mathbf{V}$ is a right-adjoint-right-inverse [left-adjoint-left-inverse] of $\mathbf{S}_D$ [$\mathbf{S}_n$]. Hence $\mathbf{S}_D$ [$\mathbf{S}_n$] embeds $\mathbb{T}$ as a full, coreflective [reflective] subcategory of $\mathbb{T}_D$.

PROOF. We detail only the adjunction $\mathbf{S}_D \dashv \mathbf{V}$, the adjunction $\mathbf{V} \dashv \mathbf{S}_n$ being left to the reader. The Major and Minor Diagrams of §0 are established via the following two steps.

CLAIM I. *$\mathbf{V}$ has a left adjoint whose action on objects coincides with $\mathbf{S}_D$.* It suffices to show:

$$\forall (X,L,\tau) \in |\mathbb{T}|, \exists \eta : (X,L,\tau) \longrightarrow [\mathbf{V}\mathbf{S}_D(X,L,\tau) = (X,L,\tau)] \text{ in } \mathbb{T},$$
$$\forall (f,\phi) : [(X,L,\tau) \longrightarrow \mathbf{V}(Y,M,\sigma,E) = (Y,M,\sigma)] \text{ in } \mathbb{T},$$
$$\exists!\, \overline{(f,\phi)} : [\mathbf{S}_D(X,L,\tau) = (X,L,\tau,D(\tau))] \longrightarrow (Y,M,\sigma,E) \text{ in } \mathbb{T}_D,$$
$$(f,\phi) = [\mathbf{V}(\overline{(f,\phi)}) = \overline{(f,\phi)}] \circ \eta$$

The Major Diagram looks as follows for this claim:

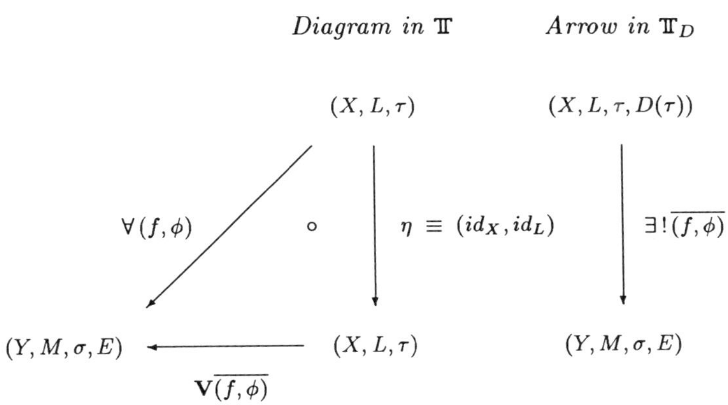

Letting $\eta = (id_X, id_L)$, $\overline{(f,\phi)}$ is forced to be (f,ϕ) in the ground category $\mathbf{SET} \times \mathbb{L}$. The only question is whether (f,ϕ) being a morphism from (X,L,τ) to (Y,M,σ) implies (f,ϕ) is a morphism from $(X,L,\tau,D(\tau))$ to (Y,M,σ,E), and this is simply the question of whether $\phi^{op}(E) \subset D(\tau)$, which is answered by noting

$$\beta \in E \Rightarrow \underline{\beta} \in \sigma \Rightarrow [\underline{\phi^{op}(\beta)} = (f,\phi)^{\leftarrow}(\underline{\beta})] \in \tau \Rightarrow \phi^{op}(\beta) \in D(\tau)$$

CLAIM II. *The action of $\mathbf{V}$'s left adjoint on morphisms coincides with that of $\mathbf{S}_D$, i.e. $\mathbf{S}_D$ is the left adjoint of $\mathbf{V}$.* The action of $\mathbf{V}$'s left adjoint on morphisms is given by the following diagrams which follow from Step I.

Diagram in $\mathbb{T}$ *Arrow in* $\mathbb{T}_D$

$$\begin{array}{ccc} (X_1, L_1, \tau_1) & \xrightarrow{(f,\phi)} & (X_2, L_2, \tau_2) \\ {\scriptstyle (id_{X_1}, id_{L_1})}\big\downarrow & \circ & \big\downarrow{\scriptstyle (id_{X_2}, id_{L_2})} \\ (X_1, L_1, \tau_1) & \xrightarrow[\mathbf{V}[!\,\overline{(id_{X_2}, id_{L_2}) \circ (f,\phi)}]]{} & (X_2, L_2, \tau_2) \end{array} \qquad \begin{array}{c} (X_1, L_1, \tau_1, D(\tau_1)) \\ \big\downarrow{\scriptstyle !\,\overline{(id_{X_2}, id_{L_2}) \circ (f,\phi)}} \\ (X_2, L_2, \tau_2, D(\tau_2)) \end{array}$$

The assertion of II immediately follows since $\overline{(id_{X_2}, id_{L_2}) \circ (f, \phi)} = \mathbf{S}_D(f, \phi)$.

The remaining claims of the theorem now follow, in connection with which see the dual of Exercise IV.4.4 of [Mac Lane 1971]. □

2.7 REMARK. Because of the uniqueness of adjoints (up to isomorphism), the adjunctions given above completely fix the role of $\mathbf{V}$ in relating $\mathbb{T}$ to $\mathbb{T}_D$, and both adjunctions yield the same philosophical conclusion, which we now state in accordance with the criteria set forth in [Hedrlin-Pultr-Trnkova 1967] and especially in [Rodabaugh 1983]: we should view $\mathbb{T}_D$ as a "slight", but bona fide generalization of $\mathbb{T}$. This seemingly "slight" difference turns out to be fundamentally important in building representations of stratified and partially stratified spaces, i.e. in defining a sobriety consistent with stratification or partial stratification, as we shall see in §6 below.

2.8 DEFINITION (**The Hutton Functor**). Let $\mathcal{H}$ denote the Hutton framework [Hutton 1980] for fuzzy topology and put $\mathbf{H} : \mathbb{T} \longrightarrow \mathcal{H}$ by $\mathbf{H}(X, L, \tau) = (L^X, \tau)$, $\mathbf{H}(f, \phi) = (f, \phi)^{\leftarrow}$. This essentially comes from [Hutton 1980] and [Erceg 1980].

2.9 REMARK. This functor can be quite useful in analyzing those properties of spaces which are purely lattice-theoretic; in this respect it is rather like the Ω functor of [Johnstone 1982] (about which we say more in §4). Now $\mathbf{H}$ is not able to say that $\mathcal{H}$ generalizes poslat topology for two synergistic reasons: $\mathbf{H}$ is **not** an embedding; but $\mathcal{H}$ **embeds** into $\mathbb{T}$ as the full subcategory of all singleton spaces with Hutton algebra bases (a Hutton algebra or **fuzzy lattice** [Hutton 1980] is a completely distributive deMorgan algebra (1.2.5.9)), as seen in our next result.

2.10 THEOREM (**The Singleton Functor**). Let $\mathbf{1}$ represent a singleton set. Put $\mathbf{F_s}$: $\mathcal{H} \longrightarrow \mathbb{T}$ by $\mathbf{F_s}(L, \tau) = (1, L, \tau^1)$ and $\mathbf{F_s} = (id, \langle \phi^{op} \rangle^{op})$. Then $\mathbf{F_s}$ is a morphism-invertible isomorphism (2.14 below) of $\mathcal{H}$ with a full subcategory (denoted) $\mathbb{T}_s$ of $\mathbb{T}$; this conclusion remains true if we replace $\mathcal{H}$ by $\mathcal{H}_{\mathbf{CSLF}}$ (3.8 below) or $\mathbb{L}^2$ (3.3 below). In fact, $\mathbf{F_s} \dashv \mathbf{H}|_{\mathbb{T}_s} \dashv \mathbf{F_s}$.

PROOF. This is straightforward and left to the reader (cf. [Rodabaugh 1983]). □

2.11 REMARK. Because of the criteria alluded to in Remark 2.7, $\mathcal{H}$ is a special, very important case, of $\mathbb{T}$. A more thorough analysis of the relationship between $\mathcal{H}$ and $\mathbb{T}$ is given in [Eklund 1984–* * **]; in particular, Eklund shows $\mathbb{T}$ morphisms factor through Chang-Goguen morphisms and Hutton morphisms (under the hypotheses that bases are Hutton algebras (2.9).

2.12 DEFINITION (**Subcategories of $\mathbb{T}$**).

(1) $\mathbb{T}(L)$ is the full subcategory of $\mathbb{T}$ each of whose objects has the same basis L.
(2) $\mathbb{T}(L, id)$ is the subcategory of $\mathbb{T}(L)$ each of whose morphisms has second component id_L. A morphism $(f, id_L) \in \mathbb{T}(L, id)$ is often written (f, id).
(3) $\mathbb{T}_k$ is the full subcategory of $\mathbb{T}$ each of whose objects is **stratified**, i.e. the topology of each object contains all constant maps — the concept is due to [Lowen 1976] and the term is due to [Pu-Liu 1980a, 1980b].
(4) $\mathbb{T}_k(L)$ and $\mathbb{T}_k(L, id)$ are subcategories of $\mathbb{T}_k$ defined *à la* (1,2).

2.13 DEFINITION (**Basis and Subbasis**). Given $(X, L, \tau) \in |\mathbb{T}|$, then $S \subset L^X$ is a **subbasis** for τ iff τ is the smallest topology containing S, in which case we write $\tau = \langle\langle S \rangle\rangle$, and $B \subset L^X$ is a **basis** for τ iff $\forall u \in \tau,\ \exists B_u \subset B,\ u = \bigvee B_u$, in which case we write $\tau = \langle B \rangle$. These definitions are due to [Warren 1978]. If we want to pass from S to τ via a basis $B \equiv$ the collection of all finite infima of members of S, then we need $L \in |\mathbf{FRM}|$; however, for arbitrary $L \in |\mathbf{CSLF}|$ we have "subbasic" continuity implies continuity [Rodabaugh 1986–7, second edition].

2.14 DEFINITION (**Morphism-Invertibility**). Let $\mathbf{F} : \mathcal{C} \longrightarrow \mathcal{D}$ be a functor and let $\mathcal{A}$ be a subcategory of $\mathcal{C}$. Then $\mathbf{F}|_{\mathcal{A}}$ is **morphism-invertible** iff $\forall f \in \mathcal{C},\ \mathbf{F}(f) \in \mathbf{F}^{\rightarrow}(\mathcal{A}) \Leftrightarrow f \in \mathcal{A}$. *Sequens*, the data $\mathbf{F}, \mathcal{A}$, and $\mathbf{F}^{\rightarrow}(\mathcal{A})$ are explicitly defined, with $\mathcal{C}, \mathcal{D}$, and the extension of $\mathbf{F}$ to $\mathcal{C}$ left to the reader. We are typically concerned with a given morphism from a ground category being fuzzy continuous iff its classical counterpart is classically continuous. See [Rodabaugh 1983].

2.15 THEOREM (**Embeddings into $\mathbb{T}$ and Stratification**).

(1) $\mathbf{TOP} \equiv \mathbb{T}(\{0,1\}, id)$ and $\mathbf{TOP} \hookrightarrow \mathbb{T}(L, id)$ via two functors, both denoted $\mathbf{G}_\chi$, defined respectively by $\mathbf{G}_\chi(X, \mathcal{T}) = (X, \{0,1\}, \chi(\mathcal{T}))$, $\mathbf{G}_\chi(f) = (f, id)$, and $\mathbf{G}_\chi(X, \mathcal{T}) = (X, L, \chi(\mathcal{T}))$, $\mathbf{G}_\chi(f) = (f, id)$, where $\chi(\mathcal{T}) = \{\chi_U : U \in \mathcal{T}\}$.
(2) $\forall L \in |\mathbf{CSLF}|$, the category of [Chang 1968, Goguen 1973] [[Lowen 1976]] is isormorphic to $\mathbb{T}(L, id)$ [$\mathbb{T}_k(L, id)$].
(3) The inclusion functor $\hookrightarrow: \mathbb{T}_k \to \mathbb{T}$ has a faithful right adjoint $\mathbf{G}_k$, such that $\hookrightarrow \dashv$

$\mathbf{G}_k$ is an epicoreflection, defined by $\mathbf{G}_k(\tau) = \tau \vee \{\underline{\alpha} : \alpha \in L\} \equiv \langle\!\langle \tau \cup \{\underline{\alpha} : \alpha \in L\}\rangle\!\rangle$, $G_k(X, L, \tau) = (X, L, \mathbf{G}_k(\tau))$, $G_k(f, \phi) = (f, \phi)$.

Outline of Proof. (1,2) are and well-known (see [Rodabaugh 1983]). To satisfy the Major Diagram (§0) for (3), let $(X, L, \tau) \in |\mathbb{T}_k|$. Then $(\hookrightarrow \circ \mathbf{G}_k)(X, L, \tau) = (X, L, \tau)$, and it can be seen that (id, id) serves as the unit of an adjunction in which the left adjoint to $\mathbf{G}_k$ acts as $\hookrightarrow$ on objects. That this left adjoint also coincides with $\hookrightarrow$ on morphisms, i.e. that the Minor Diagram (§0) is satisfied, is left to the reader. □

2.16 Remark.

(1) The category $\mathbb{T}_k$ includes stratification with change of basis. Note $\mathbb{T}(L)$ and $\mathbb{T}_k(L)$ allow change of basis since the ϕ in a morphism is unspecified. The adjunction $\hookrightarrow \dashv G_k$ restricts to the subcategories defined above in 2.12.
(2) Because of 2.6(3), $\mathbf{G}_k$ is called the **stratification functor**.

2.17 Definition (**Phi-Saturated Topologies**). For $L \in |\mathbf{CSLF}|$ and $\phi \in \mathbb{L}(L, L)$, the category $\mathbb{T}_\phi(L, \phi)$ comprises all objects (X, L, τ) in $\mathbb{T}(L)$ which satisfy

$$\forall b \in L^X,\ b \in \tau \Leftrightarrow \langle \phi^{op} \rangle (b) \in \tau$$

together with all morphisms of $\mathbb{T}$ whose second component is ϕ. The category $\mathbb{T}_\phi(L, \phi)$ internalizes a **fixed change-of-basis**.

2.18 Remark. Note that $\mathbb{T}_\phi(L, \phi)$ is not a subcategory of $\mathbb{T}$ if $|L| \geq 3$: the identity morphism of (X, L, τ) in $\mathbb{T}$ is (id,id), but the identity morphism in $\mathbb{T}_\phi(L, \phi)$ is (id, ϕ). But we may regard $\mathbb{T}_\phi(L, \phi)$ as a subcategory of $\mathbb{T}$ because of the next result.

2.19 Theorem (**Embedding of $\mathbb{T}_\phi(L, \phi)$ into $\mathbb{T}$**). Let $\mathbf{F}_\phi : \mathbb{T}_\phi(L, \phi) \longrightarrow \mathbb{T}(L, id)$ by

$$\mathbf{F}_\phi(X, L, \tau) = (X, L, \tau),\ \mathbf{F}_\phi(f, \phi) = (f, id)$$

And let $\mathbf{G}_\phi : \mathbb{T}(L, id) \longrightarrow \mathbb{T}_\phi(L, \phi)$ by

$$\mathbf{G}_\phi(\tau) = \tau_\phi \equiv \bigcap \{\sigma \subset L^X : (X, L, \sigma) \in |\mathbb{T}_\phi(L, \phi)|,\ \tau \subset \sigma\},$$

$$\mathbf{G}_\phi(X, L, \tau) = (X, L, \tau_\phi),\ \mathbf{G}_\phi(f, id) = (f, \phi)$$

Then $\mathbf{F}_\phi$ is a morphism-invertible isomorphism of $\mathbb{T}_\phi(L, \phi)$ onto a full subcategory of $\mathbb{T}(L, id)$, $\mathbf{G}_\phi$ is a full, object-onto (hence dense) functor of $\mathbb{T}(L, id)$ onto $\mathbb{T}_\phi(L, \phi)$, $\mathbf{F}_\phi \dashv \mathbf{G}_\phi$, and this adjunction is an epicoreflection.

Outline of Proof. We verify that $\mathbf{F}_\phi$ and $\mathbf{G}_\phi$ are functors, and that $\mathbf{F}_\phi \dashv \mathbf{G}_\phi$; other details and claims can be done by the reader.

CLAIM I: $\mathbf{F}_\phi : \mathbb{T}_\phi(L,\phi) \longrightarrow \mathbb{T}(L,id)$ *is a functor.* The main detail worth checking is that $\mathbf{F}_\phi$ takes morphisms to morphisms. Let $(f,\phi) : (X,L,\tau) \longrightarrow (Y,L,\sigma)$ be an arrow in $\mathbb{T}_\phi(L,\phi)$; we must show that $\mathbf{F}_\phi(f,\phi) \equiv (f,id)$ is an arrow in $\mathbb{T}$ from (X,L,τ) to (Y,L,σ), i.e. that $\forall\, v \in \sigma,\ (f,id)^{\leftarrow}(v) \in \tau$. Let $v \in \sigma$. Since (f,ϕ) is continuous, $(f,\phi)^{\leftarrow}(v) \in \tau$, i.e. we have

$$\langle \phi^{op} \rangle (v \circ f) = \phi^{op} \circ (v \circ f) = \phi^{op} \circ v \circ f \in \tau$$

But $(X,L,\tau) \in |\,\mathbb{T}_\phi(L,\phi)|$ and $\langle \phi^{op} \rangle (v \circ f) \in \tau$ imply

$$(f,id)^{\leftarrow}(v) \equiv id^{op} \circ v \circ f = v \circ f \in \tau$$

CLAIM II: $\mathbf{G}_\phi : \mathbb{T}(L,id) \longrightarrow \mathbb{T}_\phi(L,\phi)$ *takes objects to objects.* We must verify that $\forall\,(X,L,\tau) \in |\,\mathbb{T}\,(L,id)|$, it is the case that $(X,L,\tau_\phi) \in |\,\mathbb{T}_\phi(L,\phi)|$. We first observe that $\tau \subset L^X$ and $(X,L,L^X) \in |\,\mathbb{T}_\phi(L,\phi)|$, so that the collection $\{\sigma \subset L^X : (X,L,\sigma) \in |\,\mathbb{T}_\phi(L,\phi)|,\ \tau \subset \sigma\}$ is non-empty. Hence the topology

$$\tau_\phi \equiv \bigcap \{\sigma \subset L^X : (X,L,\sigma) \in |\,\mathbb{T}_\phi(L,\phi)|,\ \tau \subset \sigma\}$$

is meaningful. It must be now shown that $(X,L,\tau_\phi) \in |\,\mathbb{T}_\phi(L,\phi)|$, i.e. shown that $\forall\, b \in L^X,\ b \in \tau_\phi \Leftrightarrow \langle \phi^{op} \rangle (b) \in \tau_\phi$; but this is trivial given the definition of τ_ϕ and left to the reader.

CLAIM III: $\mathbf{G}_\phi$: $\mathbb{T}(L,id) \longrightarrow \mathbb{T}_\phi(L,\phi)$ *takes morphisms to morphisms.* Let (f,id) : $(X,L,\tau) \longrightarrow (Y,L,\sigma)$ be given. We proceed in two steps: first, we show that (f,id) is a morphism from (X,L,τ_ϕ) to (Y,L,σ_ϕ); and then we show that (f,ϕ) is a morphism from (X,L,τ_ϕ) to (Y,L,σ_ϕ). For the first step, note $\tau \subset \tau_\phi$, and so (f,id) is continuous from (X,L,τ_ϕ) to (Y,L,σ). Put

$$\sigma^* = \{v \in L^X : (f,id)^{\leftarrow}(v) \in \tau_\phi\}$$

Trivially, $\sigma \in \sigma^*$ and (f,id) is a morphism from (X,L,τ_ϕ) to (Y,L,σ^*). Now we claim that $(Y,L,\sigma^*) \in |\,\mathbb{T}_\phi(L,\phi)|$: letting $b \in L^Y$, we have

$$\begin{aligned} b \in \sigma^* &\Leftrightarrow (f,id)^{\leftarrow}(b) \in \tau_\phi \\ &\Leftrightarrow [(f,id)^{\leftarrow}(\langle \phi^{op} \rangle (b)) = (\phi^{op} \circ b) \circ f = \langle \phi^{op}((f,id)^{\leftarrow}(b)) \in \tau_\phi] \\ &\Leftrightarrow \langle \phi^{op} \rangle (b) \in \sigma^* \end{aligned}$$

Since $\sigma \subset \sigma^*$, then $\sigma_\phi \subset \sigma$ But (f,id) is therefore continuous from (X,L,τ_ϕ) to (Y,L,σ_ϕ). The second step, showing why (f,ϕ) is an arrow between these spaces, is quite easy: given $v \in \sigma_\phi$, we have

$$(f,id)^{\leftarrow}(v) = v \circ f \in \sigma_\phi \Leftrightarrow (f,\phi)^{\leftarrow}(v) = \langle \phi^{op} \rangle (v \circ f) \in \sigma_\phi$$

CLAIM IV: $\mathbf{G}_\phi$ *is a functor.* We leave this to the reader.

CLAIM V: $\mathbf{F}_\phi \dashv \mathbf{G}_\phi$. To verify the Major Diagram (§0), let $(X, L, \tau) \in |\mathbb{T}_\phi(L, \phi)|$, and note $\mathbf{G}_\phi\mathbf{F}_\phi(X, L, \tau) = (X, L, \tau_\phi) = (X, L, \tau)$. The unit will turn out to be (id, ϕ). To see this let $(Y, L, \sigma) \in |\mathbb{T}(L, id)|$ and continuous $(f, \phi) : (X, L, \tau) \longrightarrow \mathbf{G}_\phi(Y, L, \sigma) = (Y, L, \sigma_\phi)$ be given. The proof of the second step of Claim III shows that (f, id) is continuous from (X, L, τ) to (Y, L, σ_ϕ); and since $\sigma \subset \sigma_\phi$, we have (f, id) is continuous from (X, L, τ) to (Y, L, σ). It follows that (f, id) is the unique solution of the equation $(f, \phi) = \mathbf{G}_\phi(f, id) \circ (id, \phi)$, i.e. the Major Diagram is verified. It is straightforward to check that the Minor Diagram is satisfied; this is left to the reader. □

2.20 REMARK. Because of 2.19, $\mathbf{G}_\phi$ is called the **phi-ification functor**.

§3. Lattice-Theoretic Frameworks

It is important to describe the lattice-theoretic categories which "capture" or "model" the lattice-theoretic data of poslat topological frameworks, i.e. that play the same role with respect to a given notion of poslat topology that **LOC** plays with respect to **TOP**. The relationship between a poslat topological framework and its lattice-theoretic counterpart is functorial in a "Stone-like" way, and the particular functors relating the poslat topological frameworks given in §2 to their lattice-theoretic counterparts are described explicitly in §4-6. The purpose of this section is to intuitively motivate and rigorously define these lattice-theoretic frameworks, and in the process show that **each** is a functor category of the form $\mathbb{L}^{\mathbf{r}}$, where $\mathbf{r}$ is one of the categories **1, 1.5, 2, 2.5** defined below. The reader is referred to [Herrlich-Strecker 1979] for standard terminology; we shall sometimes use labeling which is more suggestive, but non-standard.

3.1 DEFINITION (**The Categories 1, 1.5, 2, 2.5, 3**). In the following definitions we refer only to morphisms **other than** identity morphisms.

(1) The category **1** has exactly one object t. This object is labeled "t" since it is often thought of as a topology (crisp or fuzzy).

(1.5) The category **1.5** [Meßner-Rodabaugh ****] is the pre-ordered category $\{l, t\}$ with the trivial ordering; i.e. **1.5** has exactly two objects and no morphisms (other than identity morphisms). These objects are labeled "l" and "t" since the first object is often viewed as a base lattice of membership values for a (fuzzy) topology, and the second object is often viewed as a (fuzzy) topology based on that membership lattice.

(2) The category **2** is the pre-ordered category $\{l, t\}$ with the ordering $l \leq t$; i.e. **2** has exactly two objects and one morphism between them. The labeling of the two objects is the same as for **1.5**, except that the object l may be viewed as a fuzzy powerset, and t may be viewed as a topology contained in that powerset. The non-trivial arrow has domain l and codomain t. The following diagram represents **2**:

$$l \longrightarrow t$$

It is also convenient to consider the dual $\mathbf{2}^{op}$ represented by the following diagram:

$$l \longleftarrow t$$

(2.5) The category **2.5** [Meßner-Rodabaugh ****] is the pre-ordered category $\{l, t, d\}$ with the ordering $l, t \leq d$; i.e. **2.5** has exactly three objects and two arrows, both with codomain d. The first two objects are labeled l and t as in (1), i.e "l" suggests a base membership lattice for a (fuzzy) topology, and "t" suggests a topology based on that membership lattice; and the third object is labeled "d" to suggest a recognized degree of stratification of that topology. The category **2.5** is represented by the diagram:

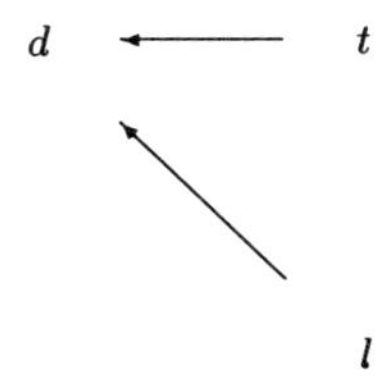

It is also convenient to consider the dual $\mathbf{2.5}^{op}$ represented by the following diagram:

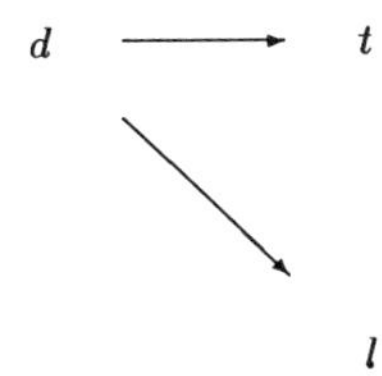

(3) The category **3** is the pre-ordered category $\{l, t, d\}$ with the ordering $l \leq t \leq d$; i.e. **3** has exactly three objects and three arrows, the three arrows being $d \longleftarrow t, t \longleftarrow l$, and $d \longleftarrow l$. The first two objects are labeled l and t as in (2), i.e "l" suggests a fuzzy powerset, and "t" may be viewed as a topology contained in that powerset; and the third object is labeled "d" to suggest a recognized collection of constant maps in that topology. The category **3** is represented by the diagram:

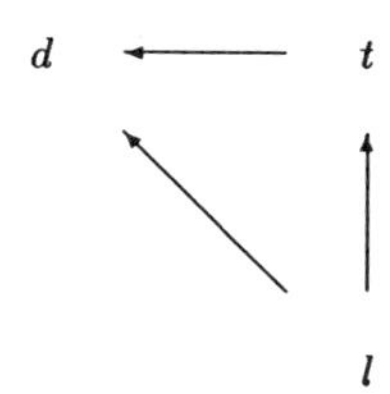

It is also convenient to consider the dual $\mathbf{3}^{op}$ represented by the following diagram.

3.2 REMARK. The categories defined in 3.1, with the exception of **3**, are clearly motivated by the diagrams implicit in the types of topology considered in [Hutton 1980] and §2. For

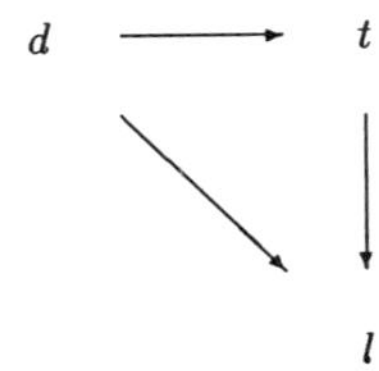

example, the categories **2** and $\mathbf{2}^{op}$ are motivated by the diagram implicit in each $(L,\tau) \in |\mathcal{H}|$, namely

$$L \hookrightarrow \tau$$

and the categories **2.5** and $\mathbf{2.5}^{op}$ are motivated by the diagram implicit in each (X, L, τ, D) $\in |\mathbb{T}_D|$, namely

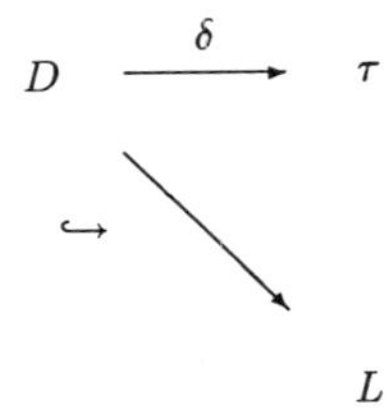

We now consider the powers of $\mathbb{L}$ with exponents taken from among the categories defined above in 3.1.

3.3 Definition. (**The Categories $\mathbb{L}^{\mathbf{r}}$**). By $\mathbb{L}^{\mathbf{r}}$) we mean, per the usual definition, a functor category with base $\mathbb{L}$ and exponent $\mathbf{r}$, where $\mathbf{r}$ is one of the categories **1, 1.5, 2, 2.5**, or **3**. In other words, an object $\mathbf{A}$ is a functor $\mathbf{A} : \mathbf{r} \longrightarrow \mathbb{L}$, and a morphism $\zeta : \mathbf{A} \longrightarrow \mathbf{B}$ is a natural transformation, and the composition of morphisms is the composition of natural transformations.

3.4 Notation. We generally draw the diagrams relevant to $\mathbb{L}^{\mathbf{r}}$ with concrete arrows (in **CSLF**), so our diagrams are drawn as if they come from the category

$$\mathbf{CSLF}^{[\mathbf{r}^{op}]} \equiv [\,\mathbb{L}^{\mathbf{r}}]^{op}$$

And in both these diagrams and all remaining text of this chapter, the components of the dual ζ^{op} of a morphism are denoted $\zeta_1, \zeta_t, \zeta_d$ bf instead of $\zeta_1^{op}, \zeta_t^{op}, \zeta_d^{op}$, respectively. Cf. 3.6 below.

3.5 Remark. $\forall$ $\mathbf{r} = \mathbf{1},\ \mathbf{1.5},\ \mathbf{2},\ \mathbf{2.5},\ \mathbf{3}$, the objects of $\mathbb{L}^{\mathbf{r}}$ are copies in $\mathbb{L}$ of the category $\mathbf{r}$, and hence, intuitively speaking, may be thought of as copies in $\mathbb{L}$ (or **CSLF**) of the diagram representing $\mathbf{r}$. This motivates the mnemonics given below in 3.6. Furthermore, the duals (in **CSLF**) of the morphisms of $\mathbb{L}^{\mathbf{r}}$ may be thought of in topological terms vis-a-vis §2:

(1) "$\zeta_1 \in$ **CSLF**" is the **ground axiom** ($\mathbf{r} \neq \mathbf{1}$).

(2) "$\zeta_t \in$ **CSLF** " is the **general continuity axiom** (all $\mathbf{r}$).
(3) "$\zeta_d \in$ **CSLF** " is the **special continuity axiom** ($\mathbf{r} = \mathbf{2.5}$ or $\mathbf{3}$).

3.6 MNEMONIC. For each $\mathbf{r}$, a commutative diagram in **CSLF** can be drawn which summarizes (via dual arrows) the structure of $\mathbb{L}^{\mathbf{r}}$, so that all subdiagrams commute. Such a **CSLF** diagram is the **mnemonic for** $\mathbb{L}^{\mathbf{r}}$. To illustrate, let $\mathbf{A}, \mathbf{B} \in |\mathbb{L}^{\mathbf{r}}|$, $\zeta : \mathbf{A} \longrightarrow \mathbf{B}$ in $\mathbb{L}^{\mathbf{r}}$, and set $\mathbf{r} = \mathbf{2.5}$. Then we have the diagram below: recall the notation of 3.4; compare with 2.4; note that it contains implicitly the mnemonics for the cases $\mathbf{r} = \mathbf{1}, \mathbf{1.5}$; and note that if the diagram were "completed" with the arrows $\mathbf{A}(t \longrightarrow l)$ and $\mathbf{B}\ (t \longrightarrow l)$, it would serve explicitly for the case $\mathbf{r} = \mathbf{3}$, and implicitly for the case $\mathbf{r} = \mathbf{2}$ (but nodes and arrows have a very different interpretation for these two cases — cf. 3.1 above).

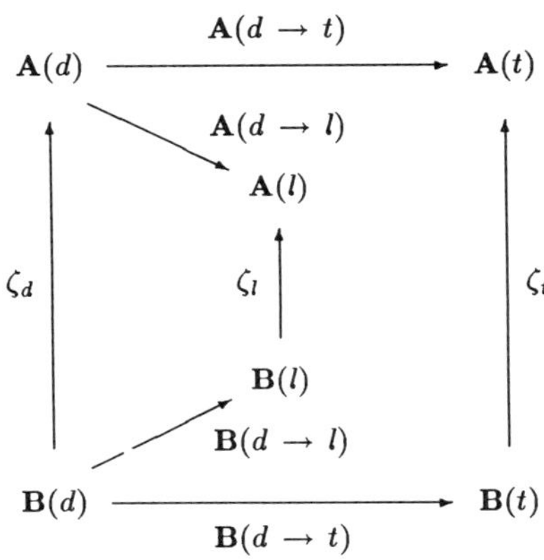

3.7 META-THEOREM (**Role of Functor Categories in Poslat Topology**). All the approaches to poslat topology discussed in [Hutton 1980] and §2 are related to (subcategories of) functor categories which are powers of $\mathbb{L}$.

(1) $\mathbb{L}^{\mathbf{1}}$ ($\equiv \mathbb{L}$) captures the topologies of the spaces in each $\mathbb{T}(L, id)$ and $\mathbb{T}_\phi(L, \phi)$;
(2) $\mathbb{L}^{\mathbf{1.5}}$ captures the topologies and the bases of each space in $\mathbb{T}$;
(3) $\mathbb{L}^{\mathbf{2}}$ captures the powersets and topologies of the objects of $\mathcal{H}$.
(4) $\mathbb{L}^{\mathbf{2.5}}$ captures the topologies, bases, and degrees of stratification for each space in $\mathbb{T}_D$.

If the first infinite distributive law is desired, then the ground may be strengthened to **SET** $\times$ **LOC**, and **LOC** would then replace $\mathbb{L}$ in (1-4).

3.8 DISCUSSION. It is of historical interest to note that the notion of functor category is implicit in Hutton's change of basis approach, though he never expresses his spaces explicitly in these terms. In particular, Hutton cannot express his viewpoint in terms of a functor category using his category **FUZLAT** as a ground for the following reasons: given a Hutton space (L, τ), the inclusion $\tau \hookrightarrow L$ need **not** be in **HUT** $\equiv$ **FUZLAT**^{op} (it may not preserve the involution), but in **CSLF**; and given a morphism $\phi \in \mathcal{H}[(L_1, \tau_1), (L_2, \tau_2)]$, $\phi^{op}|_{\tau_2}$ need

not be in **HUT**, but in **CSLF**. Since we need only have $\hookrightarrow^{op}$ and ϕ in $\mathbb{L}$, this suggests weakening **FUZLAT** to $\mathbb{L}$, redefining $\mathcal{H}$ in this setting, and calling the new category $\mathcal{H}_{\mathbf{CSLF}}$. We then would have $\mathcal{H} \hookrightarrow \mathcal{H}_{\mathbf{CSLF}} \hookrightarrow \mathbb{L}^{\mathbf{2}}$.

3.9 OPEN QUESTION. There is no topological framework in the literature which is captured lattice-theoretically by $\mathbb{L}^{\mathbf{3}}$. This suggests a new Hutton-like approach to fuzzy topology in which there are **recognized degrees of stratification**. [Eklund 1984] shows that (full) stratification is incompatible with the Tihinov Theorem based on Hutton's compactness and Hutton's product topology. We conjecture this incompatibility remains in this new Hutton framework.

3.10 QUESTION. Does each functor category consist only of those diagrams generated by the associated category of topological spaces, i.e. is each object of these functor categories **spatial**? This brings us to §4-6 below.

§4. Classical Spatiality and Sobriety

We summarize the classical Stone Representation Theorems and the underlying theory as presented in [Johnstone 1982]; the original theory is given in [Stone 1936, 1937a, 1937b]. No proofs are given — these are subsumed in the proofs of the more general, fuzzy cases presented in the next two sections.

4.1 DISCUSSION (**Set-up of the Classical Theory**). A representation theory constructs an (categorical) equivalence between some sort of topological spaces and some sort of lattice-theoretic objects. A primary issue to be addressed by such a theory is: what will be the ground set of a space constructed from a lattice-theoretic object? Such a ground set will be referred to *sequens* as the **carrier set**. Given $A \in |\mathbf{LOC}|$, the carrier set is $\mathbf{CSLF}(A, \{0,1\})$, called the set of **points of A**, and denoted $Pt(A)$. Now A should also determine a topology on $Pt(A)$, namely $\{\{p \in Pt(A) : p(a) = 1\} : a \in A\}$, and the correspondence $a \mapsto \{p \in Pt(A) : p(a) = 1\}$ thereby determines a map

$$\Phi : A \longrightarrow \mathcal{P}(Pt(A)),\ \Phi(a) = \{p \in Pt(A) : p(a) = 1\}$$

We also let Φ denote the restriction, with respect to codomain, of Φ to $\Phi^{\rightarrow}(A)$. So each A determines an object $(Pt(A), \Phi^{\rightarrow}(A)) \in |\mathbf{TOP}|$. Given an arrow $f \in \mathbf{LOC}(A, B)$, i.e. $f^{op} \in \mathbf{FRM}(B, A)$, a point $p \in Pt(A)$ determines a point $p \circ f^{op} \in Pt(B)$, and

$$\begin{array}{ccc} B & \xrightarrow{f^{op}} & A \\ & \searrow^{!p \,\circ\, f^{op}} \quad \circ & \downarrow p \\ & & \{0,1\} \end{array}$$

this correspondence determines a continuous map in

$$\mathbf{TOP}[(Pt(A), \Phi^{\rightarrow}(A)), (Pt(B), \Phi^{\rightarrow}(B))].$$

Thus is the functor **PT: LOC** $\longrightarrow$ **TOP** born, where $\mathbf{PT}(A) = (Pt(A), \Phi^{\rightarrow}(A))$ and $\mathbf{PT}(f)(p) = p \circ f^{op}$. Now the other, easier half of the story is the functor Ω : **TOP** $\longrightarrow$ **LOC**, given by $\Omega(X, \mathcal{T}) = \mathcal{T}$, $\Omega[f : (X_1, \mathcal{T}_1) \longrightarrow (X_2, \mathcal{T}_2)] = [f^{\leftarrow}|_{\mathcal{T}_2} : \mathcal{T}_2 \longrightarrow \mathcal{T}_1]^{op}$. And Ω motivates the question: given a space $(X, \mathcal{T})$, how does $(X, \mathcal{T})$ compare with $\mathbf{PT}(\Omega(X, \mathcal{T}))$, i.e. how does $(X, \mathcal{T})$ compare with $(Pt(\mathcal{T}), \Phi^{\rightarrow}(\mathcal{T}))$? A continuous map $\Psi \in$ **TOP** $[(X, \mathcal{T}), (Pt(\mathcal{T}), \Phi^{\rightarrow}(\mathcal{T}))]$ is determined by the sequence of correspondences:

$$\begin{aligned} x &\mapsto Cl\{x\} \text{ (an "irreducible" closed set)} \\ &\mapsto X - Cl\{x\} \text{ (a "prime" open set)} \\ &\mapsto \downarrow(X - Cl\{x\}) \text{ (a "prime" principal ideal of } \mathcal{T}) \\ &\mapsto p \in Pt(\mathcal{T}) \text{ by } \ker(p) = \downarrow(X - \{x\}) \end{aligned}$$

(A closed set K is **irreducible** if K is not the union of proper, nontrivial closed subsets. An open set U is **prime** if $\downarrow(U)$ is **prime**, i.e. if $\mathcal{T} - \downarrow(U)$ is a filter of $\mathcal{T}$.)

4.2 THEOREM (**Fundamental Adjunction**). $\Omega \dashv$ **PT** with unit Ψ and counit Φ^{op}.

4.3 DEFINITION (**Spatiality and Sobriety**). The locale $A \in |\mathbf{LOC}|$ is **spatial** iff Φ is injective, in which case Φ^{op} becomes a localic isomorphism of A with the topology $\Phi^{\rightarrow}(A)$ of the space $\mathbf{PT}(A)$; and the space $(X, \mathcal{T}) \in |\mathbf{TOP}|$ is **sober** iff Ψ is bijective, in which case Ψ is a homeomorhism, i.e. an isomorphism in **TOP**.

4.4 THEOREM (**Sobriety-Spatiality Representation Theorem**).

$$\forall (X, \mathcal{T}) \in |\mathbf{TOP}|, \ \Omega(X, \mathcal{T}) \ [= \mathcal{T}]$$

is spatial; and $\forall A \in |\mathbf{LOC}|$, $\mathbf{PT}(A)$ $[= (Pt(A), \Phi^{\rightarrow}(A))]$ is sober. Hence, the adjunction $\Omega \dashv$ **PT** restricts to a categorical equivalence **SOBTOP** $\approx$ **SPATLOC**, where **SOBTOP** [**SPATLOC**] is the full subcategory of **TOP** [**LOC**] of all sober [spatial] objects.

4.5 REMARK. Even though this representation is not due to Stone himself, it is the equivalence underlying his representations, and as such we speak of it as part of the classical

Stone representation theory. *Sequens*, **DLAT** [**BOOL**] is the category of distributive lattices [Boolean algebras] and lattice morphisms.

4.6 DEFINITION. (**Coherent Spaces and Stone Spaces**). A space $(X, \mathcal{T})$ is **coherent** iff it is compact, sober, and $\mathcal{T} = \langle\text{clopens}\rangle$; and $(X, \mathcal{T})$ is a **Stone space** iff it is compact, T_0, and zero-dimensional. **COHTOP** is the subcategory of **TOP** of coherent spaces and all continuous mappings whose reverse powerset operators preserve the clopens; and **STNTOP** is the full subcategory of **TOP** of all Stone spaces.

4.7 DEFINITION (**Coherent Lattices.**) $\forall A \in |\mathbf{CSLF}|$, the **finite elements of A**, $K(A)$, comprise $\{a \in A : \forall S \subset A, \vee S \geq a, \exists \text{ finite } F \subset S, \vee F \geq a\}$; and A is **coherent** iff $K(A)$ is a sublattice of A and each $a \in A$ is a join of elements of $K(A)$. The subcategory **COH$\mathbb{L}$** [**COHLOC**] of $\mathbb{L}$ [**LOC**] comprises all coherent complete lattices [locales] and all morphisms whose duals preserve the finite elements.

4.8 LEMMA. The following statements hold:

(1) $(X, \mathcal{T}) \in |\mathbf{COHTOP}| \Rightarrow \mathcal{T} \in |\mathbf{COHLOC}|$; $A \in |\mathbf{COHLOC}| \Rightarrow K(A) \in |\mathbf{DLAT}|$.
(2) $B \in |\mathbf{DLAT}| \Rightarrow \text{Idl}(B)$ (lattice of ideals of B) $\in |\mathbf{COHLOC}|$, and $\mathbf{PT}(\text{Idl}(B)) \in |\mathbf{COHTOP}|$.
(3) If $B \in |\mathbf{DLAT}|$, then $\mathbf{PT}(\text{Idl}(B))$ is Hausdorff (so in **STNTOP**) $\Leftrightarrow B \in |\mathbf{BOOL}|$.

Lemma 4.8 sets up the equivalence $\mathbf{COHLOC} \approx \mathbf{DLAT}^{op}$ and the resulting restrictions of $\mathbf{SOBTOP} \approx \mathbf{SPATLOC}$ which comprise Stone's classical representation theorems.

4.9 THEOREM (**Stone Representation Theorems**). $\mathbf{COHTOP} \approx \mathbf{DLAT}^{op}$ and $\mathbf{STNTOP} \approx \mathbf{BOOL}^{op}$.

§5. Lattice-Dependent Spatiality and Sobriety

In this and the next sections we summarize the fuzzifications of the classical Stone Representation Theorems and the underlying theory. This entails a summary of [Rodabaugh 1986, 1986–7, 1987, 1991, ****] and [Meßner-Rodabaugh ****]. An equivalent approach to fuzzifying the basic Stone adjunction can be seen in [Frith 1987] for the fixed-basis case $L = I$. See also the chapter of W. Kotzé in this volume for developments related to both the sequel and [Rodabaugh 1987].

The relevance of these two sections is manifold: the importance of the classical Stone theorems, both intrinsically and in their bearing on twentieth century mathematics (see the Introduction of [Johnstone 1982]); the role that fuzzy sets play not only in generalizing the classical theory, but in uniquely steamlining it via two explicit evaluation maps; the ability of fuzzy sets to fill **two significant deficiencies** of **TOP** indicated by the theory underlying

Stone's work (non-spatial locales have no classical space representations yet have fuzzy sober space representations, and localic products need not be classical product topologies yet are always fuzzy product topologies); and the ability of the fuzzy representation theory to address important questions internal to fuzzy topology such as the role of stratified spaces, lower-order separation axioms, the proper compactness axiom for compactifications, the significance of change of basis, etc. The Stone representation theories of this section assume topological frameworks with a fixed basis or fixed change-of-basis; the justification of $\mathbb{T}$ in this context is that it is the smallest categorical framework containing all these theories. (Such theories include, of course, the classical or crisp case, a case which is motivationally important for the fuzzy case and which was given separate treatment in the previous section.) And the theory presented in this section does not concern itself with the issue of recognized stratification. We consider these two issues — change of basis and degrees of recognized stratification — at length in the next section.

5.1 Discussion (**Set-up of the General Theory**). Let $A \in |\mathbb{L}|$ and $L \in |\mathbf{CSLF}|$. Compared to the classical theory, A plays the same role as before, except that we do not require any distributivity, i.e. we do not require $A \in |\mathbf{LOC}|$; and L plays the role classically assumed (implicitly or explicitly) by $\{0,1\}$. To be explicit, the **carrier set** is $\mathbf{CSLF}(A, L)$, called the **L-(fuzzy) points of A** and denoted $Lpt(A)$. Note the classical carrier set is captured by specifying $L = \{0,1\}$, i.e. $Pt(A) = \{0,1\}pt(A)$. Now A should determine an "L-"poslat topology on $Lpt(A)$, meaning A should determine a subcollection of the L-fuzzy powerset $L^{Lpt(A)}$ of $Lpt(A)$ which satisfies the criteria of a poslat topology, and *à la* the classical case, this topology should be the range in $L^{Lpt(A)}$ of some lattice morphism Φ_L defined on A. Thus each $a \in A$ should determine an L-fuzzy subset $[\Phi_L(a) : Lpt(A) \longrightarrow L]$ of $Lpt(A)$, and the natural way to construct the action of $\Phi_L(a)$ at each p is

$$\Phi_L(a)(p) = p(a)$$

And so the map $\Phi_L : A \longrightarrow L^{Lpt(A)}$ is in the **explicit form** of an **evaluation map**, a considerable enhancement and simplification of the classical case. Indeed, it is a meta-theorem that the explicit form of an evaluation map is obtained **if and only if** the poslat philosophy of fuzzy sets (that a subset be replaced with a mapping into a lattice) is adopted. This is immediate from $\Phi_{\{0,1\}}$ being the **unique** lifting of the classical Φ through χ in the following commutative triangle (which shows the "crisp" $\Phi_{\{0,1\}}$ in effect is the classical Φ):

$$\begin{array}{ccc} A & \xrightarrow{\Phi} & \mathcal{P}(Pt(A)) \\ & {\scriptstyle !\,\Phi_{\{0,1\}}}\searrow \quad \circ & \updownarrow \chi \\ & & \{0,1\}^{\{0,1\}pt(A)} \end{array}$$

We also let Φ_L denote the restriction, with respect to codomain, of Φ_L to $\Phi_L^{\rightarrow}(A)$. It turns out $\Phi_L \in$ **CSLF**, making $\Phi_L(A)$ a poslat topology on $Lpt(A)$, i.e. A determines $(Lpt(A), L, \Phi_L^{\rightarrow}(A)) \in |\mathbb{T}(L)|$. Given an arrow $f \in \mathbb{L}(A,B)$, i.e. $f^{op} \in \mathbf{CSLF}(B,A)$, an L-point $p \in Lpt(A)$ determines an L-point $p \circ f^{op} \in Lpt(B)$, and this in turn

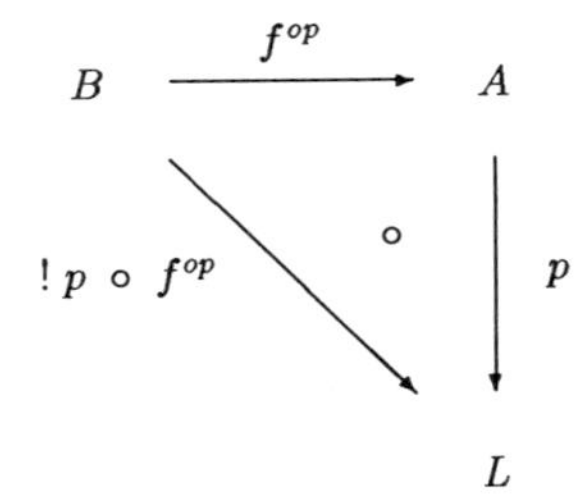

determines a morphism in $\mathbb{T}_\phi(L,\phi)[(Lpt(A), L, \Phi_L^{\rightarrow}(A)), (Lpt(B), L, \Phi_L^{\rightarrow}(B))]$; this is proved in Theorem 5.2 below. Thus is born a mapping $\mathbf{L}_\phi\mathbf{PT} : \mathbb{L} \longrightarrow \mathbb{T}_\phi(L,\phi)$ of objects and morphisms, defined by

$$\mathbf{L}_\phi\mathbf{PT}(A) = (Lpt(A), L, \Phi_L^{\rightarrow}(A)),\ \mathbf{L}_\phi\mathbf{PT}(f) = (\mathbf{LPT}^*(f), \phi) \text{ with } \mathbf{LPT}^*(f)(p) = p \circ f^{op}$$

Conditions are given below for when $\mathbf{L}_\phi\mathbf{PT}$ is indeed a functor (Theorems 5.2, 5.3). We let **LPT** denote $\mathbf{L}_\phi\mathbf{PT}$ when $\phi = id_L$; by Theorem 5.3 below, **LPT** is a functor. Note that **LPT** is a correct generalization of **PT** since $\{0,1\}\mathbf{PT}|_{\mathbf{LOC}}$ is the **unique** lifting of **PT** through $\mathbf{G}_\chi : \mathbf{TOP} \longrightarrow \mathbb{T}(\{0,1\}, id)$ (2.15) if in $\mathbb{T}(\{0,1\}id)$ we identify isomorphic objects; this follows from the lifting diagram given below for Ψ.

Now the other, easier half of the story is the construction of the functor acting as the left-adjoint to $\mathbf{L}_\phi\mathbf{PT}$ under the conditions of Theorems 5.2, 5.3. We use Ω from the classical case to also denote $\Omega : \mathbb{T} \longrightarrow \mathbb{L}$ defined by

$$\Omega(X,L,\tau) = \tau,\ \Omega[(f,\phi) : X_1, L_1, \tau_1) \longrightarrow (X_2, L_2, \tau_2)] = [(f,\phi)^{\leftarrow}|_{\tau_2} : \tau_2 \longrightarrow \tau_1]^{op}$$

We use $\mathbf{L}\Omega$ to denote the restriction of Ω to $\mathbb{T}(L, id)$, and $\mathbf{L}_\phi\Omega$ to denote the functor $\mathbf{L}\Omega \circ \mathbf{F}_\phi$ from $\mathbb{T}_\phi(L,\phi)$ to $\mathbb{L}$; note for $\phi = id_L, \mathbf{L}_\phi\Omega = \mathbf{L}\Omega$. If $L \in |\mathbf{FRM}|$, then the images of $\mathbf{L}_\phi\Omega$ and $\mathbf{L}\Omega$ are in **LOC**, in which case we use $\mathbf{L}_\phi\Omega$ and $\mathbf{L}\Omega$ to denote the restrictions of $\mathbf{L}_\phi\Omega$ and $\mathbf{L}\Omega$, with respect to codomain, to **LOC**. We note $\mathbf{L}\Omega$ is a correct generalization of the classical Ω since $\{0,1\}\Omega$ is the **unique** factorization of (classical) Ω through $\mathbf{G}_\chi : \mathbf{TOP} \longrightarrow \mathbb{T}(\{0,1\}, id)$ if in **LOC** we identify isomorphic objects; this is immediate from the diagram above giving $\Phi_{\{0,1\}}$ as the unique lifting of Φ.

Now $\mathbf{L}_\Phi\Omega$ motivates the question: given a space (X,L,τ), how does (X,L,τ) compare with $\mathbf{L}_\phi\mathbf{PT}(\mathbf{L}_\phi\Omega(X,L,\tau))$, i.e. how does (X,L,τ) compare with $(Lpt(\tau), L, \Phi_L^{\rightarrow}(\tau))$? A (fuzzy) continuous pair $(\Psi_L, \phi) \in \mathbb{T}(L,\phi)[(X,L,\tau), (Lpt(\tau), L, \Phi_L^{\rightarrow}(\tau))]$ is determined as follows (details of proof given in 5.2 below): $\Psi_L : X \longrightarrow Lpt(\tau)$ is defined by

$$\Psi_L(x) : \tau \longrightarrow L,\ \Psi_L(x)(u) = u(x)$$

And so Ψ_L is in the **explicit form** of an **evaluation map**, a considerable simplification and enhancement of the classical case. Indeed, it is a meta-theorem that the explicit form of an evaluation map is obtained **if and only if** the poslat philosophy of fuzzy sets (that a subset be replaced with a mapping into a lattice) is adopted; this is immediate from $(\Psi_{\{0,1\}}, id)$ being the **unique** lifting up to isomorphism in $\mathbb{T}(\{0,1\}, id)$ of the classical Ψ in the following commutative diagram (which shows the "crisp" $\Psi_{\{0,1\}}$ in effect is the classical Ψ):

$$
\begin{array}{ccccc}
(X,\mathcal{T}) & \xrightarrow{\ \Psi\ } & \mathbf{PT}(X,\mathcal{T}) \equiv (Pt(\mathcal{T}), \Phi^{\rightarrow}(\mathcal{T})) & \xrightarrow{\ \mathbf{G}_\chi\ } & (\{0,1\}pt(\mathcal{T}), \{0,1\}, \chi(\Phi^{\rightarrow}(\mathcal{T}))) \\
\mathbf{G}_\chi \downarrow & & \circ & & \updownarrow iso \\
(X, \{0,1\}, \chi(\mathcal{T})) & \xrightarrow[\ !\,(\Psi_{\{0,1\}},\, id)\ ]{} & & & \mathbf{LPT}(X, \{0,1\}, \chi(\mathcal{T})) \equiv (\{0,1\}pt(\chi(\mathcal{T})), \{0,1\}, \Phi^{\rightarrow}_{\{0,1\}}(\chi(\mathcal{T})))
\end{array}
$$

5.2 THEOREM (**Adjunction Conditions**). Let $\mathcal{C}$ and $\mathcal{D}$ be full subcategories of $\mathbb{T}_\phi(L,\phi)$ and $\mathbb{L}$, respectively, which satisfy both of the following conditions:

(1) $\mathbf{L}_\phi\Omega|_\mathcal{C}$ maps $\mathcal{C}$ into $\mathcal{D}$;
(2) $\mathbf{L}_\phi\mathbf{PT}$ maps $|\mathcal{D}|$ into $|\mathcal{C}|$.

Then $\mathbf{L}_\phi\mathbf{PT}|_\mathcal{D}$ is a functor, and $\mathbf{L}_\phi\Omega|_\mathcal{C} \dashv \mathbf{L}_\phi\mathbf{PT}|_\mathcal{D}$ with unit (Ψ_L, ϕ) and counit Φ_L^{op}.

PROOF. We first show that $\mathbf{L}_\phi\mathbf{PT}$ as restricted is a functor. Clearly, $\mathbf{L}_\phi\mathbf{PT}$ preserves into $\mathcal{C}$ the composition and identity morphisms of $\mathcal{D}$ if in fact $\mathbf{L}_\phi\mathbf{PT}$ maps hom-sets of $\mathcal{D}$ into the corresponding hom-sets of $\mathbb{T}_\phi(L,\phi)$. To check the latter, let an arrow $f \in \mathbb{L}(A,B)$, i.e. $f^{op} \in \mathbf{CSLF}(B,A)$, be given; it must be verified that $\mathbf{L}_\phi\mathbf{PT}(f) \in \mathbb{T}_\phi(L,\phi)[(Lpt(A), L, \Phi_L^{\rightarrow}(A)), (Lpt(B), L, \Phi_L^{\rightarrow}(B))]$, where $\mathbf{L}_\phi\mathbf{PT}(f) = (\mathbf{LPT}^*(f), \phi)$ with $\mathbf{LPT}^*(f)(p) = p \circ f^{op}$ (see Discussion 5.1 above). By inspection, $\mathbf{LPT}^*(f) : Lpt(A) \longrightarrow Lpt(B)$, so $\mathbf{L}_\phi\mathbf{PT}|_\mathcal{D}$ being a functor reduces to checking that $\mathbf{L}_\phi\mathbf{PT}(f)$ is continuous, i.e. that $(\mathbf{LPT}^*(f), \phi)^{\leftarrow}$ maps $\Phi_L^{\rightarrow}(B)$ into $\Phi_L^{\rightarrow}(A)$. Let $b \in B$ and $p \in Lft(A)$. Then

$$
\begin{aligned}
((\mathbf{LPT}^*(f), \phi)^{\leftarrow}(\Phi_L(b)))(p) &= (\phi^{op} \circ \Phi_L(b)) \circ \mathbf{LPT}^*(f))(p) \\
&= \phi^{op}(\Phi_L(b)((\mathbf{LPT}^*(f))(p))) \\
&= \phi^{op}(\Phi_L(b)(p \circ f^{op})) \\
&= \phi^{op}(p(f^{op}(b))) \\
&= \phi^{op}(\Phi_L(f^{op}(b))(p))
\end{aligned}
$$

Thus on $Lpt(A)$, we have

$$\mathbf{L}_\phi\mathbf{PT}(f)^{\leftarrow}(\Phi_L(b)) = \langle\phi^{op}\rangle(\Phi_L(f^{op}(b))$$

But $\Phi_L(f^{op}(b)) \in \Phi_L^{\rightarrow}(A)$, and since $\mathbf{L}_\phi\mathbf{PT}(A)$ is assumed to be in $|\mathcal{C}| \subset |\,\mathbb{T}_\phi(L,\phi)|$, then $\langle\phi^{op}\rangle(\Phi_L(f^{op}(b)) \in \Phi_L^{\rightarrow}(A)$, i.e. $\mathbf{L}_\phi\mathbf{PT}(f)^{\leftarrow}(\Phi_L(b)) \in \Phi_L^{\rightarrow}(A)$.

We now show the Major Diagram (§0) of the claimed adjunction is satisfied. Let a space $(X, L, \tau) \in |\mathcal{C}|$ be given. Then choose the candidate for the unit η to be the

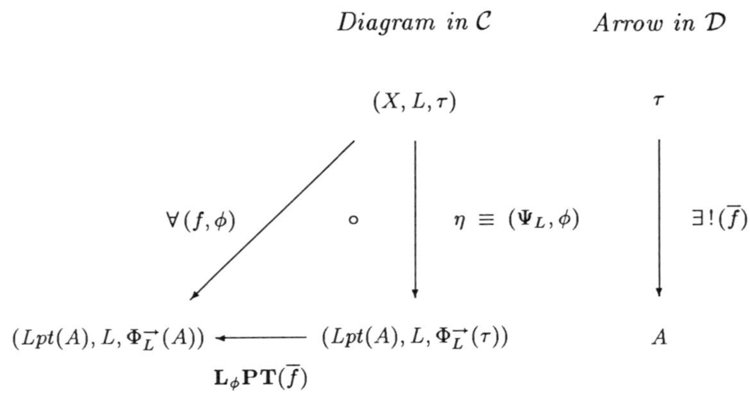

pair (Ψ_L, ϕ). To establish the continuity of (Ψ_L, ϕ), let $v \in \tau$ and $x \in X$, and note

$$\begin{aligned}(\Psi_L,\phi)^{\leftarrow}(\Phi_L(v))(x) &= \phi^{op}(\Phi_L(v)(\Psi_L(x))) = \phi^{op}(\Phi_L\phi(v)(\Psi_L(x))) \\ &= \phi^{op}(\Psi_l(x)(v)) = \phi^{op}(v(x)) = \langle\phi^{op}\rangle(v)(x)\end{aligned}$$

It follows that on X we have

$$(\Psi_L,\phi)^{\leftarrow}(\Phi_L(v)) = \langle\phi^{op}\rangle(v)$$

Since $(X, L, \tau) \in |\mathcal{C}| \subset |\,\mathbb{T}_\phi(L,\phi)|$, then $(\Psi_L,\phi)^{\leftarrow}(\Phi_L(v)) \in \tau$, and (Ψ_L, ϕ) is continuous. To finish the Major Diagram, let $(f,\phi) : (X, L, \tau) \longrightarrow \mathbf{L}_\phi\mathbf{PT}(A)$ be an arrow in $\mathcal{C}$ for some $A \in |\mathcal{D}|$. Noting $f : X \longrightarrow Lpt(A)$ and $\Phi_L : A \longrightarrow L^{Lpt(A)}$, put

$$\overline{f}^{op} : A \longrightarrow \tau,\ \overline{f}^{op}(a) = \Phi_L(a) \circ f$$

By the fullness of $\mathcal{D}$, $\overline{f} \in \mathcal{D}$. To see that (in $\mathcal{C}$) the commutivity

$$(f,\phi) = \mathbf{L}_\phi\mathbf{PT}(\overline{f}) \circ (\Psi_L, \phi)$$

holds, it suffices, given that $\mathbf{L}_\phi\mathbf{PT}(\overline{f}) \circ (\Psi_L,\phi) = (\mathbf{LPT}^*(\overline{f}) \circ \Psi_L, \phi)$ in $\mathcal{C}$, to show

$$f = \mathbf{LPT}^*(\overline{f}) \circ \Psi_L$$

Let $x \in X$ and $a \in A$. Then

$$\begin{aligned}(\mathbf{LPT}^*(\overline{f})(\Psi_L(x)))(a) &= \Psi_L(x)(\overline{f}^{op})(a)) = \Psi_L(x)(\Phi_L(a) \circ f) \\ &= \Phi_L(a)(f(x)) = f(x)(a)\end{aligned}$$

which shows $f = \mathbf{LPT}^*(\overline{f}) \circ \Psi_L$. Finally, $\overline{f}$ uniquely generates the desired commutivity in $\mathcal{C}$, for if g is a possibly different arrow generating the commutivity in $\mathcal{C}$, then those computations above which can be applied to both $\overline{f}$ and g for arbitrary x and a say that

$$\Psi_L(x)(\overline{f}^{op}(a)) = \Psi_L(x)(g^{op}(a)),\ \overline{f}^{op}(a)(x) = g^{op}(a)(x)$$

and hence $\overline{f} = g$. This completes the proof of the Major Diagram.

Finally, we must verify the Minor Diagram (§0) for the claimed adjunction. Let $(f, \phi) : (X_1, L, \tau_1) \longrightarrow (X_2, L, \tau_2)$ be given in $\mathcal{C}$. Then we have the composition

$$(\Psi_L \circ f, \phi) = (\Psi_L, \phi) \circ (f, \phi) : (X_1, L, \tau_1) \longrightarrow (Lpt(\tau_2), L, \Phi_L^{\rightarrow}(\tau_2))$$

By the Major Diagram, $\exists ! [\ \overline{\Psi_L \circ f} : \tau_1 \longrightarrow \tau_2] \in \mathcal{D}$ satisfying the equation

$$(\Psi_L, \phi) \circ (f, \phi) = \mathbf{L}_\phi\mathbf{PT}(\overline{\Psi_L \circ f}) \circ (\Psi_L, \phi)$$

$$\left[\begin{array}{ccccc} (X_1, L_1, \tau_1) & \xrightarrow{(f,\phi)} & (X_2, L_2, \tau_2) & & \tau_1 \\ {\scriptstyle (\Psi_L,\phi)}\big\downarrow & \circ & \big\downarrow{\scriptstyle (\Psi_L,\phi)} & & \big\downarrow{\scriptstyle !\ \overline{\Psi_L \circ f}} \\ \mathbf{L}_\phi\mathbf{PT}(\tau_1) & \xrightarrow[\mathbf{L}_\phi\mathbf{PT}(\overline{\Psi_L \circ f})]{} & \mathbf{L}_\phi\mathbf{PT}(\tau_2) & & \tau_2 \end{array}\right]$$

$$\Longrightarrow \mathbf{L}_\phi\Omega(f, \phi) = \overline{\Psi_L \circ f}$$

It must be shown that $\overline{\Psi_L \circ f} = \mathbf{L}_\phi\Omega(f, \phi)$. Given $v \in \tau_2$ and $x \in X_1$, the definition of $\overline{\Psi_L \circ f}$ in the proof of the Major Diagram implies:

$$\begin{aligned}(\ \overline{\Psi_L \circ f}^{op}(v))(x) &= (\Phi_L(v) \circ (\Psi_L \circ f))(x) = \Phi_L(v)(\Psi_L(f(x))(v)) \\ &= \Psi_L(f(x))(v) = v(f(x)) = ((f, id)^{\leftarrow}(v))(x) \\ &= (\mathbf{L}\Omega(\mathbf{F}_\phi(f, \phi))(v)) = (\mathbf{L}_\phi\Omega(f, \phi)(v))(x)\end{aligned}$$

This concludes the proof of $\mathbf{L}_\phi\Omega|_{\mathcal{C}} \dashv \mathbf{L}_\phi\mathbf{PT}|_{\mathcal{D}}$ and the theorem. □

5.3 DEFINITION (**Fuzzy Spatiality and Sobriety**). Let $L \in |\mathbf{CSLF}|$. Then $A \in |\mathbb{L}|$ is **L-spatial** iff Φ_L is injective; a space (X, L, τ) is (**classically or crisply**) **topological** iff τ is $\{0,1\}$-spatial. The space $(X, L, \tau) \in |\mathbb{T}|$ is **sober** $[S_0, T_0]$ iff Ψ_L is bijective [surjective, injective]. **LSPAT**$\mathbb{L}$ [**SOB**$\mathbb{T}$, **SOB**$\mathbb{T}_\phi(L,\phi)$, $\mathbf{S_0}\mathbb{T}_\phi(L,\phi)$, $\mathbf{T_0}\mathbb{T}_\phi(L,\phi)$, etc] is the full subcategory of $\mathbb{L}$ [$\mathbb{T}$, $\mathbb{T}_\phi(L,\phi)$, etc] of all L-spatial [sober, S_0, T_0] objects.

5.4 LEMMA. *A is L-spatial $\Leftrightarrow$ Φ_L^{op} is a $\mathbb{L}$ isomorphism of A with the topology $\Phi_L^{\rightarrow}(A)$ of the space $\mathbf{L}_\phi\mathbf{PT}(A)$. And if L_1 is a complete sublattice of L_2, then L_1-spatiality $\Rightarrow$ L_2-spatiality; so that classical spatiality, or $\{0,1\}$-spatiality, always implies L-spatiality $\forall\, L \in |\mathbf{CSLF}|$. Finally, (X, L, τ) is sober $\Leftrightarrow$ (Ψ_L, ϕ) is a $\mathbb{T}_\phi(L,\phi)$ isomorphism.*

PROOF. The proofs are straightforward; the last statement can be easily proven directly — it is found in more general form in [Rodabaugh 1986, 1986–7, ****]. □

5.5 DEFINITION. Let $\mathcal{C}$ be a subcategory of $\mathcal{A}$. Then $\langle\mathcal{C}\rangle$ denotes the smallest full subcategory of $\mathcal{A}$ containing $\mathcal{C}$.

5.6 THEOREM (**Fundamental Classes of "Stone" Adjunctions**). *The hypotheses of 5.2 are satisfied if any of the following sets of conditions hold:*

(1) $L \in |\mathbf{CSLF}|$, $\phi \in \mathbb{L}(L,L)$, $\mathcal{C} = \mathbf{S_0}\mathbb{T}_\phi(L,\phi)$, $\mathcal{D} = \langle L_\phi\Omega(\mathbf{S_0}\mathbb{T}_\phi(L,\phi))\rangle$.
(2) $L \in |\mathbf{CSLF}|$, $\phi \in \mathbb{L}(L,L)$, $\mathcal{C} = \mathbf{SOB}\,\mathbb{T}_\phi(L,\phi)$, $\mathcal{D} = \langle \mathbf{L}_\phi\Omega(\mathbf{SOB}\,\mathbb{T}_\phi(L,\phi))\rangle$.
(3) $L \in |\mathbf{CSLF}|$, $\mathcal{C} = \mathbb{T}(L, id)$, $\mathcal{D} = \mathbb{L}$.
(4) $L \in |\mathbf{FRM}|$, $\mathcal{C} = \mathbb{T}(L, id)$, $\mathcal{D} = \mathbf{LOC}$.
(5) $L = \{0,1\}$, $\mathcal{C} = \mathbb{T}(\{0,1\}, id) \equiv \mathbf{TOP}$, $\mathcal{D} = \mathbf{LOC}$.

*Furthermore, (3) [(4)] can be strengthened to say that $\mathbb{T}(L, id)$ is adjunctive to $\mathbb{L}$ [**LOC**] via $\mathbf{L}\Omega \dashv \mathbf{LPT}$ iff $L \in |\mathbf{CSLF}|$ [$|\mathbf{LOC}|$].*

OUTLINE OF PROOF. We refer the reader to [Rodabaugh 1986-7, ****] for (1-5); we merely note 5.13 below is needed for (1,2), and (4) $\Rightarrow$ (5). For the last statement of the theorem, we outline the localic case: sufficiency comes from (2); and necessity comes from observing that [$\mathbf{L}\Omega(1, L, \tau \equiv L^1) \in |\mathbf{LOC}|$] $\Rightarrow$ [$L^1 \in |\mathbf{LOC}|$] $\Rightarrow$ [$L \in |\mathbf{LOC}|$], where $\mathbf{1}$ is a singleton set. □

5.7 COROLLARY. *The classical $\Omega \dashv \mathbf{PT}$ holds; i.e. the class of adjunctions $\mathbf{L}_\phi\Omega \dashv \mathbf{L}_\phi\mathbf{PT}$ includes $\Omega \dashv \mathbf{PT}$.*

PROOF. Invoke 5.6(5) and the diagrams of 5.1 above. □

5.8 REMARK.

(1) These facts—(Ψ_L, ϕ) and Φ_L^{op} are the respective unit and counit of $\mathbf{L}_\phi\Omega \dashv \mathbf{L}_\phi\mathbf{PT}$, and $\mathbf{L}_\phi\Omega \dashv \mathbf{L}_\phi\mathbf{PT}$ includes $\Omega \dashv \mathbf{PT}$ — are confirmation, in addition to that of the diagrams

of 5.1, that Ψ_L generalizes Ψ and Φ_L generalizes Φ.

(2) Historically speaking, 5.2, 5.3, 5.4, 5.6, and 5.15 (below) were first announced in [Rodabaugh 1986] and proofs first given in [Rodabaugh 1986-7]; then 5.2, 5.3, 5.4, 5.6(4) for the case $L = [0,1] \subset \mathbb{R}$ and $\phi = id$ were subsequently and independently stated and proved in [Frith 1987, Chapter 6]. Further, an axiomatization for the existence of a general class of "crisp" dualities (including the special cases of topological spaces and locales, probabilistic spaces and Boolean algebras, etc.) was given in [Rodabaugh 1987], with indications on how such an axiomatization should carry over to the fuzzy case (*à la* 5.2-5.6 and 5.15) given implicitly both there and in [Rodabaugh 1986-7, Subsection 3.2]; developments along this line can be seen in the chapter of W. Kotzé in this volume.

(3) What is the role of stratification in all of this? The full resolution of this question is given in the next section, for which we need the following results.

5.9 THEOREM (**Stratification of 5.2**). Let $L \in |\mathbf{CSLF}|$, $\phi^{op} \in \mathbb{L}(L,L)$, $\mathbf{L}_\phi\Omega$ be restricted to a full subcategory $\mathcal{C}_k$ of $\mathbb{T}_{\phi k}(L,\phi)$ and $\mathbf{G}_k\mathbf{L}_\phi\mathbf{PT}$ (see 2.15) be restricted to a full subcategory $\mathcal{D}_k$ of $\mathbb{L}$ such that $\mathbf{L}_\phi\Omega(\mathcal{C}_k) \subset \mathcal{D}_k$ and $\mathbf{G}_k\mathbf{L}_\phi\mathbf{PT}(|\mathcal{D}_k|) \subset |\mathcal{C}_k|$. Then $\mathbf{G}_k\mathbf{L}_\phi\mathbf{PT}$ becomes a functor, and $\mathbf{L}_\phi\Omega \dashv \mathbf{G}_k\mathbf{L}_\phi\mathbf{PT}$ with unit (Ψ_L,ϕ) and counit^{op} being the assignment $A \mapsto \Phi_L^{\rightarrow}(A) \vee \{\underline{\alpha} : \alpha \in L\}$.

PROOF. The proof is virtually a twin of the proof of 5.2 and is essentially left to the reader. We make a few comments concerning the slight divergence from 5.2. Note that $\forall A \in |\mathcal{D}|$, $\forall f : A \longrightarrow B$ in $\mathcal{D}_k$, we have

$$\mathbf{G}_k\mathbf{L}_\phi\mathbf{PT}(A) = (Lpt(A), L, \Phi_L^{\rightarrow}(A) \vee \{\underline{\alpha} : \alpha \in L\}),$$

$$\mathbf{G}_k\mathbf{L}_\phi\mathbf{PT}(f) = \mathbf{L}_\phi\mathbf{PT}(f)$$

Hence for $b \in B$, the proof of 5.2 yields

$$\begin{aligned}(\mathbf{G}_k\mathbf{L}_\phi\mathbf{PT}(f))^{\leftarrow}(\Phi_L(b)) &= (\mathbf{L}_\phi\mathbf{PT}(f))^{\leftarrow}(\Phi_L(b)) \\ &= \phi^{op} \circ \Phi_L(f^{op}(b)) \\ &= \langle\phi^{op}\rangle(\Phi_L(f^{op}(b)))\end{aligned}$$

which is an element of

$$\langle\phi^{op}\rangle^{\rightarrow}(\Phi_L^{\rightarrow}(A)) \subset \langle\phi^{op}\rangle^{\rightarrow}(\Phi_L^{\rightarrow}(A) \vee \{\underline{\alpha} : \alpha \in L\}) \subset \Phi_L^{\rightarrow}(A) \vee \{\underline{\alpha} : \alpha \in L\}$$

the latter inclusion holding since $\mathbf{G}_k\mathbf{L}_\phi\mathbf{PT}(A) \in |\mathcal{C}_k| \subset |\mathbb{T}_{\phi k}(L,\phi)| \subset |\mathbb{T}_\phi(L,\phi)|$. Also note for a constant map in $\mathbf{G}_k\mathbf{L}_\phi\mathbf{PT}(B)$, we have that

$$(\mathbf{G}_k\mathbf{L}_\phi\mathbf{PT}(f))^{\leftarrow}(\underline{\beta}) = \langle\phi^{op}\rangle(\underline{\beta} \circ \mathbf{LPT}^*(f)) = \phi^{op} \circ \underline{\beta} \circ \mathbf{LPT}^*(f)$$

is a constant map in $\mathbf{G}_k\mathbf{L}_\phi\mathbf{PT}(A)$. So $(\mathbf{G}_k\mathbf{L}_\phi\mathbf{PT}(f))^{\leftarrow}$ preserves the subbasis of $\mathbf{G}_k\mathbf{L}_\phi\mathbf{PT}(B)$; and as noted in 2.13, this is enough for the continuity of $\mathbf{G}_k\mathbf{L}_\phi\mathbf{PT}(f)$. Similar adjustments provide the proof of the continuity of (Ψ_L,ϕ) for the stratified case. The proof of 5.9 now follows from that of 5.2. □

5.10 THEOREM (**Stratification of 5.6**). The hypotheses of 5.9 are satisfied if any of the following sets of conditions hold:

(1) $L \in |\mathbf{CSLF}|$, $\mathcal{C}_k = \mathbb{T}_k(L, id)$, $\mathcal{D}_k = \mathbb{L}$.
(2) $L \in |\mathbf{FRM}|$, $\mathcal{C}_k = \mathbb{T}_k(L, id)$, $\mathcal{D}_k = \mathbf{LOC}$.
(3) $L \in \{0,1\}$, $\mathcal{C}_k = \mathbb{T}_k(\{0,1\}, id) \equiv \mathbf{TOP}$, $\mathcal{D}_k = \mathbf{LOC}$.

Furthermore, (1) [(2)] can be strengthened to say that $\mathbb{T}_k(L, id)$ is adjunctive to $\mathbb{L}$ [**LOC**] via $\mathbf{L}\Omega \dashv \mathbf{G}_k\ \mathbf{LPT}$ iff $L \in |\mathbf{CSLF}|$ [$|\mathbf{LOC}|$].

5.11 REMARK. To answer the suggestion of [Höhle 1986], the distinction between the Chang-Goguen approach and the Lowen approach is not that the former is adjunctive to **LOC** (if the underlying lattice is a frame) and the latter is not, for this is also true of the latter. So what is missing in the stratified approach? First, there is no meaningful stratification of 5.6(1,2) for most lattices L and endomorphisms ϕ^{op}. Second, for most lattices the adjunction $\mathbf{L}_\phi\Omega \dashv \mathbf{G}_k\mathbf{L}_\phi\mathbf{PT}$ **never** restricts to a categorical equivalence even if $\phi = id_L$; thus no representation or compactification theory à la Stone can be constucted in $\mathbb{T}_k(L, id)$ using $\mathbf{L}_\phi\Omega$ or $\mathbf{L}\Omega$. Proofs for these and similar results are given in the next section.

5.12 LEMMA. $\forall (X, L, \tau) \in |\mathbb{T}|$, $\Omega(X, L, \tau) \equiv \tau$ is L-spatial.

PROOF. Let $u \neq v$ in τ; then $\exists\, x_0 \in X,\ u(x_0) \neq v(x_0)$. Put $p = \Psi_L(x_0) \in Lpt(\tau)$. Then

$$\Phi_L(u)(p) = p(u) = \Psi_L(x_0)(u) = u(x_0) \neq v(x_0) = \Psi_L(x_0)(v) = p(v) = \Phi_L(v)(p)$$

Thus $\Phi_L(u) \neq \Phi_L(v)$ on $Lpt(\tau)$, i.e. Φ_L is injective on τ. So τ is L-spatial. □

5.13 LEMMA. If $L \in |\mathbf{CSLF}|$ and $A \in |\mathbb{L}|$, then $\mathbf{L}_\phi\mathbf{PT}(A) \equiv (Lpt(A), L, \Phi_L^{\rightarrow}(A))$ is sober.

PROOF. To satisfy 5.3, we must show that $\Psi_L : Lpt(A) \longrightarrow Lpt(\Phi_L(A))$ is bijective. Let $p, q \in Lpt(A)$ with $p \neq q$. Then $\exists\, a \in A,\ p(a) \neq q(a)$, i.e. $\exists\, \Phi_L(a) \in \Phi_L^{\rightarrow}(A)$ such that

$$\Psi_L(p)(\Phi_L(a)) = \Phi_L(a)(p) = p(a) \neq q(a) = \Phi_L(a)(q) = \Psi_L(q)(\Phi_L(a))$$

which shows $\Psi_L(p) \neq \Psi_L(q)$. Thus Ψ_L is injective. To show the surjectivity of Ψ_L, let $q \in Lpt(\Phi_L^{\rightarrow}(A))$ and put $p = q \circ \Phi_L$. Clearly $p \in \mathbf{CSLF}(A, L) = Lpt(A)$. Furthermore, $\forall\, a \in A$, we have

$$\Psi_L(p)(a) = \Phi_L(a)(p) = p(a) = q(\Phi_L(a))$$

so that $\Psi_L(p) = q$, establishing surjectivity. □

5.14 REMARK. The results 5.2, 5.6, 5.12, and 5.13 above not only generalize, in a broadscale way, their classical counterparts as found in [Johnstone 1982], but they are much easier, detail by detail, to prove. These results, combined with the adjunctions of 5.2 and 5.6, have many consequences *sequens*. More importantly, we begin to see the practical significance of fuzzy sets taken in the non-model-theoretic or poslat sense: only with fuzzy sets do we have the simplicity and convenience of the explicit form of evaluation maps for the unit and the counit^{op} of the Stone paradigm $\mathbf{L}_\phi\Omega \dashv \mathbf{L}_\phi\mathbf{PT}$.

5.15 THEOREM (**Sobriety-Spatiality Representation Theorems**). Let $L \in |\mathbf{CSLF}|$.

(1) $\forall \phi \in \mathbb{L}(L,L)$, $\mathbf{SOB}\,\mathbb{T}_\phi(L,\phi) \approx \langle \mathbf{L}_\phi\Omega(\mathbf{SOB}\,\mathbb{T}_\phi(L,\phi))\rangle$.
(2) If $\phi = id_L$, then $\langle \mathbf{L}_\phi\Omega(\mathbf{SOB}\,\mathbb{T}_\phi(L,\phi))\rangle \approx \mathbf{LSPAT}\,\mathbb{L}$ via $\hookrightarrow \dashv \mathbf{L}\Omega \circ \mathbf{LPT}$.
(3) $\mathbf{SOB}\,\mathbb{T}(L,id) \approx \mathbf{LSPAT}\mathbb{L}$ via $\mathbf{L}\Omega \dashv \mathbf{LPT}$.
(4) If $L \in |\mathbf{FRM}|$, then $\mathbf{SOB}\,\mathbb{T}(L,id) \approx \mathbf{LSPATLOC}$ via $\mathbf{L}\Omega \dashv \mathbf{LPT}$.
(5) If $L = \{0,1\}$, then $\mathbf{SOBTOP} \approx \mathbf{SPATLOC}$ via $\Omega \dashv \mathbf{PT}$.

PROOF. *Ad(1).* This is a consequence of 5.4, 5.6(2), 5.12, and 5.13: these say that the adjunction of 5.6(2) has a unit each of whose components is an isomorphism in $\mathbb{T}_\phi(L,\phi)$ and a counit each of whose components is an isomorphism in $\mathbb{L}$, i.e. the adjunction is an equivalence of categories.

Ad(2). It suffices to show that the inclusion functor $\langle \mathbf{L}_\phi\Omega(\mathbf{SOB}\ \mathbb{T}_\phi(L,\phi))\rangle \hookrightarrow \mathbf{LSPAT}\mathbb{L}$ induces an adjunction which is an equivalence, and in which the right adjoint is $\mathbf{L}\Omega \circ \mathbf{LPT}$. To satisfy the dual of the Major Diagram (§0), i.e. determine the counit, let $A \in |\mathbf{LSPAT}\mathbb{L}|$, and note the arrow $\Phi_L^{op} : \Phi_L^{\rightarrow}(A) \rightarrow A$ is in $\mathbf{LSPAT}\mathbb{L}$. Now if $\xi : \tau \rightarrow A$ is in $\mathbf{LSPAT}\mathbb{L}$, where $\tau \in |\langle \mathbf{L}_\phi\Omega(\mathbf{SOB}\,\mathbb{T}_\phi(L,\phi))\rangle|$, then $\xi^* \equiv \xi^{op} \circ \Phi_L^{-1}$ gives the unique $\xi^* : \tau \rightarrow \Phi_L^{\rightarrow}(A)$ which factors ξ through Φ_L^{op}, where Φ_L^{-1} is the inverse of Φ_L in $\mathbf{SET}$ with domain $\Phi_L^{\rightarrow}(A)$; i.e. Φ_L^{op} is the counit. To satisfy the dual of the Minor Diagram (§0), i.e. determine that the right adjoint of $\hookrightarrow$ induced by the dual of the Major Diagram agrees with $\mathbf{L}\Omega \circ \mathbf{LPT}$ on arrows as well as on objects, let $v : A \longrightarrow B$ be in $\mathbf{LSPAT}\ \mathbb{L}$. Then $v \circ \Phi_L^{op} : \Phi_L^{\rightarrow}(A) \longrightarrow B$ is in $\mathbf{LSPAT}\ \mathbb{L}$, so the arrow from $\Phi_L^{\rightarrow}(A)$ to $\Phi_L^{\rightarrow}(B)$ induced from the dual of the Major Diagram is:

$$((v \circ \Phi_L^{op}) \circ \Phi_L^{-1})^{op} = (\Phi_L \circ v^{op} \circ \Phi_L^{-1})^{op}$$

where Φ_L^{-1} is the inverse of Φ_L in $\mathbf{SET}$ with domain $\Phi_L^{\rightarrow}(B)$. Using the calculation

$$(\mathbf{LPT}(v))^{\leftarrow}(\Phi_L(b)) = \Phi_L(v^{op}(B))$$

one can verify that the induced arrow referred to above is precisely $\mathbf{L}\Omega(\mathbf{LPT}\ (v))$, thus satisfying the dual of the Minor Diagram. Next, one can compute the unit to be $(\Phi_L^{-1})^{op}$, where Φ_L^{-1} is the inverse in $\mathbf{SET}$ with domain the range of Φ_L. This adjunction is an equivalence: since all components of Φ_L are isomorphisms to their ranges, then so are all components of the counit and unit.

Ad(3). This follows from (1,2) via the adjunction $\hookrightarrow \circ\ \mathbf{L}\Omega \dashv \mathbf{LPT} \circ L\Omega \circ \mathbf{LPT}$, which reduces to the equivalence given by $\mathbf{L}\Omega \dashv \mathbf{LPT}$ since (Ψ_L, id), from the space $\mathbf{LPT}(A)$

to the space $\mathbf{LPT}(\mathbf{L}\Omega(\mathbf{LPT}(A)))$ is natural and always an isomorphism in $\mathbb{T}(L, id)$. This equivalence also follows directly from 5.4, 5.6(3), 5.12, and 5.13.

Ad(4). Note $L \in |\mathbf{FRM}|$ implies $\mathbf{L}\Omega(X, L, \tau) \in |\mathbf{LOC}|$, and so the equivalence of (3) restricts as needed.

Ad(5). In the equivalence of (4), set $L = \{0, 1\}$ (as done several times before). □

5.16.1 REMARK (**First Deficiency of TOP Filled by Fuzzy Topology**). Theorem 5.15 is much more general than 4.4, and its proof (essentially 5.12 and 5.13) much simpler and easier because of the (explicit form of) the evaluation maps! More importantly, 5.15 fills a deficiency of **TOP** implicitly indicated by 4.4. The localic product $\mathcal{T}(\mathbf{Q}) \times_\ell \mathcal{T}(\mathbf{Q})$ of the usual topology on the rationals with itself is a non-spatial locale [Johnstone 1982] in the classical sense, and hence it can have no topological space representation. Setting $L = \mathcal{T}(\mathbf{Q}) \times_\ell \mathcal{T}(\mathbf{Q})$, it is not hard to prove [Rodabaugh 1986–7, ****] that $\mathcal{T}(\mathbf{Q}) \times_\ell \mathcal{T}(\mathbf{Q})$ is L-spatial and hence has the non-trivial L-sober space representation $\mathbf{LPT}(\mathcal{T}(\mathbf{Q}) \times_\ell \mathcal{T}(\mathbf{Q}))$. More generally, the fuzzy theory furnishes at least one sober space representation for **each** locale (indeed each object of $\mathbb{L}$). This is due to the fuzzy theory allowing degrees of spatiality, with $\{0, 1\}$-spatiality (or crisp or classical spatiality) the strongest degree (see 5.4). Thus all complete lattices and locales are topologies: some of them classical topologies (with crisp spatiality), some of them fuzzy topologies (with "non-crisp" degrees of spatiality). If an apologetic for the Stone Representation Theorem for Boolean algebras be that it gives representations for algebras missed by the Lindenbaum-Tarski Representation Theorem (cf. Introduction of [Johnstone 1982]), then the analogous apologetic for fuzzy topology must be made since it gives representations for all complete Heyting algebras or locales, which **TOP cannot** do. To sum up, **each complete lattice (and locale) is the fuzzy topology of some fuzzy sober space.**

5.16.2 REMARK (**Second Deficiency of TOP Filled by Fuzzy Topology**). The previous remark uses the example $\mathcal{T}(\mathbf{Q}) \times_\ell \mathcal{T}(\mathbf{Q})$ of a locale which cannot be a classical topology, but which is a sober fuzzy topology. *À fortiori*, $\mathcal{T}(\mathbf{Q}) \times_\ell \mathcal{T}(\mathbf{Q})$ is a localic product which is not a classical product topology. We show it is a fuzzy product topology, as defined in [Goguen 1973] and [Wong 1974], by outlining a proof for an arbitrary localic product $A \equiv \times_\ell A_\gamma$ with projections $\{\pi_\gamma\}_\gamma$; see [Rodabaugh 1986–7, ****] for more details. Now A is a coproduct in **FRM** with injections the frame morphisms $\{\pi_\gamma^{op}\}_\gamma$, each of which can be shown to be an isomorphism in **FRM** onto its image, thus making that image a subframe of A. Thus for each index γ, we put

$$B_\gamma = (\pi_\gamma^{op})^{\rightarrow}(A_\gamma)$$

As in 5.16.1, choose $L = A$; and also choose $X = \mathbf{1}$ (the singleton set), and put $\tau_\gamma = B_\gamma^{\mathbf{1}}$. Then B_γ being a subframe of L guarantees that $(X, L, \tau_\gamma) \in |\mathbb{T}(L, id)|$. Since each $\tau_\gamma \approx B_\gamma \approx A_\gamma$ in **FRM**, the universality of $\times_\ell$ implies that

$$A = \times_\ell A_\gamma \approx \times_\ell \tau_\gamma \text{ (in } \mathbf{LOC} \text{ or } \mathbf{FRM})$$

and since A can be shown to be L-spatial (cf. 5.16.1), the universality of Φ_L from the adjunction $\mathbf{L\Omega} \dashv \mathbf{LPT}$ can be invoked to show that $\times_\ell \tau_\gamma$ is L-spatial. It can then be proven that

$$\times_\ell \tau_\gamma \approx \times_\gamma \Phi_L^{\rightarrow}(\tau_\gamma)$$

the right-hand side denoting the poslat topology of the Goguen-Wong (categorical) product of $\{\mathbf{LPT}(\tau_\gamma)\}_\gamma$ in $\mathbb{T}(L, id)$. Thus $A \approx \times_\ell \tau_\gamma$ is a fuzzy product topology. So each localic product, in particular $\mathcal{T}(\mathbf{Q}) \times_\ell \mathcal{T}(\mathbf{Q})$, is captured by fuzzy topology as a product topology, which **TOP cannot** do. To sum up, **each localic product is a fuzzy product topology.**

5.16.3 OPEN QUESTION. Does there exist an object $A \in |\mathbf{CSLF}|$ which is not classically spatial, i.e. not $\{0,1\}$-spatial, but for which there exists a complete sublattice L strictly between $\{0,1\}$ and A, such that A is L-spatial? As indicated in 5.16.1 and 5.16.2 above, A is L-spatial for $L = A$.

5.17 DEFINITION (**Fuzzy Coherent and Fuzzy Stone Spaces**). A space (X, L, τ) is **coherent** iff it is C-compact (compact in the sense of [Chang 1968]), sober, and τ is coherent in the sense of 4.7; and (X, L, τ) is a **Stone space** iff it is coherent and HR-regular (regularity of [Hutton-Reilly 1980]). **COH** $\mathbb{T}$ [**COH** $\mathbb{T}_\phi(L, \phi)$] is the subcategory of $\mathbb{T}$ [$\mathbb{T}_\phi(L, \phi)$] of coherent spaces and all fuzzy continuous mappings whose reverse powerset operators preserve finite elements (as defined in 4.7) of the (second) topology; and **STN** $\mathbb{T}$ [**STN** $\mathbb{T}_\phi(L, \phi)$] is the full subcategory of $\mathbb{T}$ [$\mathbb{T}_\phi(L, id)$] of all Stone spaces.

5.18 REMARK. The definitions given in 5.17 are motivated by these results for the classical setting: a space is coherent iff it is sober and its topology is a coherent lattice, and a space is Stone iff it is coherent and regular.

The next lemma sets up the appropriate restrictions of the class of equivalences **SOB** $\mathbb{T}(L, id) \approx$ **LSPAT** $\mathbb{L}$ to obtain general classes of Stone representation theorems.

5.19 LEMMA. The following statements hold:

(1) **COH** $\mathbb{T} \hookrightarrow$ **SOB** $\mathbb{T}$, and $\forall L \in |\mathbf{CSLF}|$, **COHLOC** $\hookrightarrow$ **LSPATLOC**.
(2) Ω maps **COH** $\mathbb{T}$ into **COH** $\mathbb{L}$; and $\forall L \in |\mathbf{DCSLF}|$, $\mathbf{L\Omega}$ maps **COH** $\mathbb{T}(L, id)$ into **COHLOC**, (where **DCSLF** is the full subcategory of **CSLF** of all (finitely) distributive objects).
(3) $\forall L \in |\mathbf{CSLF}|$, **LPT** maps **COHLOC** into **COH** $\mathbb{T}(L, id)$.
(4) **STN** $\mathbb{T} \hookrightarrow$ **COH** $\mathbb{T}$, if the ground be strengthened to **SET** $\times$ $\mathbf{CBOOL}^{op}$, (where **CBOOL** is the full subcategory of **BOOL** of all complete Boolean algebras).
(5) $\forall L \in |\mathbf{CBOOL}|$, the following hold: C-compactness $\Leftrightarrow$ H-compactness (compactness in the sense of [Hutton 1980]); and if $B \in |\mathbf{DLAT}|$, then $\mathbf{LPT}(Idl(B))$ is HR-regular (and so in **STN** $\mathbb{T}(L, id)$) $\Leftrightarrow$ $B \in |\ \mathbf{BOOL}|$.

OUTLINE OF PROOF. (Full details are given in [Rodabaugh 1986–7, ****].)

Ad(1). $\mathbf{COH}\,\mathbb{T} \hookrightarrow \mathbf{SOB}\,\mathbb{T}$ is clear by definition. Now the second inclusion holds trivially w.r.t. morphisms; for the object case, we use the result that each coherent locale is classically spatial [Johnstone 1982], and classical spatiality implies L-spatiality for each $L \in |\mathbf{CSLF}|$ by 5.4.

Ad(2). The assertion for Ω is obvious, and the assertion for $\mathbf{L}\Omega$ is immediate from the following result implicit in [Johnstone 1982] (cf. [Rodabaugh 1986, 1986–7, ****]): each $A \in |\mathbf{DCSLF}|$ is coherent $\Leftrightarrow$ A is isomorphic to the locale of ideals of some $B \in |\mathbf{DLAT}|$.

Ad(3). Let $A \in |\mathbf{COHLOC}|$. Then $\mathbf{LPT}(A) = (Lpt(A), L, \Phi_L^{\rightarrow}(A))$ is sober by 5.13; by the argument given in Ad(1) above, A is L-spatial, so Φ_L is an isomorphism, and the coherence of A implies the coherence of $\Phi_L^{\rightarrow}(A)$; and so $\mathbf{LPT}(A)$ is coherent. Now let $f \in \mathbf{COHLOC}(A,B)$. Recall $\mathbf{LPT}(f) \in \mathbb{T}(L,id)(\mathbf{LPT}(A),\mathbf{LPT}(B))$ and that $\forall b \in B$, $\mathbf{LPT}(f)^{\leftarrow}(\Phi_L(b)) = \Phi_L(f^{op}(b))$: set $\phi = id_L$ in the proof of 5.2. Since all components of Φ_L are isomorphisms on A and B (coherence $\Rightarrow$ L-spatiality), the following implications hold: $\Phi_L(b) \in K(\Phi_L^{\rightarrow}(B)) \Rightarrow b \in K(A) \Rightarrow f^{op}(b) \in K(A)$ ($f \in \mathbf{COHLOC}(A,B)$ is a coherent morphism) $\Rightarrow \Phi_L(f^{op}(b)) \in K(\Phi_L^{\rightarrow}(A))$. Summing up, $\mathbf{LPT}(f) \in \mathbf{COH}\,\mathbb{T}(\mathbf{LPT}(A), \mathbf{LPT}(B))$.

Ad(4). We need only check that morphisms between Stone spaces are coherent. Let $(f,\phi) : (X,L,\tau) \longrightarrow (Y,M,\sigma)$ be a continuous morphism, where $(X,L),(Y,M)$, and (f,ϕ) are from the ground category $\mathbf{SET} \times \mathbf{CBOOL}^{op}$. Continuity implies that $(f,\phi)^{\leftarrow}(\sigma) \subset \tau$. Let $'$ denote the Boolean complementation on each of L and M, and let $\tau' = \{u' : u \in \tau\}$ and $\sigma' = \{v' : v \in \sigma\}$. It can be shown that this stronger ground implies that $(f,\phi)^{\leftarrow}(\sigma') \subset \tau'$, and hence that $(f,\phi)^{\leftarrow}(\sigma \cap \sigma') \subset (\tau \cap \tau')$. Finally, one can prove for fuzzy Stone spaces that $K(\tau) = \tau \cap \tau'$ and $K(\sigma) = \sigma \cap \sigma'$.

Ad(5). The equivalence of C-compactness with H-compactness can be shown to follow from the Boolean complementation. To see that the "fuzzy spectrum" $\mathbf{LPT}(Idl(B))$ is HR-regular $\Leftrightarrow$ $B \in |\mathbf{BOOL}|$, we make a few prelimniary comments: for $B \in |\mathbf{DLAT}|$, $B \approx K(Idl(B))$; the coherence of $Idl(B)$ implies $Idl(B)$ is L-spatial, i.e. $Idl(B) \approx \Phi_L^{\rightarrow}(Idl(B))$, so $B \approx K(\Phi_L^{\rightarrow}(Idl(B)))$; $\mathbf{LPT}(Idl(B))$ is the image under $\mathbf{LPT}$ of a coherent locale, and so is coherent, hence C-compact, and therefore H-compact. To see that $\mathbf{LPT}(Idl(B))$ being HR-regular $\Rightarrow$ $B \in |\,\mathbf{BOOL}|$, note the former is both H-compact and HR-regular; it can be shown that this implies that $K(\Phi_L^{\rightarrow}(Idl(B)))$, and hence B, is a Boolean algebra. Now for the converse direction, B is Boolean $\Rightarrow$ $K(\Phi_L^{\rightarrow}(Idl(B)))$ is Boolean, and the coherence of $\mathbf{LPT}(Idl(B))$ says $K(\Phi_L^{\rightarrow}(Idl(B)))$ is a sublattice of $\Phi_L^{\rightarrow}(Idl(B))$, which in turn is a sublattice of $L^{Lpt(Idl(B))}$, i.e. $K(\Phi_L^{\rightarrow}(Idl(B)))$ is a Boolean subalgebra of $L^{Lpt(Idl(B))}$, and this can be shown to imply the HR-regularity of $\mathbf{LPT}(Idl(B))$. □

5.20 Theorem (**Classes of Stone Representation Theorems**). If $L \in |\mathbf{DCSLF}|$, then $\mathbf{COH}\,\mathbb{T}(L,id) \approx \mathbf{DLAT}^{op}$; and if $L \in |\mathbf{CBOOL}|$, then $\mathbf{STN}\,\mathbb{T}(L,id) \approx \mathbf{BOOL}^{op}$. For $L = \{0,1\}$, the classical representations $\mathbf{COHTOP} \approx \mathbf{DLAT}^{op}$ and $\mathbf{STNTOP} \approx$

$\mathbf{BOOL}^{op}$ follow.

PROOF. If $L \in |\mathbf{DCSLF}|$, 5.19(1-3) restrict $\mathbf{SOB}\mathbb{T}(L, id) \approx \mathbf{LSPAT}\mathbb{L}$ to obtain the equivalence $\mathbf{COH}\,\mathbb{T}\,(L, id) \approx \mathbf{COHLOC}$, and invoking the classical equivalence $\mathbf{COHLOC} \approx \mathbf{DLAT}^{op}$, the fuzzy representation $\mathbf{COH}\,\mathbb{T}\,(L, id) \approx \mathbf{DLAT}^{op}$ is at hand. Now 5.19(4-5), for $L \in |\mathbf{CBOOL}|$, restrict the representation $\mathbf{COH}\,\mathbb{T}\,(L, id) \approx \mathbf{DLAT}^{op}$ to obtain the fuzzy representation $\mathbf{STN}\,\mathbb{T}\,(L, id) \approx \mathbf{BOOL}^{op}$. □

5.21 REMARK.

(1) The equivalence of HR-regularity of a space with the localic regularity (in the sense of [Johnstone 1982]) of its (fuzzy) topology, when the base lattice is from **CBOOL**, is established in [Rodabaugh 1986–7, ****]. This shows that for such base lattices, a fuzzy Stone space can be equivalently defined by picking one of C-compactness or H-compactness in the definition of coherence, and then picking one of HR-regularity or localic regularity (of the topology). In [Rodabaugh 1986–7, ****] is a detailed comparison of localic and "traditional" fuzzy separation axioms; also see [Rodabaugh 1991] for a philosophical critique of fuzzy separation axioms.
(2) Only the fuzzy case reveals the lattice-theoretic properties of $\{0, 1\}$ needed for Stone's original representation theorems: $\mathbf{COHTOP} \approx \mathbf{DLAT}^{op}$ [$\mathbf{STNTOP} \approx \mathbf{BOOL}^{op}$] rests on $\{0, 1\}$ being in $|\mathbf{DLAT}|$ [$|\mathbf{BOOL}|$].
(3) Applications of this section to the existence of fuzzy Stone-Cech compactifications and compactification reflectors are abstracted in [Rodabaugh 1986, 1991] and fully given [Rodabaugh 1986–7, ****].

§6. Lattice-"Independent" Spatiality and Sobriety

In this section we attempt to build a Stone representation theory which allows for both change of basis and recognized degrees of stratification. The motivation for change of basis should be clear, so we motivate the latter issue, and in so doing motivate the categories $\mathbb{T}_D$ and $\mathbb{L}_d \equiv \mathbb{L}^{2.5}$ defined in §2 and §3, respectively. We begin by showing that for most bases, stratified spaces are totally isolated from the representation theories constructed in the previous section: this philosophically justifies the existence and study of non-stratified spaces as well as the search for a representation theory compatible with stratification. Full details of this section are given in [Meßner-Rodabaugh ****].

6.1 LEMMA. $\forall A \in |\mathbb{L}|$, $\mathbf{L}_\phi\mathbf{PT}(A)$ has a constant open set $\underline{\alpha} \Leftrightarrow \exists\, a \in A$, $\forall p \in Lpt(a)$, $p(a) = \alpha$.

PROOF. $\underline{\alpha} \in \Phi_L^{\rightarrow}(A) \Leftrightarrow \exists\, a \in A,\ \Phi_L(a) = \underline{\alpha} \Leftrightarrow \exists\, a \in A,\ \forall p \in Lpt(A),\ \Phi_L(a)(p) = \alpha \Leftrightarrow \exists\, a \in A,\ \forall p \in Lpt(A),\ p(a) = \alpha$. □

6.2 LEMMA. Let $L \in |\mathbf{CSLF}|$, $A \in |\mathbb{L}|$ such that $Lpt(A) \neq \emptyset$, and $\phi \in \mathbb{L}(L, L)$.

(1) If $|\mathbf{CSLF}(L, L)| > 1$, then $\mathbf{L}_\phi\mathbf{PT}(A)$ is not stratified.

(2) If $[\forall\, \alpha \in L,\ \exists\, v \in \mathbf{CSLF}(L, L),\ v^{op}(\alpha) \neq \alpha]$ holds, then $\mathbf{L}_\phi\mathbf{PT}(\mathrm{A})$ is **completely non-stratified**, i.e. contains **no** constant open sets other than $\underline{0}$ and $\underline{1}$.

(3) If $|L| > 2$ and $A - \{0, 1\}$ consists solely of prime elements, then $\mathbf{L}_\phi\mathbf{PT}(A)$ is completely non-stratified.

PROOF. *Ad(1).* By assumption, $\exists\, v \in \mathbf{CSLF}(L, L) - \{id_L\}$, and so $\exists\, \alpha \in L - \{0, 1\}$, $v^{op}(\alpha) \neq \alpha$. Our goal is to show that the topology of $\mathbf{L}_\phi\mathbf{PT}(A)$ is not stratified: it suffices to show that the non-trivial constant open set $\underline{\alpha}$ is not in the topology, and by 6.1 it therefore suffices to show that $\forall\, a \in A$, $\exists\, p \in Lpt(A)$, $p(a) \neq \underline{\alpha}$. Let a $\in A$. Since $Lpt(A) \neq \emptyset$, $\exists\, p \in Lpt(A)$. If $p(a) \neq \alpha$, we are done. If $p(\underline{\alpha}) = \alpha$, put $q : A \longrightarrow L$ by $q = v^{op} \circ p$; this is a composition in $\mathbf{CSLF}$, so $q \in \mathbf{CSLF}(A, L)$, i.e. $q \in Lpt(A)$. The proof of (1) is completed by observing that $q(a) = v^{op}(p(a)) = v^{op}(\alpha) \neq \alpha$.

Ad(2). Similar to the proof of (1).

Ad(3). Let $\alpha \in L - \{0, 1\}$. Our goal is to show via 6.1 that $\underline{\alpha}$ is not open in $\mathbf{L}_\phi\mathbf{PT}(A)$.

First, let $a \in A - \{0, 1\}$, and put $p_a : A \longrightarrow L$ by $p_a \equiv 0$ on $\downarrow(a)$ and $p_a \equiv 1$ on $A - \downarrow(a)$. Since $p_a(a) = 0 \neq \alpha$, this case is completed by verifying that $p_a \in \mathbf{CSLF}(A, L) \equiv Lpt(A)$. The preservation of meets and joins over the empty set is assured since $p_a(1) = 1$ and $p_a(0) = 0$. If $p_a(b \wedge c) = 1$, then $b \wedge c \notin \downarrow(a)$, so $b \notin \downarrow(a)$ and $c \notin \downarrow(a)$; hence $p_a(b) = 1, p_a(c) = 1$, and so $p_a(b) \wedge p_a(c) = 1$. And if $p_a(b \wedge c) = 0$, then $b \wedge c \in \downarrow(a)$; hence W.L.O.G. $b \in \downarrow(a)$ (since a is prime), so $p_a(b) = 0$ and $p_a(b) \wedge p_a(c) = 0$. For joins, if $p_a(\bigvee_\gamma b_\gamma) = 1$, then some $b_{\gamma_0} \notin \downarrow(a)$, so $\bigvee_\gamma p_a(b_\gamma) \geq p_a(b_{\gamma_0}) = 1$; and if $p_a(\bigvee_\gamma b_\gamma) = 0$, then each $b_\gamma \in \downarrow(a)$, and so each $p_a(b_\gamma) = 0$ and $\bigvee_\gamma p_a(b_\gamma) = 0$. Thus p_a preserves arbitrary $\bigvee$ and finite $\wedge$ and is in $Lpt(A)$.

Second, if $a \in \{0, 1\}$, let p_a be any member of $Lpt(A)$ (which is nonempty); then $p_a(0) = 0 \neq \alpha \neq 1 = p_a(1)$. The requirement of 6.1 is now met. □

Before proceeding further, we give a result showing that the various conditions or antecendents of 6.2 are not very restrictive at all.

6.3 LEMMA.

(1) $\forall\, (X, L, \tau) \in |\mathbb{T}|$ with $X \neq \emptyset$, $Lpt(\tau) \neq \emptyset$.

(2) $\forall$ non-empty $A \in |\mathbf{LSPAT}\,\mathbb{L}|$, $Lpt(A) \neq \emptyset$.

(3) The antecedent of 6.2(1) holds if the antecedent of 6.2(2) holds.

(4) The antecedent of 6.2(2) holds if any of the following are true of $L \in |\mathbf{CSLF}|$:

  (a) $0 \in L^b$ (see [Rodabaugh 1982]: $L^b = \{\alpha \in L : \alpha < \beta,\ \alpha < \gamma \Rightarrow \alpha < \beta \wedge \gamma\}$).

  (b) L is a chain (e.g. $L = [0, 1] \subset \mathbb{R}$).

  (c) L is a diamond with lower handle (e.g. the topological square of the Sierpinski topology) or string of diamonds with lower handle.

(d) $L \in |\mathbf{COHLOC}|$.
(e) L is an atomic Boolean algebra.
(f) L is a reflective diamond (see [Rodabaugh 1988b]).

(5) The antecedent of 6.2(3) is satisfied if A is the four-point diamond.

OUTLINE OF PROOF. *Ad(1).* If $x \in X$, then $\Psi_L(x) \in Lpt(\tau)$.

Ad(2). We use several results of §5. Because of L-spatiality, $A \approx \Phi_L^{\rightarrow}(A)$, so $A \neq \emptyset \Rightarrow \Phi_L^{\rightarrow}(A) \neq \emptyset$. Now $\mathbf{LPT}(A)$ is categorically isomorphic in $\mathbb{T}(L, id)$ to $\mathbf{LPT}(\Phi_L^{\rightarrow}(A))$ (because of the universality of the unit (Ψ_L, id)), so $Lpt(A)$ is bijective with $Lpt(\Phi_L^{\rightarrow}(A))$. Using (1), $\Phi_L^{\rightarrow}(A) \neq \emptyset \Rightarrow Lpt(\Phi_L^{\rightarrow}(A)) \neq \emptyset \Rightarrow Lpt(A) \neq \emptyset$.

Ad(3). Trivial.

Ad(4)(a). Put $v^{op} : L \longrightarrow L$ by $v^{op}(0) = 0$ and $v^{op} \equiv 1$ on $L - \{0\}$. It can be shown that $v^{\{op\}}$ preserves arbitrary $\bigvee$, the $[0 \in L^b]$ condition guarantees v^{op} preserves finite $\wedge$, and v^{op} moves each $\alpha \in L - \{0, 1\}$.

Ad(4)(b–c). Each such lattice satisfies the $[0 \in L^b]$ condition.

Ad(4)(d). If $L \in |\mathbf{COHLOC}|$, then $L \approx Idl(B)$, for some $B \in |\mathbf{DLAT}|$. And in $Idl(B)$, all non-empty ideals dominate $\downarrow(0)$, which in turn dominates the empty ideal $\emptyset$. Hence $\emptyset \in (Idl(B))^b$.

Ad(4)(e). The required v can be constructed by means of the Lindenbaum-Tarski Representation Theorem [Johnstone 1982] for complete atomic Boolean algebras.

Ad(4)(f). The required v can be constructed by gluing together the given **CSLF** morphisms between the facets of the diamond.

Ad(5). This is trivial. □

6.4 OPEN QUESTION. Characterize those $L \in |\mathbf{CSLF}|$ satisfying $|\mathbf{CSLF}|(L, L)| > 1$.

6.5 LEMMA. Stratification is preserved by all categorical isomorphisms in $\mathbb{T}$ and $\mathbb{T}_\phi(L, \phi)$; indeed, each isomorphism in $\mathbb{T}$ and $\mathbb{T}_\phi(L, \phi)$ preserves each constant fuzzy open set.

PROOF. A morphism (f, ϕ) in the ground category $\mathbf{SET} \times \mathbb{L}$ is an isomorphism in $\mathbb{T} \Leftrightarrow f$ is a bijection in $\mathbf{SET}$, ϕ is an isomorphism in $\mathbb{L}$ (hence ϕ^{op} is an isomorphism in $\mathbf{CSLF}$), and each of (f, ϕ) and $(f^{-1}, ((\phi^{op})^{-1})^{op})$ are continuous. The proof that isomorphisms in $\mathbb{T}$ preserve constant open fuzzy sets now follows using the argument given at the end of the proof of Claim I in the proof of 2.6. As for the $\mathbb{T}_\phi(L, \phi)$ case, a morphism (f, ϕ) in the ground category is an isomorphism $\Leftrightarrow f$ is a bijection and each of (f, ϕ) and (f^{-1}, ϕ) are continuous. But given such an isomorphism (f, ϕ) in $\mathbb{T}_\phi(L, \phi)$, its image (f, id) under $\mathbf{F}_\phi$ (2.19 above) is an isomorphism in $\mathbb{T}$ between the same domain and codomain, and so the $\mathbb{T}$ case applies to say that the one space is stratified iff the other is stratified. □

6.6 THEOREM (**Justification of Non-Stratified Spaces**). Under each set of conditions of 6.2 and 6.3(4,5), no object of $\mathbb{T}$ is both sober and stratified. Furthermore, under any of the conditions of 6.2(2,3) or 6.3(4,5), each sober space is completely non-stratified, i.e. has only $\underline{0}$ and $\underline{1}$ as its constant open fuzzy sets.

PROOF. A space (X, L, τ) is sober $\Leftrightarrow$ it is isomorphic in $\mathbb{T}$ via (Ψ_L, id) to $\mathbf{LPT}(\tau) \equiv (Lpt(\tau), L, \Phi_L^{\rightarrow}(\tau))$ (5.3). The latter cannot be stratified (6.2), and (Φ_L, id) preserves this lack (6.5). A similar proof obtains for the second statement. □

6.7 THEOREM (**Non-Existence of Stratified Representations and Compactifications**).

(1) The adjunctions $\mathbf{L}_\phi\Omega \dashv \mathbf{G}_k\mathbf{L}_\phi\mathbf{PT}$ of 5.9 (and 5.10(1,2)) never restrict to an equivalence of (sub)categories under the conditions of 6.2(1).
(2) If $\mathbf{J} : |\mathbb{L}| \longrightarrow |\mathbb{T}_{\phi k}(L, \phi)|$ is any mapping such that $\mathbf{J}$ is a functor and $\mathbf{L}_\phi\Omega \dashv \mathbf{J}$ when $L_\phi\Omega$ and $\mathbf{J}$ are respectively restricted to subcategories $\mathcal{C}_k, \mathcal{D}_k$ satisfying the conditions of 5.9 — then the adjunction $\mathbf{L}_\phi\Omega \dashv \mathbf{J}$ never restricts to an equivalence of respective subcategories of $\mathcal{C}_k$ and $\mathcal{D}_k$ under the conditions of 6.2(1).

PROOF. *Ad(1).* The unit of $\mathbf{L}_\phi\Omega \dashv \mathbf{G}_k\mathbf{L}_\phi\mathbf{PT}$ is formally the same as for $\mathbf{L}_\phi\Omega \dashv \mathbf{L}_\phi\mathbf{PT}$, namely (Ψ_L, ϕ). Assume that $\mathbf{L}_\phi\Omega \dashv \mathbf{G}_k\mathbf{L}_\phi\mathbf{PT}$ can be restricted to an equivalence of respective (non-empty) subcategories. Then given a space (X, L, τ) in the first of these subcategories, $(\Psi_L, \phi) : (X, L, \tau) \longrightarrow \mathbf{G}_k\mathbf{L}_\phi\mathbf{PT}(\tau)$ is an isomorphism in $\mathbb{T}_\phi(L, \phi)$. Now $\mathbf{G}_k\mathbf{L}_\phi\mathbf{PT}(\tau)$ is stratified; and since (Ψ_l, ϕ) is an isomorphism, $\Psi_L : X \longrightarrow Lpt(\tau)$ is bijective, hence (X, L, τ) is sober (5.3). But 6.5, applied to (Φ_L^{-1}, ϕ), says (X, L, τ) is stratified. So (X, L, τ) is both sober and stratified, a contradiction to 6.6.

Ad(2). The assumptions guarantee that $\mathbf{L}_\phi\Omega \dashv \mathbf{G}_k\mathbf{L}_\phi\mathbf{PT}$ and $\mathbf{L}_\phi\Omega \dashv \mathbf{J}$ hold simultaneously on the same pair of categories $\mathcal{C}_k$ and $\mathcal{D}_k$. By various results of [Herrlich-Strecker 1979], especially 26.7, we have: $\forall (X, L, \tau) \in |\mathcal{C}_L|$, $\exists!$ isomorphism $(f, \phi) \in \mathbb{T}_\phi(L, \phi)$ from $\mathbf{J}(\tau)$ to $\mathbf{G}_k\mathbf{L}_\phi\mathbf{PT}(\tau)$ factoring the unit (Ψ_L, ϕ) of $\mathbf{L}_\phi\Omega \dashv \mathbf{G}_k\mathbf{L}_\phi\mathbf{PT}$ through the unit of $\mathbf{L}_\phi\Omega \dashv \mathbf{J}$. Now suppose $\mathbf{L}_\phi\Omega \dashv \mathbf{J}$ restricts to an equivalence on respective subcategories $\mathcal{C}' \hookrightarrow \mathcal{C}_k$, $\mathcal{D}' \hookrightarrow \mathcal{D}_k$. Then for a space $(X, L, \tau) \in |\mathcal{C}'|$, the unit of $\mathbf{L}_\phi\Omega \dashv \mathbf{J}$ from (X, L, τ) to $\mathbf{J}(\tau)$ is an isomorphism in $\mathbb{T}_\phi(L, \phi)$, which means that $(\Psi_L, \phi) : (X, L, \tau) \longrightarrow \mathbf{G}_k\mathbf{L}_\phi\mathbf{PT}(\tau)$ is now a composition of isomorphisms in $\mathbb{T}_\phi(L, \phi)$, and so is an isomorphism in $\mathbb{T}_\phi(L, \phi)$. Using 6.5, a contradiction to 6.6 now obtains just as in the proof of (1). □

6.8 REMARK. To sum up, there can be **no** representation or compactification theory which uses stratified spaces **and** $\mathbf{L}_\phi\Omega$ or $\mathbf{L}\Omega$. Thus a new approach is needed to build a representation theory using stratified spaces, and this brings us to the categories $\mathbb{T}_D$ and $\mathbb{L}_d \equiv \mathbb{L}^{2.5}$. In the development which follows, we also attempt to incorporate change-of-basis beyond the fixed-change-of-basis seen in the previous section.

6.9 QUESTION. In view of the foregoing, what sort of lattice-theoretic objects, if any, do stratified spaces represent? The key is recognizing that any constant maps originally present in a space must be explicitly recognized by the category of spaces, and this is the motivation for $\mathbb{T}_D$. It will appear from the remainder of the section that the "sober" objects of $\mathbb{T}_D$ represent the "spatial" objects of the functor category $\mathbb{L}^{2.5}$. This secures then a place for stratified spaces in our next representation theory: the (completely) stratifed spaces in $\mathbb{T}_D$ have the form (X, L, τ, L), there are sober objects of this shape in $\mathbb{T}_D$, and they represent those objects of $\mathbb{L}^{2.5}$ for which $d = l$.

6.10 DISCUSSION. **(Set-up of the General Theory)**. To emphasize the connection with $\mathbb{T}_D$, $\mathbb{L}^{2.5}$ is now denoted $\mathbb{L}_d$. Let $\mathbf{A} \in |\mathbb{L}_d|$. The carrier set, denoted $Pt(\mathbf{A})$, is defined as follows:

$$Pt(\mathbf{A}) = \{p \in \mathbf{CSLF}(\mathbf{A}(t), \mathbf{A}(1)) : \mathbf{A}(d \longrightarrow l) = p \circ \mathbf{A}(d \longrightarrow t)\}$$

It is the set of all **CSLF** morphisms making the following triangle **commute**, where we abbreviate $\mathbf{A}(d \longrightarrow t)$ $[\mathbf{A}(d \longrightarrow l)]$ by $\mathbf{A}dt[\mathbf{A}dl]$:

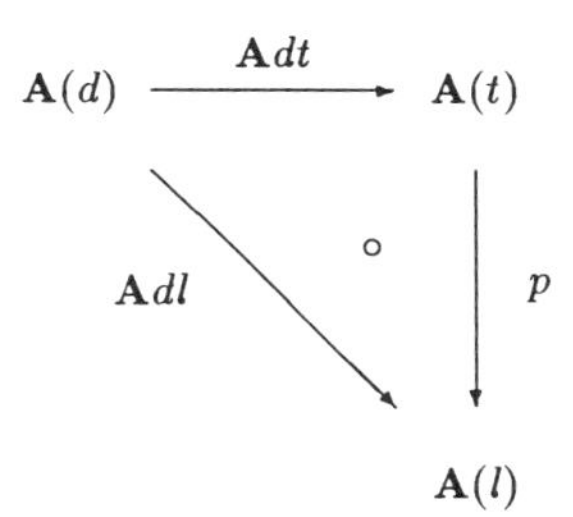

This commutivity is precisely what preserves degree of stratification in the following development. Put $\Phi : \mathbf{A}(t) \longrightarrow \mathbf{A}(l)^{Pt(\mathbf{A})}$ by

$$\Phi(a)(p) = p(a)$$

Then $\Phi \in \mathbf{CSLF}$ and we have a space

$$(Pt(\mathbf{A}), \mathbf{A}(l), \Phi^{\rightarrow}(\mathbf{A}), \mathbf{A}(d \longrightarrow l)^{\rightarrow}(\mathbf{A}(d)))$$

in $\mathbb{T}_D$. Continuing to use $\mathbf{A}(d \longrightarrow l)$ and Φ to denote the above maps restricted to their range, a functor $\mathbf{A}$ is **spatial** iff $\mathbf{A}(d \longrightarrow l)$ and Φ are injective. Now let (X, L, τ, D) in $\mathbb{T}_D$ be given, and construct a corresponding $\mathbf{A}$ as follows:

$$\mathbf{A}(l) = L, \ \mathbf{A}(t) = \tau, \ \mathbf{A}(d) = D, \ \mathbf{A}(d \longrightarrow l) = \hookrightarrow, \ \mathbf{A}(d \longrightarrow t) = \delta$$

Then $\mathbf{A}$ is spatial (proof of 6.12 below) **and** $D = \mathbf{A}(d \longrightarrow l)^{\rightarrow}(\mathbf{A}(d))$, i.e. (X, L, τ, D) has the **same degree of stratification** as $(Pt(\mathbf{A}), \mathbf{A}(l), \Phi^{\rightarrow}(\mathbf{A}), \mathbf{A}(d \rightarrow 1)^{\rightarrow}(\mathbf{A}(d)))$. Furthermore we can define a map $\Psi : X \longrightarrow Pt(\mathbf{A})$ by

$$\Psi(x) : \tau \longrightarrow L,\ \Psi(x)(u) = u(x)$$

and say that a space is **sober** iff Ψ is bijective.

We construct a functor $\Lambda : \mathbb{T}_D \longrightarrow \mathbb{L}_D$ by

$$\Lambda(X, L, \tau, D) = \mathbf{A} \text{ defined in the second preceding display,}$$

and

$$\Lambda(f, \phi) = \zeta \text{ with } \zeta_l = \phi^{op},\ \zeta_d = \phi^{op}|_{D_2}, \text{ and } \zeta_t = (f, \phi)^{\leftarrow}$$

And an action $\Sigma : |\mathbb{L}_d| \longrightarrow |\mathbb{T}_D|$ is given by

$$\Sigma(\mathbf{A}) = (Pt(\mathbf{A}), \mathbf{A}(l), \Phi^{\rightarrow}(\mathbf{A}), \mathbf{A}(d \longrightarrow l)^{\rightarrow}(\mathbf{A}(d)))$$

The symbol "Λ" ["Σ"] means a "lattice-like-object producing functor" ["space-object producing functor"].

To construct an action on the morphisms of $\mathbb{L}_d$ which is functorial and leads to adjunction and representation, it is both **necessary and sufficient** to restrict $\mathbb{T}_D$ and $\mathbb{L}_d$ to the subcategories $\mathbb{IT}_D$ and $\mathbb{IL}_d$: $|\mathbb{IT}_D| = |\mathbb{T}_D|$ and for each morphism (f, ϕ) in $\mathbb{IT}_D$, ϕ^{op} is an isomorphism in **CSLF**, i.e. ϕ may be taken as a concrete arrow in **CSLF**, namely the functional inverse of ϕ^{op}; and $|\mathbb{IL}_d| = |\mathbb{L}_d|$ and for each morphism ζ in $\mathbb{IL}_d$, ζ_l is an isomorphism in **CSLF**, so that ζ_l^{op} is a concrete arrow in **CSLF**, namely the functional inverse of ζ_l. Now let $\zeta \in \mathbb{IL}_d(\mathbf{A}, \mathbf{B})$ be given; it behooves us to construct an action $\Sigma^*(\zeta) : Pt(\mathbf{A}) \longrightarrow Pt(\mathbf{B})$. For $p \in Pt(\mathbf{A})$,

$$\Sigma^*(\zeta)(p) = \zeta_l^{op} \circ p \circ \zeta_t$$

Then $\Sigma(\zeta) : \Sigma(\mathbf{A}) \longrightarrow \Sigma(\mathbf{B})$ by

$$\Sigma(\zeta) = (\Sigma^*(\zeta), \zeta_l^{op})$$

Therefore the action of Σ on morphisms lives on the commutative completions of the diagram:

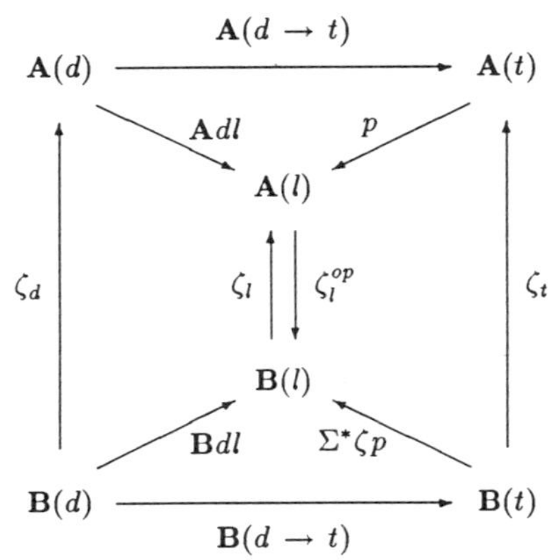

6.11 THEOREM (**Fundamental Adjunction**). Let Λ, Σ denote the respective restrictions to the subcategories $\mathbb{T}_D$ and $\mathbb{L}_d$ of the actions of Λ and Σ defined above. Then Σ is a functor, and $\Lambda \dashv \Sigma$ with unit (Ψ, id) and counit ε, the latter given by:

$$\varepsilon \equiv \langle \varepsilon_d = \mathbf{A}(d \longrightarrow l),\ \varepsilon_l = id_{\mathbf{A}(l)},\ \varepsilon_t = \Phi \rangle$$

where the arrows $\mathbf{A}(d \longrightarrow l)$ and Φ are restricted to their set-theoretic ranges.

OUTLINE OF PROOF. We discuss a series of claims, a more complete discussion of which is given in [Meßner-Rodabaugh ****]. to the reader.

CLAIM I. $\Sigma^*[\zeta(p)] \in Pt(\mathbf{B})$. We need only check that

$$\mathbf{B}(d \longrightarrow l) = \Sigma^*[\zeta(p)] \circ \mathbf{B}(d \longrightarrow t)$$

Starting with the right side, we have

$$\begin{aligned} \Sigma^*[\zeta(p)] \circ \mathbf{B}(d \longrightarrow t) &= \zeta_l^{op} \circ p \circ \zeta_t \circ \mathbf{B}(d \longrightarrow t) \\ &= \zeta_l^{op} \circ p \circ \mathbf{A}(d \longrightarrow t) \circ \zeta_d \\ &= \zeta_l^{op} \circ \mathbf{A}(d \longrightarrow l) \circ \zeta_d \\ &= \zeta_l^{op} \circ \zeta_l \circ \mathbf{B}(d \longrightarrow l) \\ &= \mathbf{B}(d \longrightarrow l) \end{aligned}$$

CLAIM II. $\Sigma(\zeta) \in \mathbb{T}_D(\Sigma(\mathbf{A}), \Sigma(\mathbf{B}))$ *and* Σ *is a functor.* Referring to 2.2, the ground axiom is assured by Claim I. As for general continuity, let $\Phi(b) \in \Phi^{\rightarrow}(\mathbf{B}(t))$ and $p \in Pt(\mathbf{A})$. Then

$$\begin{aligned} (\Sigma(\zeta))^{\leftarrow}(\Phi(b))(p) &= (\Sigma^*(\zeta), \zeta_l^{op})^{\leftarrow}(\Phi(b))(p) \\ &= \zeta_l(\Phi(b)(\Sigma^*(\zeta)(p))) \\ &= \zeta_l((\Sigma^*(\zeta)(p))(b)) \\ &= \zeta_l(\zeta_l^{op}(p(\zeta_t(b)))) \\ &= \Phi(\zeta_t(b))(\zeta_l \circ \zeta_l^{op} \circ p) \\ &= \Phi(\zeta_t(b))(p) \end{aligned}$$

i.e., $(\Sigma(\zeta))^{\leftarrow}(\Phi(b)) = \Phi(\zeta_t(b)) \in \Phi^{\rightarrow}(\mathbf{A}(t))$. And for special continuity, it suffices to show that

$$cod[\zeta_l|_{\mathbf{B}(d \longrightarrow l)^{\rightarrow}(\mathbf{B}(d))}] = \mathbf{A}(d \longrightarrow l)^{\rightarrow}(\mathbf{A}(d))$$

and for this, note $\forall\, \beta \in \mathbf{B}(d)$, $\zeta_l(\mathbf{B}(d \longrightarrow 2)(\beta)) = \mathbf{A})d \longrightarrow 1)(\zeta_d(\beta))$ and $\zeta_d(\beta) \in \mathbf{A}(d)$. It can now be checked that Σ preserves composition and identities and is a functor.

CLAIM III. *The Major Diagram* (§0) *for* $\Lambda \dashv \Sigma$ *holds, i.e.* Σ *has a left adjoint whose action on objects coincides with that of* Λ. It suffices to show:

$$\forall (X, L, \tau, D) \in |\mathbb{T}_D|,\ \exists\, \eta : (X, L, \tau, D) \longrightarrow \Sigma\Lambda(X, L, \tau, D),$$

$$\forall (f, \phi) : (X, L, \tau, D) \longrightarrow \Sigma(\mathbf{B}),\ \exists ! \zeta : \Lambda(X, L, \tau, D) \longrightarrow \mathbf{B},$$
$$(f, \phi) = \Sigma(\zeta) \circ \eta$$

Recall $\Lambda(X, L, \tau, D) = \mathbf{A}$, where

$$\mathbf{A}(1) = L,\ \mathbf{A}(t) = \tau,\ \mathbf{A}(d) = D,\ \mathbf{A}(d \longrightarrow t) = \delta, \mathbf{A}(d \longrightarrow l) = \hookrightarrow$$

and then recall

$$\Sigma(\mathbf{A}) = (Pt(\mathbf{A}), \mathbf{A}(l), \Phi^{\rightarrow}(\mathbf{A}(t)), \mathbf{A}(d \longrightarrow l)^{\rightarrow}(\mathbf{A}(d))) = (Pt(\mathbf{A}), L, \Phi^{\rightarrow}(\tau), D)$$

The following diagrams summarize what we must show to justify the claim, in which diagrams the morphism abbreviations are as in the diagrams of 6.10.

Diagram in $\mathbb{T}_D$

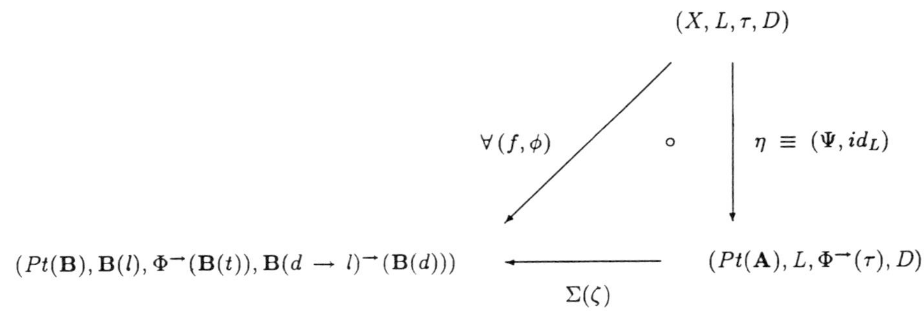

Arrow in $\mathbb{L}_d$

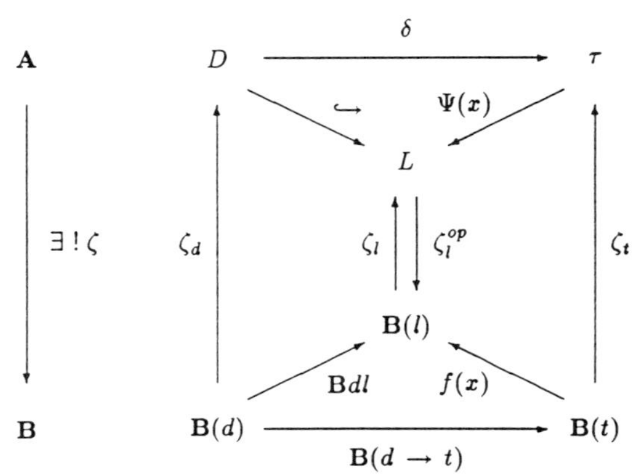

Let $x \in X$ and recall $\Psi(x) : \tau \longrightarrow L$ by $\Psi(x)(u) = u(x)$. As in 5.1, $\Psi(x) \in \mathbf{CSLF}$. We now show $\Psi(x) \in Pt(\mathbf{A})$, i.e. that $\mathbf{A}(d \longrightarrow l) = \Psi(x) \circ \mathbf{A}(d \longrightarrow t)$. Letting $\alpha \in D$, we have

$$\begin{aligned}\Psi(x)(\mathbf{A}(d \longrightarrow t)(\alpha)) &= \Psi(x)(\delta(\alpha)) = \Psi(x)(\underline{\alpha}) \\ &= \underline{\alpha}(x) = \hookrightarrow (\alpha) \\ &= \mathbf{A}(d \longrightarrow 1)(\alpha)\end{aligned}$$

Now we show (Ψ, id_L) is a morphism in $\mathbb{T}_D$. Let $\Phi(v) \in \Phi^{\rightarrow}(\tau)$ and $x \in X$. Then

$$(\Psi, id_L)^{\leftarrow}(\Phi(v))(x) = id_L^{-1}(\Phi(v)(\Psi(x))) = \Psi(x)(v) = v(x)$$

i.e. $(\Psi, id_L)^{\leftarrow}(\Phi(v)) = v \in \tau$. Thus we set

$$\eta \equiv (\Psi, id_L) : (X, L, \tau, D) \longrightarrow (Pt(\mathbf{A}), L, \Phi^{\rightarrow}(\tau), D)$$

and claim η is the desired unit. So let $(f, \phi) : (X, L, \tau, D) \longrightarrow \Sigma(\mathbf{B})$. We must find a unique ζ such that:

$$(f, \phi) = \Sigma(\zeta) \circ (\Psi, id_L) \tag{*}$$

Now $\Sigma(\zeta)$, if such a ζ exists, is the pair $(\Sigma^*(\zeta) \circ (\Psi, \zeta_l^{op})$, so that $(*)$ becomes

$$(f, \phi) = (\Sigma^*(\zeta) \circ \Psi,\ \zeta_l^{op})$$

So if $(*)$ should be true, then ζ_l is determined uniquely as ϕ^{op}. To determine ζ_d and ζ_t, refer to the diagrams above. Imposing commutivity on the left square of the $\mathbb{L}_d$ diagram implies

$$\hookrightarrow \circ\, \zeta_d = \zeta_l \circ \mathbf{B}(d \longrightarrow l)$$

Let $\beta \in \mathbf{B}(d)$. Then

$$\hookrightarrow (\zeta_d(\beta)) = \zeta_l(\mathbf{B}(d \longrightarrow l)(\beta)),\ \zeta_d(\beta) = \phi^{op}(\mathbf{B}(d \longrightarrow l)(\beta))$$

i.e. we must choose $\zeta_d = \phi^{op} \circ \mathbf{B}(d \longrightarrow l)$. Now requiring $f = \Sigma^*(\zeta) \circ \Psi$ means that if $x \in X,\ b \in \mathbf{B}(t)$, then

$$\begin{aligned}f(x)(b) &= (\Sigma^*(\zeta)(\Psi(x)))(b) \\ &= (\zeta_l^{op} \circ \Psi(x) \circ \zeta_t)(b) \\ &= \phi(\Psi(x)(\zeta_t(b))) \\ &= \phi(\zeta_t(b)(x)\end{aligned}$$

so that

$$\Phi(b)(f(x)) = \phi(\zeta_t(b)(x)),$$

$$\phi^{op}(\Phi(b)(f(x))) = \phi^{op}(\phi(\zeta_t(b)(x))) = \zeta_t(b)(x)$$

forcing $\zeta_t = \phi^{op} \circ \Phi(\ \ ) \circ f$. Now each of ζ_d, ζ_l, ζ_t is in **CSLF** , and the choice

$$\zeta = \langle \zeta_d \equiv \phi^{op} \circ \mathbf{B}(d \longrightarrow l),\ \zeta_l \equiv \phi^{op},\ \zeta_t \equiv \phi^{op} \circ \Phi(\ \ ) \circ f \rangle$$

makes the diagram in $\mathbb{IL}_d$ commute; and since $\zeta_l = \phi^{op}$ is an isomorphism, we now have $\zeta \in \mathbb{IL}_d$. But this choice also satisfies $(*)$. Finally, this discussion establishes the uniqueness of ζ w.r.t. $(*)$.

CLAIM IV. *The Minor Diagram (§0) for $\Lambda \dashv \Sigma$ holds, i.e. the action of Σ's left adjoint on morphisms coincides with that of Λ, making Λ the left adjoint of Σ.* The proof of Claim III shows the action of Σ's left adjoint on morphisms is given by the following diagrams, in which $\mathbf{A} = \Lambda(X_1, L_1, \tau_1, D_1)$ and $\mathbf{B} = \Lambda(X_2, L_2, \tau_2, D_2)$:

Diagram in $\mathbb{IT}_D$ *Arrow in $\mathbb{IL}_d$*

$$\begin{array}{ccc} (X_1, L_1, \tau_1, D_1) & \xrightarrow{(f,\phi)} & (X_2, L_2, \tau_2, D_2) \\ (\Psi, id_{L_1}) \Big\downarrow & \circ & \Big\downarrow (\Psi, id_{L_2}) \\ (Pt(\mathbf{A}), L_1, \Phi^{\rightarrow}(\tau_1), D_1) & \xrightarrow[\Sigma_D[!\,\overline{(\Psi, id_{L_2}) \circ (f,\phi)}]]{} & (Pt(\mathbf{B}), L_2, \Phi^{\rightarrow}(\tau_2), D_2) \end{array}
\qquad
\begin{array}{c} \mathbf{A} \\ \Big\downarrow\ !\,\overline{(\Psi, id_{L_2}) \circ (f,\phi)} \\ \mathbf{B} \end{array}$$

where the proof of Claim III identifies $\overline{(\Psi, id_{L_2}) \circ (f, \phi)}$, the action of Σ's left adjoint on (f, ϕ), as the arrow

$$\langle \zeta_d \equiv \phi^{op} \circ \mathbf{B}(d \longrightarrow l),\ \zeta_l \equiv \phi^{op},\ \zeta_t \equiv \phi^{op} \circ \Phi(\ \ ) \circ \Psi \circ f \rangle$$

But the action of Λ on (f, ϕ) is the arrow

$$\Lambda(f, \phi) = \langle \zeta_d \equiv \phi^{op}|_{D_2},\ \zeta_l \equiv \phi^{op},\ \zeta_t \equiv (f, \phi)^{\leftarrow}|_{\tau_2} \rangle$$

To show these are the same, we need only show that they are the same in the ζ_d and ζ_t components. For the ζ_d question, let $\beta \in \mathbf{B}(d) = D_2$. Then

$$\phi^{op}(\mathbf{B}(d \longrightarrow l)(\beta)) = \phi^{op}(\delta_2(\beta)) = \phi^{op}(\beta) = \phi^{op}|_{D_2}(\beta)$$

so $\phi^{op} \circ \mathbf{B}(d \longrightarrow l) = \phi^{op}|_{D_2}(\beta)$. And for the ζ_t question, let $x \in X_1,\ v \in \tau_2$. Then

$$\phi^{op}(\Phi(v)(\Psi(f(x)))) = \phi^{op}(\Psi(f(x))(v)) = \phi^{op}(v(f(x)))$$

so that $\phi^{op} \circ \Phi(\ \ ) \circ \Psi \circ f = (f, \phi)^{\leftarrow}|_{\tau_2}(\ \ )$, concluding the proof of Claim IV and the adjunction $\Lambda \dashv \Sigma$.

Claim V. *The counit of* $\Lambda \dashv \Sigma$ *has the claimed coordinates*

$$\langle \varepsilon_d \equiv \mathbf{A}(d \longrightarrow l),\ \varepsilon_l \equiv id_{\mathbf{A}(l)},\ \varepsilon_t \equiv \Phi \rangle$$

We determine the counit from the universality of the unit. Note the following diagrams follow from (the proofs of) Claims III and IV.

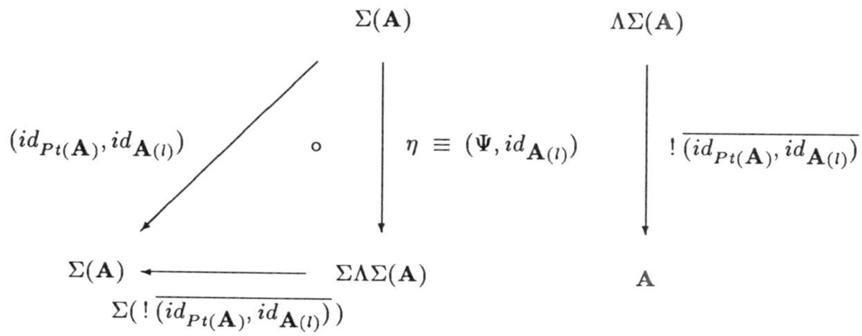

Now we claim the unique arrow $\overline{id_{Pt(\mathbf{A})}, id_{\mathbf{A}(l)})}$ is the counit. Recall an adjunction $\mathbf{F} \dashv \mathbf{G}$ builds a bijection between $Morph(a, \mathbf{G}(b))$ and $Morph(\mathbf{F}(a), b)$. By Theorem IV.1.1 of [Mac Lane 1971, p.80], the bijection is given by $[a \longrightarrow \mathbf{G}(b)] \longleftrightarrow [\varepsilon \circ \mathbf{F}(a \longrightarrow \mathbf{G}(b))]$, so that

$$[\mathbf{G}(b) \ \overrightarrow{{}_{id_{\mathbf{G}(b)}}} \mathbf{G}(b)] \longmapsto [\varepsilon_b \circ \mathbf{F}(id_{\mathbf{G}(b)})] = [\varepsilon_b \circ id_{\mathbf{F}(\mathbf{G}(b))}] = [\varepsilon : \mathbf{F}(\mathbf{G}(b)) \longrightarrow b]$$

Hence ε is the arrow $\overline{(id_{Pt(\mathbf{A})}, id_{\mathbf{A}(l)})}$. But from the proof of Claim III, we have that this arrow must be

$$\langle \zeta_d \equiv id^{op}_{\mathbf{A}(1)} \circ \mathbf{A}(d \longrightarrow l),\ \zeta_l \equiv id^{op}_{\mathbf{A}(l)},\ \zeta_t \equiv id^{op}_{\mathbf{A}(l)} \circ \Phi(\ \ ) \circ id_{Pt(\mathbf{A})} \rangle$$

which is clearly the arrow of Claim V. □

6.12 Theorem (**Sobriety-Spatiality Representation Theorem**). Sobriety is equivalent to the unit (Ψ, id) being an isomorphism in $\mathbb{T}_D$, spatiality is equivalent to the counit ε being an isomorphism in $\mathbb{L}_d$. Accordingly the adjunction $\Lambda \dashv \Sigma$ restricts to the categorical equivalence $\mathbf{SOB}\mathbb{T}_D \approx \mathbf{SPAT}\mathbb{L}_d$, where $\mathbf{SOB}\mathbb{T}_D$ [$\mathbf{SPAT}\mathbb{L}_d$] is the full subcategory of $\mathbb{T}_D$ [$\mathbb{L}_d$] of all sober [spatial] objects.

Outline of Proof. The first statement of the theorem regarding sobriety and spatiality is fairly straightforward from the descriptions of the unit and counit given in the proof of the 6.11 (see [Meßner-Rodabaugh ****]). The rest of the theorem rests on the adjunction

$\Lambda \dashv \Sigma$, two claims which we discuss below, and the fact that these claims show that the adjunciton restricts to the subcategories $\mathbf{SOB}\ \mathbb{T}_D$ and $\mathbf{SPAT}\ \mathbb{L}_d$.

CLAIM I. *If* $\mathbf{A} \in |\mathbb{L}_d|$, *then* $\Sigma(\mathbf{A})$ *is sober; hence* Σ *is a functor from* $\mathbb{L}_d$ *to* $\mathbf{SOB}\mathbb{T}_D$. We have to show $\Psi : \Sigma(\mathbf{A}) \longrightarrow \Sigma\Lambda\Sigma(\mathbf{A})$ is bijective. For injectivity, let $p_1, p_2 \in Pt(\mathbf{A})$, $p_1 \neq p_2$. Then $\exists\, a \in \mathbf{A}(t)$, $p_1(a) \neq p_2(a)$. Since $\Phi(a) \in \Phi(\mathbf{A}(t))$,

$$\Psi(p_1)(\Phi(a)) = p_1(a) \neq p_2(a) = \Phi(p_2)(\Phi(a))$$

And for surjectivity, consider the following diagram in $\mathbb{L}_d$, labeled with the usual morphism abbreviations,

Arrow in $\mathbb{L}_d$

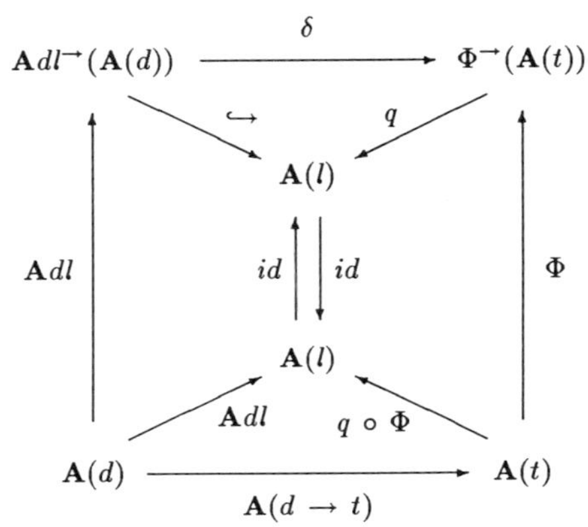

and let $q \in Pt(\Lambda\Sigma(\mathbf{A}))$. We have to find a point $p \in Pt(\mathbf{A})$, $\Psi(p) = q$. Choose $p \equiv q \circ \Phi$. Clearly $p \in \mathbf{CSLF}(\mathbf{A}(t), \mathbf{A}(l))$ and we note

$$\begin{aligned} p \circ \mathbf{A}(d \longrightarrow t) &= q \circ \Phi \circ \mathbf{A}(d \longrightarrow t) \\ &= q \circ \delta \circ \mathbf{A}(d \longrightarrow l) \\ &= \hookrightarrow \circ\, \mathbf{A}(d \longrightarrow l) \\ &= \mathbf{A}(d \longrightarrow l) \end{aligned}$$

i.e. $p \in Pt(\mathbf{A})$. Now let $a \in \mathbf{A}(t)$. Then

$$\Psi(p)(\Phi(a)) = p(a) = (q \circ \Phi)(a) = q(\Phi(a))$$

i.e. $\Psi(p) = q$.

CLAIM II. For (X, L, τ, D) in $\mathbb{T}_D$, $\Lambda(X, L, \tau, D)$ is spatial; hence Λ is a functor from $\mathbb{T}_D$ to $\mathbf{SPAT}\mathbb{L}_d$. Let $(X, L, \tau, D) \in |\mathbb{T}_D|$, and recall from 6.10 that

$$\Lambda(X, L, \tau, D) = \mathbf{A} = \left\{ \begin{array}{l} \mathbf{A}(1) = L,\ \mathbf{A}(t) = \tau,\ \mathbf{A}(d) = D \\ \mathbf{A}(d \longrightarrow l) = \hookrightarrow,\ \mathbf{A}(d \longrightarrow t) = \delta \end{array} \right\}$$

Note $\mathbf{A}(d \longrightarrow l)$ is an injection, so we only have to check that Φ is injective from $\mathbf{A}(t)$ to $\Phi(\mathbf{A}(t))$. Let $u, v \in \mathbf{A}(t)$, $u \neq v$. Then $\exists x \in X$, $u(x) \neq v(x)$, i.e. $\Phi(u)(\Psi(x)) \neq \Phi(v)(\Psi(x))$. Since $\Psi(x) \in Pt(\mathbf{A})$, we conclude $\Phi(u) \neq \Phi(v)$. □

6.13 EXAMPLES.

(1) (**Application of Degree of Stratification**). Theorem 6.12 implies that each meet-semilattice diagram with nodes in **CSLF** (as in 3.1(2.5)) which is spatial, has a sober fuzzy space representation with a degree of stratification. In particular, such diagrams with $d = l$ have sober fuzzy space representations which are fully stratified (cf. 6.9).

(2) (**Application of Degree of Stratification**). This section succeeds in providing a categorical setting for sobriety which is compatible with stratification. The fuzzy real line $\mathbb{R}(I)$ of [Gantner-Steinlage-Warren 1978], with base $L = I = [0,1] \subset \mathbb{R}$, is one of the most important spaces in fuzzy topology. It is completely non-stratified, so its stratification $\mathbf{G}_k(\mathbb{R}(I))$ is important, and it is non-sober in the sense of §5 [Meßner ****], so its "soberification" $\mathbf{LPT}(\mathbf{L}\Omega(\mathbb{R}(I)))$ is important. The latter is also completely non-stratified according to 6.2 and 6.3, and the situation is unchanged if we take the soberification of the stratification $\mathbf{LPT}(\mathbf{L}\Omega(\mathbf{G}_k(\mathbb{R}(I))))$, such a space also being completely non-stratified. But if we apply the soberification of this section to the stratification, then the resulting space $\Sigma(\Lambda(\mathbf{G}_k(\mathbb{R}(I))))$ is both sober in the sense of this section and fully stratified.

(3) (**Existence of Non-Represented Objects**). To see that the equivalence of 6.12 does not extend to all of $\mathbb{T}_D$ and $\mathbb{L}_d$, it suffices to exhibit examples of nonsober spaces in $\mathbb{T}_D$ and examples of nonspatial diagrams in $\mathbb{L}_d$. As pointed out in [Johnstone 1982], any infinite set X with the cofinite topology $\mathcal{T}_{co}$ is a classically nonsober object of **TOP**. It follows $(X, \{0,1\}, \{\chi_U : U \in \mathcal{T}_{co}\}, \{0,1\})$ is a nonsober object of $\mathbb{T}_D$. Also, the localic product $\mathcal{T}(\mathbf{Q}) \times_\ell \mathcal{T}(\mathbf{Q})$ referred to in 5.16 above is classically nonspatial. It follows that the functor $\mathbf{A}$ defined by

$$\mathbf{A}(l) = \{0,1\} = \mathbf{A}(d),\ \mathbf{A}(t) = \mathcal{T}(\mathbf{Q}) \times_\ell \mathcal{T}(\mathbf{Q}),\ \mathbf{A}(d \longrightarrow l) = \hookrightarrow = \mathbf{A}(d \longrightarrow t)$$

is a nonspatial object of $\mathbb{L}_d$. Thus $\mathbf{A}$ is an object of $\mathbb{L}_d$ having no space representation. This brings us to our final discussion.

6.14 DISCUSSION. (**Comparison of Representation Theories of Sections 5 and 6**).

(1) Clearly the classical theory is a subtheory of both theories of §5 and §6. Philosophically speaking, the classical theory and the theory of §6 are somewhat alike: each comprises **one** basic sobriety-spatiality representation based based on **one** fundamental adjunction, and so each is a unified theory; each is compatible with stratification (the classical case vacuously); and each fails to provide space representations for each possible lattice-theoretic object. On the other hand, the theory of §5 is really a schemum of theories, each indexed by a base of membership values, and virtually each theory is incompatible with stratification; but this diversity has the advantage of providing space representations for each possible lattice-theoretic object and capturing topologically all localic

products; and the fixed-change-of-basis allowed in the fundamental adjunctions of §5 cannot be captured by the adjunction of §6.

(2) The issues of change of basis and stratification need not be linked. One can easily fix a basis, require the commutivity condition of 6.10, and have a fixed-basis theory compatible with stratification of spaces over that base. And one can make a unified theory in the manner of this section without degrees of stratification and the commutivity condition of 6.10: throughout this section replace $\mathbb{T}_D$ by $\mathbb{T}$, $\mathbb{L}_d$ by $\mathbb{L}_l \equiv \mathbb{L}^{1.5}$, and define subcategories $\mathbb{TT}$ and $\mathbb{LL}_l$ replacing $\mathbb{TT}_D$ and $\mathbb{LL}_d$, respectively; the proofs are similar and simpler to those of these section.

(3) It is clear that in both §5 and §6 only a limited change of basis is permitted. It would appear that change of basis **within** a given representation theory is a delicate issue on which more work is needed. Profs. Höhle and Stout have suggested to this author the possible connection with the model-theoretic issue of geometric morphisms discussed in [Johnstone 1977].

(4) The theory of §5 possesses generalizations of the classical Stone Representation Theorems for $\mathbf{DLAT}^{op}$ and $\mathbf{BOOL}^{op}$, and the reader may notice these are missing in the theory of §6. In both the classical case and §5, the appropriate restriction of a sobriety-spatiality representation to a coherence-distributivity (or Stone-Boolean) representation hinges on the (classical) fact that (for locales) coherence ⇒ spatiality. This implication in turn rests unavoidably on the Prime Ideal Theorem, a result which is essentially linked to the carrier sets defined in §4 and §5, but unrelated to the carrier sets defined in §6—the difficulty is caused by the very commutivity condition of 6.10 which makes possible the compatibility of sobriety and stratification; and the difficulty obviously remains if we choose a fixed-basis theory compatible with stratification as described in (2) above (since the same commutivity condition is involved). Of course, one can rewrite the Stone Representation Theorems of §5 (5.20) in the language of §6 by specifying $D = \{0,1\}$ and $\mathbf{A}(d) = \{0,1\}$, but this is trivial. We thus pose an **open question**: are there non-trivial analogues of the Stone Representation Theorems (for distributive lattices and Boolean algebras) which result from restrictions of the equivalence $\mathbf{SOB}\mathbb{TT}_D \approx \mathbf{SPAT}\mathbb{LL}_d$.

§7. Final Remarks and Acknowledgements

To this author, space representations of lattice-theoretic objects is a very intriguing and appealing idea, and it has been most rewarding to see the role played by fuzzy sets and fuzzy topologies taken in the poslat sense. However, a chapter of even this restricted scope is impossible without significant mathematical and logistical support, which the author would now like to acknowledge.

I am most grateful to Prof. Höhle for many helpful discussions during recent years of the issues and ideas addressed herein. Deepest appreciation is due Prof. Klement of Linz and the Program Committee of the annual International Seminar on Fuzzy Set Theory for supporting my work and for the role this Seminar has played in shaping my ideas over the years. Special acknowledgement is given Profs. Johnstone and Mac Lane for their insightful, instructive criticisms of the preprint versions subsumed under [Rodabaugh ****]; these criticisms directly benefitted this tutorial. My chairman Prof. Klein is thanked for a supportive teaching schedule during the 1990–1 academic year, as is my secretary Karen DeMatteo for her patience and help. My final thanks go to my wife Cathy and daughters Rachel, Hannah, and Julia for the personal sacrifices they made in order for this chapter to be written.

Part III:

Applications and Related Topics in Logic and Topology

Chapter 8:

Pointless Metric Spaces and Fuzzy Spaces

S. E. Rodabaugh et al. (eds.), Applications of Category Theory to Fuzzy Subsets, 235–244.

POINTLESS METRIC SPACES AND FUZZY SPACES

G. Gerla

§0. Introduction

Recall some definitions and results given in [Gerla 1989]. A **pointless pseudometric space**, briefly **p-p-m-space**, is any structure $\mathcal{R} = (R, \leq, \delta, |\,|)$ where $(R, \leq)$ is a partial order and $|\,| : R \longrightarrow [0, \infty]$, $\delta : R \times R \longrightarrow [0, \infty)$ are functions such that for every $x, y, z \in R$

$$x \geq y \Rightarrow |x| \geq |y|; \tag{A1}$$

$$x \geq y \Rightarrow \delta(y, z) \geq \delta(z, x); \tag{A2}$$

$$\delta(x, x) = 0; \tag{A3}$$

$$\delta(x, y) \leq \delta(x, z) + \delta(z, y) + |z|. \tag{A4}$$

We call **regions** the elements of R, **inclusion** the relation $\leq$, **distance** between two regions x and y the number $\delta(x, y)$ and **diameter** of x the number $|x|$. Notice that A2 is equivalent to saying that δ is symmetric and decreasing i.e.

$$\delta(x, y) = \delta(y, x) \text{ and } x \leq x', y \leq y' \Rightarrow \delta(x, y) \geq \delta(x', y')$$

where x, y, x', y' are in R.

From A1–A4 it follows that if a region z is included both in x and y then $\delta(x, y) = 0$, in particular, if $x \leq y$, then $\delta(x, y) = 0$. Thus, the existence of a minimum in $(R, \leq)$ would entail that δ is constantly equal to zero. Likewise the pseudometric spaces, every (nonempty) subset of a p-p-m-space defines a p-p-m-space.

If x and y are bounded regions, i.e. $|x| \neq \infty$ and $|y| \neq \infty$, we can consider the value $e(x, y) = \delta(x, y) + |x|/2 + |y|/2$. It is immediate that

$$e(x, y) = e(y, x); \; e(x, x) = |x|; \; e(x, y) \leq e(x, z) + e(z, y)$$

We call a **pointless metric space**, briefly **p-m-space**, a p-p-m-space $\mathcal{R} = (R, \leq, \delta, |\,|)$ such that

$$(|x| \geq |y| \text{ and } \delta(z,x) \leq \delta(y,z) \text{ for every } z \in R) \Rightarrow x \geq y \tag{A5}$$

In particular, in a p-m-space two regions x and y with equal diameters and the same distances (i.e. $\delta(x,z) = \delta(y,z)$ for every $z \in R$) are equal. The connection between p-p-m-space and p-m-spaces is similar to the connection between metric spaces and pseudometric spaces. Indeed, it is possible to associate to every p-p-m-space a p-m-space via an obvious notion of **quotient**.

The theory of the pointless pseudometric (metric) spaces generalizes the theory of the pseudometric (metric) spaces; namely, the pseudometric (metric) spaces are the p-p-m-spaces (p-m-spaces) for which every region is an atom with diameter equal to zero (in this case $\leq$ becomes the identity relation).

If R is equal to a class of nonempty subsets of a pseudometric space (M, d), $\leq$ is the inclusion relation, and δ and $|\ |$ are the usual distance and diameter functions defined by

$$\delta(X,Y) = \inf\{d(x,y) \mid x \in X, y \in Y\},\ |X| = \sup\{d(x,y) \mid x, y \in X\} \tag{1}$$

then we obtain a p-p-m-space. We call **canonical** any space of this type. A canonical p-p-m-space satisfying A5 arises in interval analysis by setting R equal to the set of closed intervals of the real line. The class $\mathcal{P}_0(E)$ of the nonempty subsets of a Euclidean space E is a p-p-m-space in which A5 is not satisfied. Indeed, an open ball x and its closure $\bar{x}$ are examples of two different regions such that $|x| = |\bar{x}|$ and $\delta(z,x) = \delta(z,\bar{x})$ for every nonempty set z.

The following is an example of non canonical p-p-m-space. Let X and Y be two nonempty sets and denote by $F(X,Y)$ the class of partial functions from X to Y. If $f \in F(X,Y)$ we denote by D_f the domain of f. Given a finitely additive measure $\mu : \mathcal{P}(X) \to [0,1]$, we set, for every $f, g \in F(X,Y)$.

$$\delta(f,g) = \mu(\{x \in X \mid x \in D_f \cap D_g \text{ and } f(x) \neq g(x)\});\ |f| = \mu(\{x \in X \mid x \notin D_f\}) \tag{2}$$

Moreover, we define an order relation $\leq$ by

$$f \leq g \iff D_g \subseteq D_f \text{ and } g(x) = f(x),\ \forall x \in D_g. \tag{3}$$

In a sense, $|f|$ a measure of the indeterminateness of f, $f \leq g$ means that f is an extension of g. With respect to this order relation the atoms are the totally defined functions, the empty function is the maximum.

In [Gerla 1989] we prove the following proposition.

PROPOSITION 1. Let X and Y be two nonempty sets, $\mu : \mathcal{P}(X) \longrightarrow [0,1]$ a finitely additive measure and δ, $|\ |$, and $\leq$ defined by (2) and (3). Then for every nonempty class $\mathcal{C} \subseteq F(X,Y)$ of partial functions $(\mathcal{C}, \leq, \delta, |\ |)$ is a p-p-m-space.

In the case $Y = \{0,1\}$ the elements of $F(X,Y)$ may be interpreted as **partially defined** subsets of X. Interesting classes of partially defined subsets related to mathematical logic can be found in [Gerla 1990b].

§1. Fuzzy Sets in a Metric Space

Let X be a set; then a **fuzzy subset** of X is any map $s : X \longrightarrow [0,1]$. A fuzzy subset $s : X \longrightarrow [0,1]$ is **crisp** if $s(x) \in \{0,1\}$ for every $x \in X$; we identify the crisp fuzzy subsets of X with the subsets of X via the characteristic functions. Given two fuzzy subsets s and s', we set $s \leq s'$ provided that $s(x) \leq s'(x)$ for every $x \in X$. With respect to $\leq$, the class $F(X)$ of the fuzzy subsets of X is a lattice extending the lattice of the subsets of X. The following equations define the main operations in $F(X)$:

$$(s \wedge s')(x) = \inf\{s(x), s'(x)\},\ (s \vee s')(x) = \sup\{s(x), s'(x)\},\ ({\sim} s)(x) = 1 - s(x)$$

We say that a fuzzy subset s is **nonempty** if there is $x \in X$ such that $s(x) = 1$. If $\alpha \in [0,1)$, the **open α-cut** $O(s,\alpha)$ of s is the set $\{x \in X \mid S(X) > \alpha\}$, while, if $\alpha \in (0,1]$, the set $C(s,\alpha) = \{x \in X \mid s(x) \geq \alpha\}$ is called the **closed α-cut**.

Let (M,d) be a pseudometric space, denote by $F_0(M)$ the class of the nonempty fuzzy subsets of M, and set, for every $s, s' \in F_0(M)$

$$\delta^*(s,s') = \int_0^1 \delta(O(s,\alpha), O(s',\alpha))\, d\alpha;\ |s|^* = \int_0^1 |O(s,\alpha)|\, d\alpha. \tag{4}$$

where δ and $|\ |$ are defined by (1) and $\int_0^1 |O(s,\alpha)|\, d\alpha = \infty$ whenever there is $\alpha \in (0,1)$ such that $|O(s,\alpha)| = \infty$. Notice that both the integrands in (4) are monotone functions of α so that they are Riemann-integrable.

PROPOSITION 2. The structure $\mathcal{R} = (F_0(M), \leq, \delta^*, |\ |^*)$ is a p-p-m-space extending the canonical space $(\mathcal{P}_0(M), \leq, \delta, |\ |)$.

PROOF. A1, A2 and A3 are immediate. Let s, s', t be nonempty fuzzy subsets; then by integrating both sides of the inequality

$$\delta(O(s,\alpha), O(s',\alpha)), \leq \delta(O(s,\alpha), O(t,\alpha)) + \delta(O(t,\alpha), O(s',\alpha)) + |O(t,\alpha)|$$

we obtain A4. □

Since any subset of a p-p-m-space is a p-p-m-space, any class of nonempty fuzzy subsets of a pseudometric space defines a p-p-m-space. For example, the class of the fuzzy numbers, is a p-p-m-space extending the p-m-space of the interval numbers. Recall that a **fuzzy number** is defined as a nonempty fuzzy subset of the real line whose open cuts are intervals. These numbers generalize the interval numbers and therefore the real numbers.

§2. Entropy, Diameters, Distances

If S is a set and μ a measure on an algebra A of subsets of S, then it is well known that the equality $d(\alpha,\beta) = \mu(X \bigtriangledown Y)$, where $X \bigtriangledown Y = (X - Y) \cup (Y - X) = X \cup Y - X \cap Y$ is the symmetric difference, defines in A a structure of pseudometric space (A, d). In this section we will examine how the pseudodistance d could be extended to a suitable class of fuzzy subsets of S. Now, the simmetric difference of two fuzzy subsets s and s' is defined by $s \bigtriangledown s' = (s \wedge \sim s') \vee (s' \wedge \sim s)$, and a natural extension μ^* of μ to the class $\mathcal{A}$ of the measurable fuzzy subsets is obtained by $\mu^*(s) = \int s\, d\mu$. So, the first way we have to extend the distance d to $\mathcal{A}$ is given by setting $\delta(s,s') = \mu^*(s \bigtriangledown s')$. Unfortunately, in such a way we do not obtain a pseudometric space; indeed $\delta(s,s) = \int (s \wedge \sim s)\, d\mu$ is different from zero, in general. On the other hand, since μ^* is a valuation in the lattice $\mathcal{A}$, the equality $\delta(s,s') = \mu^*(s \vee s') - \mu^*(s \wedge s')$ defines a pseudometric in $\mathcal{A}$ [Birkhoff 1940].

There are more interesting possibilities, related to the concept of measure of entropy, if we admit the extension of (A, d) to be a p-p-m-space rather than a pseudometric space. To this purpose, recall that the **sharpness relation** $\leq_s$ is defined in [0,1] by setting $\alpha \leq_s \beta$ if and only if

$$\beta < 1/2 \Rightarrow \alpha \leq \beta \text{ and } \beta > 1/2 \Rightarrow \alpha \geq \beta.$$

This relation is extended to the class $F(S)$ of the fuzzy subsets of S by setting $f \leq_s g$ provided that $f = g$ or $f(x) \leq_s g(x)$ for every $x \in S$. The maximum in $F(S)$ with respect to $\leq_s$ is the completely undefined fuzzy subset $i : S \longrightarrow [0,1]$, i.e. the fuzzy set such that $i(x) = 1/2$ for every $x \in S$. The atoms coincide with the classical subsets of S. An **entropy** is a map $h : \mathcal{A} \longrightarrow [0,1]$ increasing with respect to $\leq_s$ such that $e(i) = 1$ and $e(s) = 0$ for every classical subset s. The entropies are measures of the "degree of vagueness" of the fuzzy subsets [DeLuca and Termini 1972].

Now, a way of extending (A, d) in a p-p-m-space in which the order relation is the sharpness relation is furnished by the class $F(S, \{0,1\})$ of the partial functions from S into $\{0,1\}$ equipped with the structure of p-p-m-space as seen in Section 1. Indeed, the partial functions from S into $\{0,1\}$ may be identified with the fuzzy subsets of S with values in $\{0, 1/2, 1\}$. Namely, we associate to every partial function $f : S \longrightarrow \{0,1\}$ its extension $\bar{f}$ obtained by setting $\bar{f}(x) = 1/2$ if f is not defined in x. Thus, Proposition 1 says that the class of the measurable fuzzy subsets with values in $\{0, 1/2, 1\}$ defines a p-p-m-space extending (A, d). It is also immediate that the "inclusion" relation defined by (3) is the sharpness relation.

In order to consider the whole class $\mathcal{A}$ of measureable fuzzy subsets, let α be an element of $[1/2, 1)$. We say that the fuzzy subsets s **and** t **α-fight over** x provided that either $s(x) > \alpha$ and $t(x) < 1 - \alpha$ or $s(x) < 1 - \alpha$ and $t(x) > \alpha$; moreover we say that s is **α-undetermined** in x provided that $1 - \alpha \leq s(x) \leq \alpha$. The following proposition holds.

PROPOSITION 3. For every $\alpha \in [1/2, 1)$ the structure $(\mathcal{A}, \leq_s, \delta_\alpha, |\ |_\alpha)$ defined by

$$\delta_a(s,t) = \mu(\{x \in S \mid s \text{ and } t\ \alpha\text{-fight over } x\}) \quad s,t \in \mathcal{A}; \tag{5}$$

$$|s|_\alpha = \mu(\{x \in S \mid s\ \alpha\text{–undetermined in } x\}\quad s \in \mathcal{A}. \tag{6}$$

is a p-p-m-space extending the pseudometric space (A, d).

PROOF. A1 and A3 are immediate. To prove A2, that is $s \leq_s s'$ implies $\delta(s,t) \geq \delta(t,s')$, observe that $s \leq_s s'$ entails that if s' and t α-fight over x then also s and t α-fight over x. To prove A4, i.e. $\delta(s,s') \leq \delta(s,t) + \delta(t,s') + |t|$, we have to prove that $\{x \in S \mid s$ and $s'\alpha$-fight over $x\}$ is contained in the union of the sets $\{x \in S \mid s$ and t α-fight over $x\}$, $\{x \in S \mid s'$ and t α-fight over $x\}$ and $\{x \in S/1 - \alpha \leq t(x) \leq \alpha\}\}$.

Now, if s and s' α-fight over x, for example $s(x) > \alpha$ and $s'(x) < 1 - \alpha$, and x is not in $\{x \in S \mid 1 - \alpha \leq t(x) \leq \alpha\}\}$, then either $t(x) > \alpha$ or $t(x) < 1 - \alpha$. In the first case t and s' α-fight over x, in the latter s and t α-fight over x. Likewise we proceed if $s'(x) > \alpha$ and $s(x) < 1 - \alpha$. This concludes the proof. □

An extension of (A, d) with no reference to a prefixed "level" α is given by the following proposition.

PROPOSITION 4. If we set

$$\delta(s,s') = 2\int_{\frac{1}{2}}^{1} \mu(C(s \bigtriangledown s', \alpha))\, d\alpha;\quad |s| = 2\int s \wedge \sim s\ d\mu, \tag{7}$$

then the structure $(\mathcal{A}, \leq_s, \delta, |\ |)$ is a p-p-m-space extending (A, d).

PROOF. Since $(\mathcal{A}, \leq_s, \delta_\alpha, |\ |_\alpha)$ is a p-p-m-space for every $\alpha \in [1/2, 1)$, it is immediate that if we define δ and $|\ |$ by

$$\delta(s,s') = 2\int_{\frac{1}{2}}^{1} \delta_\alpha(s,s')\, d\alpha;\qquad |s| = 2\int_{\frac{1}{2}}^{1} |s|_\alpha\, d\alpha$$

then we obtain a p-p-m-space $(\mathcal{A}, \leq_s, \delta, |\ |)$. On the other hand, since s and s' α-fight over x if and only if either $s(x) \wedge \sim s'(x) > \alpha$ or $s'(x) \wedge \sim s(x) > \alpha$, i.e. if and only if $s \bigtriangledown s'(x) > \alpha$, we have

$$\delta_\alpha(s,t) = \mu(O(s \bigtriangledown t, \alpha)). \tag{8}$$

Likewise, since s is α-undetermined in x if and only if $(s \wedge \sim s)(x) \geq 1 - \alpha$, we have

$$|s|_\alpha = \mu(C(s \wedge \sim s, 1 - \alpha)). \tag{9}$$

and $\int_{\frac{1}{2}}^{1} |s|_\alpha\, d\alpha = \int_{\frac{1}{2}}^{1} \mu(C(s \wedge \sim s, 1 - \alpha))\, d\alpha = \int_{0}^{\frac{1}{2}} \mu(C(s \wedge \sim s, \alpha))\, d\alpha = \int s \wedge \sim s\ d\mu.$ □

Notice that the diameter defined in (7) is a well known example of entropy.

Are there other possible extensions of the pseudometric space (A, d)? Now, a rather general method to build up examples of p-p-m-space in fuzzy set theory is the following. We call a **continuous p-p-m-space** in $[0,1]$ any p-p-m-space $([0,1], \leq', \delta', |\ |')$ such that δ'

and $|\ |'$ are measurable maps. Then it is possible to extend $\leq'$, δ', and $|\ |'$ to $\mathcal{A}$ by putting, for every $s,t \in \mathcal{A}$,

$$s \leq' t \Longleftrightarrow s(x) \leq' t(x),\ \forall x \in S;\ \delta(s,t) = \int \delta'(s,t)\,d\mu;\ |s|(x) = \int |s|'\,d\mu \tag{10}$$

where $\delta'(s,t)$ and $|s|'$ are the fuzzy subsets defined by $(s'(s,t))(x) = \delta'(s(x),t(x))$ and $|s|'(x) = |s(x)|'$.

PROPOSITION 5. Let μ be a probability measure on S, $\mathcal{A}$ the class of measurable fuzzy subsets of S and $([0,1],\leq',\delta',|\ |')$ a continuous p-p-m-space. The structure $(\mathcal{A},\leq',\delta',|\ |')$ defined by (10) is a p-p-m-space.

PROOF. Obvious. □

It is not too difficult to find examples to continous p-p-m-spaces. For example, if we want $\leq'$ to be the sharpness relation, then we have to find a function δ' such that $\delta'(\alpha,\beta) = 0$ whenever $\alpha \leq_s \beta$ or $\beta \leq_s \alpha$. This means that we have to define δ' only in the case $\alpha \wedge \beta < 1/2 < \alpha \vee \beta$. Moreover, δ' must be decreasing with respect to $\leq_s$ and therefore increasing with respect to the distance $|\alpha - \beta|$. This suggests to consider the function

$$\delta'(\alpha,\beta) = \begin{cases} |\alpha-\beta|, & \text{if } \alpha \wedge \beta < 1/2 < \alpha \vee \beta \\ 0, & \text{otherwise} \end{cases}$$

It is immediate that δ' satisfies A2 and A3. So, it remains to find a diameter satisfying A1 and A4 i.e. a function $|\ |'$ increasing in $[0,1/2]$ decreasing in $[1/2,1]$ and satisfying

$$|\alpha|' \geq \bigvee\{\delta'(x,y) - \delta'(x,\alpha) - \delta'(\alpha,y) \mid x,y \in [0,1]\} \tag{11}$$

Since $\delta'(x,1/2) = \delta'(1/2,y) = 0$, then (11) says that $|1/2|' \geq 1$. If $\alpha > 1/2$, since it is not restrictive to assume that $x < 1/2 < y$, then (11) is equivalent to

$$|\alpha|' \geq \bigvee\{y - x - \alpha + x \mid x < 1/2 < y\} = 1 - \alpha$$

If $\alpha < 1/2$ then (11) becomes

$$|\alpha|' \geq \bigvee\{y - x - y + \alpha \mid a < 1/2 < y\} = \alpha$$

In conclusion, $|\ |'$ is required to be increasing in $[0,1/2]$, decreasing in $[1/2,1]$ such that $|1/2|' \geq 1$ and $|\alpha|' \geq \alpha \wedge \sim\alpha$ if $\alpha \neq 1/2$. Examples are obtained putting $|\alpha|' = 2\cdot(\alpha \wedge \sim\alpha)$ or $|\alpha|' = 4\alpha - 4\alpha^2$.

§3. The Points as Derived Concept

In this section we will define the points via suitable sequences of regions. Namely, we call a **Cauchy sequence** of a p-p-m-space $\mathcal{R}$ every sequence $(p_n)_{n\in N}$ of bounded regions such that

$$\forall\,\varepsilon > 0,\ \exists\,\nu \in N,\ \forall\,h \geq \nu,\ \forall\,k \geq \nu,\ e(p_h, p_k) < \varepsilon \tag{12}$$

It is immediate to prove that (12) is equivalent to

$$\forall\,\varepsilon > 0,\ \exists\,\nu \in N,\ \forall\,n \in N,\ e(p_\nu, p_{\nu+n}) < \varepsilon \tag{13}$$

or to

$$\lim |p_n| = 0 \text{ and } \forall\,\varepsilon > 0,\ \exists\,\nu,\ \delta(p_h, p_k) < \varepsilon,\ \forall\,h \geq \nu,\ \forall\,k \geq \nu \tag{14}$$

Decreasing sequences with vanishing diameters are examples of Cauchy sequences. Indeed in this case, $\delta(p_h, p_k) = 0$ for every h and k and (14) is satisfied. Axioms A1–A5 do not imply the existence of Cauchy sequences in a p-p-m-space but we can assure their existence adding some new axiom. For example, we can assume that every region contains arbitrarily small regions.

$$\forall\,\varepsilon > 0,\ \forall\,r \in R,\ \exists\ r' \leq r,\ |r'| < \varepsilon \tag{A6}$$

It follows from A6 that for every region r there is a decreasing sequence of subregions of r with vanishing diameters and therefore a Cauchy sequence contained in r. If $\mathcal{C}$ is the class of the finite partial functions from X into Y, where $|Y| \geq 2$, then the p-p-m-space defined in Proposition 1 satisfies A6. The same holds for the space defined in Proposition 2. In the canonical p-p-m-space of the open balls with diameters greater than 1, A6 is not satisfied, obviously.

Assume A6. Let S be the set of Cauchy sequences and define $d : S \times S \longrightarrow [0, \infty)$ by setting, for every pair $p = (p_n)_{n\in N}$ and $q = (q_n)_{n\in N}$ of elements of S,

$$d(p, q) = \lim \delta(p_n, q_n) = lim\ e(p_n, q_n) \tag{15}$$

Then it is easy to prove that (S, d) is a pseudometric space. We denote by $M(\mathcal{R}) = (\mathbb{P}, d)$ the metric space obtained as a quotient of (S, d) modulo the relation $\equiv$ defined by $p \equiv q$ if and only if $d(p, q) = 0$. Moreover we call a **point** every element of $\mathbb{P}$; as a consequence, a point P is a class $[p] = \{q \in S \mid d(q, p) = 0\}$, and $d : \mathbb{P} \times \mathbb{P} \rightarrow [0, \infty)$ is defined putting $d([p], [q]) = d(p, q)$.

A smaller metric space can be defined by considering only the class S_1 of decreasing sequences of bounded regions with vanishing diameters. We denote by $M_1(\mathcal{R})$ the metric space so obtained. $M_1(\mathcal{R})$ is different from $M(\mathcal{R})$, in general. As an example, if $\mathcal{R}$ is a metric space, then $M(\mathcal{R})$ is its completion, while $M_1(\mathcal{R})$ coincides with R. If $\mathcal{R}$ is the canonical p-p-m-space of the open intervals of the rational number set, then $M_1(\mathcal{R})$ is equal to $M(\mathcal{R})$ and both are equal to the real line.

The following proposition is proven in [Gerla 1989].

PROPOSITION 6. If $\mathcal{R}$ is a p-p-m-space satisfying A6, then $M(\mathcal{R})$ is the completion of $M_1(\mathcal{R})$.

We conclude this section by emphasizing that the construction of the metric space $M(\mathcal{R})$ shows that **it is possible to give a foundation of the geometry without the primitive "point"**. Indeed, it is well known that a large class of geometries may be defined adding suitable axioms to the axioms of metric spaces. So, by imposing suitable axioms to the metric space $M(\mathcal{R})$ we obtain an axiomatization of these geometries. More general information about point-free approaches to the geometry can be found in [Gerla 1990a].

§4. Categorical Features

Let $\mathcal{R}$ and $\mathcal{R}'$ be two p-p-m-spaces whose sets of regions we denote by R and R', respectively. A basic question in the theory of p-p-m-spaces is to give a good notion of "function" (equivalently, of "morphism"). Now, since we consider regions as a tool to give an "approximate representation" of points, it is very natural to require that a morphism is a function $f : R \longrightarrow R'$ such that the equality

$$M(f)([(p_n)_{n\in N}]) = [(f(p_n))_{n\in N}],\ (p_n)_{n\in N} \in S \tag{16}$$

defines a function $M(f)$ from $M(\mathcal{R})$ to $M(\mathcal{R}')$, i.e. a point-function. By using the language of category theory, we have to look for a definition of morphism such that:

- the class of p-p-m-spaces becomes a category;
- the extension of the operator M given by (16) becomes a functor from this category into the category of the metric spaces.

For example, we can consider the following class of morphisms. Given two p-p-m-spaces $\mathcal{R}$ and $\mathcal{R}'$, we say that a function $f : R \longrightarrow R'$ is **uniformly continuous** if

$$\forall \varepsilon > 0,\ \exists \delta > 0,\ [e(x,y) < \delta \Rightarrow e(f(x), f(y)) < \varepsilon] \tag{17}$$

PROPOSITION 7. A map $f : R \rightarrow R'$ is uniformly continuous if and only if

$$\lim e(x_n, y_n) = 0 \Rightarrow \lim e(f(x_n), f(y_n)) = 0. \tag{18}$$

for every pair $(x_n)_{n\in N}$ and $(y_n)_{n\in N}$ of sequences of regions.

PROOF. Assume (17) and $\lim e(x_n, y_n) = 0$. Then, since

$$\forall \varepsilon > 0, \exists \delta > 0, [e(x,y) < \delta \Rightarrow e(f(x), f(y)) < \varepsilon]$$

and

$$\forall\, \delta > 0,\, \exists\, \nu \in N,\, [n \geq \nu \Rightarrow e(x_n, y_n) < \delta]$$

we have that $\forall\, \varepsilon > 0,\ \exists\, \nu \in N,\ (n \geq \nu \Rightarrow e(f(x_n)f(y_n)) < \varepsilon$; i.e. $\lim e(f(x_n), f(y_n)) = 0$.

Assume (18) while (17) is not satisfied, then

$$\exists\, \varepsilon > 0,\ \forall\, k \exists\, x_k y_k e(x_k, y_k) < 1/k \text{ and } e(f(x_k), f(y_k)) \geq \varepsilon.$$

Consequently, $\lim e(x_k, y_k) = 0$ while $\lim e(f(x_k), f(y_k)) = 0$ is false. Thus (18) entails (17) and this completes the proof. □

PROPOSITION 8. The class of p-p-m-spaces equipped with the uniformly continuous maps defines a category **PPMS**. Moreover, the extension of M via (16) is a functor from **PPMS** into the category of metric spaces.

PROOF. The proposition is an immediate consequence of (17) and (18). As an example, we prove that $M(f)$ is uniformly continuous. Now, for every $\varepsilon > 0$, let δ be as in (17) and assume $d(P,Q) < \delta$ where P and Q are elements of $M(\mathcal{R})$. Then, since $d(P,Q) = \lim e(p_n, q_n)$, $\exists\, \nu \in N$ such that $e(p_n, q_n) < \delta$ for every $n \geq \nu$ and therefore $e(f(p_n), f(q_n)) < \varepsilon$ for every $n \geq \nu$. Thus, $d(f(P), f(Q)) = \lim e(f(p_n), f(q_n)) < \varepsilon$.

The just proposed notion of morphism lays itself open to criticism because it is too restrictive. For example, in the case $\mathcal{R}$ is the metric space of the rational numbers, the function $y = x^2$ is not uniformly continuous in spite of the fact that it can be extended to the real line $M(\mathcal{R})$. Moreover, it should be desirable that also the partially defined functions fell within the notion of morphism. Indeed these functions are basic tools for mathematical analysis and computability theory. So, to find morphisms in p-p-m-space theory is an **open question**.

Chapter 9:

Fuzzy Unit Interval and Fuzzy Paths

S. E. Rodabaugh et al. (eds.), Applications of Category Theory to Fuzzy Subsets, 245–256.

FUZZY UNIT INTERVAL AND FUZZY PATHS

F. Klawonn

§0. Introduction

In this chapter we demonstrate that it is impossible to give a reasonable definition of a product of fuzzy paths, if we consider a fuzzy path as a continuous mapping from the L-fuzzy unit interval to an L-fuzzy topological space. To obtain this result we describe the product of paths as the universal arrow of a pushout. On the other hand we show (using a Boolean valued model) how to obtain a reasonable definition of the product of fuzzy paths.

§1. The L-Fuzzy Unit Interval

Let L be a complete lattice with least element $\mathbf{0}$ and greatest element $\mathbf{1} \neq \mathbf{0}$ satisfying the infinite distributive law (i.e. $\forall\, \alpha \in L,\ \forall S \subseteq L : \alpha \wedge \bigvee S = \bigvee \{\alpha \wedge \sigma \mid \sigma \in S\}$). Let $' : L \longrightarrow L$ be an order reversing involution.

The elements of L^X are called L-fuzzy sets [Goguen 1967], where X is an arbitrary set. If $\tau \subseteq L^X$ is closed under arbitrary suprema and finite infima (and therefore the constant mappings $\mathbf{0}_X$ and $\mathbf{1}_X$ belong to τ), τ is an L-fuzzy topology on X [Goguen 1973], and the pair (X, τ) is called L-fuzzy topological space. An L-fuzzy topological space is stratified iff for each $\alpha \in L$ the constant mapping α_X belongs to τ. Let (X, τ) and (Y, σ) be two L-fuzzy topological spaces. A mapping $\varphi : X \longrightarrow Y$ is continuous iff $\forall\, M \in \sigma : \varphi^{-1}(M) \in \tau$, where the L-fuzzy set $\varphi^{-1}(M)$ in X is defined by $M \circ \varphi$. The category of L-fuzzy topological spaces is denoted by **L-FTS**, the category of stratified L-fuzzy topological spaces by **s-L-FTS**. The objects of **L-FTS**, **s-L-FTS** are L-fuzzy topological spaces, stratified L-fuzzy topological spaces, respectively. Morphisms are continuous mappings between (stratified) L-fuzzy topological spaces.

The L-fuzzy unit interval $I(L)$ [Hutton 1975] is the set of all equivalence classes $[\lambda]$, where $\lambda : \mathbb{R} \longrightarrow L$ is monotone decreasing, $\lambda(0) = \mathbf{1}$ and $\lambda(1+) = \mathbf{0}$, and $\mu \in [\lambda]$ iff $\lambda(t-) = \mu(t-)$ for each $t \in \mathbb{R}$. The mapping $\iota_{I,L} : I \longrightarrow I(L)$, $t \mapsto [\lambda_t]$, where

$$\lambda_t : \mathbb{R} \longrightarrow L, \quad s \mapsto \begin{cases} \mathbf{1}, s \leq t \\ \mathbf{0}, s > t \end{cases}$$

is an embedding of the unit interval $I = [0,1]$ into the L-fuzzy unit interval.

$I(L)$ can be endowed with the canonical L-fuzzy topology $\tau_{I(L)}$ induced by the subbasis $\{L_t \mid t \in \mathbb{R}\} \cup \{R_t \mid t \in \mathbb{R}\}$, where $L_t([\lambda]) = (\lambda(t-))', R_t([\lambda]) = \lambda(t+)$. The stratified L-fuzzy topology on $I(L)$, which has $\{L_t \mid t \in \mathbb{R}\} \cup \{R_t \mid t \in \mathbb{R}\} \cup \{\alpha_X \mid \alpha \in L\}$ as a subbasis, is denoted by $\tau^s_{I(L)}$.

An **L-fuzzy path** (**s-L-fuzzy path**) is a morphism γ in **L-FTS** (**s-L-FTS**) with domain $(I(L), \tau_{I(L)})$, $((I(L), \tau^s_{I(L)}))$. $\gamma([\lambda_0])$, $\gamma([\lambda_1])$ is called the **initial point**, **terminal point** of γ, respectively.

§2. The Product of Paths as the Universal Arrow of a Pushout

Let $\gamma_i : I \longrightarrow X$ $(i = 1,2)$ be two paths such that $\gamma_1(1) = \gamma_2(0)$ holds. The mapping

$$\gamma_1 \cdot \gamma_2 : I \longrightarrow X, \; t \mapsto \begin{cases} \gamma_1(2t), & 0 \leq t \leq \frac{1}{2} \\ \gamma_2(2t-1), & \frac{1}{2} \leq t \leq 1 \end{cases}$$

is called the **product of the paths γ_1 and γ_2**.

The product of two paths can be described as the universal arrow of a pushout:

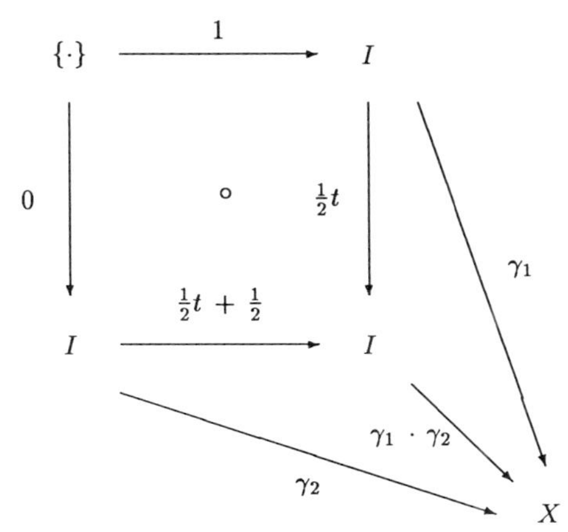

§3. The Product of L-Fuzzy Paths

We consider the pushout for L-fuzzy paths. Let $I(L)+I(L)=(I(L)\times\{1\})\cup(I(L)\times\{2\})$ be the disjoint union of two L-fuzzy unit intervals. We identify $([\lambda_1],1)$ and $([\lambda_0],2)$ in $(I(L)+I(L)$ via the equivalence relation $\approx$ on $I(L)+I(L)$, which is defined by $([\lambda],i)\approx([\mu],j)\Leftrightarrow([\lambda]=[\mu]$ and $i=j)$ or $([\lambda]=[\lambda_0]$ and $i=2$ and $[\mu]=[\lambda_1]$ and $j=1)$ or $([\lambda]=[\lambda_1]$ and $i=1$ and $[\mu]=[\lambda_0]$ and $j=2)$. Let $[[\lambda],i]$ denote the equivalence class of $([\lambda],i)$. The quotient space $I(L)\ \#\ I(L)=I(L)+I(L)/_{\approx}$ is the one-point union of two L-fuzzy unit intervals. We define $\kappa_i : I(L)\to I(L)\ \#\ I(L),[\lambda]\mapsto[[\lambda],i]$ $(i=1,2)$.

PROPOSITION 1.

$$\begin{array}{ccc} \{\cdot\} & \xrightarrow{[\lambda_1]} & I(L) \\ {\scriptstyle[\lambda_0]}\downarrow & \circ & \downarrow{\scriptstyle\kappa_1} \\ I(L) & \xrightarrow[\kappa_2]{} & I(L)\,\sharp\,I(L) \end{array}$$

is a pushout square in **L-FTS (s-L-FTS)**, if $I(L)$ is endowed with the L-fuzzy topology $\tau_{I(L)}$ (stratified L-fuzzy topology and $\tau^s_{I(L)}$) and $I(L)\ \#\ I(L)$ is endowed with the final L-fuzzy topology induced by $\{\kappa_1,\kappa_2\}$.

If we assume that there is a reasonable definition of the product of L-fuzzy paths (s-L-fuzzy paths), then there exists a pushout square in **L-FTS (s-L-FTS)** in the form of

$$\begin{array}{ccc} \{\cdot\} & \xrightarrow{[\lambda_1]} & I(L) \\ {\scriptstyle[\lambda_0]}\downarrow & \circ & \downarrow \\ I(L) & \longrightarrow & I(L) \end{array}$$

There are two reasonable constraints for the product of two $(s-)L$-fuzzy paths:

(i) The initial point of the first path is the initial point of the product path.
(ii) The terminal point of the second path is the terminal point of the product path.

Therefore, Proposition 1 would imply, that there is a canonical homeomorphism φ : $I(L)\longrightarrow I(L)\ \#\ I(L)$ from the L-fuzzy unit interval $(I(L),\tau_{I(L)})$ (stratified L-fuzzy unit interval $(I(L),\tau^s_{I(L)})$) to the one-point union of two L-fuzzy unit intervals (stratified L-fuzzy unit intervals). Conditions (i) and (ii) imply (*) $\varphi([\lambda_0]) = [[\lambda_0], 1]$ and (**) $\varphi([\lambda_1]) = [[\lambda_1], 2]$.

The following theorem shows, that these constraints can be satisfied only in the classical case of ordinary topological spaces, where $L = \{\mathbf{O}, \mathbf{1}\}$ holds.

THEOREM 1. (a) If $L \neq \{\mathbf{O}, \mathbf{1}\}$ holds, then there is no bijective morphism $\varphi : I(L) \longrightarrow I(L) \# I(L)$ in **L-FTS** from the L-fuzzy unit interval $(I(L), \tau_{I(L)})$ to the one-point union of two L-fuzzy unit intervals, such that (*) and (**) are satisfied.

(b) If $L \neq \{\mathbf{O},\mathbf{1}\}$ holds, then there is no bijective morphism $\varphi : I(L) \longrightarrow I(L) \# I(L)$ in **s-L-FTS** from the stratified L-fuzzy unit interval $(I(L), \tau^s_{I(L)})$ to the one-point union of two stratified L-fuzzy unit intervals, such that (*) and (**) are satisfied.

PROOF. We will give the proof for the case of stratified L-fuzzy topological spaces only. By omitting the constants α_i we obtain the proof for L-fuzzy topological spaces. Let $\alpha \in L$, $\mathbf{O} \neq \alpha \neq \mathbf{1}$. Let $\varphi : I(L) \longrightarrow I(L) \# I(L)$ be a bijective mapping. Obviously, $L_s \wedge L_t = L_s$ and $R_s \wedge R_t = R_t$ hold for $0 < s \leq t < 1$. Therefore, an element of $\tau^s_{I(L)}$ can be represented in the following way:

$$\bigvee_{i \in J} (L_{s_i} \wedge R_{t_i} \wedge \alpha_i), \text{ where } \alpha_i \in L$$

We show: if $M \in \tau^s_{I(L)}$ such that $M([\lambda_{\mathbf{O}}]) > \mathbf{O}$ or $M([\lambda_{\mathbf{1}}]) > \mathbf{O}$, then $M([\lambda_\alpha]) > \mathbf{O}$ holds, where

$$\lambda_\beta(t) = \begin{cases} \mathbf{1}, & t < 0 \\ \beta, & 0 \leq t \leq 1 \quad \text{for } \beta \in L. \\ \mathbf{O}, & t > 1 \end{cases}$$

Let $M \in \tau^s_{I(L)}$ such that $M([\lambda_{\mathbf{O}}]) > \mathbf{O}$, $M = \bigvee_{i \in J}(L_{s_j} \wedge R_{t_i} \wedge \alpha_i)$. $M([\lambda_{\mathbf{O}}]) > \mathbf{O}$ implies

$$(\bigvee_{j \in J_0} (L_{s_j} \wedge \alpha_j))([\lambda_{\mathbf{O}}]) = \mathbf{1}, \text{ where } J_0 = \{i \in J \mid t_j < 0 \text{ and } s_j > 0\}$$

We have $\mathbf{1} = (\bigvee_{j \in J_0}(L_{s_j} \wedge \alpha_j))([\lambda_{\mathbf{O}}]) = \bigvee_{j \in J_0}(L_{s_j}([\lambda_{\mathbf{O}}]) \wedge \alpha_j) = \bigvee_{j \in J_0} \alpha_j$. Therefore, $\mathbf{O} < \alpha' = \alpha' \wedge \bigvee_{j \in J_0} \alpha_j = \bigvee_{j \in J_0}(\alpha' \wedge \alpha_j) = \bigvee_{j \in J_0}(L_{s_j}([\lambda_\alpha]) \wedge \alpha_j) \leq M([\lambda_\alpha])$.

Analogously, we can show that $M([\lambda_{\mathbf{1}}]) > \mathbf{O}$ implies $M([\lambda_\alpha]) \geq \alpha > \mathbf{O}$. Let $[[\lambda], i] = \varphi([\lambda_\alpha])$. We define an open L-fuzzy set N of $I(L) \# I(L)$

$$N = \begin{cases} \kappa_2(R_{\frac{1}{2}}), & \text{if } i = 1 \\ \kappa_1(L_{\frac{1}{2}}), & \text{if } i = 2 \end{cases}$$

Obviously, $N([[\lambda], i]) = \mathbf{O}$ and $\big(N([[\lambda_{\mathbf{O}}, 1]]) = \mathbf{1}$ or $N([[\lambda_{\mathbf{1}}, 2]]) = \mathbf{1}\big)$ holds. This implies $(\varphi^{-1}(N))([\lambda_\alpha]) = \mathbf{O}$ and $\big((\varphi^{-1}(N))([\lambda_{\mathbf{O}}]) = \mathbf{1}$ or $(\varphi^{-1}(N))([\lambda_{\mathbf{1}}]) = \mathbf{1}\big)$. So N is an open L-fuzzy set of $I(L) \# I(L)$, but $\varphi^{-1}(N)$ is not open in $I(L)$, and therefore φ is not continuous. □

§4. A Boolean Valued Model of the Unit Interval

Using a Boolean valued model of the unit interval we show in the following how to obtain a reasonable definition of the product of fuzzy paths.

DEFINITION. Let $(\Omega, \mathcal{A}, P)$ be a probability space and let $\mathbb{B} = \mathcal{A}/_{[P=0]}$ be the quotient algebra of the σ-field $\mathcal{A}$ modulo the σ-ideal $[P = 0]$ of events in $\mathcal{A}$ of P-measure zero. $\mathbb{B}$ is a complete Boolean algebra.

DEFINITION (**$\mathbb{B}$-underdeterminate set** [Höhle 1985]). Let X be an arbitrary set. An **equality relation** on X is a mapping $E : X \times X \longrightarrow \mathbb{B}$ equipped with the following properties.

(E1)	$E(x,x) = \mathbf{1}$	(**Total Existence**)
(E2)	$E(x,y) = E(y,x)$	(**Symmetry**)
(E3)	$E(x,y) \wedge E(y,z) \leq E(x,z)$	(**Transitivity**)

If E is an equality relation on X then the order pair (X, E) is said to be a **$\mathbb{B}$-underdeterminate set.**

DEFINITION (**Category $\mathbb{B}$US** [Höhle 1985]). The objects of the category **$\mathbb{B}$US** are $\mathbb{B}$-underdeterminate sets, and morphisms $(X,E) \xrightarrow{\varphi} (Y,F)$ are ordinary mappings $\varphi : X \times Y \longrightarrow \mathbb{B}$ satisfying the following axioms:

(M1)	$\varphi(x,y) \wedge E(x,t) \leq \varphi(t,y)$	(**Extensionality**)
	$\varphi(x,y) \wedge F(y,s) \leq \varphi(x,s)$	
(M2)	$\varphi(x,y) \wedge \varphi(x,s) \leq F(y,s)$	(**Singleton Property**)
(M3)	$\bigvee\{\varphi(x,y) \mid y \in Y\} = \mathbf{1}$	(**Total Definiteness**)

The composition of two morphisms $(X,E) \xrightarrow{\varphi} (Y,F)$ and $(Y,F) \xrightarrow{\psi} (Z,G)$ is given by $(\psi \circ \varphi)(x,z) = \bigvee\{\varphi(x,y) \wedge \psi(y,z) \mid y \in Y\}$.

Note that the category **$\mathbb{B}$US** is a weak topos [Höhle 1985].

DEFINITION. A fuzzy set on a $\mathbb{B}$-underdeterminate set (X, E) is an ordinary mapping $f : X \longrightarrow \mathbb{B}$ such that $f(x) \wedge E(x,y) \leq f(y)$. A **topology** on (X, E) is a set τ of fuzzy sets on (X, E) equipped with the following properties:

$$\text{(i) if } \alpha \in \mathbb{B} \text{ then } \alpha_X \in \tau, \quad \text{(ii) } f, g \in \tau \Rightarrow f \wedge g \in \tau, \quad \text{(iii) } \mathcal{M} \subseteq \tau \Rightarrow \bigvee \mathcal{M} \in \tau.$$

Let τ and τ' be topologies on (X, E) and (Y, F), respectively. Let φ be a morphism from (X, E) to (Y, F). φ is called **continuous** if for all $f' \in \tau'$, $\varphi^{-1}(f') \in \tau$, where $\varphi^{-1}(f')$ is given by $(\varphi^{-1}(f'))(x) = \bigvee\{\varphi(x, y) \wedge f'(y) | y \in Y\}$.

DEFINITION. $\hat{I}$ denotes the set of all random variables $\xi : \Omega \longrightarrow \mathbb{R}$ such that $P(\xi^{-1}(I)) = 1$ holds. Statements like $\xi = \eta$ or $\xi \leq \eta$ are given the obvious truth values as follows: $[\![\xi = \eta]\!] = \{\omega \in \Omega \mid \xi(\omega) = \eta(\omega)\}/_{[P=0]}$ and $[\![\xi \leq \eta]\!] = \{\omega \in \Omega \mid \xi(\omega) \leq \eta(\omega)\}/_{[P=0]}$. $(\hat{I}, \hat{E})$ is a $\mathbb{B}$-underdeterminate set where we define $\hat{E}(\xi, \eta) = [\![\xi = \eta]\!]$. $(\hat{I}, \hat{E})$ is called a **Boolean valued model** of the unit interval. (compare [Scott 1967]). We define a topology $\hat{\tau}$ on $(\hat{I}, \hat{E})$ by taking as a subbasis $\{(\leftarrow, \xi) \mid \xi \in \hat{I}\} \cup \{(\xi, \rightarrow) \mid \xi \in \hat{I}\} \cup \{\alpha_x \mid \alpha \in \mathbb{B}\}$ where $(\leftarrow, \xi)(\zeta) = [\![\zeta < \xi]\!]$ and $(\xi, \rightarrow)(\zeta) = [\![\xi < \zeta]\!]$. The fuzzy set (ζ, η) is defined by $(\zeta, \eta)(\xi) = [\![\zeta < \xi < \eta]\!]$.

THEOREM 2.

$$\begin{array}{ccc} \{\cdot\} & \xrightarrow{\;c_1\;} & \hat{I} \\ {\scriptstyle c_0}\downarrow & \circ & \downarrow{\scriptstyle \iota_1} \\ \hat{I} & \xrightarrow[\;\iota_2\;]{} & \hat{I} \end{array}$$

is a pushout square in the category $\mathbb{B}\mathbf{US}$, where

$$c_i(\cdot, \xi) = [\![\xi = i]\!] \qquad (i = 0, 1)$$
$$\iota_1(\xi, \eta) = [\![\tfrac{1}{2}\xi = \eta]\!]$$
$$\iota_2(\xi, \eta) = [\![\tfrac{1}{2}\xi + \tfrac{1}{2} = \eta]\!]$$

PROOF. Obviously, c_0 and c_1 are morphisms of the category $\mathbb{B}\mathbf{US}$. We have to check the axioms (M1), (M2), and (M3) for ι_1.

(M1).

$$\begin{aligned} \iota_1(\xi, \zeta) \wedge [\![\xi = \zeta]\!] &= \{\omega \in \Omega \mid \frac{1}{2}\xi(\omega) = \eta(\omega)\}/_{[P=0]} \wedge \{\omega \in \Omega \mid \xi(\omega) = \zeta(\omega)\}/_{[P=0]} \\ &\leq \{\omega \in \Omega \mid \frac{1}{2}\zeta(\omega) = \eta(\omega)\}/_{[P=0]} = [\![\frac{1}{2}\zeta = \eta]\!], \\ \iota_1(\xi, \eta) \wedge [\![\eta = \zeta]\!] = [\![\frac{1}{2}\xi = \eta]\!] \wedge [\![\eta = \zeta]\!] &\leq [\![\frac{1}{2}\xi = \zeta]\!] = \iota_1(\xi, \zeta) \end{aligned}$$

(M2). $\iota_1(\xi, \eta) \wedge \iota_1(\xi, \zeta) = [\![\frac{1}{2}\xi = \eta]\!] \wedge [\![\frac{1}{2}\xi = \zeta]\!] \leq [\![\eta = \zeta]\!]$.

(M3). $\bigvee\{\iota_1(\xi, \eta) \mid \eta \in \hat{I}\} = \bigvee\{\, [\![\frac{1}{2}\xi = \eta]\!] \mid \eta \in \hat{I}\} \geq [\![\frac{1}{2}\xi = \frac{1}{2}\xi]\!] = \mathbf{1}$.

Therefore, ι_1 is a morphism of the category $\mathbb{B}\mathbf{US}$. The proof for ι_2 runs analogously.

Let (X, E) be a $\mathbb{B}$-underdeterminate set and let $\hat{I} \xrightarrow{\gamma_i} X$ $(i = 1, 2)$ be morphisms from $\hat{I}$ to X such that $\gamma_1 \circ c_1 = \gamma_2 \circ c_0$ holds. We define a morphism $\hat{I} \xrightarrow{\gamma_1 \cdot \gamma_2} X$ from $\hat{I}$ to the $\mathbb{B}$-underdeterminate set (X, E) by

$$(\gamma_1 \cdot \gamma_2)(\xi, x) := ([\![\xi \leq \tfrac{1}{2}]\!] \wedge \gamma_1(2\underline{\xi}, x)) \vee ([\![\xi \geq \tfrac{1}{2}]\!] \wedge \gamma_2(2\bar{\xi} - 1, x))$$

where

$$\underline{\xi}(\omega) := \min\{\xi(\omega), \tfrac{1}{2}\} \text{ and } \bar{\xi}(\omega) := \max\{\xi(\omega), \tfrac{1}{2}\}$$

We have to show, that $\gamma_1 \cdot \gamma_2$ is a morphism and satisfies the conditions $\gamma_i = (\gamma_1 \cdot \gamma_2) \circ \iota_i (i = 1, 2)$. Furthermore, $\gamma_1 \cdot \gamma_2$ must be uniquely determined by γ_1 and γ_2. Note that $[\![\xi \leq \tfrac{1}{2}]\!] = [\![\xi = \underline{\xi}]\!]$ and $[\![\xi \geq \tfrac{1}{2}]\!] = [\![\xi = \bar{\xi}]\!]$.

(M1).

$$\gamma_1 \cdot \gamma_2(\xi, x) \wedge [\![\xi = \zeta]\!] =$$

$$([\![\xi \leq \frac{1}{2}]\!] \wedge \gamma_1(2\underline{\xi}, x) \wedge [\![\xi = \zeta]\!]) \vee ([\![\xi \geq \frac{1}{2}]\!] \wedge \gamma_2(2\bar{\xi} - 1, x) \wedge [\![\xi = \zeta]\!]) \leq$$

$$([\![\zeta \leq \frac{1}{2}]\!] \wedge \gamma_1(2\underline{\xi}, x) \wedge [\![2\underline{\xi} = 2\underline{\zeta}]\!]) \vee ([\![\zeta \geq \frac{1}{2}]\!] \wedge \gamma_2(2\bar{\xi} - 1, x) \wedge [\![2\bar{\xi} - 1 = 2\bar{\zeta} - 1]\!])$$

$$\leq \gamma_1 \cdot \gamma_2(\zeta, x)$$

so that $\gamma_1 \cdot \gamma_2(\xi, x) \wedge E(x, y) \leq \gamma_1 \cdot \gamma_2(\xi, y)$ obviously holds.

(M2).

$$\gamma_1 \cdot \gamma_2(\xi, x) \wedge \gamma_1 \cdot \gamma_2(\xi, y) =$$

$$([\![\xi = \underline{\xi}]\!] \wedge \gamma_1(2\underline{\xi}, x) \wedge \gamma_1(2\underline{\xi}, y)) \vee ([\![\xi = \bar{\xi}]\!] \wedge \gamma_2(2\bar{\xi} - 1, x) \wedge \gamma_2(2\bar{\xi} - 1, y))$$

$$\vee([\![\xi = \underline{\xi}]\!] \wedge \gamma_1(2\underline{\xi}, x) \wedge [\![\xi = \bar{\xi}]\!] \wedge \gamma_2(2\bar{\xi} - 1, y))$$

$$\vee([\![\xi = \bar{\xi}]\!] \wedge \gamma_2(2\bar{\xi} - 1, x) \wedge [\![\xi = \underline{\xi}]\!] \wedge \gamma_1(2\underline{\xi}, y))$$

$$\leq E(x, y) \vee ([\![\xi = \frac{1}{2}]\!] \wedge \gamma_1(2\underline{\xi}, x) \wedge \gamma_2(2\bar{\xi} - 1, y)) \vee ([\![\xi = \frac{1}{2}]\!] \wedge \gamma_2(2\bar{\xi} - 1, x) \wedge \gamma_1(2\underline{\xi}, y))$$

$$\leq E(x, y) \vee (\gamma_1(1, x) \wedge \gamma_2(0, y)) \vee (\gamma_1(1, y) \wedge \gamma_2(0, x))$$

We have $(\gamma_1 \circ c_1)(\cdot, z) = \bigvee\{\gamma_1(\xi, z) \wedge ([\![\xi = 1]\!] \mid \xi \in \hat{I}\} \leq \bigvee\{\gamma_1(1, z) \mid \xi \in \hat{I}\} = \gamma_1(1, z) \leq (\gamma_1 \circ c_1)(\cdot, z)$ and analogously $(\gamma_2 \circ c_0)(\cdot, z) = \gamma_2(0, z)$. Because of $\gamma_1 \circ c_1 = \gamma_2 \circ c_0$ we obtain $\gamma_1(1, z) = \gamma_2(0, z)$, which implies $\gamma_1(1, x) \wedge \gamma_2(0, y) = \gamma_1(1, x) \wedge \gamma_1(1, y) \leq E(x, y)$.

(M3).

$$\bigvee\{\gamma_1 \cdot \gamma_2(\xi, x) \mid x \in X\} =$$

$$([\![\xi \leq \frac{1}{2}]\!] \wedge \bigvee\{\gamma_1(2\underline{\xi}, x) \mid x \in X\}) \vee ([\![\xi \geq \frac{1}{2}]\!] \wedge \bigvee\{\gamma_2(2\bar{\xi} - 1, x) \mid x \in X\}) =$$

$$[\![\xi \leq \frac{1}{2}]\!] \vee [\![\xi \geq \frac{1}{2}]\!] = \mathbf{1}$$

We have to show that $\gamma_1 = (\gamma_1 \cdot \gamma_2) \circ \iota_1$ holds.

$$\gamma_1(\xi, x) \leq ([\![\tfrac{1}{2}\xi \leq \frac{1}{2}]\!] \wedge \gamma_1(2 \cdot \underline{\tfrac{1}{2}\xi}, x)) \vee ([\![\tfrac{1}{2}\xi \geq \tfrac{1}{2}]\!] \wedge \gamma_2(2 \cdot \overline{\tfrac{1}{2}\xi} - 1, x)) = (\gamma_1 \cdot \gamma_2)(\tfrac{1}{2}\xi, x)$$

$$= \bigvee\{(\gamma_1 \cdot \gamma_2)(\eta, x) \wedge [\![\tfrac{1}{2}\xi = \eta]\!] \mid \eta \in \hat{I}\} = ((\gamma_1 \cdot \gamma_2) \circ \iota_1)(\xi, x) = (\gamma_1 \cdot \gamma_2)(\tfrac{1}{2}\xi, x)$$

$$= \gamma_1(\xi, x) \vee ([\![\xi = 1]\!] \wedge \gamma_2(0, x)) = \gamma_1(\xi, x) \vee ([\![\xi = 1]\!] \wedge \gamma_1(1, x)) \leq \gamma_1(\xi, x)$$

Analogously, we obtain $\gamma_2 = (\gamma_1 \cdot \gamma_2) \circ \iota_2$.

Finally, we have to show that for any morphism $\gamma : \hat{I} \longrightarrow X$ such that $\gamma_i = \gamma \circ \iota_i$ holds for $i = 1, 2$, we can derive $\gamma = \gamma_1 \cdot \gamma_2$. We obtain

$$\gamma_1(\xi, x) = (\gamma \circ \iota_1)(\xi, x) = \bigvee\{\gamma(\eta, x) \wedge [\![\frac{1}{2}\xi = \eta]\!] \mid \eta \in \hat{I}\} = \gamma(\frac{1}{2}\xi, x)$$

and analogously we obtain $\gamma_2(\xi, x) = \gamma(\frac{1}{2}\xi + \frac{1}{2}, x)$. Now note

$$\begin{aligned} \gamma_1(2\underline{\xi}, x) \wedge [\![\xi = \underline{\xi}]\!] &\leq \bigvee\{\gamma_1(\eta, x) \wedge [\![\frac{1}{2}\eta = \xi]\!] \mid \eta \in \hat{I}\} \\ &= \bigvee\{\gamma(\frac{1}{2}\eta, x) \wedge [\![\frac{1}{2}\eta = \xi]\!] \eta \in \hat{I}\} \\ &\leq \gamma(\xi, x) \end{aligned}$$

and analogously $\gamma_2(2\bar{\xi} - 1, x) \wedge [\![\xi = \bar{\xi}]\!] \leq \gamma(\xi, x)$, and therefore $(\gamma_1 \cdot \gamma_2)(\xi, x) \leq \gamma(\xi, x)$ holds. Also note

$$\gamma(\xi, x) = (\gamma(\xi, x) \wedge [\![\xi = \underline{\xi}]\!] \wedge [\![\tfrac{1}{2} \cdot 2\underline{\xi} = \xi]\!]) \vee (\gamma(\xi, x) \wedge [\![\xi = \bar{\xi}]\!] \wedge [\![\tfrac{1}{2} \cdot (2\bar{\xi} - 1) + \tfrac{1}{2} = \xi]\!])$$

$$\leq (\gamma(\tfrac{1}{2} \cdot 2\underline{\xi}, x) \wedge [\![\xi = \underline{\xi}]\!]) \vee (\gamma(\tfrac{1}{2} \cdot (2\bar{\xi} - 1) + \tfrac{1}{2}, x) \wedge [\![\xi = \bar{\xi}]\!]) = (\gamma_1 \cdot \gamma_2)(\xi, x) \quad \square$$

PROPOSITION 2. ι_1 and ι_2 are continuous morphisms.

PROOF. It is sufficient to prove that the inverse images of fuzzy sets of the subbasis of $\hat{\tau}$ are open.

$$\iota_1^{-1}((\zeta, \rightarrow))(\xi) = \bigvee\{\, [\![\frac{1}{2}\xi = \eta]\!] \wedge [\![\zeta < \eta]\!] \mid \eta \in \hat{I}\,\} =$$

$$[\![\zeta < \frac{1}{2}\xi]\!] = [\![2\underline{\zeta}, < \xi]\!] = (2\underline{\zeta}, \rightarrow)(\xi),$$

$\iota_1^{-1}((\leftarrow, \zeta)) = (\leftarrow, 2\bar{\zeta} - 1)$. For ι_2 we obtain similar results. $\square$

PROPOSITION 3. $\hat{\tau}$ is the final topology on $\hat{I}$ induced by ι_1 and ι_2.

PROOF. Because of Proposition 2 we only have to show that a fuzzy set u on $\hat{I}$ is open, if $\iota_1^{-1}(u)$ and $\iota_2^{-1}(u)$ are open. We define the image of a fuzzy set f under a morphism

$\varphi : X \longrightarrow Y$ by $(\varphi f)(y) = \bigvee\{\varphi(x,y) \wedge f(x) \mid x \in X\}$. Let u be a fuzzy set on $\hat{I}$ such that $u_1 = \iota_1^{-1} u \in \hat{\tau}$ and $u_2 = \iota_2^{-1} \in \hat{\tau}$ hold.

$$(\iota_1 u_1 \vee \iota_2 u_2)(\xi) =$$

$$\bigvee_\eta \bigvee_\zeta ((\llbracket \tfrac{1}{2}\eta = \xi \rrbracket \wedge \llbracket \tfrac{1}{2}\eta = \zeta \rrbracket \wedge u(\zeta)) \vee (\llbracket \tfrac{1}{2}\eta + \tfrac{1}{2} = \xi \rrbracket \wedge \llbracket \tfrac{1}{2}\eta + \tfrac{1}{2} = \zeta \rrbracket \wedge u(\zeta)))$$

$$\leq u(\xi)$$

and by choosing $\eta = 2\underline{\xi}, \eta = 2\bar{\xi} - 1$, respectively, and $\zeta = \xi$ we obtain $(\iota_1 u_1 \vee \iota_2 u_2)(\xi) \geq u(\xi)$. If $v \in \hat{\tau}$, then

$$v = \bigvee_{\zeta:(\zeta,\rightarrow)\leq v} (\zeta,\rightarrow) \vee \bigvee_{\eta:(\leftarrow,\eta)\leq v} (\leftarrow,\eta) \vee \bigvee_{\zeta,\eta:(\zeta,\eta)\leq v} (\zeta,\eta)$$

because $(\zeta_1,\eta_1) \wedge (\zeta_2,\eta_2) = (\max\{\zeta_1,\zeta_2\}, \min\{\eta_1,\eta_2\})$, and constant fuzzy sets can be represented in the following way. Let $\alpha \in \mathbb{B}$, $\alpha = M/_{[P=0]}$. $\chi_M : \Omega \longrightarrow \mathbb{R}$ denotes the characteristic function of M. Then $\alpha_{\hat{I}} = (\leftarrow, \chi_M) \vee (\chi_{\Omega - M}, \rightarrow)$. Representing u_1 and u_2 in this way and taking into account that

$$u = \iota_1 u_1 \vee \iota_2 u_2,\ \iota_1((\zeta,\eta)) = (\frac{1}{2}\zeta, \frac{1}{2}\eta),\ \iota_1((\leftarrow,\eta)) = (\leftarrow, \frac{1}{2}\eta),$$

$$\iota_2((\zeta,\eta)) = (\frac{1}{2}\zeta + \frac{1}{2}, \frac{1}{2}\eta + \frac{1}{2}),\ \iota_2((\zeta,\rightarrow)) = (\frac{1}{2}\zeta + \frac{1}{2}, \rightarrow)$$

hold, it remains to be proved that $\bigvee_{\zeta:(\zeta,\rightarrow)\leq u_1} \iota_1((\zeta,\rightarrow)) \vee \bigvee_{\eta:(\leftarrow,\eta)\leq u_2} \iota_2((\leftarrow,\eta)) \in \hat{\tau}$ holds.

Therefore, it is sufficient to show that

$$\bigvee_{\zeta:(\zeta,\rightarrow)\leq u_1} \iota_1((\zeta,\rightarrow)) \ \vee \bigvee_{\eta:(\leftarrow,\eta)\leq u_2} \iota_2((\leftarrow,\eta)) \tag{1}$$

$$= \bigvee_{\zeta:(\zeta,\rightarrow)\leq u_1,\ \eta:(\leftarrow,\eta)\leq u_2} (\tfrac{1}{2}\zeta, \tfrac{1}{2}\eta + \tfrac{1}{2}) \tag{2}$$

holds.

$$\iota_1((\zeta,\rightarrow))(\xi) = \bigvee_\eta (\llbracket \zeta < \eta \rrbracket \wedge \llbracket \frac{1}{2}\eta = \xi \rrbracket) =$$

$$\bigvee_\eta (\llbracket \frac{1}{2}\zeta < \frac{1}{2}\eta \rrbracket \wedge \llbracket \frac{1}{2}\eta = \xi \rrbracket \wedge \llbracket \xi = \underline{\xi} \rrbracket) \leq$$

$$\llbracket \frac{1}{2}\zeta < \underline{\xi} \rrbracket \wedge \llbracket \xi = \underline{\xi} \rrbracket$$

and by choosing $\eta = 2\underline{\xi}$ we obtain $\iota_1((\zeta,\rightarrow))(\xi) \geq \llbracket \frac{1}{2}\zeta < \underline{\xi} \rrbracket \wedge \llbracket \xi = \underline{\xi} \rrbracket$. Analogously, $\iota_2((\leftarrow,\eta))(\xi) = \llbracket \xi < \frac{1}{2}\eta + \frac{1}{2} \rrbracket \wedge \llbracket \xi = \bar{\xi} \rrbracket$.

$$(\iota_1((\zeta,\rightarrow)) \vee \iota_2((\leftarrow,\eta)))(\xi) =$$

$$(\llbracket \frac{1}{2}\zeta < \xi \rrbracket \vee (\llbracket \xi < \frac{1}{2}\eta + \frac{1}{2} \rrbracket) \wedge (\llbracket \frac{1}{2}\zeta < \xi \rrbracket \vee \llbracket \xi = \frac{1}{2} \rrbracket g) \wedge (\llbracket \xi < \frac{1}{2}\eta + \frac{1}{2} \rrbracket \wedge \llbracket \xi = \frac{1}{2} \rrbracket)$$

$$= (\llbracket \frac{1}{2}\zeta < \xi \rrbracket \vee \llbracket \xi < \frac{1}{2}\eta + \frac{1}{2} \rrbracket) \wedge (\llbracket \xi = \frac{1}{2} \rrbracket \vee \llbracket \frac{1}{2}\zeta < \xi < \frac{1}{2}\eta + \frac{1}{2} \rrbracket)$$

$$= (\llbracket \xi = \frac{1}{2} \rrbracket \wedge \llbracket \zeta < 1 \rrbracket) \vee (\llbracket \xi = \frac{1}{2} \rrbracket \wedge \llbracket \eta > 0 \rrbracket) \vee \llbracket \frac{1}{2}\zeta < \xi < \frac{1}{2}\eta + \frac{1}{2} \rrbracket$$

Applying this result to (1) it remains to be proved that

$$\llbracket \xi = \tfrac{1}{2} \rrbracket \wedge (\bigvee_{\zeta:(\zeta,\rightarrow)\leq u_1} \llbracket \zeta < 1 \rrbracket \ \vee \bigvee_{\eta:(\leftarrow,\eta)\leq u_2} \llbracket \eta > 0 \rrbracket) \leq (2) \tag{3}$$

$\bigvee_{\zeta:(\zeta,\rightarrow)\leq u_1} \llbracket \zeta < 1 \rrbracket = (\bigvee_{\zeta:(\zeta,\rightarrow)\leq u_1} (\zeta,\rightarrow))(1) = u_1(1)$, because $u_1 \in \hat{\tau}$, and analogously $\bigvee_{\eta:(\leftarrow,\eta)\leq u_2} \llbracket \eta > 0 \rrbracket = u_2(0)$. $u_1(1) = \bigvee\{\llbracket \frac{1}{2} = \vartheta \rrbracket \wedge u(\vartheta) \mid \vartheta \in \hat{I}\} = u(\frac{1}{2}) = u_2(0)$. This reduces (3) to

$$\llbracket \xi = \tfrac{1}{2} \rrbracket \wedge u(\tfrac{1}{2}) \tag{4}$$

$$(2) \geq \llbracket \xi = \tfrac{1}{2} \rrbracket \wedge (\bigvee_{\zeta:(\zeta,\rightarrow)\leq u_1} \llbracket \zeta < 1 \rrbracket \ \wedge \bigvee_{\eta:(\leftarrow,\eta)\leq u_2} \llbracket \eta > 0 \rrbracket) = (4) \quad \square$$

COROLLARY 1. Let (X, E) be a $\mathbb{B}$-underdeterminate set and let τ be a topology on (X, E). Let $\gamma_1 : (\hat{I}, \hat{E}) \longrightarrow (X, E)$ be continuous morphisms $(i = 1, 2)$. Then $\gamma_1 \cdot \gamma_2$ as defined in the proof of Theorem 2 is continuous.

EXAMPLE 1. card $\Omega = 1$ implies $\mathbb{B} = \{\mathbf{0}, \mathbf{1}\}$. In this case the only truth values are true and false. The Boolean valued model then coincides with the usual notion of the unit interval.

EXAMPLE 2. If the lattice L of Section 1 is equal to the unit interval I, then we obtain Lowen's definition of the fuzzy unit interval [Lowen 1985]. In this example we discuss a Boolean valued model of the unit interval which is adequate for Lowen's fuzzy unit interval. Let $\Omega = \mathbb{R}$, $\mathcal{A} = \mathcal{B}(\mathbb{R})$ (the Borel sets in $\mathbb{R}$), and let P denote the Borel measure on $\mathbb{R}$. Let ξ and η be two random variables concentrated on the unit interval. $\llbracket \xi = \eta \rrbracket = \mathbf{1}$ holds, iff the corresponding distribution functions F_ξ and F_η are equal. Therefore, we may consider the set $\mathcal{F}$ of all distribution functions concentrated on the unit interval instead of the set $\hat{I}$, because we cannot distinguish random variables ξ and η satisfying $\llbracket \xi = \eta \rrbracket = \mathbf{1}$ in our Boolean valued model. If we consider the set $\mathcal{F}^* = \{1 - F \mid F \in \mathcal{F}\}$ as a subset of all monotone decreasing functions $\lambda : \mathbb{R} \longrightarrow L = I$, satisfying $\lambda(0) = \mathbf{1}$ and $\lambda(1+) = \mathbf{0}$ for each equivalence class $[\lambda]$ there is a unique $F^* \in \mathcal{F}^*$ such that F^* is a representative of the of the equivalence class $[\lambda]$.

Therefore, the Boolean valued model of the unit interval, which we obtain in the case of $(\mathbb{R}, \mathcal{B}(\mathbb{R}), P)$, can be identified with Lowen's fuzzy unit interval, if we forget about the fuzzy equality in the Boolean valued model of the unit interval.

§5. Conclusions

Although it is possible to define the product of paths in the case of a Boolean valued model there are still difficulties in handling the more general case of Heyting valued sets. Therefore, considering fuzzy equality does not solve the problem of defining fuzzy paths in general. A reasonable definition for the product of fuzzy paths is essential for a homotopy theory for fuzzy topological spaces. We have shown that the fuzzy unit interval cannot be used in a canonical way as a basis for fuzzy homotopy theory.

Chapter 10:

Lattice Morphisms, Sobriety, and Urysohn Lemmas

S. E. Rodabaugh et al. (eds.), Applications of Category Theory to Fuzzy Subsets, 257–274.

LATTICE MORPHISMS, SOBRIETY, AND URYSOHN LEMMAS

W. Kotzé

§0. Introduction

C. H. Dowker and Dona Papert established a relationship between continuous functions and frame maps in their 1967 paper on Urysohn's Lemma. This can be improved upon and extended to the fuzzy situation. The role of sobriety is emphasized.

The author is indebted to Steve Rodabaugh who showed the author the way onto the path of sobriety, and this chapter is dedicated to him. Bernhard Banaschewski from McMaster University, Canada, is another good friend whose comments resulted in significant improvements. Grants from the F.R.D., Pretoria, and Rhodes University are greatly appreciated.

§1. Lattices and Lattice Morphisms

If property A_i denotes one of

A_1 : closed under finite suprema
A_2 : closed under countable suprema
A_3 : closed under arbitrary suprema

and property B_j denotes one of

B_1 : closed under finite infima
B_2 : closed under countable infima
B_3 : closed under arbitrary infima

then an $\mathbf{A}_i\mathbf{B}_j$**-lattice** is one which has properties A_i and B_j for a fixed i and j (1, 2 or 3).

Such lattices will have a least element $\underline{0}$ and a largest element $\underline{1}$. We can also require an A_iB_j-lattice to satisfy the appropriate distributive law of, e.g. for i = 3 and j = 1, $a \wedge (\bigvee b_i) = \bigvee a \wedge b_i$ (i over an arbitrary index set); in other words, finite infima distribute over arbitrary suprema. However, in this paper, distributivity plays a role only in section 7.

So a frame is an A_3B_1-lattice and a σ-frame an A_2B_1-lattice.

The morphisms between two A_iB_j-lattices (with i and j fixed as the same in both) are the maps μ which preserve properties A_i and B_j, e.g. for $i = 3, j = 1$, the morphisms are the frame maps which satisfy $\mu(\bigvee a_i) = \bigvee \mu a_i$ for all subfamilies $\{a_i\}$ of the first lattice, and $\mu(\wedge b_j) = \wedge \mu b_j$ for all finite subfamilies $\{b_j\}$.

$\mathbf{CA}_i\mathbf{B}_j$ will denote the category of A_iB_j-lattices and A_iB_j-morphisms.

If I is an A_iB_j-lattice, then I^X (the set of maps from a set of X into I) is again an A_iB_j-lattice (or an object from $\mathbf{CA}_i\mathbf{B}_j$) under the definitions $(f \vee g)(x) = f(x) \vee g(x)$ and $(f \wedge g)(x) = f(x) \wedge g(x)$.

EXAMPLES 1.1.

(i) A topology T on a set X is an A_3B_1-sublattice (subframe) of 2^X.

(ii) A fuzzy topology (I-topology) F on a set X is an A_3B_1-sublattice (subframe) of I^X where I is an A_3B_1-lattice. Originally I was taken to be the unit interval. An I-topological space is denoted by (X, F, I) and the $\underline{0}$ and $\underline{1}$ in it are respectively the functions on X which are identical on X to the $\underline{0}$ and $\underline{1}$ in I. Clearly any topological space (X, $T, 2$) is an I-topological (fuzzy topological) space.

(iii) If (X, T) is a topological space, the collection of F_σ in X is an A_2B_1-sublattic (σ-subframe) of 2^X. The corresponding extension of the concept to a σ-frame in (X, F, I) is obvious.

Given a map $f : X \longrightarrow Y$ and I an A_iB_j-lattice, we obtain maps between I^X and I^Y as follows (the so-called Zadeh extension principle):

DEFINITION 1.2. If f is a map from a set X into a set Y, then for $v \in I^Y$, $f^{\leftarrow}(v)$ is defined by $f^{\leftarrow}(v)(x) = v(f(x))$, and for $u \in I^X$, $f^{\rightarrow}(u)$ is defined by

$$f^{\rightarrow}(u)(y) = \begin{cases} \bigvee u(x)_{x \in f^{\leftarrow}\{y\}} & \text{if } f^{\leftarrow}\{y\} \neq \emptyset \\ 0 & \text{if } f^{\leftarrow}\{y\} = \emptyset. \end{cases}$$

In the special case when $I = 2$, this corresponds with the classical notions of image and pre-image of "crisp" sets under a map.

EXAMPLES 1.3.

(i) If $f : (X, T_1) \longrightarrow (Y, T_2)$ is a continuous map between two topological spaces, then $f^{\leftarrow}$ is a frame map from T_2 into T_1, and if $f : (X, F_1, I) \longrightarrow (Y, F_2, I)$ is "fuzzy continuous" (i.e. $f^{\leftarrow}(v) \in F_1$ for every $v \in F_2$), then $f^{\leftarrow}$ is a frame map from F_2 into F_1.

(ii) If $f : (X, T_1) \longrightarrow (Y, T_2)$ is a function of class 1 (see e.g. [Kotzé 1984]), then $f^{\leftarrow}$ is a σ-frame morphism from the σ-frame T_2 into the σ-frame of F_σ in X.

We have on the other hand the following result from [Dowker and Papert 1967].

THEOREM 1.4. If (X,T_1) and (Y,T_2) are two topological spaces with (Y,T_2) Hausdorff, and $\mu : T_2 \longrightarrow T_1$ is a frame map, then there exists a unique continuous function $f : X \longrightarrow Y$ such that $f^{\leftarrow} = \mu$.

This result can be considerably generalised as well as improved, as it turns out that sobriety of (Y, T_2) is sufficient (and in a sense also necessary).

§2. Sobriety

To accommodate the reader we briefly summarise the elements of sobriety. (For more details see e.g. [Johnstone 1982], [Rodabaugh 1986–7].)

Given a topological space (X,T), pt$T$ (or spectrum of T) will denote the collection of all frame maps p from T into the frame $\{0,1\}$ (the topology on a singleton space).

Define $\Psi : X \longrightarrow \text{pt}T$ as $\Psi(x)(u) = \chi_u(x)$, the characteristic function of u, for $x \in X$ and $u \in T$. Clearly Ψ is injective iff (X,T) is T_0, and if (X,T) is Hausdorff, then Ψ is a bijection (see [Johnstone 1982], Chapter 2). We say (X,T) is **sober** iff Ψ is a bijection. (So Hausdorff $\Rightarrow$ sober $\Rightarrow T_0$). We put a topology on pt$T$ as follows:

$$\Phi : T \longrightarrow \mathcal{P}(\text{pt}T) \text{ as } \Phi(u) = \{p \in \text{pt}T : p(u) = 1\}$$

Then $\{\Phi(u) : u \in T\}$ is a topology on pt$T$ under which Ψ becomes continuous and open since $\Psi^{\leftarrow}(\Phi(u)) = u$ and $\Psi^{\rightarrow}(u) = \Phi(u)$. Thus if (X,T) is sober, Ψ is a homeomorphism. Note that $\Phi(u)$ as a subset of pt$T$ can also be considered as a map from pt$T$ into $\{0,1\}$ where

$$\begin{aligned}\Phi(u)(p) &= \begin{cases} 1 & \text{if } p(u) = 1 \\ 0 & \text{if } p(u) = 0 \end{cases} \\ &= p(u).\end{aligned}$$

[Rodabaugh 1986–7] generalised these considerations to fuzzy topological (I-topological) spaces (X,F,I) as in Example 1.1 (ii) above. In this case pt$F$ is defined as the set of all frame maps p from F into I, and $\Psi : X \longrightarrow \text{pt}F$ as $\Psi(x)(u) = u(x)$ for $x \in X$ and $u \in F$. Note that:

2.1 Ψ is an injection iff $x_1 \neq x_2 \Rightarrow$ there exists a $u \in F$ such that $u(x_1) \neq u(x_2)$. We call this property of F, T_0.

2.2 Ψ is a surjection iff for every $p \in \text{pt}F$ there exists an $x \in X$ such that for every $u \in F, p(u) = u(x)$.

We say (X,F,I) is **sober** iff Ψ is a bijection. An I-topology is generated on pt F as follows: For $u \in F, \Phi(u)$ is a "fuzzy subset" of pt$F$, i.e. $\Phi(u)$: pt$F \longrightarrow I$ through

$\Phi(u)(p) = p(u)$. Then $\{\Phi(u) : u \in F\}$ is an I-topology on pt$F$ under which Ψ becomes fuzzy continuous since

$$\begin{aligned}\Psi^{\leftarrow}(\Phi(u))(x) &= \Phi(u)(\Psi(x)) \text{ (from Definition 1.2)}\\ &= \Psi(x)(u)\\ &= u(x)\end{aligned}$$

and fuzzy open since $\Psi^{\rightarrow}(u)(p) = \bigvee_{x\in\Psi^{\leftarrow}(p)} u(x) = p(u) = \Phi(u)(p)$ or $\Psi^{\rightarrow}(u) = \Phi(u)$. So once again, if (X, F, I) is sober then Ψ is a fuzzy homeomorphism.

We can extend these concepts even further to the general categories $\mathbf{CA}_i\mathbf{B}_j$. Let I be an $\mathrm{A}_i\mathrm{B}_j$-lattice with $\underline{0}$ and $\underline{1}$ and F an $\mathrm{A}_i\mathrm{B}_j$-sublattice of I^X. pt F will be the set of all $\mathrm{A}_i\mathrm{B}_j$-morphisms of F into I (i.e. pt$F$ = mor(F, I)). Again $\Psi : X \longrightarrow$ pt$F$ is defined as $\Psi(x)(u) = u(x)$ for $x \in X$ and $u \in F$. (X, F, I) is **sober** iff Ψ is a bijection.

If $\Phi(u)$: pt$F \longrightarrow I$ is defined as $\Phi(u)(p) = p(u)$, then $\{\Phi(u) : u \in F\}$ is an $\mathrm{A}_i\mathrm{B}_j$-sublattice of the $\mathrm{A}_i\mathrm{B}_j$-lattice $I^{\mathrm{pt}F}$. Once again $\Psi^{\leftarrow}(\Phi(u)) = u$ for $u \in F$, in other words $\Psi^{\leftarrow}$ is an $\mathrm{A}_i\mathrm{B}_j$-morphism from $\{\Phi(u) : u \in F\}$ into F. Furthermore we also have that $\Psi^{\rightarrow}(u) = \Phi(u)$.

2.3 Important examples of I-topological spaces are the fuzzy real line $\mathbb{R}(I)$ and the fuzzy unit interval $[0,1](I)$. (See [Hutton 1975]). G. Meßner proved (in an as yet unpublished doctoral dissertation with the Johannes Kepler Universität, Linz) that these spaces are sober iff I is Boolean.

It might be worthwhile to try to better understand the meaning of sobriety, at least for some special cases, through the following considerations.

DEFINITION 2.4. An **ideal** $\mathcal{J}$ of a lattice L (with $\underline{0}$ and $\underline{1}$) is a subset of L such that

(i) $\underline{0} \in \mathcal{J}$ and $a \in \mathcal{J}, b \in \mathcal{J} \Rightarrow a \vee b \in \mathcal{J}$;
(ii) $\mathcal{J}$ is a lower set, i.e. $a \in \mathcal{J}$ and $b \leq a \Rightarrow b \in J$.

If $a \in L$, the subset $\downarrow(a) = \{b \in L : b \leq a\}$ is an ideal of L, called the **principal ideal** generated by a.

DEFINITION 2.5. A **filter** $\mathcal{F}$ of a lattice $\mathbf{L}$ is a subset of L such that

(i) $\underline{1} \in \mathcal{F}$ and $a \in \mathcal{F}, b \in \mathcal{F} \Longrightarrow$ a $\wedge b \in \mathcal{F}$
(ii) $\mathcal{F}$ is an upper set, i.e. $a \in \mathcal{F}$ and $b \geq a \Rightarrow b \in \mathcal{F}$.

Clearly $\uparrow(a) = \{b \in L : b \geq a\}$ is a filter in L.

PROPOSITION 2.6. Let $\mathcal{J}$ be an ideal of a lattice L. Then the following conditions are equivalent:

(i) The complement of $\mathcal{J}$ in L is a filter,
(ii) $\underline{1} \notin \mathcal{J}$ and $a \wedge b \in \mathcal{J} \Rightarrow a \in \mathcal{J}$ or $b \in \mathcal{J}$,

(iii) $\mathcal{J}$ is the kernel of a lattice morphism $\varphi : L \longrightarrow I$ where I is a lattice with $\underline{0}$ and $\underline{1}$ and with $\underline{0}$ as a prime element of I (i.e. $a \wedge b = \underline{0} \Rightarrow a = \underline{0}$ or $b = \underline{0}$). (cf. [Johnstone 1982] Proposition 2.2, [Rodabaugh 1986–7] Lemma 5.4.1).

PROOF. (i) $\Leftrightarrow$ (ii): is obvious.

(ii) $\Rightarrow$ (iii): Define $\varphi : L \longrightarrow [0,1]$ as $\varphi(a) = \underline{0}$ if $a \in \mathcal{J}$ and $\varphi(a) = \underline{1}$ otherwise. Then it can be easily checked (using Definition 2.4 (ii) and Definition 2.5 (ii)) that φ is a lattice morphism.

(iii) $\Rightarrow$ (ii): Let $\mathcal{J} = \text{Ker } \varphi$ with φ as in (iii). Then clearly $\underline{1} \notin \mathcal{J}$. If $a \wedge b \in \text{Ker } \varphi$, then $\varphi(a) \wedge \varphi(b) = \varphi(a \wedge b) = \underline{0}$. Since $\underline{0}$ is prime in I, at least one of $\varphi(a)$ or $\varphi(b)$ is zero, i.e. at least one of a or b is in $\mathcal{J}$. □

An ideal satisfying the condition of Proposition 2.6 is called a **prime ideal**, and an element a of a lattice is a **prime element** if $\downarrow(a)$ is a prime ideal. Thus an element a is prime iff $b \wedge c \leq a \Longrightarrow b \leq a$ or $c \leq a$.

DEFINITION 2.7. In a lattice L an element a is **irreducible** (in L) iff $b \vee c \geq a \Longrightarrow b \geq a$ or $c \geq a$.

PROPOSITION 2.8. In a lattice L with an order reversing involution $a \in L$ is a prime element in L iff a' is irreducible in L'.

PROOF. Follows from de Morgan's law. □

Now consider (X, F, I) with I a complete lattice with an order reversing involution and with $\underline{0}$ a prime element and F an A_3B_j-sublattice of I^X. We also have an order reversing involution in I^X ($a'(x) \equiv a(x)'$). Then F' (the involutes of members of F) is an A_jB_3-sublattice of I^X.

If $p \in \text{pt}F$, by Proposition 2.6, its kernel, $p^{-1}(\underline{0})$, is a prime ideal. But since p preserves arbitrary suprema, $p^{-1}(\underline{0}) = \downarrow(\bigvee\{a : a \in p^{-1}(\underline{0})\})$, so it is a principal ideal. So for each $p \in \text{pt}F$ there exists a prime element in F, and hence (by Proposition 2.8) an irreducible element in F'. Conversely, given a prime element $a \in F$ (or an irreducible element $a' \in F'$), i.e. $\downarrow(a)$ is a principal prime ideal, then $\downarrow(a)$ is the kernel of a $p \in \text{pt}F$ (by Proposition 2.6).

If we now consider a $p \in \Psi(X)$, where Ψ is the map of $X \longrightarrow \text{pt}F$ above, the question arises, what is the irreducible element of F' identified with it? This is easy to see since for such a p, there exists an $x_1 \in X$ such that $p = \Psi(x_1)$, and so Ker $p = \downarrow(\bigvee\{u \in F : p(u) = \underline{0}\}) = \downarrow(\bigvee\{u \in F : \Psi(x_1)(u) = u(x_1) = \underline{0}\})$, and so the associated irreducible element in F' is $\bigwedge\{u' \in F' : u'(x_1) = \underline{1}\}$ i.e. the smallest member of F' which is $\underline{1}$ at x_1 — let's call it the F-**closure** of x_1.

Note that the F-closure of a singleton x_1 in X is clearly irreducible in F' since $\underline{1}$ is irreducible in the I we consider. Thus sober implies (Ψ a bijection) that the only irreducible members of F' are the F-closures of the singletons in X. So a T_0 space is sober iff the only irreducible members of F' are the F-closures of the singletons in X.

From this it is clear that in the case (X, F, I) with I other than $\{0, 1\}$, the relationship between Hausdorff and sobriety is not as simple as in the classical case. Whatever notion might be (or has been) proposed for (X, F, I) to be Hausdorff, $F = I^X$ will be such. In $F = I^X$, the only irreducible members in $F' = F$ need not only be the F-closures of the singletons in X. In fact, unless each $\alpha \neq \underline{0}$ in I is reducible, the "fuzzy points"

$$u(x) = \begin{cases} \alpha & x = x_1, \quad \underline{0} < \alpha \leq \underline{1} \\ \underline{0} & x \neq x_1 \end{cases}$$

will be irreducible.

§3.

A strengthening of Theorem 1.4 can now be obtained as a special case of the following result.

THEOREM 3.1. (X, F_1, I) and (Y, F_2, I) are two I-topological spaces with (Y, F_2, I) sober. If $\mu : F_2 \longrightarrow F_1$ is a frame map, there exists a unique fuzzy continuous function $f : X \longrightarrow Y$ such that $f^{\leftarrow} = \mu$.

And in the case of σ-frames:

THEOREM 3.2. Given (X, F_1, I) and (Y, F_2, I) with I a σ-frame and F_1 and F_2 σ-frames in I^X and I^Y respectively, with (Y, F_2, I) sober. If μ is a σ-frame morphism from F_2 to F_1, then there exists a unique $f : X \longrightarrow Y$ such that $f^{\leftarrow} = \mu$.

These two results are special cases of:

THEOREM 3.3. Given (X, F_1, I) and (Y, F_2, I) with I an $\mathrm{A}_i\mathrm{B}_j$-lattice and F_1 and F_2 A_iB_j-sublattices of I^X and I^Y respectively, with (Y, F_2, I) sober. If $\mu \in \mathrm{mor}(F_2, F_1)$, then there exists a unique $f : X \longrightarrow Y$ such that $f^{\leftarrow} = \mu$.

PROOF. So we have $\Psi_X : X \longrightarrow \mathrm{pt}F_1$ and $\Psi_Y : Y \longrightarrow \mathrm{pt}F_2$ with the latter a bijection. Define $G : \mathrm{pt}F_1 \longrightarrow \mathrm{pt}F_2$ as $G(p) = p \circ \mu$ for $p \in \mathrm{pt}F_1$. So we have the diagram

$$\begin{array}{ccc} pt\,F_1 & \xrightarrow{\quad G \quad} & pt\,F_2 \\ \Psi_X \Big\uparrow & & \Big\downarrow \Psi_X^{-1} \\ X & & Y \end{array}$$

and therefore define $f : X \longrightarrow Y$ as $f = \Psi_Y^{-1} \circ G \circ \Psi_X$. We check that $f^{\leftarrow} = \mu$ as $\mathrm{A}_i\mathrm{B}_j$-morphisms — in other words if $v \in F_2$ and $x \in X$, do we have $f^{\leftarrow}(v)(x) = \mu(v)(x)$? But

$$\begin{aligned} f^{\leftarrow}(v)(x) &= v(f(x)) \text{ (from Definition 1.2)} \\ &= v\left(\Psi_Y^{-1}(G(\Psi_X(x)))\right) \\ &= v\left(\Psi_Y^{-1}(\Psi_X(x) \circ \mu)\right) \end{aligned}$$

Now if $\Psi_Y^{-1}(q) = y \quad (q \in \mathrm{pt}F_2)$, then $\Psi_Y(y) = q$, i.e. for every $v \in F_2, q(v) = \Psi_Y(y)(v) = v(y)$ by definition of Ψ_Y. So for $y = \Psi_Y^{-1}(\Psi_X(x) \circ \mu)$ we have $v(y) = \Psi_X(x)(\mu(v)) = \mu(v)(x)$. Hence $f^{\leftarrow}(v)(x) = \mu(v)(x)$ for every $v \in F_2$ and every $x \in X$. Thus $f^{\leftarrow} = \mu$.

Uniqueness of f: Suppose we also have $\hat{f} : X \longrightarrow Y$ such that $\hat{f}^{\leftarrow} = \mu$. We must show that $f = \hat{f}$, or that $\hat{f} = \Psi_Y^{-1} \circ G \circ \Psi_X$. Now $f(x) = \Psi_Y^{-1}(G(\Psi_X(x))) = \Psi_Y^{-1}(\Psi_X(x) \circ \mu)$. And again $\Psi_Y^{-1}(\Psi_X(x) \circ \mu) = y$ iff for every $v \in F_2, v(y) = \mu(v)(x) = \hat{f}^{\leftarrow}(v)(x) = v(\hat{f}(x))$. But this implies, since Ψ_Y is injective (see 2.1) that $y = \hat{f}(x)$. Hence $f = \hat{f}$. □

The sobriety of (Y, F_2, I) is also necessary in the following sense:

THEOREM 3.4. If (Y, F_2, I) is T_0 (i.e. Ψ is an injection — see 2.1) and for every (X, F_1, I) and every $\mathrm{A}_i\mathrm{B}_j$-morphism $\mu : F_2 \longrightarrow F_1$, there exists an $f : X \longrightarrow Y$ such that $f^{\leftarrow} = \mu$, then (Y, F_2, I) is sober.

PROOF. Since (Y, F_2, I) is T_0, we have that Ψ_Y is an injection. We therefore only have to prove that Ψ_Y is surjective, i.e. we have to prove the characterisation as is expressed in 2.2. Now take $\mathrm{X} = \{x\}$ with $F_1 = I^{\{x\}}$ and any $\mu : F_2 \longrightarrow F_1$. Then by hypothesis there exists an $f : X \longrightarrow Y$ such that $f^{\leftarrow} = \mu$. Call $f(\{x\}) = f(x) = y$. Then for any $v \in F_2$,

$$\begin{aligned} \mu(v)(x) &= f^{\leftarrow}(v)(x) \\ &= v(f(x)) \text{ (from Definition 1.2)} \\ &= v(y). \end{aligned}$$

Identify $I^{\{x\}}$ with I. Then the $\mathrm{A}_i\mathrm{B}_j$-morphisms $\mu : F_2 \longrightarrow F_1$ are identified with the $p \in \mathrm{pt}F_2$, and after that identification $\mu(v)(x) = p(v)$. We therefore have that for every $p \in \mathrm{pt}F_2$, there exists a $y \in Y$ such that for every $v \in F_2$, $p(v) = v(y)$. Thus Ψ_Y is surjective. □

On the other hand we can drop the T_0 requirement in Theorem 3.4:

THEOREM 3.5. If for every (X, F_1, I) and every $\mathrm{A}_i\mathrm{B}_j$-morphism $\mu : F_2 \longrightarrow F_1$, there exists a unique $f : X \longrightarrow Y$ such that $f^{\leftarrow} = \mu$, then (Y, F_2, I) is sober.

PROOF. If (Y, F_2, I) is not T_0 we can shuffle the points of Y with the same values for all $v \in F_2$ around without any effect on the frame of the latter, but violating the unicueness of f. □

§4.

Let us now look at the category $\mathbf{CA_2B_4}$ of lattices which are closed under countable suprema and property B_4, which is closedness under complementation (order reversing involution).

If I is an A_2B_4-lattice (an object from $\mathbf{CA_2B_4}$), then I^X is an A_2B_4-lattice under the definition $f'(x) \equiv f(x)'$. If F is a A_2B_4-sublattice of I^X, then $\text{pt}F = \text{mor}(F, I)$ are the maps from F to I which preserve countable suprema and are **autoconjugate**, i.e. maps μ such that $\mu(\bigvee a_n) = \bigvee \mu(a_n)$ and $\mu(a') = \mu(a)'$. Again $\Psi : X \longrightarrow \text{pt}F$ is defined as $\Psi(x)(u) = u(x)$ for $x \in X$ and $u \in F$. Then $\Psi(x)(u') = u'(x) = u(x)'$. $(X, \mathbf{F}, I)$ is **sober** iff Ψ is a bijection.

If $\Phi(u) : \text{pt}F \longrightarrow I$ is defined as $\Phi(u)(p) = p(u)$ then $\{\Phi(u) : u \in F\}$ is an A_2B_4-sublattice of the A_2B_4-lattice $I^{\text{pt}F}$ — e.g. $\Phi(u')(p) = p(u') = p(u)' = [\Phi(u)(p)]' = \Phi(u)'(p)$, hence $\Phi(u') = \Phi(u)'$. Furthermore $\Psi^{\leftarrow}(\Phi(u)) = u$ for $u \in F$, in other words $\Psi^{\leftarrow}$ is an A_2B_4-morphism from $\{\Phi(u) : u \in F\}$ into F.

DEFINITION 4.1. A **σ-algebra** A on X is an A_2B_4-sublattice of 2^X (with involution set-theoretic complementation). (X, A) is called a **measurable space**. A function $f : X \longrightarrow Y$, where (X, A_1) and (Y,A_2) are measurable spaces, is **measurable** iff $f^{\leftarrow} \in \text{mor}(A_2, A_1)$.

Once more we can deduce from a counterpart of Theorem 3.3 for A_2B_4-lattices that:

THEOREM 4.2. Given two measurable spaces (X, A_1) and (Y, A_2) with (Y, A_2) sober, then if $\mu \in$ mor (A_2, A_1) there exists a unique measurable function $f : X \longrightarrow Y$ such that $f^{\leftarrow} = \mu$.

§5.

Theorems 3.3 and 4.2 can also be derived from a general result in category theory (see e.g. [Banaschewski and Bruns 1988], Lemma 1):

THEOREM 5.1. If $\Omega : \mathbf{P} \longrightarrow \mathbf{S}$ and $\Sigma : \mathbf{S} \longrightarrow \mathbf{P}$ are contravariant functors between categories $\mathbf{P}$ and $\mathbf{S}$, adjoint on the right, with adjunctions $\Psi_P : P \longrightarrow \Sigma\Omega P$ and $\Psi_S : S \longrightarrow \Omega\Sigma S$ for all $P \in \mathbf{P}$ and $S \in \mathbf{S}$, then for $P_1, P_2 \in \mathbf{P}$ and for any $\mu : \Omega P_2 \longrightarrow \Omega P_1$ (so $\Sigma\mu : \Sigma\Omega P_1 \longrightarrow \Sigma\Omega P_2$) there exists a unique $f : P_1 \longrightarrow P_2$ ($f \in$ mor $\mathbf{P}$) such that $\Omega f = \mu$ provided Ψ_{P_2} is an isomorphism — in fact $f = \Psi_{P_2}^{-1} \circ \Sigma\mu \circ \Psi_{P_1}$.

Now let I be an A_iB_j-lattice (for some $i = 1,2,3$ and $j = 1,\ldots,4$) with F_X an A_iB_j-sublattice of I^X. Let $\mathbf{P}$ be the category of all triples (X, F_X, I) with morphisms the maps $f : X \longrightarrow Y$ such that $f^{\leftarrow} : F_Y \longrightarrow F_X$. Let $\mathbf{S}$ be the category $\mathbf{CA}_i\mathbf{B}_j$. Then we have a contravariant functor $\Omega : \mathbf{P} \longrightarrow \mathbf{S}$, $\Omega(X,F,I) = F$, $\Omega(f) = f^{\leftarrow}$ for $f \in \text{mor}\mathbf{P}$. We have another contravariant functor $\Sigma : \mathbf{S} \longrightarrow \mathbf{P}$, $\Sigma L = \text{mor}(L, I) = \text{pt}L$ (or spectrum of L). $F_{\Sigma L}$ is the A_iB_j-sublattice of $I^{\Sigma L}$ consisting of the evaluation maps $\{\Phi(u) : u \in L\}$ where $\Phi(u)(p) = p(u)$ for $p \in \text{pt}L$. If μ is an A_iB_j-morphism from $L \longrightarrow M$, $\Sigma\mu : \Sigma M \longrightarrow \Sigma L$ is defined as $\Sigma\mu(p) = p \circ \mu$ and $(\Sigma\mu)^{\leftarrow}$ of Definition 1.2 is a map from $\mathbf{F}_{\Sigma L}$ to $\mathbf{F}_{\Sigma M}$ — for $p \in \Sigma\ \mathbf{M}, (\Sigma\mu)^{\leftarrow}\Phi(u)(p) = \Phi(u) \circ \Sigma\mu(p)$ (Definition 1.2) $= \Phi(u)(p \circ \mu) = p \circ \mu(u) = \Phi(\mu(u))(p)$. So $(\Sigma\mu)^{\leftarrow}\Phi(u) = \Phi(\mu(u))$ which is a member of $F_{\Sigma M}$.

THEOREM 5.2. Ω and Σ are adjoint from the right. (Cf. [Frith 1987] Theorem 6.10 for the frame case.)

PROOF. Let $f \in \text{mor}((X,F),(\Sigma L, F_{\Sigma L}))$ and define for $u \in L$ and $x \in X$, $\bar{f}(u)(x) = f(x)(u)$. Then $\bar{f} \in \text{mor}(L,\ \Omega(X,F))$ — letting $u \in L$, then $\Phi(u) \in F_{\Sigma L}$ and so $f^{\leftarrow}(\Phi(u)) = \Phi(u) \circ f \in F$.

But $\Phi(u)_0 f(x) = f(x)(u) = \bar{f}(u)(x)$, so $\bar{f}(u) \in F$. The morphism properties appropriate to the particular $\mathbf{CA}_i\mathbf{B}_j$ category are easy to check. On the other hand, let $g \in \text{mor}(L, \Omega(X,F))$ and define for $u \in L$, $x \in X$, $\tilde{g}(x)(u) = g(u)(x)$. Then $\tilde{g} \in \text{mor}((X,F),$ $(\Sigma, L, F_{\Sigma L}))$, e.g. for $x \in X$ and $u, v \in \Sigma L$,

$$\begin{aligned}\tilde{g}(x)(u \wedge v) &= g(u \wedge v)(x)\\ &= g(u) \wedge g(v)(x)\\ &= g(u)(x) \wedge g(v)(x)\\ &= \tilde{g}(x)(u) \wedge \tilde{g}(x)(v), \quad \text{etc.}\end{aligned}$$

Furthermore, consider $\Phi(u) \in F_{\Sigma L}(u \in L)$, then $\tilde{g}^{\leftarrow}(\Phi(u))(x) = \Phi(u)_0\tilde{g}(x) = \tilde{g}(x)(u)$ $= g(u)(x)$ or $\tilde{g}^{\leftarrow}(\Phi(u)) = g(u)$ which belongs to F. Also $\tilde{\bar{f}}(x)(u) = \bar{f}(u)(x) = f(x)(u)$ and $\bar{\tilde{g}}(u)(x) = \tilde{g}(x)(u) = g(u)(x)$. Hence $\tilde{\bar{f}} = f$ and $\bar{\tilde{g}} = g$. We therefore have that $\text{mor}((X,F),(\Sigma L, F_{\Sigma L})) \cong \text{mor}(L, \Omega, (X,F))$. The naturality conditions are easily checked. □

The adjunction units are $\eta_L : L \longrightarrow \Omega\Sigma L$, $\eta_L(u) = \Phi(u)$ and $\Psi_X : X \longrightarrow \Sigma\Omega X$, $\Psi_X(x)(u) = u(x)$. Also $\eta_L(u)f(x) = \Phi(u)(f(x)) = f(x)(u)$ and $\Psi_X(x)g(x) = g(u)(x)$. Again, (X,F,I) is sober iff Ψ_X is an isomorphism. It now follows that we can deduce Theorem 3.3 (and in a similar way also Theorem 4.2) from Theorem 5.1 and the above.

§6. Frame Spatiality

A frame L is **spatial** iff the adjuntion unit $\eta_L : L \longrightarrow \Omega\Sigma L$ is a bijection (isomorphism) i.e. when it is the image under Ω of some I-topological space and the frame morphisms $L \longrightarrow I$ separate the elements of L.

Now for any I-topological space (X, F, I), $\Omega(X, F, I) = F$ is spatial since

$$\begin{aligned} \eta_L(u) = \eta_L(v) &\Rightarrow \Phi(u) = \Phi(v) \\ &\Rightarrow p(u) = p(v) \quad \text{for all } p \in \text{pt } \mathbf{F} \\ &\Rightarrow u(x) = v(x) \quad \text{for all } \ x \in X \\ &\Rightarrow u \ = v \end{aligned}$$

We deduce from Theorem 5.1 that

THEOREM 6.1. For two I-topological spaces (X, F_1, I) and (Y, F_2, I) and any continuous $g : \Sigma F_1 \longrightarrow \Sigma F_2$, there exists a unique frame morphism $\mu : F_2 \longrightarrow F_1$ such that $\Sigma\mu = g$.

Thus if (X, F_1, I) is sober and $f : X \longrightarrow Y$ is continuous, we have $g = \Psi_Y \circ f \circ \Psi_X^{-1}$ a continuous function from ΣF_1 to ΣF_2

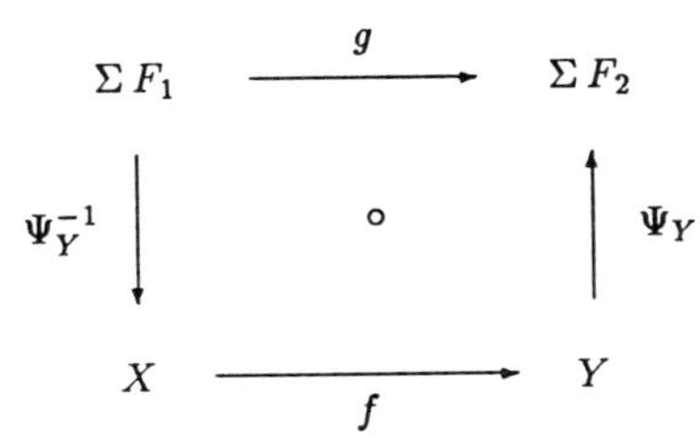

and so by Theorem 6.1, there exists a unique frame morphism $\mu : F_2 \longrightarrow F_1$ such that $\Sigma\mu = g$. In fact $\mu = \eta_{F_1}^{-1} \circ \Omega g \circ \eta_{F2}$

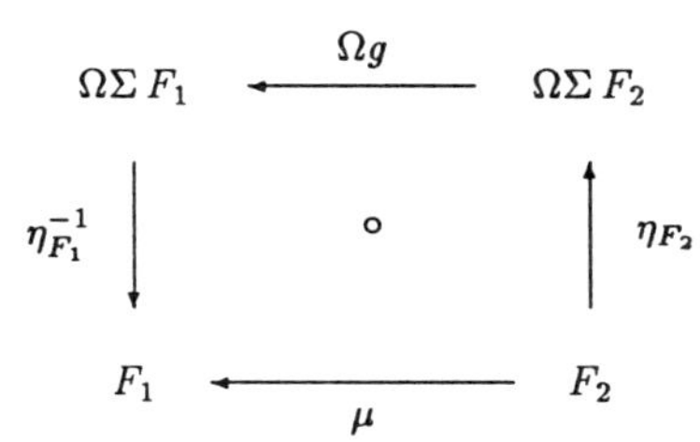

which is simply $f^{\leftarrow}$: for $v \in F_2$,

$$\begin{aligned}\mu(v) &= \eta_{F_1}^{-1} \circ \Omega g \circ \eta_{F_2}(v) \\ &= \eta_{F_1}^{-1} \circ \Omega g \circ \Phi_2(v) \\ &= \eta_{F_1}^{-1}(\Psi_X^{\rightarrow} f^{\leftarrow} \Psi_Y^{\leftarrow})\Phi_2(v) \\ &= \eta_{F_1}^{-1}(\Psi_X^{\rightarrow} f^{\leftarrow}(v)) \\ &= \eta_{F_1}^{-1}(\Phi(f^{\leftarrow}(v))) \text{ (from the observation just before 2.3 above)} \\ &= f^{\leftarrow}(v)\end{aligned}$$

§7. Urysohn Lemmas

There are several versions and generalisations of Urysohn's Lemma in the literature. One of the more recent is that of B. Hutton for fuzzy topological spaces. The remainder of this chapter relates some of these by considering quite general lattices with a normality condition. The associated lattice which is considered to be the canonical one is the lattice of open subsets of $[0, 1]$. This is not the one Hutton considered, but under certain conditions the Hutton case is included.

L will, in what follows, denote an $\mathrm{A}_i\mathrm{B}_j$-lattice ($i, j = 1, 2$ or 3) with an order reversing involution. (So the de Morgan laws hold). From now on we also require the appropriate $\mathrm{A}_i\mathrm{B}_j$ distributive law. (See Section 1 above.) L_1 will denote an $\mathrm{A}_i\mathrm{B}_j$-sublattice of the $\mathrm{A}_i\mathrm{B}_j$-lattice L. Of course, L_1', the involutes of members of L_1, will also be in L.

PROPOSITION 7.1. Requiring that the order reversing involution be an "orthocomplement" i.e. for each $a \in L, a \vee a' = \underline{1}$ and $a \wedge a' = \underline{0}$ (in other words the algebra generated by $L_1 \cup L_1'$ is a Boolean algebra) is equivalent to the requirement that for all $a, b \in L_1$, $a \wedge b = \underline{0} \Leftrightarrow a \leq b'$.

PROOF. (i) Suppose that $a \vee a' = \underline{1}$ and $a \wedge a' = \underline{0}$. Then $a \leq b' \Rightarrow a \wedge b \leq b' \wedge b = 0$ and $a \wedge b = \underline{0} \Longrightarrow a = a \wedge (b \vee b') = (a \wedge b) \vee (a \wedge b') = a \wedge b' \leq b'$.

(ii) Take $b' = a$. Then $a \leq a \Longrightarrow a \wedge a' = \underline{0} \Leftrightarrow a' \vee a = \underline{1}$. □

PROPOSITION 7.2. If $(L, \underline{0}, \underline{1}, ')$ and $(M, \underline{0}, \underline{1}, ')$ are two $\mathrm{A}_i\mathrm{B}_j$-lattices with both involutions orthocomplements, then an A_iB_j-morphism $\mu : L \longrightarrow M$ is autoconjugate and monotone on L.

PROOF. Elementary. □

EXAMPLES 7.3.

(i) $L = \mathcal{P}(X)$; X a set; $L_1 = T$, a topology on X (so T is a frame or an A_3B_1-lattice); L_1' the closed subsets of X.

(ii) $L = I^X$; X a set; I a complete lattice with $\underline{0}$ and $\underline{1}$ satisfying Birkhoff's first distributive law (i.e. a complete Heyting algebra),and with an order reversing involution; $L_1 = F$, a fuzzy topology (I-topology) on X (so F is a frame); L_1' the fuzzy-closed subsets of X.

(iii) $L = \mathcal{P}(X)$; X a topological space; L_1 the collection of F_σ in X (a σ-frame or an A_2B_1-lattice); L_1' the collection of G_δ in X.

DEFINITION 7.4. A subfamily B of an A_iB_j-lattice L is a **base** for L if every element in L can be expressed as the supremum of members of a subfamily (for $i = 3$); or a countable subfamily (for $i = 2$), or finite subfamily (for $i = 1$) of B.

7.5 DIFFERENT CONCEPTS OF NORMALITY

Our concept of normality for L_1 (an A_iB_j-sublattice of L) will be

DEFINITION 7.5.1. L_1 is **normal** (in L) iff for every $u, v \in L_1$ with $u' \leq v$ (equivalently $v' \leq u$), there exist $g, h \in L_1$ such that $u' \leq g \leq h' \leq v$.

How does this compare with some of the other notions of normality in the literature? e.g.

SPEED 7.5.2 [Speed 1973]. L_1 is **S-normal** iff for every $a, b \in L_1$ with $a \wedge b = \underline{0}$, there exist $c, d \in L_1$ such that $a \wedge c = b \wedge d = \underline{0}$ and $c \vee d = \underline{1}$.

DOWKER AND PAPERT 7.5.3 [Dowker and Papert 1967] L_1 is **D-P normal** iff for every $u, v \in L_1$ with $u \vee v = \underline{1}$ there exist $g,\ h \in L_1$ such that $u \vee g = \underline{1},\ v \vee h = \underline{1},\ g \wedge h = \underline{0}$.

Clearly, if in L we have $a \wedge b = \underline{0} \Leftrightarrow a \leq b'$ (as happens in $L = \mathcal{P}(X), X$ a set) then L_1 normal $\Leftrightarrow$ L_1 D-P normal $\Leftrightarrow$ L_1' S-normal. Furthermore, for ordinary topological spaces (X, T), (X, T) is a normal topological space $\Leftrightarrow$ the frame T is normal in $\mathcal{P}(X)$ $\Leftrightarrow$ T is D-P normal $\Leftrightarrow$ the family of closed sets is S-normal.

(X, T) is F_σ-normal (see [Speed 1973]) $\Leftrightarrow$ the frame F_σ in X is normal (D-P normal) in $\mathcal{P}(X)$ $\Leftrightarrow$ the family of G_δ in X is S-normal. However, if in L we only have $a \wedge b = 0 \Rightarrow a \leq b'$ (as in the case $L = I^X$ — see Example 7.3 (ii) above), our concept of normality (Definition 7.5.1) cannot be compared with D-P normality and S-normality. (Of course, the latter two are still comparable).

Furthermore, for fuzzy topological (I-topological or generalised topological) spaces (X, F, I), (X, F, I) is normal (see [Hutton 1975]) $\Leftrightarrow$ the frame F is normal in I^X.

7.6 DIFFERENT URYSOHN LEMMAS

Below, $T_{\mathbb{R}}$ denotes the ordinary topology on $\mathbb{R}$ and $T_{[0,1]}$ the corresponding relative topology on $[0,1]$.

THE CLASSICAL URYSOHN LEMMA 7.6.1. The topological space (X,T) is normal iff $u,v \in T$ with $u \vee v = \underline{1} \Rightarrow$ there exists a frame map $\mu : T_{[0,1]} \to T$ such that $u \geq \mu(0,1]$ and $v \geq \mu[0,1)$.

DOWKER-PAPERT VERSION 7.6.2. [Dowker and Papert 1967] If a frame L_1 is D-P normal, then $u,v \in L_1$ with $u \vee v = \underline{1} \Rightarrow$ there exists a frame map $\mu : T_{\mathbb{R}} \longrightarrow L_1$ such that $u \geq \mu(\mathbb{R} \setminus \{0\})$ and $v \geq \mu\,(\mathbb{R} \setminus \{1\})$.

SPEED VERSION 7.6.3. [Speed 1973] A σ-frame L_1 in $L = \mathcal{P}(X)$ is normal/D-P normal iff $u,v \in L_1$ with $u \vee v = \underline{1} \Rightarrow$ there exists a σ-frame map $\mu : T_{[0,1]} \longrightarrow L_1$ such that $u \geq \mu(0,1]$ and $v \geq \mu[0,1)$.

This version has the following result (see [Kotzé 1984]) as a special case: the topological space (X,T) is F_σ-normal iff $u,v \in \{F_\sigma\}$, $u \vee v = \underline{1} \Rightarrow$ there exists a σ-frame map $\mu : T_{[0,1]} \longrightarrow \{F_\sigma\}$ such that $u \geq \mu(0,1]$ and $v \geq \mu[0,1)$.

HUTTON VERSION 7.6.4. [Hutton 1975] If the I-topological space (X,F,I) is normal, then for $u,v \in F$ with $u' \leq v \Rightarrow$ there exists a frame map $\mu : \{R_t, L_t\} \longrightarrow F$ such that $u \geq \mu(R_0)$ and $v \geq \mu(L_1)$ where $\{R_t, L_t\}$ indicates the frame (topology on the fuzzy unit interval) generated by R_t and L_t.

NOTE 7.7.

(i) The maps μ of 7.6.1, 7.6.3 and 7.6.4 are in fact all autoconjugate since they are the inverses of functions.
(ii) The "only if" in 7.6.1 and 7.6.3 follows from the fact that $\mathbf{T}_{[0,1]}$ is sober, and an "only if" will be valid in 7.6.4 if I is Boolean. (Theorem 3.1 above.)

LEMMA 7.8. Let L_1 be an $\mathrm{A}_i\mathrm{B}_j$-sublattice of L as above. Then L_1 is normal iff $u,v \in L_1$ with $u' \leq v$, there exists a **scale** $\{g_p : p \in \mathbf{Q}\} \bigcup \{h_q : q \in \mathbf{Q}\} \subset L_1$ such that $p > q \Rightarrow g_p \leq g_q$, $h_p \geq h_q$, $h'_p \leq g_q$, and $v' \leq g_p \leq h'_p \leq u$ for all $p \in \mathbf{Q}$ (equivalently $u' \leq h_p \leq g'_p \leq v$) where $\mathbf{Q}$ is a countable set dense in $(0,1)$.

PROOF. $\Leftarrow$: Trivial.

$\Rightarrow$: This follows the usual inductive construction which is contained in some form in the proofs of 7.6.1 – 7.6.4. Since L_1 is normal, we have that there exist $g_{r_1}, h_{r_1} \in L_1$ such that $v' \leq g_{r_1} \leq h'_{r_1} \leq u$ $(r_1 \in \mathbf{Q})$. Suppose we have constructed $\{g_{r_1}, \ldots, g_{r_n}, h_{r_1}, \ldots, h_{r_n}\} \subset L_1$ such that $v' \leq \ldots \leq g_p \leq h'_p \leq g_q \leq h'_q \leq \ldots \leq u$, where $p,q \in \{r_1 \ldots r_n\} (p > q)$, then again by normality there exist $g_{r_{n+1}}, h_{r_{n+1}}$ such that $h'_p \leq g_{r_{n+1}} \leq h_{r_{n+1}} \leq g_q$ where r_{n+1} is chosen $> q$. $\square$

LEMMA 7.9. Given L_1 and L_2 frames (in lattices such as L) with B a base for L_1 and φ a map from B into L_2. If φ satisfies

(i) $\varphi \underline{1}_{L_1} = \varphi\{\bigvee b : b \in B\} = \underline{1}_{\mathbf{L}_2}$, and
(ii) if $\{b_j\}$ is a finite subset of B, $\{c_i\} \subset B$ with $\wedge b_j \leq \vee c_i$, then $\wedge \varphi\, b_j \leq \vee \varphi\, c_i$;

then φ can be extended to a frame map $\mu : L_1 \longrightarrow L_2$. If also φ is defined on B' by $\varphi(b') = \varphi(b)'$, then there exists an extension μ such that $\mu(a') = \mu(a)'$ for all a $\in L_1$ (autoconjugate). (Cf. [Dowker and Papert 1967].)

PROOF. Define for $h \in L_1$, $\mu h = \bigvee\{\varphi b : b \leq h,\ b \in B\}$. Then μ is clearly an extension of φ and μ is monotone. If $\{h_j, j = 1 \dots k\}$ is a finite subfamily of L_1, then

$$\begin{aligned}\wedge \mu h_j &= \bigvee_{a \leq h_1} \varphi a \wedge \bigvee_{b \leq h_2} \varphi b \wedge \dots \wedge \bigvee_{c \leq h_k} \varphi c \\ &= \bigvee_{a \leq h_1} \dots \bigvee_{c \leq h_k} \varphi a \wedge \varphi b \wedge \dots \wedge \varphi c \text{ (by the frame distributive law).}\end{aligned}$$

Since $a \wedge \dots \wedge c \leq \wedge h_j = \bigvee_{b \leq \wedge h_j} b$, we have from (ii) that φ a $\wedge \dots \wedge \varphi c \leq \bigvee_{b \leq \wedge h_j} \varphi b = \mu(\bigwedge h_j)$. Thus $\bigwedge \mu h_j \leq \mu(\bigwedge h_j)$. The reverse inequality is trivially true, and so $\bigwedge \mu h_j = \mu(\bigwedge h_j)$. In case $\{h_j\}$ is empty, this is still true since $\mu \underline{1}_{L_1} = \bigvee_{b \leq \underline{1}} \varphi b = \varphi \underline{1}_{\mathbf{L}_1} = \underline{1}_{\mathbf{L}_2}$ from (i).

For any family $\{h_i\}, \mu \vee h_i = \bigvee_{b \leq \bigvee h_i} \varphi b$. When $b \leq \bigvee h_i = \bigvee_i \bigvee\{c : c \leq h_i,\ c \in B\}$, then from (ii) $\varphi b \leq \bigvee_i \bigvee_c \varphi c = \bigvee_i \mu h_i$. Hence $\mu) \bigvee h_i) \leq \bigvee \mu h_i$. Since the reverse inequality is again trivially true we have that $\mu(\bigvee h_i) = \bigvee \mu h_i$. Thus μ is a frame map from L_1 into L_2. If $\varphi(b') = \varphi(b)'$, define μ on L_1' as $\mu h' = \mu(\bigwedge\{b' : b' \geq h'\}) \equiv \bigwedge\{\varphi b' : b' \geq h'\}$. Then $\mu h' = (\mu h)'$ for all $h \in L_1$. □

THEOREM 7.10. L_1 is a normal frame in L iff $u, v \in L_1$ with $u' \leq v \Rightarrow$ there exists a frame map $\mu : T_{[0,1]} \longrightarrow L_1$ such that $u \geq \mu(0,1]$ and $v \geq \mu[0,1)$, $\mu(a') = \mu(a)'$, and μ is monotone on $T_{[0,1]} \cup T'_{[0,1]} = \{\text{sets}\} \cup \{\text{closed sets}\}$ on $[0,1]$.

PROOF. $\Leftarrow$: For $0 < t < 1$ we have $u \geq \mu(0,1] \geq \mu[t,1]$, hence $\mu' \leq \mu[0,t) \leq \mu[0,t] \leq \mu[0,1) \leq v$ and $\mu[0,t) \in L_1$ and $\mu[0,t] \in L_1'$. Thus L_1 is normal.

$\Rightarrow$: We will use the scale of Lemma 7.8 to define a map φ on a base for $\mathbf{T}_{[0,1]}$ and then extend it to a frame map μ on $\mathbf{T}_{[0,1]}$ through Lemma 7.9. So we define for $0 \leq x < 1$, $0 < y \leq 1$, $\varphi(x,1] \equiv \bigvee_{p>x} g_p$, $\varphi[0,y) \equiv \bigvee_{p<y} h_p$ and for $x < y$, $\varphi(x,y) \equiv \varphi(x,1] \wedge \varphi[0,y)$. Furthermore define $\varphi[0,x] = \varphi((x,1]') \equiv (\varphi(x,1])' = \bigwedge_{p>x} g_p'$, $\varphi[y,1] = \varphi([0,y)') \equiv (\varphi[0,y))' = \bigwedge_{p<y} h_p'$, $\varphi[0,1] \equiv \underline{1}$, $\varphi(\phi) = \underline{0}$. Note that $\varphi[0,y] \geq \varphi[0,y)$ and $\varphi[x,1] \geq \varphi(x,1]$ because $\bigwedge_{p>y} g_p' \geq \bigvee_{q<y} h_q$ and $\bigwedge_{q<x} h_q' \geq \bigvee_{p>x} g_p$ since for $p > q$, $g_p' \geq h_q$.

Also $\varphi[0,y_1) \geq \varphi[0,y_2]$ for $y_1 > y_2$: if $y_1 > y_2$ there exist $p < y_1$, $q > y_2$ such that $p > q$ and so $h_p \geq g_q'$. Thus $\bigvee_{p<y_1} h_p \geq g_q' \geq \bigwedge_{q>y_2} g_q'$.

We now show that φ satisfies the conditions required by Lemma 7.9. Of course condition (i) is satisfied by definition of φ. φ is clearly monotone on $B = \{(x,y) : 0 \leq x < y \leq 1\}$

and because of the frame distributive law we have that $\bigwedge \varphi(x_j, y_j) = \varphi(\bigcap(x_j, y_j))$ and $\vee\varphi(x_i, y_i) = \varphi(\bigcup(x_i, y_i))$ for all non-empty finite families $\{b_j = (x_j, y_j)\}$ and all families $\{c_i = (x_i, y_i)\}$ in B. Thus if $\bigcap_j b_j \leq \bigcup_i c_i$, we have that $\bigwedge \varphi b_j \leq \bigvee \varphi c_i$. Even if the family $\{b_j\}$ is empty (i.e. $\bigwedge b_j = [0,1]$) and $\vee c_i = [0,1]$ we have $\bigwedge \varphi b_j \leq \bigvee \varphi c_i$ since $\varphi[0,1] = \underline{1}$. Thus by Lemma 7.9, φ can be extended to a frame map $\mu : T_{[0,1]} \longrightarrow L_1$ such that $\mu(a') = a'$ and is monotone on $T_{\{0,1\}} \cup T'_{[0,1]}$ (since φ is such on $B \cup B'$). Clearly $\mu[0,1) = \bigvee_{p<1} h_p \leq v$ and $\mu(0,1] = \bigvee_{p>0} g_p \leq u$. □

Lemma 7.9 has the following counterpart for σ-frames:

LEMMA 7.11. *Given L_1 and L_2 σ-frames (in L) with B a base for L_1 and φ a map from B into L_2. If φ satisfies*

(i) $\varphi\underline{1}_{L_1} = \varphi(\bigvee\{b_n : b_n \in B\}) = \underline{1}_{L_2}$, and
(ii) if $\{b_j\}$ is a finite subset of B, $\{c_n\}$ a countable subfamily of B with $\bigwedge b_j \leq \bigvee c_n$, then $\wedge\varphi b_j \leq \vee\varphi c_n$;

then φ can be extended to a σ-frame map $\mu : L_1 \longrightarrow L_2$. If also φ is defined on B' by $\varphi(b') = \varphi(b)'$ then there exists an extension μ such that $\mu(a') = \mu(a)'$ for all a $\in L_1$ (autoconjugate).

PROOF. As that of Lemma 7.9, in other words define for h $\in L_1$, $\mu h = \bigvee\{\varphi b_n : b_n \leq h,\ b_n \in B\}$ etc. □

THEOREM 7.12. *L_1 is a normal σ-frame iff $u, v \in L_1$ with $u' \leq v \Rightarrow$ there exists a σ-frame map $\mu : T_{[0,1]} \longrightarrow L_1$ such that $u \geq \mu(0,1]$ and $v \geq \mu[0,1)$, $\mu(a') = \mu(a)'$ and μ is monotone on $T_{[0,1]} \cup T'_{[0,1]} = \{\text{open sets}\} \cup \{\text{closed sets}\}$ on $[0,1]$.*

PROOF. Similar to that of Theorem 7.10. □

Lemma 7.9 clearly has its counterparts for all $\mathrm{A}_i\mathrm{B}_j$-lattices ($i, j = 1, 2$ or 3):

LEMMA 7.13. *Given L_1 and L_2 two $\mathrm{A}_i\mathrm{B}_j$-lattices with B a base for L_1 and φ a map from B into L_2 which satisfies*

(i) $\varphi\underline{1}_{L_1} = \varphi(\bigvee\{b : b \in B\}) = \underline{1}_{L_2}$ and
(ii) $\bigwedge b_j \leq \bigvee c_i \Rightarrow \bigwedge \varphi b_j \leq \vee\varphi c_i$;

then φ can be extended to an $\mathrm{A}_i\mathrm{B}_j$-morphism from L_1 to L_2.

In condition (i), the supremum should be interpreted as a countable or finite one in case of $\mathrm{A}_2\mathrm{B}_j$ or $\mathrm{A}_1\mathrm{B}_j$ lattices respectively.

In condition (ii), $\{b_j\}$ should be interpreted as being finite/countable/without countability restriction depending on whether we are dealing with two $\mathrm{A}_i\mathrm{B}_j$-lattices with $j =$

1/2/3 respectively, and $\{c_i\}$ as being finite/countable/without countability restriction depending on whether we are dealing with two A_iB_j-lattices with $i = 1/2/3$ respectively.

Again we can deduce from Lemma 7.13 as a general Urysohn Theorem:

THEOREM 7.14. L_1 is a normal (Definition 7.5.1) A_iB_j-sublattice of an A_iB_j-lattice L iff $u, v \in L_1, u' \leq v \Rightarrow$ there exists an autoconjugate and monotone A_iB_j-morphism μ from the A_iB_j-lattice generated by $T_{[0,1]} \cup T'_{[0,1]}$ into L_1 such that $u \geq \mu(0,1]$ and $v \geq \mu[0,1)$.

NOTE 7.15.

(i) If the involution of L is an orthocomplement, i.e. $a \wedge b = 0 \Leftrightarrow a \leq b'$, then in view of Proposition 7.2, Theorem 7.10 simply reads: L_1 is a normal frame in L iff $u, v \in L$, with $u \vee v = \underline{1} \Rightarrow$ there exists a frame map $\mu : T_{[0,1]} \longrightarrow L_1$ such that $u \geq \mu(0,1]$ and $v \geq \mu[0,1)$.

(ii) Theorem 7.10 is the counterpart of the Dowker-Papert result for frames in lattices with an order reversing involution, and includes the Dowker-Papert result if this involution satisfies $a \wedge b = 0 \Longleftrightarrow a \leq b'$.

(iii) Theorem 7.12 generalises that of Speed (7.6.3 above).

NOTE 7.16. Theorem 7.10 obviously translates for I-topological spaces (X, F, I) with F a frame in I^X to:

(1) An I-topological space is normal iff $u, v \in F$ with $u' \leq v \Rightarrow$ there exists a frame map $\mu : T_{[0,1]} \longrightarrow F$ such that $u \geq \mu(0,1]$ and $v \geq \mu[0,1)$, $\mu(a') = \mu(a)'$, and μ is monotone on $T_{[0,1]} \cup T'_{[0,1]}$.

(2) If however I is a Boolean algebra (in other words, its involution is an orthocomplement) it reduces to (Note 7.15 (i) above): (X, F, I) is normal iff $u, v \in F$ with $u \vee v = \underline{1} \Rightarrow$ there exists a frame map $\mu : T_{[0,1]} \longrightarrow F$ such that $u \geq \mu(0,1]$ and $v \geq \mu[0,1)$.

This latter statement is equivalent to the Hutton result (7.6.4 above) for I a Boolean algebra in view of Note 7.7 (ii) and the fact that Hutton proved ([Hutton 1975], Theorem 3) that if I has an orthocomplement, then there exists a bijective frame map $\varphi : T_{[0,1]} \longrightarrow \{R_t, L_t\}$. Indeed in terms of the terminology of that proof $\varphi(0,1](\lambda) = \varphi(0,+\infty)(\lambda) = \lambda(0+) \wedge \lambda(+\infty-)' = \lambda(0+) = R_0(\lambda)$ and $\varphi[0,1)(\lambda) = \varphi(-\infty,1) = \lambda(-\infty+) \wedge \lambda(1-)' = \lambda(1-)' = L_1(\lambda)$. Now take $\mu = \tilde{\mu}_0\varphi$ where $\tilde{\mu}$ is the frame map of Hutton (7.6.4).

Such a situation arises in the case of lower semi-continuous set-valued functions: If X and Y are (ordinary) topological spaces, then

DEFINITION 7.17. A set valued function $f : X \longrightarrow \mathcal{P}(Y)$ is **lower semi-continuous** (respectively **upper semi-continuous**) iff for every G open (closed) in Y, $f^{\leftarrow}G = \{x \in X : f(x) \cap G \neq \emptyset\}$ is open (closed) in X. (Note that this differs from Kuratowski's [Kuratowski 1970] definition in so far as that we do not restrict the range of f to the closed subsets of Y.)

Now in the lattice $\mathcal{P}(Y)$

$$\begin{aligned} \bigwedge f_i(x) &= \bigcap f_i(x) \text{ and } \bigvee f_i(x) = \bigcup f_i(x) \\ \text{and } (\bigwedge f_i)^{\leftarrow} G &= \{x \in X : \bigcap_i f_i(x) \cap G \neq \emptyset\} \\ &= \bigcap_i \{x \in X : f_i(x) \cap G \neq \emptyset\} \\ (\bigvee f_i)^{\leftarrow} G &= \{x \in X : \bigcup_i f_i(x) \cap G \neq \emptyset\} \\ &= \bigcup_i \{x \in X : f_i(x) \cap G \neq \emptyset\} \end{aligned}$$

It is then clear that

PROPOSITION 7.18. The collection of all lower semi-continuous functions is a subframe of $P(y)^x$, i.e. is an I-topology on X where $I = \mathcal{P}(Y)$.

Since $\mathcal{P}(Y)$ is Boolean we have by 7.16 above:

THEOREM 7.19. The collection F of lower semi-continuous functions $X \longrightarrow \mathcal{P}(Y)$ is normal iff $f, g \in F$ with $f(x) \cup g(x) = Y$ for all $x \in X \Rightarrow$ there exists a frame map $\mu : T_{[0,1]} \longrightarrow F$ such that $f \geq \mu(0,1]$ and $g \geq \mu[0,1)$.

Chapter 11:

The Topological Modification of the L-fuzzy Unit Interval

S. E. Rodabaugh et al. (eds.), Applications of Category Theory to Fuzzy Subsets, 275–305.

THE TOPOLOGICAL MODIFICATION OF THE L-FUZZY UNIT INTERVAL

T. Kubiák

§0. Introduction

In 1975, Hutton invented the L-fuzzy unit interval $I(L)$, a very fruitful concept which enabled him to characterize the normal L-fuzzy topological spaces in terms of Urysohn's type lemma and to introduce a very efficient complete regularity axiom for L-fuzzy spaces (see [Hutton 1975 1977]). Although the appropriateness of $I(L)$, as a space which plays the same role in fuzzy topology as the real unit interval plays in general topology, has been proved by a number of other authors (e.g. [Liu 1983], [Liu and Luo 1989], [Rodabaugh 1983b, 1988a], [Kubiak 1987, 1986]), the structure of this canonical L-fuzzy space is not yet well-understood. A survey of the results on the L-fuzzy unit interval that have been obtained up to 1982 was made by [Rodabaugh 1982a]. That paper, however, contains more open questions than results.

The present chapter has been stimulated by some of those questions concerning the properties of the natural L-fuzzy topology of $I(L)$ and its topological modification. As a matter of fact, we do not solve herein, in principle, any of those questions (unless the underlying lattice L is a complete chain). Strictly speaking, we do that but with all the involved concepts redefined (including a-compactness, strong compactness, ultracompactness, etc; only (c1)-connectedness of [Lowen 1981b] and 1-connectedness of [Rodabaugh 1982b] need not be redefined). After [Gantner *et al.* 1978] those notions involve the strict less-than relation. This relation is useful when L is a complete chain. In the context of a non-linearly ordered lattice, however, this relation is very far from being a useful tool. One illustration of this statement is that there is really no significant result in the literature involving the relation $<$ and proved without the restriction to the subsets $L^c = \{x \in L : x$ is comparable to each member of $L\}$, $L^b = \{x \in L : x < y$ and $x < z$ imply $x < y \wedge z\}$, or $L^a = L^b \cap L^c$ (all that is not very far removed from the hypothesis that L is itself a chain); see for instance [Gantner *et al.* 1978] and [Rodabaugh 1980]. One natural way to overcome these restrictions is to replace the relation $<$ by the relation $\not\geq$. This idea of course is not new. The relation $\not\geq$ has been successfully used by some workers of the Chinese school (see [Zhao 1987] and the papers of Liu and Luo just mentioned).

Here is an example which shows how one can take advantage of replacing $<$ by $\not\geq$. An L-fuzzy topological space X is called a-compact in [Gantner *et al.* 1978] (resp., (pf, a)-compact in [Rodabaugh 1982a]), where a is an element of the designated lattice, if whenever $\mathcal{V}$ is a family of open L-fuzzy sets of X such that for each $x \in X$ there exists $U \in \mathcal{V}$ with $U(x) > a$ (resp., $\bigvee \mathcal{V} > \mathbf{a}$), there is a finite subfamily of $\mathcal{V}$ with the same property. Note that a-compactness is equivalent to (pf, a)-compactness if $a \in L^c$. A result of [Gantner *et al.* 1978] states that $I(L)$ is a-compact provided that $a \in L^a$. After rewriting both these compactness notions in terms of the relation $\not\leq$ (cf. 2.3), $I(L)$ becomes a-compact for each $a \in L$ (see 4.1-4.5 and 7.6), and a-compactness is then equivalent to (pf, a)-compactness for each $a \in L$, too. It is also worth mentioning that although these two sorts of a-compactness generally differ, the corresponding statements about $I(L)$ are comparable and that of [Gantner *et al.* 1978] trivially follows from the second one. (The same situation will take place after possibly re-proving the other results originally related to the strict less-than relation and the set L^c and its subsets.)

Besides revisiting some compactness notions, the objective of this chapter necessitates a generalization of the Lowen functors ω and ι to a lattice L (endowed with some topology) other than the real unit interval with its right-hand topology. We will do this in a manner different from that of [Rodabaugh 1982, 1988a], where the role which the unit interval played in Lowen's work is taken by an arbitrary complete lattice with a topology generated by the subbase $\{\uparrow(a) \setminus \{a\} : a \in L\}$ (thus, Rodabaugh's approach involves the strict less-than relation and emphasizes connections with all those L-fuzzy topological notions that are defined in terms of this relation). However the link between the so extended Lowen functors remains valid if L is a complete chain (note that ω has to be at least a right inverse of ι and possibly a left adjoint of it). To be compatible with the concepts redefined in terms of $\not\leq$ (as well as for other reasons), our choice is a complete lattice with its upper topology (it is the one generated by the sets $L \setminus \downarrow(a)$). If L is completely distributive (or more generally, if it is complete and each element is an inf of prime elements), then all the categorical aspects of the original Lowen functors remain valid. This is the approach initiated in [Kubiak 1988a].

There is another direction in which we can generalize the Lowen functors. It has been available for many years in [Gierz *et al.* 1980] and has been recently resurrected by [Warner 1990]. She replaced the unit interval by a continuous frame with its Scott topology (cf. 3.11(2)). It is evident how to define the functor ι (which is not present in her paper) as to be compatible with her extended ω. Then one can show (the proofs of Section 3 go over this case) that all the features of Lowen's machinery remain valid. It deserves to be pointed out that Warner's approach and the one of [Kubiak 1988b], i.e. that we shall use herein, differ as long as we talk only about open L-fuzzy sets and not about the closed ones, that is when the underlying lattice is not supposed to have an order-reversing involution. If it does, then a continuous frame becomes completely distributive, and its Scott topology becomes the upper topology. Furthermore, note that the definition of the natural L-fuzzy topology of $I(L)$ needs to possess an order-reversing involution (even if closed L-fuzzy sets are thrown away). This along with what we said above justifies the presence of the complete distributive law in this paper. As each completely distributive lattice is continuous, we shall extensively use the so-called **way below** relation which proves to be yet another relation that can be taken to play the role of the strict less-than relation.

The present chapter is a much expanded and improved version of the unpublished preprint [Kubiak 1988b], which in turn extends [Kubiak 1988a] from the lattice being the unit interval or a complete chain to an arbitrary completely distributive lattice. Both these papers were incomplete since they did not establish a precise description of the topological modification of $I(L)$. In the present chapter we close this gap.

The contents of this chapter are arranged as follows. In Section 1 we present a brief review of some relevant basic concepts concerning continuous lattices and completely distributive ones. Section 2 collects information concerning L-fuzzy topological spaces. Section 3 discusses how to create a proper setting for a generalization of Lowen functors. It is not our aim in this section to study these functors for their own sake, therefore it contains only those results and observations that we shall need in what follows. In Section 4 we redefine some L-fuzzy topological axioms of compactness as to be in tune with our new ultracompactness axiom and with the new Lowen functor ω. Also, we simplify the concept of N-compactness of [Zhao 1987] (and call it Ch-compactness) and extend the fuzzy compactness of [Lowen 1976] to a general lattice setting (the latter is done in a manner similar to but different from that of [Rodabaugh 1982b]). An evidence of the appropriateness of all those new definitions is provided by the following: ultracompactness $\Rightarrow$ Ch-compactness $\Rightarrow$ strong compactness (= a-compactness for each $a \in L$) $\Rightarrow$ fuzzy compactness, which was previously known for the case when L is the real unit interval (see [Lowen 1978] and [Wang 1983]). Note also that all the proposed axioms coincide with the ordinary compactness axiom via the generalized functor ω (good extensions in the sense of [Lowen 1978]), and clearly via the functor taking open sets to their characteristic functions. In Section 5 we introduce a lattice-valued variant of the Helly space, i.e. the subspace of the cube L^I consisting of all order-reversing mappings with L carrying the interval topology. In Section 6 we deal with a quotient of the Helly space resulting via identifying its members having the same right-hand limits at each point. The underlying quotient set is clearly the L-fuzzy unit interval and the resulting quotient topology (which we call the natural topology of $I(L)$) is compact Hausdorff and connected (the latter holds if L is connected in its interval topology). The L-fuzzy topology topologically generated from the natural topology of $I(L)$ proves to be strictly finer than the natural L-fuzzy topology of $I(L)$ if $L \neq \{0,1\}$. A general conclusion is that the L-fuzzy unit interval has all the properties that are good extensions of the properties of the natural topology and that are preserved under weakening of an L-fuzzy topology. On the other hand, the natural topology of $I(L)$ is precisely the topological modification of the natural L-fuzzy topology (= the initial topology for all the open L-fuzzy sets and L with the upper topology). In particular, this shows that $I(L)$ is ι-Hausdorff (ultra-Hausdorff in the customary non-adequate terminology). In spite of the fact that everything becomes clear after Section 6, in Section 7 we list all the properties of $I(L)$ (and of the L-fuzzy Tychonoff cube) which follow via the information about the natural topology of $I(L)$. We also record a theorem of [Kubiak 1990a] that the topological modification of $I(L)$ is an absolute extensor for normal topological spaces provided L is the unit interval.

An earlier version of this chapter (whose abstract is [Kubiak 1989]) was read at the

Eleventh International Seminar on Fuzzy Set Theory, Applications of Category Theory to Fuzzy Subsets, held in Linz, September 11-15, 1989. The author would like to thank the organizers, Professors E. P. Klement and S. E. Rodabaugh, for the opportunity of presenting it. The chapter was completed and presented in its final form at Rhodes University in Grahamstown, during a leave from Adam Mickiewicz University of Poznań. Special thanks go to Professor W. Kotzé. Thanks go also to the various members of his Seminar on Fuzzy Topology, notably Dr. J. J. Chadwick who contributed a counterexample I had been looking for. The author gratefully acknowledges support through the South African Council for Scientific and Industrial Research and the Hugh Kelly Fellowship from Rhodes University.

§1. Some Background in Lattices

Our general source of reference is [Gierz *et al.* 1980]. For an easy reference we shall reproduce here some definitions, notation and results that are relevent for our purposes.

Completeness of each lattice $L = (L, \leq)$ is assumed from the beginning. The bounds of any lattice will be denoted by 0 and 1. The two-point lattice will be denoted by $2 = \{0, 1\}$. The real unit interval is denoted by I. To each L there is associated the opposite lattice L^{op} in which $a \leq b$ iff $b \leq a$ in L. If $A \subset L$, then we write $\bigvee A = \bigvee\{a \in L : a \in A\}$ and $\bigwedge A = \bigwedge\{a \in L : a \in A\}$. For any element $a \in L$ we let $\downarrow(a) = \{x \in L : x \leq a\}$ and $\uparrow(a) = \{x \in L : a \leq x\}$.

On each complete lattice L one defines a new transitive relation $\ll$ as follows: $x \ll y$ (one says that x is **way below** y) whenever for any $A \subset L$ with $y \leq \bigvee A$ there exists a finite subset $B \subset A$ with $x \leq \bigvee B$. The relation $x \ll y$ implies $x \leq y$. Also, $0 \ll x$ always. We shall also make unacknowledged use of the following two properties: $u \leq x \ll y \leq z$ implies $u \ll z$; $x \ll z$ and $y \ll z$ imply $x \vee y \ll z$. For any element $a \in L$ we write $\Uparrow(a) = \{x \in L : a \ll x\}$ and $\Downarrow(a) = \{x \in L : x \ll a\}$. Note that in I the way below relation becomes the strict less-than relation. Also note that in a complete lattice $x \ll y$ may never hold unless $x = 0$.

A **continuous lattice** is a complete lattice L in which $a = \bigvee \Downarrow(a)$ for each $a \in L$ or, equivalently, in which one has the following property:

$$\text{(C)}\quad a \not\leq b \text{ implies there exists } x \in L \text{ with } x \ll a \text{ and } x \not\leq b$$

A **completely distributive** lattice is a complete lattice L in which

$$\text{(CD)}\quad \bigwedge\{\bigvee a_{st} : t \in T_s\} : s \in S\} = \bigvee\{\bigwedge\{a_{sf(s)} : s \in S\} : f \in \Pi_{s \in S} T_s\}$$

for every family $\{a_{st} : s \in S,\ t \in T_s\} \subset L$.

If L is completely distributive, so is L^{op}, i.e. the dual of (CD) is valid in L. Every complete chain is completely distributive. Every completely distributive lattice is continuous.

An element $p \in L$ is called **prime** whenever $x \wedge y \leq p$ implies $x \leq p$ or $y \leq p$. The set of all primes is denoted PRIME L. An element q is called **coprime** if it is prime in $L^{\rm op}$ (that is, if $q \leq x \vee y$ implies $q \leq x$ or $q \leq y$). All the coprimes in L are denoted COPRIME L. We say that the set PRIME L is **order generating** iff each $a \in L$ is the inf of primes, i.e. if $a = \bigwedge(\uparrow(a) \cap \text{PRIME } L)$.

We record the following facts from [Gierz *et al.* 1980]:

THEOREM 1.1. (I-3.14, *op. cit.*). In each continuous distributive lattice the primes order generate.

In fact, by I-3.13, *op. cit.*, each complete lattice L in which PRIME L is order generating is necessarily a frame, i.e. it satisfies the infinite distributive law

$$\text{(ID)} \quad a \wedge \bigvee B = \bigvee\{a \wedge b : b \in B\} \text{ for each } a \in L \text{ and } B \subset L$$

Thus each continuous distributive lattice is a continuous frame.

THEOREM 1.2. (I-3.15, *op. cit.*). For L a complete lattice the following statements are equivalent:

(1) L is continuous and $a = \bigvee\{\downarrow(a) \cap \text{ COPRIME } L\}$ for all $a \in L$,
(2) L is completely distributive,
(3) L is distributive and both L and $L^{\rm op}$ are continuous.

We now review quickly some of the most important among the natural topologies existing on a complete lattice L:

(1) The **upper topology** SUP(L) generated by the subbasic sets $L \setminus \downarrow(a)$.
(2) The **lower topology** INF(L) generated by the subbasic sets $L \setminus \uparrow(a)$.
(3) The **interval topology** INT(L) = SUP(L) $\vee$ INF(L), that is the one generated by SUP(L) $\cup$ INF(L).

For each complete lattice L the family $\{\Uparrow(a) : a \in L\}$ is a base of a topology (one easily sees it is closed under finite intersections). For L a continuous lattice we have another two topologies:

(4) The **Scott topology** $\sigma(L)$ generated by the basic sets $\Uparrow a$.
(5) The **Lawson topology** $\lambda(L) = \sigma(L) \vee$ INF(L).

For the definition of the Scott topology on an arbitrary complete lattice, see [Gierz *et al.* 1980, page 99]. We shall mention it in 3.10 but not use it explicitly.

The following holds on account of IV-2.31 and III-3.23 of [Gierz *et al.* 1980]:

THEOREM 1.3. If L is a completely distributive lattice, then $\sigma(L)$ = SUP(L) and, thus, $\lambda(L)$ = INT(L).

REMARK 1.4. We shall also refer to the following facts. Each complete lattice is compact in the interval topology ([Johnstone 1982, page 280]). If L is continuous, then $(L, \lambda(L))$

is compact Hausdorff ([Gierz *et al.* 1980, III-1.10]). In particular, by 1.3, each completely distributive lattice is compact Hausdorff in its interval topology (for a direct proof see [Johnstone 1982, page 280]).

We say that L has a **quasi-complementation** provided there exists an order-reversing map $' : L \longrightarrow L$ which is involutory, i.e. $a'' = a$ for each $a \in L$. Following [Rodabaugh 1986-7], we shall call $(L, \ ')$ a **de Morgan algebra** (note that in the usual terminology distributivity is imposed as well). In each complete de Morgan algebra one has the complete de Morgan laws: $(\bigvee A)' = \bigwedge\{a' : a \in A\}$ and its dual. Also, $0' = 1$ and $1' = 0$. When $L = I$, the quasi-complementation is understood as $a' = 1 - a$ for each $a \in I$. Note that the concept of a quasi-complementation is fairly special. It is not in general an internal notion, that is, it cannot be generally defined in terms of the ordering. In a lattice there may exist one or more quasi-complementations, or none at all (the latter may even be the case of an infinite chain). If L has a quasi-complementation, then clearly $L \cong L^{\mathrm{op}}$. By 1.2 we thus immediately have the following:

COROLLARY 1.5. Each continuous and distributive de Morgan algebra is completely distributive.

§2. Basic Concepts of Fuzzy Topology

All the information mentioned here is standard. We will therefore not cite references. The purpose of this section is just to establish notation and some terminology.

For a given de Morgan algebra $(L, \ ')$ and a non-empty set X, $(L^X, \ ')$ is a complete de Morgan algebra of all maps from X to L under pointwisely defined ordering and quasi-complementation. Members of L^X are called **L-fuzzy sets** (or just **L-sets**) in X. The constant member of L^X whose constant value is $a \in L$ is denoted by **a**. Note that $2^X = \{1_A : A \subset X\} \subset L^X$ (1_A denotes the characteristic function of A).

With each map $f : X \longrightarrow Y$ there is associated the complete lattice homomorphism $(\cdot)f : L^Y \longrightarrow L^X$ acting by the composition with f. This extends the usual inverse image map since $1_A f = 1_{f^{-1}(A)}$ for every $A \subset Y$.

The pair (X,T) is an **L-fuzzy topological space** (**L-fuzzy space** or just **L-space**) and T is an **L-fuzzy topology** (or **L-topology**) on X if $T \subset L^X$ is closed under arbitrary sups and finite infs, the sups and the infs understood in the usual pointwise order of L^X (so **0** and **1** belong to T). Members of T are called **open**, and their quasi-complements are called **closed**. We write T' for the family of all closed L-sets. If $\mathbf{a} \in T$ for each $a \in L$, then (X,T) is called **stratified**.

A mapping $f : (X,T) \longrightarrow (Y,S)$ is called **continuous** whenever $Uf \in T$ for each $U \in S$. All L-fuzzy spaces together with continuous mappings form a category called **TOP**(L) (with composition of mappings as composition of morphisms). The most straightforward embedding of **TOP** into **TOP**(L) is the one which assigns to each (X,T) in $|$**TOP**$|$ the

L-fuzzy space $(X, \chi(T))$ with $\chi(T) = \{1_U : U \in T\}$, and which preserves underlying mappings.

The set of all L-topologies on a set X is a complete lattice under inclusion as ordering. For $\mathcal{F}$ a family of L-topologies, $\bigvee \mathcal{F}$ is **generated** by $\bigcup \mathcal{F}$, that is $\bigvee \mathcal{F}$ is the inf (= intersection) of all those L-topologies on X that contain $\bigcup \mathcal{F}$. Also, we say that $S \subset L^X$ **generates** an L-topology T on X, and write $T = \ll S \gg$ (the same notation applies for ordinary topologies), if T is the inf of those L-topologies on X that contain the family S. The following was also observed in [Rodabaugh 1986-7], second edition.

REMARK 2.1. With some abuse of terminology one calls S (as well as $\bigcup \mathcal{F}$) a **subbase** of T (of $\bigvee \mathcal{F}$). Note that if L is not a frame, then the family of all finite infs of members of S need not be a **base** of T, i.e. T is not generally the collection of all sups of those finite infs of members of S.

For (Y, S) an L-space, the weakest L-topology on X making $f : X \longrightarrow Y$ continuous is denoted $f^{\leftarrow}(S)$ $(= \{Uf : U \in S\})$. The same notation applies for ordinary topologies. If $\{(X_\gamma, T_\gamma)\}_\gamma$ is a family in $|\,\mathbf{TOP}(L)\,|$, then $(\Pi_\gamma X_\gamma,\ \Pi_\gamma T_\gamma)$, where $\Pi_\gamma T_\gamma = \bigvee_\gamma \pi_\gamma^{\leftarrow}(T_\gamma)$, is the categorical product (as usual, π_γ denotes the γ-th projection). Also, for (X, T) in $|\,\mathbf{TOP}(L)\,|$ and $A \subset X$, the set of all restrictions $\{U \mid A : U \in T\}$ is an L-topology on A called the **subspace L-topology** and is denoted by T_A. The same notation will be used to denote an ordinary subspace topology. For T an L-topology on X, $T^c = T \vee \{\mathbf{a} : a \in L\}$ is called the **stratification** of T.

In addition to those remarks in 2.1 about the terms "subbase" and "base" under absence of (ID), it is noteworthy that characterizing continuity in terms of the inverse images of members of a set that generates the L-topology of the codomain does not require the underlying lattice to be a frame (the wrong thinking that it does, was the only reason that [Kubiak 1987] has been written down under the assumption that the underlying lattice is a frame and not an arbitrary complete lattice). It appears that completeness suffices as stated in the next proposition; this result appears without proof in the second edition of [Rodabaugh 1986-7], and its proof was shown to me by U. Höhle in Linz (1989).

PROPOSITION 2.2. Let L be a complete lattice, let (X, T) and (Y, S) be in $|\,\mathbf{TOP}(L)\,|$, and let $S = \ll S_0 \gg$. Then $f : (X, T) \longrightarrow (Y, S)$ is continuous iff $Vf \in T$ for each $V \in S_0$.

PROOF. We must show only the *if* part. Since $S_1 = \{U \in L^Y : Uf \in T\}$ is the strongest L-topology on Y making f continuous from (X, T) to Y, hence $S_0 \subset S_1$, and thus $S \subset S_1$. Therefore f is continuous from (X, T) to (Y, S). □

NOTATION 2.3. If $A \in L^X$ and $a \in L$, then $A \not\leq \mathbf{a}$ iff $A(x) \not\leq a$ for **all** $x \in X$ (i.e. $A \not\leq \mathbf{a}$ does not mean the negation of $A \leq \mathbf{a}$). A similar convention applies to $A \not\geq \mathbf{a}$.

Finally note that if $A \in L^X$ and $a \in L$, then $[A \not\leq \mathbf{a}] = A^{-1}(L \setminus \downarrow(a)$, $[A \gg \mathbf{a}] = A^{-1}(\uparrow(a))$ etc. As usual , if (X, T) and (Y, S) are in $|\,\mathbf{TOP}\,|$, then $C((X, T), (Y, S))$ is the collection of all continuous mappings between them.

§3. Lowen Functors

The objective of this section is to extend the Lowen functors for lattices other than complete chains, in particular the unit interval, and to show their basic properties which we shall use in what follows. Our discussion will show that completely distributive lattices with their upper topologies are undoubtedly the appropriate candidates to play the role of $(I, \mathrm{SUP}(I))$, especially when the underlying lattice has to be a de Morgan algebra. Our reference for categorical notions and results is [Mac Lane 1971].

DEFINITION 3.1. Let L be a complete lattice. For each $(X,T) \in |\mathbf{TOP}|$ define an L-fuzzy space $(X, \omega_L(T))$ with

$$\omega_L(T) = \ll C((X,T),(L,\mathrm{SUP}(L))) \gg$$

Given $(Y,S) \in |\ \mathbf{TOP}(L)|$ we define a topological space $(Y, \iota_L(S))$ with

$$\iota_L(S) = \bigvee\{V^{\leftarrow}(\mathrm{SUP}(L)) : V \in S\}.$$

PROPOSITION 3.2. For each complete lattice L, $\omega_L : \mathbf{TOP} \longrightarrow \mathbf{TOP}(L)$ is a functor which takes each (X,T) to $(X,\omega_L(T))$ and which leaves mappings unchanged, and ι_L : $\mathbf{TOP}(L) \longrightarrow TOP$ is a functor which preserves mappings too, and takes each (Y,S) to $(Y, \iota_L(S))$.

PROOF. That $f : (X,\omega_L(T)) \longrightarrow (X_1,\omega_L(T_1))$ in $\mathbf{TOP}(L)$ whenever $f : (X,T) \longrightarrow (X_1,T_1)$ in $\mathbf{TOP}$ follows on account of 2.2, by the fact that each subbasic open L-set of $\omega_L(T_1)$ composed with f is subbasic open in $\omega_L(T)$. The case of ι_L is trivial as well. For if $f : (Y,S) \longrightarrow (Y_1,S_1)$ in $\mathbf{TOP}(L)$ and U is subbasic open in $\iota_L(S_1)$, i.e. $U = [V \not\leq \mathbf{a}]$ for some $V \in S_1$ and $a \in L$, $f^{-1}(U) = [Vf \not\leq \mathbf{a}] \in \iota_L(S)$ since $Vf \in S$. Thus $f : (Y,\iota_L(S)) \longrightarrow (Y,\iota_L(S_1))$ in $\mathbf{TOP}$. □

Recall that if $L = I$ one has the original functors introduced by [Lowen 1976]. An important feature of this case is that $\omega_I(T)$ consists then precisely of all continuous maps from (X,T) to $(I,\ \mathrm{SUP}(I))$, i.e. of all lower semi-continuous maps from X to I. This is, for instance, crucial in proving $\iota_I\omega_I = 1_{TOP}$. For an arbitrary complete lattice L, $C((X,T),(L,\ \mathrm{SUP}(L)))$ is closed under arbitrary sups, but not necessarily under finite infs. One sufficient condition for the latter is that L is completely distributive, since the inf operation is then continuous for the upper topology. This fact is implicit in the proof of Proposition VII-1.12 of [Johnstone 1982] (asserting that a completely distributive L is a topological lattice for the interval topology). We wish to record the proof here, as it refers only to (CD). Note that the sup operation is continuous with respect to the upper topology for any complete lattice (obvious and irrelevant for our purposes).

PROPOSITION 3.3. Let L be a completely distributive lattice. Then $\wedge : (L,\mathrm{SUP}(L)) \times (L,\mathrm{SUP}(L)) \longrightarrow (L,\mathrm{SUP}(L))$ is continuous.

PROOF (as in [Johnstone 1982]). For a given $a \in L$, let $s(a) = \{(b_0, b_1) \in L^2 : b_0 \wedge b_1 \leq a\}$. For each $f \in 2^{s(a)}$ we write $y(f) = \bigvee\{b_{f(b_0,b_1)} : (b_0, b_1) \in s(a)\}$. By the dual of (CD) we then have that

$$\begin{aligned} a &= \bigvee_{(b_0,b_1) \in s(a)} b_0 \wedge b_1 \\ &= \bigvee_{(b_0,b_1) \in s(a)} \bigwedge_{i \in 2} \pi_i(b_0, b_1) \\ &= \bigwedge_{f \in 2^{s(a)}} \bigvee_{(b_0,b_1) \in s(a)} \pi_{f(b_0,b_1)}(b_0, b_1) \\ &= \bigwedge_{f \in 2^{s(a)}} y(f). \end{aligned}$$

Now, $b_0 \wedge b_1 \leq a$ iff either $b_0 \leq y(f)$ or $b_1 \leq y(f)$ for each $f \in 2^{s(a)}$. Indeed, if $b_0 \wedge b_1 \leq a$, then $b_{f(b_0,b_1)} \leq y(f)$ for each $f \in 2^{s(a)}$, and $b_{f(b_0,b_1)}$ is either b_0 or b_1. The converse is obvious. Therefore

$$\begin{aligned} \wedge^{-1}(\downarrow(a)) &= \{(b_0, b_1) \in L^2 : b_0 \leq y(f) \text{ or } b_1 \leq y(f) \text{ for all } f \in 2^{s(a)}\} \\ &= \bigcap_{f \in 2^{s(a)}} ((\downarrow(y(f)) \times L) \cup (L \times \downarrow(y(f))), \end{aligned}$$

which is closed in L^2 for each $a \in L$. □

PROPOSITION 3.4. With L completely distributive, $\omega_L(T)$ consists precisely of all continuous mappings from (X,T) to $(L, \mathrm{SUP}(L))$.

PROOF. We must show that $C((X,T),(L,\mathrm{SUP}(L)))$ is closed under arbitrary sups and finite infs (formed in L^X). For any $\{U_\gamma\}_\gamma \in L^X$ one has $[\bigvee_\gamma U_\gamma \not\leq \mathbf{a}] = \bigcup_\gamma [U_\gamma \not\leq \mathbf{a}]$ for each a in L (only completeness involved). Closedness under finite infs follows immediately from 3.3. □

PROPERTY 3.5. Let $(X,T) \in |\mathbf{TOP}|$ and $(Y,S) \in |\mathbf{TOP}(L)|$. Then:

(1) $S \subset \omega_L \iota_L(S)$,
(2) $T = \iota_L \omega_L(T)$ if L is completely distributive.

PROOF. We shall only show that $\iota_L \omega_L(T) \subset T$ in order to emphasize the use of 3.4. For if $V \in \omega_L(T)$, then $V^{\leftarrow}(\mathrm{SUP}(L)) \subset T$, hence the assertion follows. □

PROPOSITION 3.6. If L is completely distributive, then ω_L is a left adjoint right inverse of ι_L.

PROOF. Since both ω_L and ι_L preserve mappings, ω_L is a right inverse of ι_L by 3.5(2). Again, as these functors preserve mappings, the statement that $f : (X, \omega_L(T)) \longrightarrow (Y,S)$ in $\mathbf{TOP}(L)$ iff $f : (X,T) \longrightarrow (Y, \iota_L(S))$ in TOP yields a natural bijection between the hom-sets. To verify the *if* part of this claim, act with ω_L and use 3.5(1). For the converse act with ι_L and apply 3.5(2). □

REMARK 3.7. Complete distributivity of L is not necessary for 3.3 and, thus, for 3.4, 3.5(2) and 3.6 as well. The equality $\omega_L(T) = C((X,T),(L,\mathrm{SUP}(L)))$ still holds if L is a complete lattice in which PRIME L is order generating. Indeed, then for each $a \in L$ we have

$$\begin{aligned}\downarrow(a) &= \downarrow(\bigwedge(\uparrow(a) \cap \text{PRIME } L)) \\ &= \{b \in L : b \leq p \text{ for each prime } p \geq a\} \\ &= \bigcap\{\downarrow(p) : p \geq a \text{ and } p \in \text{PRIME } L\},\end{aligned}$$

and therefore $\mathrm{SUP}(L) = \ll \{L\setminus \downarrow(p) : p \in \text{PRIME } L\} \gg$. In particular, for the inverse image of $\downarrow(a)$ under $\wedge$ to be closed in L^2 it is sufficient that a is prime, and with a prime we clearly have $\wedge^{-1}(\downarrow(a)) = (\downarrow(a) \times L) \cup (L \times \downarrow(a))$, so that 3.3 follows (recall that in each completely distributive lattice the primes order generate). Note also that for each complete lattice L in which PRIME L is order generating, and each $(Y,S) \in |\mathbf{TOP}(L)|$ one has $\iota_L(S) = \ll\{[V \not\leq \mathbf{p}] : V \in S_0 \text{ and } p \in \text{PRIME } L\}\gg$ for every $S_0 \subset L^Y$ that generates S (a proof of this uses the fact that such an L is a frame).

PROPOSITION 3.8. If L is completely distributive, $\omega_L(T)$ is the stratification of $\chi(T)$ for each $(X,T) \in |\mathbf{TOP}|$.

PROOF. Clearly, $\chi(T)^c \subset \omega_L(T)$. For the converse observe that for each $U \in L^X$ one has the following decomposition:

$$U = \bigvee\{\mathbf{a} \wedge 1_{[U \gg \mathbf{a}]} : a \in L\}$$

(which is of course valid for L a continuous lattice and is just a point-free formulation of the statement that $U(x) = \bigvee\{a \in L : a \ll U(x)\}$ for every $x \in X$). □

PROPOSITION 3.9. Let $(L, ')$ be a completely distributive (resp., complete) de Morgan algebra. Then for each $(X,T) \in |\mathbf{TOP}|$, $C((X,T),(L,\mathrm{INF}(L)))$ is the family of all closed (resp., subbasic closed) L-sets of $(X, \omega_L(T))$.

PROOF. This follows immediately from the evident fact that the quasi-complementation is a homeomorphism between $(L,\mathrm{SUP}(L))$ and $(L,\mathrm{INF}(L))$. □

REMARK 3.10. All the statements above reduce to the original Lowen's ones if $L = I$. As has already been mentioned in the Introduction, there exist other directions in which one can extend the Lowen functors. We shall briefly make a few comments about them. A general idea is evidently to replace $\mathrm{SUP}(L)$ in 3.1 by an arbitrary topology τ_L on L such that $\tau_I = \mathrm{SUP}(I)$. Then the functoriality of 3.2 holds true as an inspection of the proof of 3.2 easily shows (for a perfect evidence of this, it would perhaps be better to replace $[V \not\leq \mathbf{a}]$ in that proof by $V^{-1}(W)$ with some W in $\mathrm{SUP}(L)$). Furthermore, the proofs of 3.5(2) (inverseness) and of 3.6 (adjunction) do not really depend on $\mathrm{SUP}(L)$ explicitly, and merely on the fact that SUP(L) with L completely distributive (or with L as in 3.7) is such that $C((X,T),(L,\mathrm{SUP}(L)))$ forms an L-fuzzy topology. Consequently, one has 3.5(2) and 3.6 for each (L,τ_L) such that $C((X,T),(L,\tau_L))$ is an L-fuzzy topology on X (note that 3.5(1) is true always). As proved recently by [Warner 1990], Proposition 3.1, this is the case of L a continuous frame and τ_L the Scott topology. This is a special case of II-4.17 of [Gierz *et al.*, 1980], which states that $C((X,T),(L,\sigma(L)))$ is a frame (with respect to the pointwise operations formed in L^X, hence an L-topology on X) provided L is a frame with $\sigma(L)$ a continuous lattice. (This is a special case, since if L is continuous, so is $\sigma(L)$ by II-1.14 of [Gierz *et al.* 1980].) Therefore the behaviour of Warner's ω_L (and the corresponding ι_L) agrees with the classical one as well (note that she did not consider the categorical aspects). Also note that 3.8 goes over to her ω_L and is actually the same as her Proposition 5.4.

REMARK 3.11.

(1) Let us now compare the ω_L of 3.1 with that of [Warner 1990]. If, in 3.1, L is complete with primes order generating (as discussed in 3.7), these two approaches are incomparable (though each continuous frame is such a lattice, SUP(L) is generally weaker than $\sigma(L)$; see II-1.17 in [Gierz *et al.* 1980]). Now if L is completely distributive, then ω_L of 3.1 coincides with that of Warner by 1.3. In particular, 3.4 therefore follows from Proposition 3.1 of [Warner 1990]. However under presence of a quasi-complementation these two approaches coincide by 1.5 and 1.3. It is to be noted here that Warner's program is, As described in the chapter of Warner in this volume [Warner 1989], to avoid closedness defined by composing open L-sets with the quasi-complementation.

(2) In fact, the frame law plays no role in [Warner 1990]. She merely uses the meet continuity (directed sups distribute over finite infs), a property implied by continuity (see [Johnstone 1982, Lemma VII. 4.1] for a short proof). Consequently, 3.4 – 3.6 and 3.8 are valid for a continuous $(L,\sigma(L))$ too. This is also the case for 3.14, 3.17 – 3.20 which follow.

(3) Note that continuity of L is not necessary for $C((X,T),(L,\sigma(L))$ to be an L-topology. One sufficient condition is the meet-continuity of L and continuity of $\sigma(L)$ (see the proof of II - 4.7 of [Gierz et al. 1980].

REMARK 3.12. Question 6.2 of [Rodabaugh 1988c] asks if one can redefine (Rodabaugh's) ω_L and ι_L (i.e. those constructed from a complete lattice with a topology generated by the subbasic sets $\uparrow(a)\setminus\{a\}$) by using prime filters of L as to have the adjunction for L other than a complete chain, and ω_L being equal to χ composed with the (stratification) functor ^{c} for each such L. This seems to be partially done by 3.6 and 3.8 (together with 3.7), since $L\setminus\downarrow(p)$ is a prime filter if p is prime.

Some of the (auxiliary) results of this chapter can be performed for complete lattices in which primes order generate. The other will involve continuity of the lattice. Further, the definition of the natural L-topology of the L-fuzzy unit interval requires a complete lattice with a quasi-complementation. All these hypotheses together with what is said in 3.10 and 3.11 bring us back to complete distributivity. Nevertheless, we shall occasionally indicate whether complete distributivity is used in its full strength. Following [Lowen 1981c, 1978], we shall use the following terminology.

DEFINITION 3.13. Members of $|\,\omega_L(\mathbf{TOP})|$ are called **topologically generated**, and members of $|\,\iota_L(\mathbf{TOP}(\mathrm{L}))|$ are called the **topological modifications** of the corresponding members of $|\mathbf{TOP}(\mathrm{L})|$. For any topological property P we shall say that $(Y,S) \in |\mathbf{TOP}(L)|$ has ι-P iff its topological modification $(Y,\iota_L(S))$ has P. We say that an L-fuzzy topological property FP is a **good extension** of a topological property P iff for every $(X,T) \in |\mathbf{TOP}|$: (X,T) has P iff $(X,\omega_L(T))$ has FP.

COROLLARY 3.14. Let L be a completely distributive lattice. The following statements hold for any topological property P:

(1) ι-P is a good extension of P,

(2) P is preserved under weakening (resp., strengthening) of a topology iff ι-P is preserved under weakening (resp., strengthening) of an L-topology.

Moreover, if FW and FS are two L-fuzzy topological properties which are preserved under, resp., weakening and strengthening of an L-topology, and which are good extensions of the topological properties W and S, resp., then:

(3) ι-W implies FW,

(4) FS implies ι-S.

PROOF. (1) is trivial by 3.5(2). For (2) note that ι_L and ω_L act monotonically on L-topologies, resp., and use (1). Statements (3) and (4) follow from (1) and 3.5(1). □

REMARK 3.15.

(1) The converse of (3) and (4) is not true in general. Counterexamples are scattered in the literature (see [Wang 1983] and [Lowen 1981b] for W = compactness and connectedness, resp., and [Lowen 1981c] for S = Hausdorffness with $L = I$).

(2) As is customary in the literature, an L-space which has ι-P is called **ultra-P**. Since **ultracompact** and **ultraconnected** are appropriate terms in the light of 3.14, we shall use them (though the same terms have different meanings in general topology).

(3) Whether or not each good extension of, say, topological compactness (or Hausdorffness) must necessarily be preserved under weakening (or strengthening) of an L-topology was a question in [Kubiak 1989]. It was clear enough that this need not be the case, but the problem was with a counterexample. A surprisingly simple counterexample is due to J. J. Chadwick. Namely, define an L-space to be Hausdorff (resp., compact or connected) iff it is topologically generated from a Hausdorff (resp., compact or connected) topological space. These concepts are trivially good extensions of the coresonding topological

concepts and are not preserved under strengthening (resp., weakening) of the L-fuzzy topology. Examples are simple to construct.

DEFINITION 3.16. Given an ordinary (resp., L-fuzzy) topological property P, a topological space (resp., L-space) (X,T) is **maximal** (resp., **minimal**) P iff every (L-)topology on X strictly finer (resp., coarser) than T is **non-**P.

PROPOSITION 3.17. Let L be a completely distributive lattice. For P a topological property, (X,T) a topological space, and (Y,S) an L-topological space, the following statements hold:

(1) (X,T) is maximal P iff $(X,\omega_L(T))$ is maximal ι-P.
(2) If (Y,S) is maximal ι-P, then it is topologically generated.

PROOF.

(1) ($\Rightarrow$) We know that if T is P, then $\omega_L(T)$ is ι-P. Assume $S \supset \omega_L(T)$ is ι-P. Thus $\iota_L(S) \supset T$, hence $\iota_L(S) = T$ since $\iota_L(S)$ is P and T is maximal P. Now $S \subset \omega_L\iota_L(S) = \omega_L(T)$, hence $S = \omega_L(T)$.
($\Leftarrow$) Let both $T_1 \supset T$ be P. Then $\omega_L(T_1)$ is ι-P and $\omega_L(T_1) \supset \omega_L(T)$. As $\omega_L(T)$ is maximal P, we have $\omega_L(T_1) = \omega_L(T)$, hence $T_1 = T$ (ω_L is injective by 3.5(2)).
(2) Always $S \subset \omega_L\iota_L(S)$. Since $\iota_L(S)$ is P, hence $\omega_L\iota_L(S)$ is ι-P, and thus $S = \omega_L\iota_L(S)$, i.e. S is topologically generated. □

The following proposition is a counterpart of Lemma 1.2 of [Lowen 1977]. As its proof is actually the one for L a continuous lattice with the Scott topology, the proposition is actually Proposition 4.1 of [Warner 1990] (cf. 3.11(2)). The proof, however, extends that of [Wang and Hu 1985], whose proof is an alternative to that of Lowen.

PROPOSITION 3.18 Let L be a completely distributive lattice. Let $(X,T) \in |\mathbf{TOP}|$ and let $f : Y \to X$. Then $\omega_L(f^{\leftarrow}(T)) = f^{\leftarrow}(\omega_L(T))$.

PROOF. Let $U \in \omega_L(f^{\leftarrow}(T))$. Then $[U \gg \mathbf{a}] \in f^{\leftarrow}(T)$ for each $a \in L$, and thus $[U \gg \mathbf{a}] = V_a f$ for some $V_a \in T$. Now

$$\begin{aligned} U &= \bigvee_{a\in L}(\mathbf{a} \wedge 1_{[U\gg\mathbf{a}]}) \\ &= \bigvee_{a\in L}(\mathbf{a} \wedge 1_{V_a} f) \\ &= (\bigvee_{a\in L}(\mathbf{a} \wedge 1_{V_a}))f, \end{aligned}$$

that is $U \in f^{\leftarrow}(\omega_L(T))$. The reverse inclusion is straightforward. □

COROLLARY 3.19. For (X,T) in $|\mathbf{TOP}|$ and $A \subset X$ one has $\omega_L(T)_A = \omega_L(T_A)$.

PROOF. By 3.18 with $Y = A$ and f the identity embedding of A into X. □

PROPOSITION 3.20. Let L be completely distributive. Then both ι_L and ω_L preserve products.

PROOF. One knows that right adjoints preserve categorical products, so ι_L does. The functor ω_L commutes with sups (i.e. $\omega_L(\bigvee_\gamma T_\gamma) = \bigvee_\gamma \omega_L(T_\gamma)$ for any family $\{T_\gamma\}_\gamma$ of L-topologies on a given set), which may be verified in exactly the same way as in Lemma 1.3 of [Lowen 1977]. This also comes from an observation (in the context of the mentioned result of Lowen, first made by [Cerruti 1981]) that sups are special colimits, and that ω_L being a left adjoint must preserve them (see [Mac Lane 1971], pp. 67, 68, and 115). This along with 3.18 yields the assertion for ω_L. □

COROLLARY 3.21. For L a completely distributive lattice and for any topological property P, P is productive in **TOP** iff ι-P is productive in **TOP**(L).

PROOF. The part ($\Rightarrow$) uses 3.20 for ι_L, and ($\Leftarrow$) uses 3.20 for ω_L and the fact that a topologically generated L-space is ι-P iff it is generated from a space which is P. □

As in [Lowen 1977], for any $a \in L \setminus \{1\}$ and $(Y, S) \in |\mathbf{TOP}(L)|$, we let

$$\iota_{L,a}(S) = \ll \{U^{-1}(L \setminus \downarrow(a)) : U \in S\} \gg .$$

PROPOSITION 3.22. The following statements hold for L a completely distributive lattice:

(1) $\iota_L(S) = \bigvee\{\iota_{L,a}(S) : a \neq 1\}$.
(2) The assignment $(Y, S) \to (Y, \iota_{L,a}(S))$ is the object part of the functor $\iota_{L,a} : \mathbf{TOP}(L) \to \mathbf{TOP}$ which preserves mappings, and which is a left inverse of ω_L.

PROOF. (1) and the functoriality of (2) are immediate from the definition. Given (X, T) in $|\mathbf{TOP}|$ we clearly have $\iota_{L,a}\omega_L(T) \subset T$ by (1) and 3.5(2). For the converse note that $U = [1_U \not\leq \mathbf{a}] \in \iota_{L,a}\omega_L(T)$ for each $U \in T$ and $a \neq 1$. This proves the second assertion of (2). □

REMARK 3.23. Note that 3.14, 3.17, 3.20-3.22 hold true for every complete lattice in which primes order generate (cf. 3.7).

§4. Axioms of Compactness

As mentioned in the Introduction, we shall develop several new compactness axioms for L-fuzzy spaces. Before proceeding we have to make some comments about the terminology.

We shall keep the terms **a-compactness** and **strong compactness**, though these notions coincide with those of [Gantner *et al.* 1978] and [Lowen 1978] only if L is a complete chain. We follow tradition and use the term **fuzzy compactness** for the concept that extends the fuzzy compactness of [Lowen 1976] from I-spaces to L-spaces. The only exception we make is connected with the N-compactness of [Zhao 1987] (N is for a net). After considerable simplification, it will be renamed **Ch**-compactness (**Ch** stands for Chinese). The reason for which we rename it is that N-compactness is a well-known concept of general topology being recently of rather great interest (a topological space is N-compact if it is homeomorphic to a closed subspace of some topological power of N, the space of positive integers). We shall, however, use the term **N-compactness** when speaking of what was exactly introduced by Zhao.

In the following definition, note the use of the notation of 2.3.

DEFINITION 4.1. Let L be a complete lattice and $a \in L$. An L-space (X,T) is **a-compact** iff for every $\mathcal{V} \subset T$ such that $\bigvee \mathcal{V} \not\leq \mathbf{a}$ there exists a finite subfamily $\mathcal{V}_0$ of $\mathcal{V}$ such that $\bigvee \mathcal{V}_0 \not\leq \mathbf{a}$.

PROPOSITION 4.2. For L a complete lattice, (X,T) an L-space, and $a \in L \setminus \{1\}$, the following statements are equivalent:

(1) (X,T) is a-compact.
(2) For each $\mathcal{V} \subset T$ such that for each $x \in X$ there exists $V \in \mathcal{V}$ such that $V(x) \not\leq a$ (such a $\mathcal{V}$ is called an **a-shading**) there exists a finite $\mathcal{V}_0 \subset \mathcal{V}$ which is an a-shading.
(3) $(X, \iota_{L,a}(T))$ is compact.

PROOF This is trivial. Note only that (2) $\Rightarrow$ (3) follows from the Alexander subbase lemma (for ordinary spaces). □

REMARK 4.3. If one replaces $\not\leq$ in 4.2(2) (resp., 4.1) by $>$, one has the concept of a-compactness of [Gantner *et al.* 1978] (resp., (pf, a)-compactness of [Rodabaugh 1982a], where pf = point-free). In particular, if $a \in L^c = \{x \in L : x \text{ is comparable to each element of } L\}$, then all the discussed concepts are equivalent. Note also that (2) $\Leftrightarrow$ (3) of 4.2 (with $L = I$) comes from [Lowen 1978]. In what follows, by **a**-compactness we shall mean the concept of 4.1.

The other compactness axioms will be defined in terms of closed L-sets to be in tune with Zhao's N-compactness which was defined in such a way.

DEFINITION 4.4. Let $(L, {}')$ be a complete de Morgan algebra and (X,T) an L-space. Let $a \in L$ and $\mathcal{K} \subset T'$ be arbitrary and such that $\bigwedge \mathcal{K} \not\geq \mathbf{a}$. Then (X,T) is called:

(1) Ch-compact iff $(\exists\, b \ll a)(\exists \text{ finite } \mathcal{K}_0 \subset \mathcal{K}) :\ \bigwedge \mathcal{K}_0 \not\geq \mathbf{b}$,
(2) strongly compact iff $(\exists \text{ finite } \mathcal{K}_0 \subset \mathcal{K}) :\ \bigwedge \mathcal{K}_0 \not\geq \mathbf{a}$,
(3) fuzzy compact iff $(\forall\, b \gg a)(\exists \text{ finite } \mathcal{K}_0 \subset \mathcal{K}) :\ \bigwedge \mathcal{K}_0 \not\geq \mathbf{b}$.

REMARK 4.5. The following observations are straightforward:

(1) An L-space is strongly compact iff it is a a-compact for each $a \in L$.

(2) Ch-compactness $\Rightarrow$ strong compactness $\Rightarrow$ fuzzy compactness. That none of these implications is reversible will become clear later on, after showing (see 4.15) that Ch-compactness is equivalent to N-compactness of [Zhao 1987] (recall that N-compactness requires L to be completely distributive), hence to N-compactness of [Wang 1983] whenever $L = I$, and after showing that fuzzy compactness reduces to Lowen's fuzzy compactness if $L = I$ (see 4.18). Having said this we can refer to [Wang 1983] and [Lowen 1978] for counterexamples.

We now observe that all the introduced axioms of compactness are good extensions of ordinary compactness. In the meantime we note that an ordinary topological space (X,T) is compact iff $(X,\chi(T))$ is Ch-compact, strongly compact, and fuzzy compact.

PROPOSITION 4.6. Let $(L, ')$ be a completely distributive de Morgan algebra. Then Ch-compactness, strong compactness and fuzzy compactness are good extensions of ordinary compactness.

PROOF. Assume (X,T) is a compact topological space. We first show that $(X,\omega_L(T))$ is Ch-compact. Let $a \in L$ be arbitrary, and let $\mathcal{K}$ be an arbitrary family of closed L-sets of $(X,\omega_L(T))$ such that $\bigwedge \mathcal{K} \not\geq \mathbf{a}$. Then

$$\begin{aligned}\emptyset &= (\bigwedge \mathcal{K})^{-1}(\uparrow(a)) \\ &= \bigcap_{K\in\mathcal{K}} K^{-1}(\uparrow(\bigvee \downarrow(a))) \\ &= \bigcap_{K\in\mathcal{K}} \bigcap_{b\ll a} K^{-1}(\uparrow(b)).\end{aligned}$$

By 3.9, $\omega_L(T)' = C((X,T),(L,\mathrm{INF}(L)))$, hence each $K^{-1}(\uparrow(b))$ is closed in (X,T). Thus, since (X,T) is compact, there exist $\mathcal{K}_0 = \{K_1,\ldots,K_n\} \subset \mathcal{K}$ and $b_1,\ldots,b_m \ll a$ such that

$$\begin{aligned}\emptyset &= \bigcap_{i=1}^{n} \bigcap_{j=1}^{m} K_i^{-1}(\uparrow(b_j)) \\ &= (\bigwedge \mathcal{K}_0)^{-1}(\uparrow(b)),\end{aligned}$$

where $b = b_1 \vee \ldots \vee b_m \ll a$. Thus $\bigwedge \mathcal{K}_0 \not\geq \mathbf{b}$ with $b \ll a$, which proves that $(X,\omega_L(T))$ is Ch-compact.

After 4.5(2) it remains to show that if $(X,\omega_L(T))$ is fuzzy compact, then (X,T) is compact. Let $\mathcal{F}$ be any family of closed sets of (X,T) such that $\bigcap \mathcal{F} = \emptyset$. Let $1_{\mathcal{F}} = \{1_F : F \in \mathcal{F}\}$. Then $(\bigwedge 1_{\mathcal{F}})^{-1}(\uparrow(a)) = \emptyset$ for each $a \neq 0$. Now, for each $b \gg a$ (thus $b \neq 0$) there exists a finite $\mathcal{F}_0 \subset \mathcal{F}$ with $(\bigwedge 1_{\mathcal{F}_0})^{-1}(\uparrow(b)) = \emptyset$. Thus $\bigcap \mathcal{F}_0 = \emptyset$. □

COROLLARY 4.7. For L-spaces with $(L, ')$ a completely distributive de Morgan algebra the following implications hold: ultracompactness $\Rightarrow$ Ch-compactness $\Rightarrow$ strong compactness $\Rightarrow$ fuzzy compactness. None of these implications is reversible in general.

PROOF. First note that all the compactness axioms are preserved under weakening of an L-fuzzy topology. Therefore ultracompactness implies Ch-compactness on account of 3.14(3). Also, cf. 4.5(2) and note that an example for N-compact $\not\Rightarrow$ ultracompact is in [Wang 1983] for $L = I$. □

We are now going to discuss the concept of N-compactness of [Zhao 1987].

REMARK 4.8. As proved by [Hutton 1977] (in fact, the same can be found in [Balbes and Dwinger 1974], Def. 12 and Th. 16(ii), pp. 246, 247), if L is a completely distributive lattice and $a \in L$, then there exists $B \subset L$ such that:

(1) $\bigvee B = a$, and
(2) if $A \subset L$ and $a \leq \bigvee A$, then for each $b \in B$ there is a $c \in A$ such that $b \leq c$.

Such a set B is called an **atomic-like decomposition** of a. Let $\beta(a)$ denote the union of all the atomic-like decompositions of a. One easily sees that $\beta(a)$ is itself an atomic-like decomposition of a (cf. [Liu and Luo 1987]). Let $\beta^*(a) = \beta(a) \cap (\text{COPRIME } L \setminus \{0\})$. As proved in [Zhao 1987], one has the following statements:

(3) $\bigvee \beta^*(a) = a$ (in fact, it is an atomic-like decomposition of a),
(4) $\beta^*(a)$ is directed iff $a \in \text{COPRIME } L \setminus \{0\}$ (recall that $A \subset L$ is directed if for each $x, y \in A$ there is $z \in A$ with $x \vee y \leq z$).

The N-compactness of [Zhao 1987] (as well as its special real unit interval case of [Wang 1983]) concerns arbitrary L-sets and not only the whole L-space as we did in 4.4. Though it is not necessary for the main purposes of this chapter to consider compactness of L-sets, we shall reproduce here the definition of Zhao in its full generality. Afterwards it will become clear how to extend 4.4 to arbitrary L-sets. This will provide a unified way of extending various compactness axioms from the whole L-space to its L-sets (see also 4.19).

After avoiding fuzzy points and some other terminology, Zhao's definition is as follows.

DEFINITION 4.9 ([Zhao 1987]). Let $(L, ')$ be a completely distributive de Morgan algebra, and (X, T) and L-space. Let $a \in \text{COPRIME } L \setminus \{0\}$. A family $\mathcal{K} \in T'$ is called an **a-R-neighbourhood family** of an $A \in L^X$ iff whenever $x \in X$ and $a \leq A(x)$, there is a $K \in \mathcal{K}$ such that $a \not\leq K(x)$. Furthermore, $A \in L^X$ is **N-compact** iff for every a-R-neighbourhood family of $\mathcal{K}$ of A there exist $b \in \beta^*(a)$ and a finite $\mathcal{K}_0 \subset \mathcal{K}$ such that $\mathcal{K}_0$ is a b-R-neighbourhood family of A. We say that (X, T) is **N-compact** if $1 (= 1_X)$ is N-compact.

REMARK 4.10. Observe that the definition of an a-R-neighbourhood family can easily be restated in a point-free form analogous to those of 4.4. Namely, $\mathcal{K}$ is an a-R-neighbourhood family of A iff $A^{-1}(\uparrow(a)) \subset X \setminus \bigcap_{K\in\mathcal{K}} K^{-1}(\uparrow(a))$, that is, iff

$$(A \wedge \bigwedge \mathcal{K})^{-1}(\uparrow(a)) = A^{-1}(\uparrow(a)) \cap (\bigwedge \mathcal{K})^{-1}(\uparrow(a)) = \emptyset.$$

We thus have the first refinement of 4.9.

DEFINITION 4.11. If (X,T) is an L-space with L a completely distributive lattice, then $A \in L^X$ is **N-compact** iff for every $a \in$ COPRIME $L \setminus \{0\}$ and every $\mathcal{K} \subset T'$ such that $A \wedge \bigwedge \mathcal{K} \not\geq \mathbf{a}$ there exist a finite $\mathcal{K}_0 \subset \mathcal{K}$ and $b \in \beta^*(a)$ such that $A \wedge \bigwedge \mathcal{K}_0 \not\geq \mathbf{b}$.

We now include some comments concerning the role which the set $\beta^*(a)$ plays. It will become apparent that some of the results of [Zhao 1987] are valid if in 4.11 L is a continuous lattice and $\beta^*(a)$ is replaced by $\Downarrow(a)$ with a an arbitrary element of L. This is just what we shall call afterwards Ch-compactness of an L-set.

COMMENT 4.12.

(1) An inspection of the proofs of results of [Zhao 1987] shows that the set $\beta^*(a)$ with a a non-zero coprime has essentially been developed for only two reasons: to have a directed subset of L, other than $\downarrow a$, which satisfies (1) of 4.8, and which reduces to the interval $(0,a)$ whenever $L = I$ (the latter in order to capture N-compactness of [Wang 1983]). Recall that the coprimality of a is essential only for $\beta^*(a)$ to be directed. That $\beta^*(a)$ satisfies 4.8(2) is used, in [Zhao 1987], merely in the proof of Theorem 6.8 characterizing N-compactness in terms of some filters.

(2) One knows that in each continuous lattice $L, \Downarrow(a)$ satisfies 4.8(1) for each $a \in L$, is directed (see the properties of $\ll$ in Section 1), and is equal to $[0,a)$ if $a \in I \setminus \{0\}$ (note that there is no problem with zero since 0-R-neighbourhoods do not exist). Therefore $\beta^*(a)$ with a a non-zero coprime may well be replaced in the whole of Zhao's paper (except Theorem 6.8) by $\Downarrow(a)$ with an arbitrary a in L. This is not the case of Theorem 6.8, since $\Downarrow(a)$ need not be an atomic-like decomposition of a, i.e. (2) of 4.8 need not hold. Also, the complete distributivity of L might be relaxed to the continuity, provided that no distributivity will have to be used (recall that distributivity in continuous de Morgan algebra implies complete distributivity), cf. 4.16. It should also be noted that there are other results in the literature which use the fact that $\beta^*(a)$ satisfies 4.8(2). As far as the author knows this is only the case of Section V in [Liu and Luo 1987]. The paper of Liu and Luo, however uses complete distributivity also for other reasons than those indicated in (1).

(3) On the other hand, the only proof which exists in the literature and in which the coprimality of a in $\beta^*(a)$ is required for other reasons than only the directness of $\beta^*(a)$ is, to the author's knowledge, the one of Lemma 3 in [Liu and Luo 1988], which is crucial in Theorem 1 asserting that the L-fuzzy unit interval is N-compact (cf. 7.6(1) and 4.15). Afterwards they used his result to develop a compactfication theory in [Liu and Luo 1989].

The following will show that what we suggested in 4.12(2) is actually to avoid coprimality from the discussion, since $\beta^*(a)$ proves to consist of all coprimes way below a.

LEMMA 4.13. Let L be a completely distributive lattice. Then the following statements hold for every $a \in L$:

(1) $\Downarrow(a) \cap (\text{COPRIME } L \setminus \{0\})$ is an atomic-like decomposition of a.
(2) $\beta^*(a) = \Downarrow(a) \cap (\text{COPRIME } L \setminus \{0\})$.

PROOF. Let $P_a = \Downarrow(a) \cap (\text{COPRIME } L \setminus \{0\})$. By 1.2(1), we have that $b = \bigvee(\downarrow(b) \cap (\text{COPRIME } L \setminus \{0\}))$ for each b in L (recall that $0 = \bigvee \emptyset$). Thus,

$$\begin{aligned} a &= \bigvee_{b \ll a} \bigvee_{q \in \downarrow(b) \cap (\text{COPRIME } L \setminus \{0\})} q \\ &= \bigvee \Big(\bigcup_{b \ll a} \downarrow(b) \cap (\text{COPRIME } L \setminus \{0\}) \Big) \\ &= \bigvee P_a \end{aligned}$$

since $\Downarrow(a) = \bigcup\{\downarrow(b) : b \ll a\}$. Now, let $A \subset L$ and $a \leq \bigvee A$. As for each $b \in P_a$ we have $b \ll a$, hence there exists a finite subset D of A such that $b \leq \bigvee D$. Since b is coprime, there exists $c \in D$ with $b \leq c$. This proves (1). Thus $P_a \subset \beta^*(a)$ as the latter is the intersection of all non-zero coprimes and $\beta(a)$ (= the union of all atimic-like decompositions of a). For the reverse inclusion observe that if $B \subset L$ satisfies 4.8(2), then $B \subset \Downarrow(a)$. Indeed, if $b \in B$ and $a \leq \bigvee A$ for $A \subset L$, then $b \leq \bigvee\{c\}$ for some $c \in A$, i.e. $b \ll a$. Thus $\beta(a) \subset \Downarrow(a)$, hence $\beta^*(a) \subset P_a$. This proves (2). □

REMARK 4.14. In the notation of [Zhao 1987], $b \lhd a$ iff $b \in \beta^*(a)$. Therefore the relation $\lhd$ is the way below relation restricted to coprimes. Additionally note that after 4.13(2), for the directness of $\beta^*(a)$ with a a non-zero coprime one may refer to V-1.9 of [Gierz *et al.* 1980]. However Zhao's argument in proving this is incomparably simpler than that of [Gierz *et al.* 1980]

The following will fill in those omissions we have made in 4.5(2) and in the proof of 4.7, where we anticipated relationships between N-compactness and Ch-compactness.

PROPOSITION 4.15. Let $(L, ')$ be a completely distributive de Morgan algebra. Then each Ch-compact L-space is N-compact. The converse holds for L a complete chain.

PROOF. Let (X,T) be Ch-compact. Given a non-zero coprime A and a family $K \subset T'$, let $\bigwedge K \geq \mathbf{a}$. Then there exist $b \ll a$ and a finite $K_0 \subset K$ such that $\bigwedge K_0 \not\geq \mathbf{b}$. Since $b \ll a$ and $a = \bigvee(\Downarrow(a) \cap (\text{COPRIME } L\setminus\{0\}))$ (see the proof of 4.13), there exists a finite subset $A \subset \Downarrow(a) \cap (\text{COPRIME } L\setminus\{0\}))$ such that $b \leq \bigvee A$. By 4.13(2) and 4.8(4) (also cf. 4.14), $\beta * (a) = \Downarrow(a) \cap (\text{COPRIME } L\setminus\{0\})$ is directed, so that $\bigvee A \leq c$ for come $c \in \beta * (a)$. Since $b \leq c$, hence $\bigwedge K_0 \not\geq \mathbf{c}$. Thus (X,T) is N-compact.

For L a chain, one has L = COPRIME L. For this case Ch-compactness is trivially equivalent to N-compactness, since $\beta * (a) = \downarrow(a)$ for each $a \in L$ by 4.13(2). □

REMARK 4.16. The assumption that L is a complete chain is not necessary for the implication N-compactness ⇒ CH-compactness. In exactly the same way as in the proof of 4.6 (just byreplacing the directed set $\Downarrow(a)$ by $\beta * (a)$ with a a nonzero copirme, hence directed by 4.8(4)) one shows that N-compactness requires L to be completely distributive). Therefore N-compactness and Ch-compactness are equivalent in the class of all topologically generated L-spaces with L a complete distributive de Morgan algebra.

REMARK 4.17. In the proof of 4.6, the requirement that $\omega_L(T)' = C((X,T),(L,\mathrm{INF}(L)))$ is not necessary if one has an Alexander's type subbase theorem for Ch-compactness of L-spaces with L a continuous de Morgan algebra. Theorem 5.1 of [Zhao 1987] provides an Alexander's type subbase theorem for N-compactness. Its proof, however, does not go over to Ch-compactness with L a continuous de Morgan algebra, as it implicitly uses the frame law (when representing an open L-set as a sup of finite infs of elements of a given subbase; also cf. 2.1).

REMARK 4.18. It is now perfectly clear how to extend 4.4 to arbitrary L-sets. Let $(L, ')$ be a complete de Morgan algebra, (X,T) and L-space, and $A \in L^X$. Given $a \in L$ and $\mathcal{K} \subset T'$ with $A \wedge \bigwedge \mathcal{K} \not\geq \mathbf{a}$, we say A is **Ch-compact** (resp., **strongly compact, fuzzy compact**) if the condition of 4.4(1) (resp., (2), (3)) is replaced by $A \wedge \bigwedge \mathcal{K}_0 \not\geq \mathbf{b}$ (resp., $\not\geq \mathbf{a}$, $\not\geq \mathbf{b}$). It is also clear that 4.15 is valid if A is used in place of 1_X. A study of these compactness notions is pursued in [Kubiak 1990b].

REMARK 4.19. Assume $L = I$. Then fuzzy compactness of an I-set, when rephrased in terms of open I-sets, reads as follows: A is fuzzy compact iff for each $a \in I$ and each $\mathcal{V} \subset T$ such that $A' \vee \bigvee \mathcal{V} > \mathbf{a}$ and for each $b < a$ there exists a finite subfamily $\mathcal{V}_0 \subset \mathcal{V}$ such that $A' \vee \bigvee \mathcal{V}_0 > \mathbf{b}$. Since $a \in I$ is arbitrary, it is easy to check that we may replace the relation above by $\geq$, preserving the less-than relation between b and a. We thus have the concept of **f-compactness** due to [Chadwick 1989] which reduces to the fuzzy compactness (of the entire space) in the sense of [Lowen 1976] if $A = 1_X$ (as therefore 4.4(3) does if $L = I$). Note also that, unlike the case $L = I$, we cannot restate 4.4 or 4.18 in terms of open L-sets by just acting with the quasi-complementation since the latter need not reverse the way below relation (unless L is a complete chain without members isolated from below).

Up to this point we were concerned about some generalities relevant to our principal purpose of this chapter. What follows is our main objective.

§5. The Helly Space of L-Valued Functions

Throughout the following we consider complete lattices equipped with their interval topologies. Also, from now on we think of the set L^I as a product of I copies of $(L, \mathrm{INT}(L))$.

DEFINITION 5.1. Let L be a complete lattice. Let $H_L = \{f \in L^I : f \text{ is order-reversing}\}$. We call $(H_L, \mathcal{H}_L)$ the **Helly space** where $\mathcal{H}_L$ is the subspace topology induced from the cube L^I.

Recall that the usual Helly space is the subspace of I^I consisting of all order-preserving functions. For notational reasons we shall use here the nonincreasing variant as it agrees with the definition of the L-fuzzy unit interval. In fact, if $\phi : L \cong L^{\mathrm{op}}$, then $(H_L, \mathcal{H}_L)$ is homeomorphic to the subspace of L^I consisting of all order-preserving functions (by taking each $f \in H_L$ to ϕf).

Now, we shall show some properties of the Helly space analogous to those that hold for $L = I$ (cf. [Steen and Seebach 1978]).

PROPOSITION 5.2. If L is completely distributive, then the Helly space is compact and Hausdorff.

PROOF. As noted in 1.4, $(L, \mathrm{INT}(L))$ is compact Hausdorff. Therefore H_L is clearly Hausdorff, and since L^I is compact it suffices to show that H_L is closed in L^I. For if $f \in L^I \setminus H_L$, then there exist $s < t$ in I such that $f(t) \not\leq f(s)$. By (C) (in Section 1), there is an $a \in L$ with $a \ll f(t)$ and $a \not\leq f(s)$, so that $U = \pi_t^{-1}(\Uparrow(a)) \cap \pi_s^{-1}(L \setminus \uparrow(a))$ is an open neighbourhood of f. Also, U is disjoint from H_L. For if $g \in U \cap H_L$, then $a \not\leq g(s)$, and since $a \ll g(t) \leq g(s)$, hence $a \leq g(s)$, a contradiction. □

REMARK 5.3. The above proof is clearly the one for L a continuous lattice with the Lawson topology (cf. 1.4) and with the Helly space constructed from $(L, \lambda(L))$. On the other hand, the complete distributivity of L in $(L, \mathrm{INT}(L))$ can be relaxed to the assumption that L is a complete Hausdorff topological lattice in the interval topology. Such a lattice has sufficient separation properties for proving 5.2 and need not be completely distributive (see [Johnstone 1982], p. 282 for details).

LEMMA 5.4. The Helly space is a topological lattice relative to the pointwise ordering from L^I, where L is completely distributive.

PROOF. Since H_L is clearly a sublattice of L^I, we must only show that $\wedge$ and $\vee$ are continuous mappings from $H_L \times H_L$ to H_L. We shall use parentheses to write down two dual statements simultaneously. For each $a \in L$ and each $t \in I$ we clearly have (only completeness of L involved)

$$\wedge[\vee]^{-1}(H_L \cap \pi_t^{-1}(\uparrow[\downarrow](a))) = (\pi_t^{-1}(\uparrow[\downarrow](a)) \times H_L) \cap (H_L \times \pi_t^{-1}(\uparrow[\downarrow](a)).$$

For the two other cases recall that SUP(L) is generated by the subbasic sets $L \setminus \downarrow(a)$ with $a \in$ PRIME L (cf. 3.7). Similarly, by 1.2(1), INF(L) is generated by the subbasic sets $L \setminus \uparrow(a)$ with $a \in$ COPRIME L. Therefore, to complete the proof is suffices to note that for every $a \in$ COPRIME L [PRIME L] and every $t \in I$ one has

$$\vee[\wedge]^{-1}(H_L \cap \pi_t^{-1}(\uparrow[\downarrow](a))) = (\pi_t^{-1}(\uparrow[\downarrow](a)) \times H_L) \cup (H_L \times \pi_t^{-1}(\uparrow[\downarrow](a)).$$

Thus, inverse images of every subbasic closed set of H_L under $\wedge$ and $\vee$ are closed in $H_L \times H_L$.

REMARK 5.5. We note that 5.4 and its proof are valid for $(L, ')$ a complete de Morgan algebra in which PRIME L is order generating, since then COPRIME $L = \{a' : a \in$ PRIME $L\}$. Consequently, in such a lattice each element is a sup of coprimes too.

PROPOSITION 5.6. For $(L, ')$ a completely distributive de Morgan algebra, the Helly space is connected whenever $(L, \mathrm{INT}(L))$ is connected. (Actually, the Helly space is pathwise connected with L playing the role of the unit interval in the usual definition of path connectedness.)

NOTE. One knows that if L is a compact topological lattice (as, recall again, each completely distributive L with INT(L) does), then L is connected in its topology iff L is order dense (see VI-5.15 of [Gierz *et al.* 1980]).

PROOF. Given $f, g \in H_L$, let $\psi : L \to H_L$ be defined by $\psi(a) = (\mathbf{a} \wedge g) \vee (\mathbf{a}' \wedge f)$ for each $a \in L$. Since the embedding $a \to \mathbf{a}$ of L into H_L is continuous, likewise the quasi-complementation with respect to INT(L), hence ψ is continuous on account of 5.4. Also $\psi(0) = f$ and $\psi(1) = g$. Therefore f and g lie in $\psi(L)$, a connected subset of H_L. Thus, H_L is (pathwisely) connected. □

REMARK 5.7. By 5.5, 5.6 is valid for each complete de Morgan algebra in which primes order generate.

§6. A Quotient of the Helly Space versus the L-Fuzzy Unit Interval

In this section we shall show that after identification of members of the Helly space having the same right-hand limits at each point, the resulting quotient space is the topological modification of the L-fuzzy unit interval. For this purpose, we first discuss briefly the concept of right and left-hand limits of members of H_L with L a complete lattice (cf. [Hutton 1975]).

For every $f \in H_L$ and every $t \in I$, let

$$f^+(t) = \bigvee\{f(s) : s > t\} \quad \text{and} \quad f^-(t) = \bigwedge\{f(s) : s < t\}.$$

Note that $f^+(1) = 0$ and $f^-(0) = 1$. The following observations are straightforward.

REMARK 6.1. Both $(\cdot)^+ : f \to f^+$ and $(\cdot)^- : f \to f^-$ are order-preserving selfmappings of H_L. The mapping $(\cdot)^-$ is an expansion (i.e. $f \leq f^-$), and $(\cdot)^+$ is a contraction (i.e. $f^+ \leq f$). Both $(\cdot)^+$ and $(\cdot)^-$ are idempotents (i.e. $f^{++} = f^+$ and $f^{--} = f^-$).

In addition to those properties we have the following (cf. [Kubiak 1986]). Of course, either 6.1 or what follows does not require any distributivity (but note that $(\cdot)^+$ and $(\cdot)^-$ are lattice homomorphisms provided, resp., L and L^{op} are frames).

LEMMA 6.2. Let L be a complete lattice, and let $f, g \in H_L$. The following statements hold:

(1) $f^{-+} = f^+$,
(2) $f^{+-} = f^-$,
(3) $f^+ \leq g^+$ iff $f^- \leq g^-$ *iff* $f^+ \leq g^-$,
(4) $f^+ = g^+$ iff $f^- = g^-$.

PROOF. (1) is true at $t = 1$. If $t < 1$, then $f^+(t) \geq f^-(s)$ for all $s \in (t, 1]$. Thus $f^+(t) \geq \bigvee\{f^-(s) : s > t\} = f^{-+}(t)$. Since $(\cdot)^-$ is an expansion and $(\cdot)^+$ is order-preserving, we get $f^{-+} \geq f^+$. The equality follows. The proof of (2) is similar. The first implication of (3) follows from the monotonicity of $(\cdot)^-$ and (2). The second one is trivial. Now, if $f^+ \leq g^-$, then $f^+ \leq g^+$ by (1) and the fact that $(\cdot)^+$ is a monotone idempotent. Finally, (4) is immediate from (3). □

DEFINITION 6.3 ([Hutton 1975]). Let L be a complete lattice. For each $f, g \in H_L$, let $f \sim g$ iff $f^+ = g^+$ (an equivalence relation). The quotient set $I(L) = H_L/\sim$ is called the **L-fuzzy unit interval** (shortly, **L-unit interval**).

NOTE. We wish to stress once again (cf. [Kubiak 1986]), that $f \sim g$ iff $f^+ = f^+$ **iff** $f^- = g^-$, since after [Hutton 1975] almost everywhere one has the definition: $f \sim g$ iff $f^+ = g^+$ **and** $f^- = g^-$.

DEFINITION 6.4. The quotient topology $\mathcal{T}_L = \mathcal{H}_L/\sim$ is called the **natural topology** of the L-unit interval.

PROPOSITION 6.5. The following statements hold for L a completely distributive lattice:
(1) $(I(L), \mathcal{T}_L)$ is compact Hausdorff.

(2) $(I(L), \mathcal{T}_L)$ is (pathwise) connected if $(L, \mathrm{INT}(L))$ is a connected de Morgan algebra.

PROOF. (1) By 5.2, $\mathcal{T}_L$ is compact as a quotient of compact topology. As the Helly space is compact and Hausdorff by 5.2, therefore by a Bourbaki's theorem (see e.g. [Willard 1970], 17N), $\mathcal{T}_L$ is Hausdorff iff $K = \{(f,g) \in H_L^2 : f \sim g\}$ is closed in H_L^2. So let $(f,g) \in H_L^2 \setminus K$. Then $f^+ \neq g^+$. Without loss of generality we may assume that $f^+ \not\leq g^+$. Thus $f^+(t) \not\leq g^+(t)$ for some $t \neq 1$. Consequently, there is an $s > t$ such that $f(s) \not\leq g^+(t)$. By (C) there exists an $a \in L$ with $a \ll f(s)$ and $a \not\leq g^+(t)$. Select u such that $t < u < s$. Then $a \not\leq g(u)$, since $g(u) \leq g^+(t)$. Therefore $(f,g) \in U = (\pi_s^{-1}(\Uparrow a) \times \pi_u^{-1}(L \setminus \uparrow a)) \cap H_L^2$, an open set in H_L^2. To show U is disjoint from K, assume $(h,k) \in U$ and $u < r < s$. Then $h^+(r) \geq h(s) \gg a$, hence $h^+(r) \gg a$, and thus $h^+(r) \geq a$. Now $k(u) \not\geq a$, and since $k^+(r) \leq k(u)$, hence $k^+(r) \not\geq a$. Therefore $h^+(r) \neq k^+(r)$, i.e. $(h,k) \notin K$. This shows that $H_L^2 \setminus K$ is open. Hence $\mathcal{T}_L$ is Hausdorff.

(2) By 5.6 and the fact that (pathwise) connectedness is preserved under quotients. □

REMARK 6.6. That the natural topology of $I(L)$ is Hausdorff will follow from 6.11 too. We included the above proof since it is actually the one for a continuous lattice with the Lawson topology (cf. 5.3). Though the proof of 6.11 will require nothing but a continuous de Morgan algebra with its Scott topology, to deduce 6.5 from 6.11 we shall have to use the fact the ι_L is a left inverse of ω_L. Then all the needed assumptions gathered together imply complete distributivity. Incidentally, note that we may also restate 5.7 in the context of 6.5(2).

DEFINITION 6.7 ([Hutton 1975]). Let $(L, {}')$ be a complete de Morgan algebra. The **natural L-fuzzy topology** $\mathcal{U}_L$ on $I(L)$ is one generated by the L-sets $R_t, L_t \in L^{I(L)}$ ($t \in I$) defined by $R_t[f] = f^+(t)$ and $L_t[f] = f^-(t)'$ for each $[f] \in I(L)$ and each $t \in I$.

REMARK 6.8. Let us recall that $e : (I, \chi(\mathcal{E})) \to (I(L), \mathcal{U}_L)$ defined by $e(t) = [1_{[0,t)}]$ is a continuous embedding, where $\mathcal{E}$ is the natural topology on I. One has $R_s e = 1_{(s,1]}$ and $L_s e = 1_{[0,s)}$ for each $s \in I$. In particular, $(I(2), \mathcal{U}_2) \cong (I, \chi(\mathcal{E}))$.

PROPOSITION 6.9. Let $(L, {}')$ be a complete de Morgan algebra. Then $\mathcal{U}_L \subset \omega_L(\mathcal{T}_L)$.

PROOF. The assertion will follow if it is shown that $R_t q, L_t q : (H_L, \mathcal{H}_L) \to (L, \mathrm{SUP}(L))$ are continuous for each $t \in I$, where q is the natural quotient mapping from H_L to $I(L)$. For each $a \in L$, $\downarrow(a)$ is a subbasic closed set in L, and

$$\begin{aligned}(L_t q)^{-1}(\downarrow(a)) &= \{f \in H_L : \bigvee_{s<t} f(s)' \leq a\} \\ &= \bigcap_{s<t} \{f \in H_L : \pi_s(f) \geq a'\} \\ &= H_L \cap \bigcap_{s<t} \pi_s^{-1}(\uparrow(a'))\end{aligned}$$

is closed in H_L. Similarly for each $a \in L$ we have

$$(R_t q)^{-1}(\downarrow(a)) = H_L \cap \bigcap_{s>t} \pi_s^{-1}(\downarrow(a))$$

is closed in H_L too. The assertion follows. □

REMARK 6.10. As proved by [Artico and Moresco 1988], for any completely distributive lattice $L \neq 2$ one has $\mathcal{U}_L \cap 2^{I(L)} = \{0,1\}$. This is presumably the most evident argument that the inclusion $\mathcal{U}_L \subset \omega_L(\mathcal{T}_L)$ is proper if $L \neq 2$. We note that one cannot simply replace SUP(L) by $\sigma(L)$ in 6.9 (and thus in 6.13 which follows), since the quasi-complementation need not reverse the way below relation.

PROPOSITION 6.11. Let $(L, ')$ be a completely distributive de Morgan algebra. Then $(I(L), \iota_L(\mathcal{U}_L))$ is Hausdorff.

PROOF. Let $[f] \neq [g]$ in $I(L)$. Then $f^+ \neq g^+$. Without loss of generality we may assume that $f^+ \not\leq g^+$. By 6.2(3), this is equivalent to $f^+ \not\leq g^-$, and thus $f^+(t) \not\leq g^-(t)$ for some $t \neq 1$. By (C), there exists $a \in L$ such that $a \ll f^+(t)$ and $a \not\leq g^-(t)$. Thus $[f] \in R_t^{-1}(\uparrow(a))$ and $[g] \in L_t^{-1}(L \setminus \downarrow(a'))$. Then $R_t^{-1}(\uparrow(a))$ and $L_t^{-1}(L \setminus \downarrow(a'))$ are in $\iota_L(\mathcal{U}_L)$ (recall that $\uparrow a \in \mathrm{SUP}(L)$ by 1.3) and are disjoint. If not, $R_t^{-1}(\uparrow(a)) \cap L_t^{-1}(L \setminus \downarrow(a'))$ would include an h with $h^-(t) \geq h^+(t) \gg a$ and $h^-(t) \not\geq a$, which is a contradiction. □

REMARK 6.12. We note that 6.11 is valid for L a continuous de Morgan algebra with its Scott topology and with $\iota_L(\mathcal{U}_L) = \bigvee\{U^{\leftarrow}(\sigma(L)) : U \in \mathcal{U}_L\}$ (recall that SUP$(L) \subset \sigma(L)$).

The following provides a complete characterization of the topological modification: of $I(L)$ with its natural L-topology.

THEOREM 6.13. If $(L, ')$ is a completely distributive de Morgan algebra, then $\iota_L(\mathcal{U}_L) = \mathcal{T}_L$.

PROOF. Since $\mathcal{U}_L \subset \omega_L(\mathcal{T}_L)$, hence $\iota_L(\mathcal{U}_L) \subset \mathcal{T}_L$ by 3.5(2). Since $\mathcal{T}_L$ is compact, so is $\iota_L(\mathcal{U}_L)$. As both $\iota_L(\mathcal{U}_L)$ and $\mathcal{T}_L$ are Hausdorff, they must agree by the well-known fact that compact Hausdorff spaces are minimal Hausdorff (and maximal compact). □

§7. Some L-Topological Properties of the L-Unit Interval

In this section, unless otherwise stated, $(L, ')$ is a completely distributive de Morgan algebra. Observations made in Section 6 allow us to formulate the following:

PROPOSITION 7.1.

(1) If L is a complete de Morgan algebra, then $\mathcal{U}_L$ as well as $\mathcal{U}_L^c$ have all the properties that are good extensions of the properties of $\mathcal{T}_L$ and that are preserved under weakening of an L-topology.
(2) For P a topological property, $\mathcal{U}_L$ and $\mathcal{U}_L^c$ have ι-P iff $\mathcal{T}_L$ has P.

PROOF. By 6.9 and 6.13, respectively. □

Recall the following two notions of connectedness in L-spaces.

DEFINITION 7.2. (1) [Lowen 1981b] An L-space (X,T) is **(c1)-connected** if no clopen (= closed and open) L-set $U > 0$ has a representation $U = V \vee W$ where both V and W are open L-sets distinct from U and such that $V \wedge W = 0$.

(2) [Rodabaugh 1982b] An L-space (X,T) is 1-**connected** if there do not exist $U, V \in T \setminus \{0,1\}$ such that $U \vee V > 0$ and $U \wedge V = 0$.

NOTE. For the concept of (c2)-connectedness, which makes sense only when $L = I$, see [Lowen 1981b].

As proved by [Lowen 1981b], both (c1)- and (c2)-connectedness are good extensions of ordinary connectedness if $L = I$. He also provided examples that these two concepts are generally weaker than ultraconnectedness (cf. 3.15(2)). That 1-connectedness is generally weaker than ultraconnectedness if $L = I$ is shown in [Kubiak 1988a]. In general lattice setting we have the following straightforward observation (for the definition of L^b see the Introduction).

PROPOSTION 7.3. If $0 \in L^b$ (L complete), then (c1)-connectedness and 1-connectedness are good extensions of ordinary connectedness.

By 6.9 (and 6.10), the following exhibits an ultracompact and ultraconnected L-topology in $I(L)$ which is strictly finer than the natural one (if $L \neq 2$). In what follows, note the use of the terminology of 3.15(2). We shall avoid statements about N-compactness (cf. 4.15).

THEOREM 7.4. The space $(I(L), \omega_L(\mathcal{T}_L))$ is:

(1) ultracompact (hence Ch-compact, strongly compact and fuzzy compact),
(2) ultraconnected provided $(L, \mathrm{INT}(L))$ is connected (hence (c1)-connected and 1-connected if $0 \in L^b$),
(3) ι-Hausdorff.

PROOF. (1) is a consequence of 6.5(1), 3.14(1) and 4.7, (2) comes from 6.5(2), 3.14(1), and 7.3 together with 3.14(3). Finally, (3) follows from 6.5(1) and 3.14(1). □

PROPOSITION 7.5. $\omega_L(\mathcal{T}_L)$ is a maximal ultracompact L-topology on $I(L)$. Consequently, it is a maximal Ch-compact, strongly compact and fuzzy compact L-topology on $I(L)$.

PROOF. Since $\mathcal{T}_L$ is maximal compact (by 6.5(1)), $\omega_L(\mathcal{T}_L)$ is maximal ultracompact by 3.17(1). As for the other compactness axioms note that they are equivalent to ultracompactness in the class of topologically generated L-spaces by 4.6. □

By the fact that all the properties listed in (1) and (2) of 7.4 are preserved under weakening of an L-topology and by 6.11, we may restate 7.4 for the case of the L-unit interval with its natural L-topology.

THEOREM 7.6. The spaces $(I(L),\mathcal{U}_L)$ and $(I(L),\mathcal{U}_L^c)$ are:

(1) ultracompact (hence Ch-compact, strongly compact and fuzzy compact),
(2) ultraconnected provided $(L,\mathrm{INT}(L))$ is connected (hence (c1)-connected and 1-connected if $0 \in L^b$),
(3) ι -Hausdorff.

REMARK 7.7.

(1) That $(I(I),\mathcal{U}_I)$ is ultracompact was first proved by [Lowen 1984]. That this is true for L a complete chain was proved in [Kubiak 1988a]. A direct proof that $(I(L),\mathcal{U}_L)$ is N-compact is given in [Liu and Luo 1988]. For a discussion about a-compactness of $I(L)$ see the Introduction.

(2) It is to be noted that $(I(L),\mathcal{U}_L)$ is known to be 1-connected if $0 \in L^b$ and without the hypothesis about connectedness of $(L,\mathrm{INT}(L))$ (see [Rodabaugh 1982b]). As for 1-connectedness and (c1)-connectedness also note that by 5.7 and 7.3, complete distributivity of L in 7.4 and 7.6 (as well as in 7.10) can be relaxed to the assumption that L is complete with primes order generating. Clearly, (c1)-connectedness (and (c2)-connectedness with $L = I$) may hold vacuously in the nonstratified case.

(3) ι-Hausdorff is the only known Hausdorff-like axiom for L-spaces which $(I(L),\mathcal{U}_L)$ satisfies (see also 7.16) and note that the last sentence in [Kubiak 1988a] is therefore false (that has already been corrected in [Kubiak 1990a]). There were previously only negative results in this direction, namely in [Liu 1983] and [Rodabaugh 1980, 1983b]. As far as the other lower separation axioms are concerned, $(I(L),\mathcal{U}_L)$ is known to be L-T_0 with L a complete lattice, i.e. for each two distinct points there exists an open L-set whose values at these points are distinct (see [Rodabaugh 1986-7, 1991]; note that this property is implicit in the proof of 6.5, which actually shows that the right-hand L-topology of $I(L)$ (i.e. the one having $\{R_t : t \in I\}$ as a base) is already L-T_0). The other T_0-like axiom holding for $(I(L),\mathcal{U}_L)$ was proposed by [Liu 1983], and is called by him sub-T_0. However, it is not hard to check that for any complete L one has the following: ι_L-Hausdorff $\Rightarrow$ L-T_0 $\Leftrightarrow$ ι_L-T_0 $\Leftrightarrow$ sub-T_0. We note that with $L = I$, the first equivalence is in Theorem 3.1 of [Lowen and Srivastava 1989], likewise the second one,

since condition (3) of the mentioned result is obviously equivalent with sub-T_0 for any complete L. One cannot claim $\iota_L(\mathcal{U}_L)$ to be Hausdorff without SUP(L) (or $\sigma(L)$, cf. 6.12) being Hausdorff, in particular for L being merely complete.

(4) As noted in 5.3 (see also 6.6 and 6.12), with the Helly space constructed from $(L, \lambda(L))$, 5.2 and 6.5(1) hold for L a continuous lattice, likewise all the statements (and their proofs) about Hausdorfness and compactness of $\mathcal{T}_L$ and $\omega_L(\mathcal{T}_L)$. This is not the case for connectivity properties, since 5.4 involves order generating of primes, hence (ID), and 5.6 uses a quase-complementation (cf. 1.5). Finally note that all the statements about compactness of $\mathcal{U}_L$ depend on 6.13, a result involving (CD) and INT(L) (cf. 6.10).

PROPOSITION 7.8. $\mathcal{U}_L$ is maximal ultracompact iff $L = 2$.

PROOF Note only that if $L = 2$, then $\mathcal{E} = \iota_2(\chi(\mathcal{E}))$ (for $\mathcal{E}$ see 6.8), hence $\chi(\mathcal{E})$ is maximal ultracompact on I since $\mathcal{E}$ is maximal compact on I (note that ι_2 is injective). Finally recall that $(I(2),\mathcal{U}_2) \cong (I, \chi(\mathcal{E}))$. □

REMARK 7.9. Since $(I(2),\mathcal{U}_2) \cong (I(2), \omega_2(\mathcal{T}_2))$ $(\cong (I, \chi(\mathcal{E})))$, hence $(I(L), \omega_L(\mathcal{T}_L))$ might be regarded as another "fuzzy version" of the real unit interval.

Let Γ be an index set. Let $I(L)^\Gamma$, $\mathcal{U}_L^\Gamma$ and $\mathcal{T}_L^\Gamma$ denote products of Γ copies of $I(L)$, $\mathcal{U}_L$ and $\mathcal{T}_L$, respectively. Then by 3.20, 6.9, 6.13 and 6.5 one has the following:

PROPOSITION 7.10. The following statements hold:

(1) $\mathcal{U}_L^\Gamma \subset \omega_L(\mathcal{T}_L^\Gamma)$.
(2) $\iota_L(\mathcal{U}_L^\Gamma) = \mathcal{T}_L^\Gamma$.
(3) The cubes $(I(L)^\Gamma, \omega_L(\mathcal{T}_L^\Gamma))$ and $(I(L)^\Gamma, \mathcal{U}_L^\Gamma)$ (and the stratification of the latter) have all the properties listed in 7.6.

REMARK 7.11. Using an embedding theorem of [Liu 1983] and the N-compactness of the L-fuzzy cube, a Stone-Čech type N-compactification theory was developed by [Liu and Luo 1988 and 1989]. All the cubes of 7.10(3) can therefore be used to develop an ultracompactification (cf. [Martin 1980b]).

REMARK 7.12. Some results about a-level spaces of $(I(L),\mathcal{U}_L)$ can be deduced from 6.13 as well (cf. 3.22), in particular some of the properties of $(I(I), \iota_{I,a}(\mathcal{U}_I))$ listed in [Lowen 1986] (for this purpose see [Berri *et al.* 1971]).

QUESTION 7.13.

(1) When is $(I(L),\mathcal{U}_L)$ in fact ultraconnected? Our argument for this was rather uneconomical: INT(L) is connected $\Rightarrow$ $\mathcal{H}_L$ is connected $\Rightarrow$ $\mathcal{T}_L$ (= a quotient of the former) is connected $\Rightarrow$ $\omega_L(\mathcal{T}_L)$ is ultraconnected $\Rightarrow$ $\mathcal{U}_L$ (= the L-topology weaker than the

former) is ultraconnected. The only advantage is that we established "ultra path connectedness".

(2) We repeat the following question: Is $(I(L), \mathcal{U}_L)$ 1*-connected? (See [Rodabaugh, 1982b] for details.) We note that $(I(L), \omega_L(\mathcal{T}_L))$ fails to be 1*-connected as $\mathcal{T}_L$ is Hausdorff (see also Remark 5 in [Kubiak 1988a]). Incidentally note that various sufficient conditions for 1*-compactness of $(I(L), \mathcal{U}_L)$ were given by [Artico and Moresco 1988].

Another observation is that $(I(I), \mathcal{U}_I)$ is an ι_I-AR(normal), i.e. that its topological modification can be substituted for the real unit interval in the Tietze Extension Theorem. We note that the Helly space $(H_I, \mathcal{H}_I)$ itself is not even an ANR(compact Hausdorff) (see [Borges 1985]). For the concept of normality in fuzzy topology see [Hutton 1975]. Also note that T_1-separation is not assumed when speaking about normal topological spaces. (That f is (T, S)-continuous will clearly mean that $f : (X, T) \to (Y, S)$ is continuous.)

THEOREM 7.14. Let (X, T) be a topological space. The following statements are equivalent:

(1) (X, T) is normal.
(2) For every closed $A \subset X$, each continuous $f : (A, T_A) \to (I(I), \mathcal{T}_I)$ has a continuous extension to the whole of X.

PROOF.

(1) $\Rightarrow$ (2): If $f : (A, T_A) \to (I(I), \mathcal{T}_I)$ is continuous then it is $(\omega_I(T_A), \omega_I(\mathcal{T}_I))$-continuous, hence $(\omega_I(T)_A, \mathcal{U}_I)$-continuous by 3.19 and 6.9. Since 1_A is closed in $(X, \omega_I(T))$ and $(X, \omega_I(T))$ is normal (see [Martin 1981] or [Wang and Hu 1985]), hence f is extendable to a continuous $F : (X, \omega_I(T)) \to (I(I), \mathcal{U}_I)$ by the fuzzy Tietze Extension Theorem (see [Kubiak 1987]). Further, F is $(\iota_I \omega_I(T), \iota_I(\mathcal{U}_I))$-continuous, and since ι_I is a left inverse to ω_I and $\mathcal{T}_I = \iota_I(\mathcal{U}_I)$, F is therefore the required extension.

(2) $\Rightarrow$ (1): Since $(I(I), \mathcal{T}_I)$ is Hausdorff, this is well-known.

REMARK 7.15. Note that 7.14 characterizes extremally disconnected spaces if "closed" in (2) is replaced by "open". See [Kubiak 1990a, 1986] for details.

We finish this section with the following comments about Hausdorffness in fuzzy topology.

REMARK 7.16. Let **F-compact** and **F-Hausdorff** denote arbitrary two good extensions of ordinary compactness and Hausdorffness which are preserved under, resp., weakening and strengthening of an L-topology (cf. 3.15(3)). Then $(I(L), \mathcal{U}_L)$ $(L \neq 2)$ is never F-compact and F-Hausdorff if one requires that an L-space which is both F-compact and F-Hausdorff is minimal F-Hausdorff and maximal F-compact (as $\mathcal{U}_L$ is strictly coarser than $\omega_L(\mathcal{T}_L)$ and the latter is F-compact and F-Hausdorff).

REMARK 7.17. An example of such F-compact and F-Hausdorff is fuzzy compact ($L = I$) and fuzzy Hausdorff in the sense of [Lowen 1981a]. It is well known that a fuzzy Hausdorff and fuzzy compact stratified I-space is topologically generated. Note that Lowen's Hausdorffness is equivalent to the one of [Pu and Liu 1980a], as mentioned in [Wuyts and Lowen 1983] but not proved. This is indeed the case (J. J. Chadwick). Now, in the presence of the Pu/Liu-Lowen T_2-axiom, ultracompactness is equivalent to fuzzy compactness (see [Wang 1983]). Therefore by 3.17(2) we get Lowen's "Compact Hausdorff fuzzy topological spaces are topological" (see [Lowen 1981c]). In passing: when [Wang 1983] talks about N-compact and T_2-I-spaces, he actually talks about the topologically generated ones. (Is [Zhao 1987] doing the same?)

REMARK 7.18. As proved by J. J. Chadwick (unpublished), fuzzy compact I-sets (see 4.19) in (Pu/Liu-Lowen) Hausdorff stratified I-spaces are closed. This is the only such circumstance that is known without additional hypotheses (cf. Proposition 2 in [Liu and Luo 1987] which states: In a stratified L-space (X,T) each Ch-compact L-set is closed provided (X,T_0) is a Hausdorff topological space where $T_0 = \{U \subset X : 1_U \in T\}$). But we cannot be satisfied if the entire space is not allowed to be fuzzy compact (unless we like topologically generated spaces). We therefore call for a good (not necessarily good extension) "fuzzy compact Hausdorff".

Chapter 12:

A Categorical Approach to Fuzzy Relational Database Theory

S. E. Rodabaugh et al. (eds.), Applications of Category Theory to Fuzzy Subsets, 307–324.

A CATEGORICAL APPROACH TO FUZZY RELATIONAL DATABASE THEORY

W. A. MacCaull

§0. Introduction

The representation of inexact or incomplete information has become an important area of database research. A number of authors have proposed a variety of generalizations to parts of classical database theory so as to be able to deal with vague or incomplete information. For the most part, the generalizations have been of one specific data model, the relational data model. One of the features of the relational data model is that its associated query language, though algebraic in nature, is equivalent to a fragment of predicate calculus, the relational calculus. Many of the generalizations use fuzzy set theory to provide an interpretation of impreciseness in relational databases. An excellent review of the literature pertaining to this approach can be found in [Raju and Majumdar 1989].

There have been a number of problems with the generalizations. First, there is disagreement as to what a fuzzy relation should be; second, as a rule, the generalized theory does not get beyond the specification of the query language; and finally, the definitions appear to be somewhat ad hoc.

In this chapter we take a categorical viewpoint to the problem of generalizing relational database theory. From this viewpoint the way to develop fuzzy relational database theory is to interpret the usual or "classical" theory in a category with more general properties than the category of **SET**. The advantage of this approach is that as a generalization of ordinary relational database theory on a set, it is a direct translation of the original theory, say, as found in [Ullman 1982]. Thus, there is a natural way to extend the definition of all of the important constructions of relational database theory. In consequence, the generalization is less ad hoc. Another advantage to this approach is that when there are two classically equivalent definitions, the choice between which of them to generalize is often more easily made, because in a categorical interpretation, one definition usually has desirable or undesirable features that the other does not have. Models for relational database theory are then interpretations of the usual theory in appropriate categories: when the theory is interpreted in the category of **SET**, it gives the classical theory, and when the theory is interpreted in the category of **FUZZY SET**, it gives fuzzy relational database theory. We believe that the categorical approach will guide further development of a practically useful fuzzy relational database theory.

One of the challanges still facing fuzzy set theory is to find appropriate categorical foundations for fuzzy sets. At the same time, much of the theory of fuzzy relational databases has not yet been developed; in fact, there is disagreement over parts that have been investigated. Indeed, this situation will probably exist until a proper categorical foundation for fuzzy sets is found. However, there are some aspects of fuzzy set theory that practioneers in fuzzy relational databases have used. Thus, it would seem that a prerequisite for the fuzzy models that we obtain, is that a special case should yield what the practioneers have found useful.

In [Raju and Majumdar 1989] we find a generalized approach for treating precise as well as imprecise data which extends results in classical relational database literature involving problems of data design. Here, we present a fuzzy relational data model for which many of the results achieved in the Raju and Majumdar paper hold. The advantages of our method will become clearer below, but in short, they can be summarized as follows: (1) we do not need the fuzzy relation EQUAL; (2) the definition of natural join of a decomposition of a relation scheme is better understood in the categorical setting, so there is no need to introduce the notion of cylindrification, which has no counterpart in the classical theory; (3) the general approach we use here can be used in another category if it turns out that the appropriate category for fuzzy set theory is not the one we consider.

A number of different categories have been proposed as the category of **FUZZY SET**. For reasons of extensionality [Johnstone 1989], it is widely felt that the equality as well as the membership predicate must be fuzzy. In [Ponasse 1988] we find a description of the various categories of fuzzy sets with valued (i.e., fuzzy) equality. There has been an extensive literature on the relationships among them [Pitts 1982; Stout 1984; Barr 1986; Johnstone 1989]. However,as the categorical foundation of fuzzy sets has not yet been decided, we shall discuss three categories: **FUZ(L)**, **L-SET**, and **CM-SET**, because of the topos–like characteristics of these categories. The idea of using topos theory and sheaf representations to generalize classical results is not new (see, for example [Mulvey 1974] and [MacCaull 1989]). Indeed, in [Höhle 1986], we find an example where categorical ideas are used to clarify the notion of fuzzy topology. The author believes that choice of appropriate fuzzy notions will be explicated by investigating the interpretation of the "crisp" notion inside categories that have topos–like properties. This approach will not only suggest that the appropriate fuzzy analogue to the "crisp" notion, but will also suggest ways to extend the fuzzy theory in ways which are compatible with the so–called "crisp" theory. The already extensive development of categorical logic should be used to further the development of fuzzy relational database theory.

The outline of the chapter is as follows: after a brief introduction to the categories of interest here, we outline and interpret the basic constructions of relational database theory in two that seem particularly suitable as foundations of fuzzy set theory. For reasons which will become clear, we then focus on one category, namely **L-SET**, and show how to evaluate queries in **L-SET**. Then we define functional dependency, prove that Armstrong's Axioms are sound and complete in **L-SET**, and then, moving on to the problems of data design, prove that a theorem, which serves as the basis of classical decomposition theory, holds in **L-SET**.

§1. Some Topos-like Categories

Let $(L, \leq)$ be a complete Heyting algebra. We begin by defining the category **FUZ(L)**. An **L-fuzzy set** is an ordinary set A equipped with a map $\in_A$: $A \longrightarrow L$. We may abuse notation by writing $\in$ in place of $\in_A$, if the set A is clear from the context. A **morphism** $f : (A, \in) \to (B, \in)$ of L-fuzzy sets is a map $f : A \longrightarrow B$ such that for all a in A, $\in(a) \leq \in(f(a))$.

Again, let $(L, \leq)$ be a complete Heyting algebra and let X be a set. An **L-equality relation** on X is a map $\delta_X : X \times X \longrightarrow L$ satisfying the following conditions:

1. $\delta_X(x, y) = \delta_X(y, x)$;
2. $\delta_X(x, y) \wedge \delta_X(y, z) \leq \delta_X(x, z)$.

Again, we usually abuse notation by writing δ in place of δ_X if the set X is clear from the context. The value $\delta(x, y)$ can be interpreted as the degree that the elements x and y are equal.

The category **L-SET** has as **objects**, pairs (X, δ_X) consisting of a set X and an **L-equality**. Its **morphisms** $F : (X, \delta_X) \longrightarrow (Y, \delta_Y)$ are ordinary maps $F : X \times Y \longrightarrow L$ satisfying:

1. $F(x, y) \wedge \delta(x, x') \leq F(x', y)$
   $F(x, y) \wedge \delta(y, y') \leq F(x, y')$
2. $F(x, y) \leq \delta(x, x') \wedge \delta(y, y')$
3. $F(x, y) \wedge F(x, y') \leq \delta(y, y')$
4. $\bigvee_{y \in Y} F(x, y) = \delta(x, x)$

It is well known that **L-SET** is a topos, so subobjects of (A, δ) can be identified with ordinary maps $s : A \longrightarrow L$ satisfying the following conditions:

1. $s(x) \wedge \delta_A(x, y) \leq s(y)$ $(*\prime)$
2. $s(x) \leq \delta_A(x, x)$.

Such a function assigns to each $x \in A$ an element $s(x)$, which we think of as the truth value of "x ∈ s" or a measure of the extent to which x belongs to the set s.

A subobject $s : A \longrightarrow L$ of A determines a monic arrow $f_s : A_s \rightarrowtail A$. A_s has the same set of elements as A but with δ-equality given by:

$$\delta_{A_s}(x, y) = s(x) \wedge s(y) \wedge \delta_A(x, y)$$

The inclusion arrow f_s has $\delta_A(f_s(x), y) = \delta_{A_s}(x, y)$. Since $s(x)$ is the same as $\delta_{A_s}(x, x)$, we may write

$$\delta_{A_s}(x, y) = \delta_{A_s}(x, x) \wedge \delta_{A_s}(y, y) \wedge \delta_A(x, y)$$

Equation $(*')$, above, may also be written

$$\delta_{A_s}(x,x) \wedge \delta_A(x,y) \leq \delta_{A_s}(y,y). \qquad (*)$$

Mimicing [Pitts 1982], we now identify the category **FUZ(L)** with a full subcategory of **L-SET**. Given a set A, let ${}^A\delta : A \times A \longrightarrow L$ be the function:

$$ {}^A\delta(a,a') = \begin{array}{ll} 1, & \text{if } a = a' \\ 0, & \text{otherwise} \end{array} $$

where 1 and 0 are the top and bottom elements of L respectively. Since ${}^A\delta$ trivially satisfies the two conditions above, $(A, {}^A\delta)$ is an L-valued set, often referred to as a **constant L-valued set on A**. An L-valued set (X, δ_A) is called **subconstant** if there is a monomorphism from it to some constant L-valued set. If $\in_A : A \longrightarrow L$ is an object of **FUZ(L)**, we obtain an L-valued set (A, δ_A) where:

$$\delta_A(x,x') = {}^A\delta(x,x') \wedge \in_A(x)$$

It is straightforward to verify that this (A, δ_A) is subconstant, and that **FUZ(L)** is equivalent to the full subcategory of **L-SET**, whose objects are the subconstant L-valued sets. Since the subcategory consists of subconstant objects, i.e., subobjects of constant objects, the category is really a category of fuzzy subsets. An excellent review of the fundamental properties of categories of **L-SET** can be found in [Goldblatt 1984].

Both the above categories have sufficient structure to interpret first order logic. However, **FUZ(L)** is not a topos, and higher order logic cannot be interpreted in it. The usual categorical interpretation of a first order language in a category is explained in [Stout 1984], or in [Fourman and Scott 1977]. This interpretation gives what is commonly known as Heyting logic. However, there is a problem: fuzzy set theory had its origins in attemps to deal with imprecision and vagueness in engineering problems of man-machine interface. Connectives, reflecting intuitive understanding of how these truth values seem to combine in the fuzzy world being described, were imposed. Different connectives proved useful in different settings. The use of Łukaciewicz implication and related negation were characteristic. An excellent comparison of fuzzy set theory and topos theory can be found in [Stout 1990]. The chart below summarizes the differences between Heyting interpretations of the connectives and one set of fuzzy interpretations commonly used for the connectives when L is the unit interval. The first would be the interpretation in **FUZ(L)** or **L-SET** for $L = [0,1]$, with the usual order. The Heyting negation is particularly pathological for real world applications. The model for intuitionistic fuzzy set theory developed by [Takeuti and Titani 1986] suffers from the same pathology as **L-SET** since it is based on intuitionistic logic.

CONNECTIVES ON $[0,1]$

	Heyting	**Fuzzy**
$a \wedge b$	$\min(a,b)$	$\max(0, a+b-1)$
$a \vee b$	$\max(a,b)$	$\min(1, a+b)$
$a \Rightarrow b$	$\max \mid \{h/h \wedge a \leq b\}$	$\min(1, 1-a+b)$
$\sim a$	$\max \mid \{h/h \wedge a = 0\}$	$1-a$

In fuzzy sets used in practical applications, there is a large variety of interpretations of **implies** ($\Rightarrow$), so there is a large number of possible implication operators, as well as a variety of interpretations for **and** ($\wedge$) and **or** ($\vee$).

It would seem then that the categories thus far consider lack some features needed for a good foundation for fuzzy set theory. Höhle has suggested that another kind of category — which he termed a weak topos — would be a better choice for a foundation. Weak toposes are described in [Höhle 1990a]. As their name suggests, they do not have all the properties of toposes, but they have a number of them which make them a good category to consider as a foundation for fuzzy set theory. In particular they allow the interpretation of higher order as well as first order logic. This already makes them better candidate for a generalized set theory than **FUZ(L)**.

We shall describe one example of the weak topos, which is the category **CM-SET**. This, more or less canonical, example of a weak topos is a generalization of **L-SET**.

Let $M = (L, *, \leq)$ be a strong monoid (cf. [Birkhoff 1967]) with the added condition:

$$\text{if } e * e = e \text{ and } e \leq \bigvee_{a \in A} a, \text{ then } e \leq \bigvee_{a \in A} a * a.$$

Examples of such are complete Heyting algebras where $*$ is $\wedge$, and the real unit interval where $*$ is a left continuous t-norm. An **M-valued equality** on a set X is a map $\delta : X \times X \longrightarrow M$ such that:

1. $\delta(x,x) * \delta(x,y) * \delta(y,y) = \delta(x,y)$
2. $\delta(x,x) \vee \delta(y,y) \leq \delta(x,y) \Rightarrow x = y$
3. $\delta(x,y) \leq \delta(y,x)$
4. $\delta(x,y) * \delta(y,z) \leq \delta(x,z)$

The ordered pair (X, δ) is an M-valued set. A **singleton** of an **M-valued set** is a map $\sigma : X \to L$ provided with properties:

1. $\sigma(x) * \delta(x,x) = \sigma(x)$
   $\sigma(x) * \bigvee_{y \in X} \sigma(y) = \sigma(x)$
2. $\sigma(x) * \delta(x,y) \leq \sigma(y)$
3. $\sigma(x) * \sigma(y) \leq \delta(x,y)$.

An M-valued set (X, δ) is said to be **complete** iff for every singleton σ of (X, δ), and for every $x \in X$, there exists a (unique) element $x_0 \in X$ such that $\sigma(x) = \delta(x_0, x)$. The category of **CM-SET** has, as objects, complete M-valued set, and as morphisms, maps $F : (X, \delta_X) \longrightarrow (Y, \delta_Y)$ with the following ("equality preserving") conditions:

1. $\delta_X(x, x') \leq \delta_Y(F(x), F(x'))$
2. $\delta_X(x, x) = \delta_Y(F(x), F(x))$.

The category of **CM-SET** is monoidal closed, but not in general cartesian closed.

One of the simplest examples of **CM-SET** uses the unit interval for M and $*$ is the t-norm T defined by:

$$T(a, b) = Max(0, a + b - 1).$$

In general, t-norms are suitable candidates for conjunctions in many valued logic: i.e., t-norms can be seen as generalized conjunctions. With each t-norm, one can associate a t-conorm which can be seen as generalized disjunction. DeMorgan laws connect the two. With each t-norm t, we define an operator $t_{\Rightarrow}$ which gives the associated implication and hence its related negation. Then the interpretation of the logical connectives gives the fuzzy connectives as described in the chart above. With some further (somewhat technical) conditions on M, which make it "locally involutary" (cf. [Höhle 1990a]), the category of **CM-SET** is a weak topos. The conditions on a locally involutary monoid insure that its underlying lattice is a complete Heyting algebra.

§2. Relational Database Theory in a Category

In order to discuss relational databases in a category, we need a many sorted language L which has sort symbols (upper case letters, e.g., A or A_i), product sorts, variable symbols of each sort, equality symbol for each sort, constant symbols of each sort, relation symbols (lower case letters, e.g., r, s or r_i), and the symbols $\vee, \wedge, \Rightarrow, \sim, \forall$, and $\exists$. We also assume that we have a unary relation symbol $\in_A$ — always simply denoted $\in$ — for each sort A, and n-ary function sorts π_i, $i = 1, \ldots, n$, for each product over n-sorts. We use the notation $x \in A$ to signify that x is a variable of sort A. We will abuse notation by identifying the symbol in the language and its interpretation in a category. Terms, atomic formulas and formula are defined as usual. A sort symbol can be thought of as an attribute. Definitions and important notions relating to "crisp" relational databases can be found in [Ullman 1982]. In particular, we are using a modification of the domain relational calculus approach, (wherein each primitive object is consider to be an element of the domain of some attribute). A relation scheme R, is a collection of attributes (sorts), i.e., $R = (A_1, \ldots, A_n)$. A relation r, relating attributes $A_1, \ldots, A_n$, is a predicate with free variables with sorts among the sorts of $A_1, \ldots, A_n$. We say $r \in R$ if the attributes of r are included in the scheme R. When we say $x \in r$, we are assuming that x is a variable of the sort which is the product of the sorts (attributes) of r. We assume that our language has a distinguished sort N, and specified relations on $N \times N$ corresponding to the arithmetic comparison relations (e.g. ">").

The fundamental operations which define relational algebra are union, set difference, cartesian product, projection and selection. For each of the five operations, there corresponds a formula in L. We now give these formulas: here, r and r' are predicates of sort A, and r_i is a predicate of the sort A_i, $i = 1, \ldots, n$, and s is a predicate of sort of $A_1 \times \ldots \times A_n$.

UNION: $r \cup r' = [x \in A \mid x \in r \vee x \in r']$.
DIFFERENCE: $r - r' = [x \in A \mid x \in r \wedge \sim (x \in r')]$.
CARTESIAN PRODUCT: $r_1 \times r_2 = [(x_1, x_2) \epsilon A_1 \times A_2 \mid \pi_1(x_1, x_2) \in r_1 \wedge \pi_2(x_1, x_2) \in r_2]$.
PROJECTION: $\pi_i s = [x_i \in A_i \mid \exists x_1 \in A_1, \ldots, \exists x_{i-1} \in A_{i-1}, \exists x_{i+1} \in A_{i+1}, \ldots, \exists x_n \in A_n, (x_1, \ldots, x_n) \in s]$. There is the obvious extension of this definition to a projection onto several factors.
SELECTION: Let $r \in R(A_1, \ldots, A_n)$ be a relation on n attributes and let $F = F(n_1, \ldots\ldots, n_k)$ be a formula involving: variables x_i of sort A_i, $i = 1, \ldots, n$, constants n_i of sort N, $i = 1, \ldots, k$ (where $n > k$), arithmetic comparison relations, as well as logical constants. Then:

$$Sel_F r(A_1, \ldots, A_n) = [(x_1, \ldots, x_n) \in A_1, \ldots, A_n \mid (x_1, \ldots, x_n) \in r \wedge$$
$$F(n_1 | \pi_1(x_1, \ldots, x_n), \ldots, n_k \mid \pi_k(x_1, \ldots, x_n))]$$

$Sel_F r(A_1, \ldots, A_n)$ is a predicate of sort $A_1 \times \ldots \times A_n$.

The other useful operations of relational database theory can be expressed in terms of the above five. However, as definitions which are equivalent using classical logic are often not equivalent using other logics, each of these operations must be examined using the logic of the category under study, to see whether a definition in terms of the five basic ones, or a more direct one, is more appropriate to use.

We are particularly interested in the natural join operation of two relations, r, and s, denoted by $r <><> s$. Suppose r has m attributes, and s has m' attributes, n of which are the same as attributes of r; suppose further that attribute $A_{j(i)}$ of r is the same as attribute $A_{j'(i)}$ of s, for $i = 1, \ldots, n$. Then let F be the formula

$$\pi_{j(1)}(x) = \pi_{j'(1)}(x') \wedge \ldots \wedge \pi_{j(n)}(x) = \pi_{j'(n)}(x'),$$

where x is a variable of sort r, and x' is a variable of sort s. Let

$$\pi_{j(1), \ldots, j'(n)}$$

be shorthand for the projection map of the $m + m'$ attributes of $r \times s$ onto the $m + m' - n$ attributes of $r \times s$, excluding those attributes of s which are the same as attributes of r. Then $r <><> s$ is the formula

$$\pi_{j'(1), \ldots, j'(n)} Sel_F(r \times s).$$

(It selects from $r \times s$ the tuples whose values of attributes $j(i)$ and $m + j'(i)$ agree to some extent, and then projects onto the appropriate $m + m' - n$ attributes.) As an example, suppose $R = R(A, B, C)$, suppose $r \in R$ and $r_1 = \pi_{12}r$ and $r_2 = \pi_{13}r$; then F is the formula $\pi_1(x) = \pi_1(x\prime)$, and $r_1 <><> r_2 = \pi_{124}Sel_F(r_1 \times r_2)$. Consequently,

$$r_1 <><> r_2 = [(a,b,c) \in A \times B \times C \mid \exists\, a' \in A((a = a') \wedge (\pi_{12}(a,b,c), (a', \pi_3(a,b,c)) \in r_1 \times r_2]$$

For a complete description of the interpretation of a language in a category with enough structure, see [Makkai and Reyes 1977]. This description is used here for **FUZ(L)** and **L-SET**; in the case of **CM-SET**, we interpret the propositional connectives as fuzzy connectives. As mentioned above, the three categories considered here have enough structure for the interpretation of first order logic.

We now list the subobjects resulting from each of the operations above for the categories **L-SET** and **CM-SET**. Since **FUZ(L)** is a full subcategory of **L-SET**, the subobjects for **FUZ(L)** are the same as for **L-SET**. The descriptions for **L-SET** are well known and can be found in [Goldblatt 1984].

A **relation scheme** R, in a category, C, is simply a collection of objects, (which will be thought of as attributes) $R = (A_1, \ldots, A_n)$. A **relation** r relating attributes $A_1, \ldots, A_n$, in a category C, is a subobject of $A_1 \times \ldots \times A_n$. We say $r \in R$ if r is a subobject of the attributes included in the scheme R. Since the attributes are objects of either **L-SET** or **CM-SET**, the notation $t \in r$ denotes that t is an element of the product of the attributes in r (to the degree $\delta(t,t)$ for the appropriate δ). Thus we may think of t as a tuple in the relation r. We let $\pi_i(t)$ denote the i–th factor of t ($i \leq n$). So far this is exactly as done for ordinary relations except we have provided an interpretation in a category with enough structure. If the category is **SET** we have an ordinary relational database, where N is the natural numbers. **L-SET** has a natural numbers object N, and relations on $N \times N$ corresponding to the arithmetic comparison relations (e.g. ">") can be interpreted in **L-SET**.

L-SET

$r \cup s$	is the set $r \cup s$ with the δ described as $\max(\delta_r(x,y), \delta_s(x,y))$;
$r - s$	is the set with the δ described as $\delta(x,y) = \delta_r(x,y)$, if $\delta_s(x,y) = 0$, $= 0$, otherwise;
$r_1 \times r_2$	is the product set $r_1 \times r_2$ with the δ described as $\delta((x,y),(x',y')) = \min(\delta_r(x,x'), \delta_s(y,y'))$.

The description of the projection is somewhat involved. Here we will consider the simple projection $\pi_1 r$. Let $j : r \rightarrowtail A \times B$ be the monomorphism into $A \times B$ corresponding to the relation r, and let π_1 be the projection mapping of $A \times B$ onto A. Then $\pi_1 r$ is the smallest subobject of A such that the following diagram commutes.

$$\begin{array}{ccc} r & \xrightarrow{\quad j \quad} & A \times B \\ \exists \downarrow & \circ & \downarrow \pi_1 \\ \pi_1 r & \xhookrightarrow{\quad\quad} & A \end{array}$$

Categorically we say $\pi_1 r$ is the image of the map j under π_1.

Using the descriptions in [Goldblatt 1984] one can show that $\pi_1 r$ is the set A with the δ described by

$$\delta_{\pi_1}(a,a) = \bigvee_{t \in r,\ \pi_1(t)=a} (\delta(t,t))$$

and

$$\delta_{\pi_1}(a,a') = \delta_{\pi_1}(a,a) \wedge \delta_{\pi_1}(a',a') \wedge \delta_A(a,a')$$

For a relation r over $R(A_1,\ldots,A_n)$, the natural extension of these ideas to $\pi_{i(1)},\ldots,\pi_{i(n)}r$, gives the general definition of projection.

REMARK: Before we discuss the δ for selection operator, we recall that if $r \rightarrowtail A$ with $\delta_r(a,a)$ given, then $\delta_r(a,a') = \delta_r(a,a) \wedge \delta_r(a',a') \wedge \delta_A(a,a')$. Thus for a subobject r of an object A, it suffices to define $\delta_r(a,a)$; $\delta_r(a,a')$ then follows.

Let r be a relation over $R(A_1,\ldots,A_n)$, and suppose the interpretations of r and $F(n_1,\ldots,n_k)$ are subobjects of $A_1 \times \ldots \times A_n$ given by the maps $j_r : r \rightarrowtail A_1 \times \ldots \times A_n$ and $j_F : F(n_1,\ldots,n_k) \rightarrowtail A_1 \times \ldots \times A_n$. Then $Sel_F r$ is the pullback of j_r along j_F. From the description of pullbacks in L–Sets in [Goldblatt 1984], it is straightforward to show that the δ of $Sel_F r$, which we will denote by $\delta_{Sel_F} r$ is described as

$$\delta_{Sel_F} r(a,a') = \delta_r(a,a') \wedge \delta_F(a,a').$$

Using this, the δ function associated with $r <><> s$ can be determined from the description of projection and selection once the common attributes of r and s are determined. For example, if r is a relation in $R(A,B)$ and s is a relation in $R'(A,C)$, and F is the formula $\pi_1(x) = \pi_1(x\prime)$, where x and x' are variables of sort $A \times B$ and $A \times C$, respectively, then $r <><> s$ is the interpretation of $\pi_{124} Sel_F(r \times s)$. The δ function associated with $r <><> s$, which we denote by δ_{RjS}, is described as follows:

$$\delta_{RjS}((a,b,c),(a,b,c)) = \vee_{a \in A} \delta_{Sel_F} r((a,b,a',c),(a,b,a',c))$$

where

$$\delta_{Sel_F} r((a,b,a',c),(a,b,a'c)) = \delta_{R\times S}((a,b,a',c),(a,b,a',c)) \wedge \delta_A(a,a')$$

and $\delta_{RjS}((a,b,c),(a',b',c'))$ can be found by referring to the Remark above. (Implicit in the above definition is the assumption that B and C have at least one global section, i.e., that there is $b \epsilon B$ and $c \in C$ such that $\delta(b,b) = 1$ and $\delta(c,c) = 1$).

CM-SET

Let us first consider the category of **CM-SET**, where the monoid is $[0,1]$, with the usual order, and $*$ is the t-norm T defined in section 1. If we interpret $\wedge$ as $*$ and $\sim a$ as $a \Rightarrow 0$, we get the following interpretation.

$r \cup s$	is the set $r \cup s$ with the δ described as $\min(1, \delta_r(x,y) + \delta_s(x,y))$;
$r - s$	is the set with the δ described as $\delta(x,y) = \max(0, \delta_r(x,y) + (1 - \delta_s(x,y)) - 1)$;
$r \times s$	is the product set $r \times s$ with the δ described as $\delta((x,y),(x',y')) = \max(0, \delta_r(x,x') + \delta_s(y,y') - 1)$.

The projection and selection operators are defined as they are in **L-SET**, except one would use the **CM-SET** interpretation of $\wedge$.

Looking at the interpretation of $r \cup s$, one is reminded of an "exclusive or" interpretation. This may or may not be warranted and depends on the situation at hand. As far as database theory is concerned, the interpretation of $r \times s$, however, is not intuitive at all. The alternative is to interpret $\wedge$ as the inf of the underlying lattice. This brings us back to the interpretation of **L-SET**. If we were to move on to a weak topos of **CM-SET**, where $M = [0,1]$, the interpretation of $\wedge$ as $*$ would suffer from the same defect, as far as database theory is concerned. So again we would interpret $\wedge$ as the inf of the lattice. For this reason, we will concentrate on the category of **L-SET** (for L the unit interval) for the remainder of this paper. This is not to say that we feel that **L-SET** is the proper categorical foundation for fuzzy sets. Rather, that it is the best available thus far, especially when dealing with formulas relating to database theory. L. N. Stout, in his chapter in this volume, has suggested that a category with a cartesian closed structure as well as a monoidal closed structure is appropriate for a proper foundation, for this will give us both the Heyting and the fuzzy connectives. This suggestion warrants further study. Another suggestion that bears looking at is that of Muir [Muir 1986] when he suggests that one should get at fuzzy theory by doing probability theory in a topos.

§3. Evaluation of the Query Language

The interpretation of the language in **L-SET** for $L = [0,1]$, allows us to evaluate all the kinds of queries that are considered in [Buckles, Petri, and Sachar 1989]. Buckles et al. introduced the notion of similarity relationship to replace the notion of identity in fuzzy relational databases. This results in a valued equality. One objectionable feature of the notion of similarity relationships is that usually, large number of similarity relationships are possible for a given fuzzy relational database, and these may not be intrinsically related to the membership relation. So in a given application, the similarity relation may be rather ad hoc. Of course **L-SET** already has a valued equality so no such extra structure is necessary. Dealing with linguistic hedges is then simply a matter of solving for elements which are equal to another up to some δ-width. This is no less ad hoc than the approach to linguistic hedges used in Buckles et al., however the formulas here are easier to read and understand than the combination of similarity relation and linguistic hedge. It is interesting to note that the L-equivalence relation in **L-SET**, wherein, we can evaluate the degree to which x is equivalent to y, written $[x \equiv y]$, which is evaluated as $\delta(x,x) \vee \delta(y,y) \Rightarrow \delta(x,y)$ is a similarity relation.

We consider the main examples found in [Buckles, Petri, and Sachar 1989].

$L =$ the interval $[0,1]$ with the usual order
$\mathcal{P} = \{\text{Players}\} = \{\text{Joe, Robin, Gary, Danny, Brian, Steve}\}$. We abbreviate the members by J, R, G, D, B, S.
$\delta_{\mathcal{P}}(x,y) = 1$ for all $x, y \in \{J, R, \ldots, S\}$.

Suppose Joe is an excellent player, Robin is good, nearly excellent player, Gary is a good player, Danny is a medium, nearly good player, and Brian and Steve are both fair players. We define the subobject of {Players} as follows:

$$\mathcal{E} = \{\text{Excellent Players}\} \text{ and : } \begin{aligned} \delta_{\mathcal{E}}(J,J) &= 1 \\ \delta_{\mathcal{E}}(R,R) &= .9 \\ \delta_{\mathcal{E}}(G,G) &= .8 \\ \delta_{\mathcal{E}}(D,D) &= .7 \\ \delta_{\mathcal{E}}(B,B) &= \delta_{\mathcal{E}}(S,S) = .5 \end{aligned}$$

Also for any $x, y \in \{J, R, G, D, B, S\}$,

$$\delta_{\mathcal{E}}(x,y) = \min\{\delta_{\mathcal{E}}(x,x), \delta_{\mathcal{E}}(y,y)\}$$

By observing that

$$\begin{aligned} \delta_{\mathcal{E}}(x,z) &= \min\{\delta_{\mathcal{E}}(x,x), \delta_{\mathcal{E}}(z,z)\} \\ &\geq \min\{\delta_{\mathcal{E}}(x,x), \delta_{\mathcal{E}}(y,y)\} \wedge \min\{\delta_{\mathcal{E}}(y,y), \delta_{\mathcal{E}}(z,z)\} \\ &= \delta_{\mathcal{E}}(x,y) \wedge \delta_{\mathcal{E}}(y,z), \end{aligned}$$

one immediately concludes that $(\mathcal{E}, \delta_{\mathcal{E}})$ is an L-Set. Suppose Joe and Robin are exceptional leaders, Gary is a competent leader, Brian and Steve are average leaders and Danny is a poor leader. We define the subobject of {Players} as follows:

$$\mathcal{L} = \{\text{Leadership Players}\} \text{ and : } \begin{aligned} &\delta_{\mathcal{L}}(J,J) \ = \delta_{\mathcal{L}}(R,R) = 1 \\ &\delta_{\mathcal{L}}(G,G) = .8 \\ &\delta_{\mathcal{L}}(B,B) = \delta_{\mathcal{L}}(S,S) = .5 \\ &\delta_{\mathcal{L}}(D,D) = .3 \end{aligned}$$

As in the definition of δ, defining $\delta_{\mathcal{L}}(x,y) = \min\{\delta_{\mathcal{L}}(x,x), \delta_{\mathcal{L}}(y,y)\}$, makes $(\mathcal{L}, \delta_{\mathcal{L}})$ into an L-Set.

We now demonstrate the two examples done in Buckles et al. We assume that the language is enriched by two binary operations on L, $+$ and $-$, and three binary operations $\delta_{\mathcal{P}}, \delta_{\mathcal{E}}$ and $\delta_{\mathcal{L}}$.

1. Find all the players whose playing ability is more or less good.

To consider x a good player $\delta_{\mathcal{E}}(x,x)$ must equal .8. We must use a linguistic hedge to translate "more or less": we shall say x is "more or less good" iff $.8 - .3 \leq \delta_{\mathcal{E}}(x,x) \leq .8 + .3$.

2. List all the players who have almost the same playing ability as Joe and almost the same leadership ability as Joe.

Again we must use a linguistic hedge; "almost" will be translated as within .2 of the value of the attribute in question. Hence to satisfy the criterion, we must find:

$$[x \mid (1 - .2 \leq \delta_{\mathcal{E}}(x,x) \leq 1 + .2) \wedge (1 - .2 \leq \delta_{\mathcal{P}}(y,y) \leq 1 + .2]$$

Generalizing, we may ask for all players who have almost the same playing ability as x and almost the same leadership as x. Then we would want to find:

$$[y \mid (\delta_{\mathcal{E}}(x,x) - .2 \leq \delta_{\mathcal{E}}(x,y) \leq \delta_{\mathcal{E}}(x,x) + .2) \wedge (\delta_{\mathcal{L}}(x,x) - .2 \leq \delta_{\mathcal{L}}(y,y) \leq \delta_{\mathcal{L}}(x,x) + .2)]$$

§4. Aspects of Data Design

We now generalize some of the classical results involving data design to the categories of interest here. Central to the design of database schemes is the idea of data dependency. If one attribute uniquely determines another, we say that there is a functional dependency of the second on the first. One studies dependencies to determine how to pick a good set of relation schemes to represent the information of interest. The reader unfamiliar with common problems of design theory such as redundancies, inconsistencies, and update anomalies should read [Ullman 1982]. In the following discussion, we use the usual notation of relational database theory.

DEFINITION 1. Let $R = R(A_1, \ldots, A_n)$ be a relation scheme. We say **X is a subset of R**, and write $X \subset R$, iff X is a subset of $\{A_1, \ldots, A_n\}$. Further, if $t \in A_1 \times \ldots \times A_n$ and $X \subset R$ we use the notation $t[X]$ to denote the **projection of t onto the factors of X**. If $X \subset R$ and $Y \subset R$, we write XY to denote the set $X \cup Y$.

DEFINITION 2. Let $R = R(A_1, \ldots, A_n)$ be a relation scheme, let X and Y be subsets of R and let $r \in R$. We say **the functional dependency of Y on X holds in r**, and write $\mathbf{X} \to \mathbf{Y}$ holds in r, iff for all t_1, t_2 such that $\delta_r(t_i, t_i) > 0$ $(i = 1, 2)$,

$$\delta(t_1[X], t_2[X]) \leq \delta(t_1[Y], t_2[Y])$$

We write $\mathbf{X} \to \mathbf{Y}$ **holds** in R iff for all $r \in R$, $X \to Y$ holds in r. If R(or r) are understood from the context, we write $X \to Y$ holds.

Given a set of data dependencies that hold on a database it is often possible to derive other dependencies that hold for the same database. It is desirable to have a sound and complete set of inference axioms for these derivations. For functional dependencies in the category of **SET**, a set of axioms known as Armstrong's Axioms have been shown to be sound and complete. These axioms are as follows.

ARMSTRONG'S AXIOMS

Let X, Y, and Z be subsets of a relation scheme R.

FD1. **Reflexivity**: If Y is a subset of X, then $X \to Y$ holds.
FD2. **Augmentation**: If $X \to Y$ holds, then $XZ \to YZ$ holds.
FD3. **Transitivity**: If $X \to Y$ and $Y \to Z$ hold, then $X \to Z$ holds.

The following inference axioms follow from the above following axioms.

FD4. **Union**: If $X \to Y$ and $Y \to Z$ hold, then $X \to YZ$ holds.
FD5. **Decomposition**: If $X \to YZ$ holds, then $X \to Y$ and $X \to Z$ hold.
FD6. **Pseudotransitivity**: If $X \to Y$ and $YW \to Z$ hold, then $XW \to Z$ holds.

We shall now show that these axioms are sound when interpreted in **L-SET**. Consider an instance r of R and suppose $\delta_r(t_1, t_1) > 0$ and $\delta_r(t_2, t_2) > 0$.

FD1. Since Y is a subset of X, clearly

$$\delta(t_1[X], t_2[X]) \leq \delta(t_1[Y], t_2[Y])$$

So $X \to Y$ holds in r

FD2. Since $X \to Y$ holds for the relation r,

$$\delta(t_1[X], t_2[Y]) \wedge \delta(t_1[Z], t_2[Z])$$

Hence,

$$\delta(t_1[X], t_2[X]) \wedge \delta(t_1[Z], t_2[Z]) \leq \delta(t_1[Y], t_2[Y]) \wedge \delta(t_1[Z], t_2[Z])$$

So, $XZ \to YZ$ holds in r.

FD3. Suppose r satisfies $X \to Y$ and $Y \to Z$. Then

$$\delta(t_1[X], t_2[X]) \leq \delta(t_1[Y], t_2[Y])$$

and

$$\delta(t_1[Y], t_2[Y]) \leq \delta(t_1[Z], t_2[Z])$$

Consequently,

$$\delta(t_1[X], t_2[X]) \leq \delta(t_1[Z], t_2[Z])$$

so $X \to Z$ holds in r.

It is easy to show axioms FD4 – FD6 are sound when interpreted in **L-SET**. Alternatively, the classical proof that FD4 – FD6 follow from FD1 – FD3 uses only intuitionistic logic, so holds in any topos, particularly in **L-SET**. □

Before we can show the completeness of the set of axioms in **L-SET**, we must make the following definition.

DEFINITION 3. Suppose F is a set of functional dependencies of a relation scheme $R = R(A_1, \ldots, A_n)$ and let W be a subset of R. Then $W+$, the closure of W with respect to F, is the set of attributes A of R such that $W \to A$ can be obtained from F using FD1 – FD6. The following lemma, which holds classically, is true in any interpretation in **L-SET**.

LEMMA 1. $W \to V$ follows from the inference axioms iff V is a subset of $W+$.

PROOF: ($\Rightarrow$) Let $V = \{A_1, \ldots, A_n\}$ and suppose that $W \to V$ follows from the inference axioms. Then for each i, $W \to A_i$ holds by FD5. So V is a subset of $W+$.

($\Leftarrow$) Suppose V is a subset of $W+$. By the above definition, for each i, $W \to A_i$ is implied by FD1 – FD6. So, by FD4, $W \to V$ holds. □

The following is a generalization of the completeness of Armstrong's Axioms in **L-SET**.

THEOREM 1. The inference axioms FD1, FD2 and FD3 form a complete set of inference axioms for functional dependencies of a relation scheme $R(A_1, \ldots, A_n)$ in **L-SET** when the following condition holds:

for each A_i in R, there exists at least one pair of elements, $a_i, b_i \in A_1$ such that $\delta(a_i, a_i) > 0$ and $\delta(b_i, b_i) > 0$ and $\delta(a_i, b_i) = 0$. Moreover, for all A_i and A_j in R $\delta(a_i, a_i) = \delta(a_j, a_j)$.

PROOF. Let F be a set of functional dependencies over R. Suppose that the functional dependency $W \to V$, with W and V subsets of R, cannot be inferred from F using the inference axioms. We shall show that there is a relation r in R for which $W \to V$ does not hold.

Let r be a relation with two tuples t and t_1 as shown:

$$t = (a_1, \ldots, a_n),$$
$$t_1[A_i] = a_i \text{ if } A_i \text{ is an element of } W+$$
$$= b_i \text{ otherwise.}$$

(where $\delta(a_i, b_i) = 0,\ i = 1, \ldots, n$).

Suppose $X \to Y$ be a functional dependency in F. We show that this functional dependency holds in r. If X is not a subset of $W+$, then for any A_i in the set $X - W+$, $\delta(t[A_i], t_1[A_i]) = \delta(a_i, b_i) = 0$. This implies that $\delta(t[X], t_1[X]) = 0$, so $X \to Y$ holds in r. If X is a subset of $W+$, then by Lemma 1, $W \to X$ and by FD3, $W \to Y$. Lemma 1 then implies that Y is a subset of $W+$. Since XY is a subset of $W+$, the definition of t and t_1 tells us that $\delta(t[X], t_1[X]) = \delta(t[Y], t_1[Y])$. Therefore $X \to Y$ holds in r.

It remains to show that the functional dependency $W \to V$ does not hold in r. We argue by contradiction. Suppose $W \to V$ holds in r. Then $\delta(t[W], t_1[W]) \leq \delta(t[V], t_1[V])$. The definition of t and t_1 then implies that V is a subset of $W+$. Lemma 1 then implies that $W \to V$ follows from F by the inference axioms. But this contradicts the assumption that $W \to V$ cannot be inferred from F using the inference axioms. Thus, $W \to V$ does not hold in r. The completeness of the inference rules follows. □

The **decomposition of a relation scheme** $R = R(A_1, \ldots, A_n)$ is its replacement by a collection $\{R_1, \ldots, R_k\}$ of subschemes, where each R_i is a subset of $\{A_1, \ldots, A_n\}$, such that the union over i of the sets of attributes in the R_i's comprises all of the attributes in R. The decomposition of relational databases is an important aspect of design theory (see [Ullman 1982]). For a decomposition to be of any value one must be able to recover the original relation r from the k projections onto the attributes in the decomposition, i.e., from the projections onto the attributes in the R_i's. The way to recover a relation from such a decomposition is to take the natural join of the projections of r onto the attributes in the R_i's. However, we must ensure that the natural join of these projections recovers (no more than) the original relation r. This idea is articulated by the following definition (see, for example [Ullman 1982]). In what follows, we shall use $\pi_{R_i}(r)$ to denote the projection of r onto the attributes of R_i, and use δ_{R_i} to denote $\delta_{\pi_{R_i}(R)}$.

DEFINITION 4. If a relation scheme R is decomposed into subschemes $R_1, \ldots, R_k$, and F is a set of functional dependencies, we say that the decomposition is a **lossless join decomposition** (with respect to F) iff for every relation r of R for which the dependencies of F hold,

$$r = \pi_{R_1}(r) <><> \pi_{R_2}(r) <><> \ldots <><> \pi_{R_k}(r)$$

Following the convention in [Ullman 1982] we use the notation $m(r)$ for the right hand side of the above equality.

As in the classical case, it is easy to prove the following lemma.

LEMMA 2. Let R be a relation scheme over **L-SET**, let $\{R_1, \ldots, R_k\}$ be a decomposition of R, and let r be a relation in R. Then r is a subobject of $m(r)$.

Another important property of a decomposition of a relation scheme R into $\{R_1, \ldots, R_k\}$ is that the set of dependencies F for R be implied by the projections onto the R_i's. Algorithms for achieving decompositions have been developed so that the decomposition is lossless, and the dependencies have been preserved (cf. [Ullman 1982]). It would be desirable to extend this, and indeed all, of classical decomposition theory to the fuzzy case. In their treatment, [Raju and Majumdar 1989] have generalized some of the classical decomposition theory. In their treatment, as well as in the classical treatment in [Ullman 1982], it was shown that one particular theorem is crucial to the extension. Classically, this theorem is stated as follows: if a functional dependency $X \to Y$ holds in a relational database $R(XYZ)$, then the decomposition $\{R_1(XY), R_2(XZ)\}$ is lossless. We generalize this theorem to **L-SET**.

THEOREM 2. Given a relation scheme $R = R(A_1, \ldots, A_n)$, over **L-SET**, and a functional dependency $X \to Y$ (where X and Y are subsets of R), which holds in R, then R has a lossless join decomposition into two components, $R_1(XY)$ and $R_2(XZ)$, where $Z = R - XY$.

PROOF. Let r be a relation over the scheme R. It satisfies the functional dependency $X \to Y$. We must show that for every $t \in A_1 \times \ldots \times A_n$,

$$\delta_r(t,t) = \delta_{m(r)}(t,t)$$

First, let $t = (x,y,z)$. By Lemma 2 we know $\delta_{m(r)}(t,t) \geq \delta_r(t,t)$. We now show that equality holds. We argue by contradiction and assume that $\delta_{m(r)}(t,t) > \delta_r(t,t)$. Hence $\delta_{m(r)}(t,t) > 0$. Now if $\delta_{m(r)}(t,t) > 0$, then from the definition of $\delta_{m(r)}$,

$$\delta_{m(r)}(t,t) = \vee_{x' \in X} \ \delta_{R_1 \times R_2}(x,y,x',z) \wedge \delta(x,x').$$

Because L is linearly ordered, there is $x' \in X$ such that

$$\begin{aligned} \delta_{m(r)}(t,t) &= \delta_{R_1 \times R_2}(x,y,x',z) \wedge \delta(x,x') \qquad (**) \\ &= \delta_{R_1}((x,y),(x,y)) \wedge \delta_{R_2}((x',z),(x',z)) \wedge \delta(x,x') \end{aligned}$$

Now:

$$\delta_{R_2}((x',z),(x',z)) = \vee_{y' \in Y} \ \delta_r((x',y',z),(x',y',z)).$$

From (**) it follows that for some $y' \in Y$ it is true that

$$\delta_r((x',y',z)(x',y',z)) \geq \delta_{m(r)}((x,y,z),(x,y,z)). \qquad (***)$$

Hence:

$$\begin{aligned} \delta_r((x,y,z),(x,y,z)) &\geq \delta_r((x',y',z),(x',y',z)) \wedge \delta_X(x,x') \wedge \delta_Y(y,y') \\ &\geq \delta_r((x',y',z),(x',y',z)) \wedge \delta_X(x,x') \\ &\geq \delta_{m(r)}((x,y,z),(x,y,z)) \wedge \delta_X(x,x') \\ &= \delta_{m(r)}((x,y,z),(x,y,z)), \end{aligned}$$

where the first inequality follows (with a little work) from (*) in Section 1, the second follows because the functional dependency $X \to Y$ holds in r, and the third follows from (***) above, and the equality follows from the definition of $\delta_{m(r)}$. This gives the desired contradiction, so the proof is finished. □

§5. Conclusions

We have presented a model for fuzzy relational databases by interpreting the usual theory in the category **L-SET**, for L the unit interval with the usual ordering. We recovered much of the ordinary theory, including a theorem of fundamental importance in data design. Though we feel that this model is superior to others which have appeared thus far, we believe that the last word has not yet been said, because the category we have used has some shortcomings. The proofs work for more general kinds of lattices L (e.g. when L is a complete, linearly ordered Heyting algebra). The challenge is to find the appropriate foundation for fuzzy set theory.

This research was supported by NSERC (Operating Grant #A8915).

Chapter 13:

Fuzzy Points and Membership

S. E. Rodabaugh et al. (eds.), Applications of Category Theory to Fuzzy Subsets, 325–334.

FUZZY POINTS AND MEMBERSHIP

M. W. Warner

§1. Introduction

One of the least satisfactory areas in the early development of fuzzy topology has been that surrounding the concept of fuzzy point. In the original classical theory, where values are taken in the closed unit interval I, it soon became apparent that, in order to build up a reasonable theory, points should be defined as fuzzy singletons while membership requires strict inequality. So crisp points, taking value 1, are excluded, and fuzzy topology would seem not to include general topology. This disturbing state of affairs was to some extent overcome by [Pu and Liu 1980a] who replaced membership by quasi-coincidence (not belonging to the complement, where belonging is taken as $\leq$), thus reinstating crisp points. More recently [Hu 1985] has drawn attention to a duality between quasi-coincidence and strict inequality membership. The duality, however, is only partial [Warner 1989]. Details of the various definitions and corresponding neighbourhood theories are clearly described by [Kerre and Ottoy 1987].

The problem has been avoided by [Hutton 1980, Lowen 1982] and others in the so-called 'pointless' approach, where either a fuzzy topological theory or a fuzzy neighbourhood theory is built up without reference to points.

Other researchers, notably the Chinese school (e.g. [Liu and Luo 1987]) concentrated on illuminating the picture by establishing a theory of L-fuzzy topology, replacing the closed unit interval by a complete lattice L, usually taken to be complete distributive. The term 'fuzzy lattice' was coined by [Luo 1987] to describe such a lattice provided with an order-reversing involution (giving a form of complementation). With respect to such an L-fuzzy lattice [Wang 1984] singles out special fuzzy points called molecules which take only union-irreducible values in L, and replaces membership with the concept of remote neighbourhood (R-neighbourhood), where the relation involved is $\not\leq$. The ensuing theory of 'topological molecular lattices' has provided a good vehicle for studying fuzzy convergence and other point-based notions such as first countability, also compactification. Most of these authors introduce L-fuzzy topologies by means of closed sets, and continue to require L to be completely distributive.

Now [Johnstone 1988] draws attention to the fact that there are good reasons for claiming that the correct lattices for the study of L-fuzzy topology are complete lattices satisfying the infinite distributive law, otherwise known as frames. The properties of the lattice L pass over to its L-fuzzy sets [Goguen 1967], so that frames ensure a well-behaved category of fuzzy sets. In fact it is sufficient to work in the category **CSLF** (complete lattices with frame morphisms) introduced by [Rodabaugh 1986-7]. The collection of such L-fuzzy subsets of a set X forms a locale, as does an L-fuzzy topology on X. L-fuzzy points should therefore be defined locale-wise by frame homomorphisms to the two-point set, and so correspond bijectively to prime elements of L (generators of prime ideals). Membership emerges in terms of $\nleq$, just as in Wang's theory. This is not surprising since molecules correspond to generators of prime filters of Wang's completely distributive lattice. These L-fuzzy points are further generalised by [Rodabaugh 1986-7] to frame morphisms $p : A \to L$ in **CSLF** and constitute an intermediate stage between these and classical crisp points.

A point-based theory of L-fuzzy topology, of which general topology is a particular case, can then be constructed without recourse to complete distributivity or order-reversing involutions. Closed L-fuzzy sets may be defined using limit points. There are enough points when L is spatial.

If, furthermore, L is taken to be a continuous frame, the Scott topology on L provides an immediate generalisation of the half-open interval topology, hence of topological generation by lower semi-continuous functions [Warner 1990]. Using a continuous spatial frame, results of general topology and of I-fuzzy topology generalise with difficulty.

§2. Definitions and Background

For convenience we give a short résumé of the main classical definitions and theory upon which this work is based. Further details and amplification may be found in [Johnstone 1982].

2.1. We shall define a **complete lattice** to be a lattice $(L, \vee, \wedge)$ with top and bottom elements denoted by $1, 0$ respectively, and possessing arbitrary joins. A complete lattice L satisfies the **infinite distributive law** if for all $a \in L$ and $S \subseteq L$,

$$a \wedge (\bigvee S) = \bigvee \{a \wedge s;\ s \in S\}. \tag{1}$$

Note that the existence of arbitrary joins implies that of arbitrary meets, and that the finite distributive law ((1) above with S a finite subset) implies its dual

$$a \vee (\bigwedge S) = \bigwedge \{a \vee s;\ s \in S\}. \tag{2}$$

But the infinite distributive law (1) does not imply an infinite version of the dual (2). This is exemplified by the complete lattice $\Omega(X)$ of open sets of a topological space X. Classically, a complete lattice L satisfies the infinite distributive law iff it is a Heyting algebra, that is

L possesses relative **pseudo-complements**, i.e. for each pair of elements (a,b) there exists an element $(a \to b)$ such that $c \leq (a \to b)$ iff $c \wedge a \leq b$. The relative pseudo-complement $(a \to 0)$ is written $\neg a$ and is called the **negation**, or **pseudo-complement**, of a.

A **Boolean algebra** is of course a (finitely) distributive complemented lattice. But a complete Boolean algebra is also a Heyting algebra.

2.2. A subset I of a lattice L is an **ideal** if (i) $0 \in I$, $a, b \in I \Rightarrow a \vee b \in I$, and (ii) I is a **lower set**, i.e. $a \in I$ and $b \leq a$ imply $b \in I$.

The **principal ideal** generated by an element a of L is $\{b \in A : b \leq a\}$ and is denoted by $\downarrow(a)$. An ideal I of L is **prime** if $1 \notin I$ and $(a \wedge b \in I) \Rightarrow$ either $a \in I$ or $b \in I$.

A subset of L satisfying axioms dual to those listed above is called **a filter** (**principal filter, prime filter**) respectively. It is not difficult to show that prime ideals and prime filters are complementary.

2.3. The category **FRM** of **frames** is the category whose objects are complete lattices satisfying the infinite distributive law (Heyting algebras), and whose morphisms are functions preserving finite meets and arbitrary joins. We shall usually avoid using the term Heyting algebra in this context since the morphisms do not preserve implication $(\to)$. As mentioned in the introduction, the category **CSLF** suffices. The category **LOC** of **locales** is the opposite of the category **FRM**. Morphisms in **LOC** will be called **continuous maps** by analogy with the functor $\Omega : \mathbf{SP} \to \mathbf{LOC}$ which sends a topological space to its lattice of open sets and a continuous map $f : X \to Y$ in **SP** to the function $f^{-1} : \Omega(Y) \to \Omega(X)$. Thus every topological space is a locale, and the question naturally arises of how, given a locale, to find a space which best approximates it. A **point** p of a locale L is therefore defined to be a continuous map $\Omega(1) = 2 \to L$; i.e. a frame homomorphism $p : L \to 2$. Such a map is known [Johnstone 1982] to be completely determined by its kernel $p^{-1}(0)$ which is a principal prime ideal. So points of L correspond bijectively to the **prime elements** of L, i.e. elements λ generating principal prime ideals $\downarrow(\lambda)$. The set of points of L will be written $pt(L)$.

Note that the elements of the principal ideal generated by a prime element of a topological space $\Omega(X)$ are the open sets to which the corresponding point p does not belong.

A locale L is said to be **spatial**, or to **have enough points** iff for all $a, b \in L, a \nleq b$, there exists a point $p \in pt(L)$ such that $p(a) = 1$ and $p(b) = 0$.

2.4. In Section 6 we discuss continuity of the lattice L. The relevant definitions, including that of the Scott topology are to be found in [Warner 1990]. We use the notation $\ll$ to denote **way-below**, and define a **continuous lattice** to be a complete lattice L in which $\ell = \bigvee\{m : m \ll \ell\}$ for every $\ell \in L$.

§3. Fuzzy Points

3.1. Throughout this work fuzziness is taken with respect to a frame L. A **fuzzy subset** A of a set X is thus defined to be a function $A : X \to L$. The completeness and infinite distributivity of L ensure that the collection of all fuzzy subsets of X is a locale $\mathcal{F}(X)$ with respect to the partial ordering $A \leq B$ iff $A(x) \leq B(x)$ for all $x \in X$, and the usual operations of meet and join:

$$(A \wedge B)(x) = A(x) \wedge B(x)$$
$$(A \vee B)(x) = A(x) \vee B(x)$$

Note that L is not required to be complemented, although it is of course pseudo-complemented.

A fuzzy topology $\mathcal{T}$ on X is a complete sublattice of $\mathcal{F}(X)$ which is therefore itself a locale. $\mathcal{F}(X)$ itself is the discrete fuzzy topology on X.

[Lowen and Wuyts 1988] have argued strongly for insisting on the inclusion of all constant fuzzy subsets, $C : X \to L$, $C(x) = C$ for all $x \in X$, as open fuzzy subsets. We shall draw specific attention to this in the final section.

3.2. Given the locale $\mathcal{F}(X)$ defined above, its fuzzy points should be frame homomorphisms $p : \mathcal{F}(X) \to \{0, 1\}$, and correspond bijectively to prime elements of $\mathcal{F}(X)$.

LEMMA. The prime elements of $\mathcal{F}(X)$ are the functions $P : X \to L$ of the form, for each $x_0 \in X$, and for each prime element λ of L,

$$P(x_0) = \lambda$$
$$P(x) = 1, \; x \neq x_0$$

PROOF. P is prime iff, for every $A, B \in \mathcal{F}(X)$ such that $A \wedge B \leq P$, then A or $B \leq P$, i.e. $A(x) \wedge B(x) \leq P(x)$ for all $x \Rightarrow A(x) \leq P(x)$ for all x or $B(x) \leq P(x)$ for all x.

Assume X is not a singleton set. Take x_i, $x_j \in X$, $x_i \neq x_j$. Let $P(x_i) = \zeta$, $P(x_j) = \eta$ where $\zeta, \eta \neq 1$, (thus presuming L not a singleton lattice). Take $A, B \in \mathcal{F}(X)$ such that $A(x_i) = \zeta$, $A(x_j) = 1$ and $B(x_i) = 1$, $B(x_j) = \eta$; $A(x_k) = B(x_k) = 0$, $k \neq i, j$. Then $A \wedge B \leq P$ but neither A nor $B \leq P$. So for every pair of points x_i, x_j either $P(x_i) = 1$ or $P(x_j) = 1$. This implies that $P(x) = 1$ for all $x \in X$ except for one point, say x_0.

Let $P(x_0) = \lambda$, $P(x) = 1$, $x \neq x_0$. Then when $x \neq x_o$, $A(x)$, $B(x)$ and $A(x) \wedge B(x)$ are always ≤ 1. And we require $A(x_o) \wedge B(x_o) \leq \lambda$ to imply $A(x_o) \leq \lambda$ or $B(x_o) \leq \lambda$, which is the requirement that λ be prime in L. Note that $\lambda \neq 1$, since 1 is not a member of a prime ideal.

If L is the singleton lattice in which $0 = 1$ there are clearly no fuzzy points since L contains no prime elements.

When X is a singleton set $\{x_o\}$ the fuzzy points are functions P which map x_0 to the prime elements of L. □

The fuzzy points of X (elements of $pt(\mathcal{F}(X))$ thus correspond bijectively to the pairs (x, λ) for all $x \in X$ and λ prime in L. The point corresponding to (x, λ) will be written $x_{(\lambda)}$.

3.3. In the case where L is the closed unit interval I with the usual partial ordering (the original lattice of fuzzy set theory) all elements are prime except 1. So the fuzzy points are $x_{(\lambda)}$ for $0 \leq \lambda < 1$, and the fuzzy point $x_{(0)}$ suggested by [Hu 1985] in his discussion of duality with Q-theory is admitted, not as taking zero values everywhere, but corresponding to the prime element of $\mathcal{F}(X)$ with value zero at x and 1 elsewhere. This is in fact the elusive crisp point. For example the (crisp) points of the locale formed by the power set $\mathcal{P}(X)$ of X correspond to the prime elements $P : X \to \{0, 1\}$ where $P(x) = 1,\ x \neq x_0,\ P(x_0) = 0$, taken over all $x_0 \in X$. In these two cases there is symmetry between meet and join since the dual infinite distributive law also holds, so (fuzzy) points may be associated with elements generating prime filters, giving the more usual definitions in which 0 and 1 are interchanged. Fuzzy points are then those of [Pu and Liu 1980a]. The question of membership is the subject matter of the next section. Meanwhile we note that when L is a complete Boolean algebra considered as a locale the prime elements are those whose complements are atoms, so that fuzzy points may in fact be identified with functions P taking zero values everywhere except at some x_0 whose value is an atom. Thus in $\mathcal{P}(X)$ the original points of X are recovered from singleton subsets which are the atoms of $\mathcal{P}(X)$.

3.4. The locale of fuzzy subsets of X is spatial iff for all $A, B \in \mathcal{F}(X)$ with $A \nleq B$, there exists a fuzzy point p such that $p(A) = 1,\ p(B) = 0$, i.e. there exists a prime element P of $\mathcal{F}(x)$ such that $B \leq P,\ A \nleq P$. This is true iff there exists a prime element λ of L and $x_0 \in X$ such that $B(x_0) \leq \lambda,\ A(x_0) \nleq \lambda$. Since $A \nleq B$ there exists $x_0 \in X$ such that $A(x_0) \nleq B(x_0)$. The condition for spatiality is therefore satisfied iff L is spatial. Thus we have

LEMMA. The locale of fuzzy sets with respect to the frame L is spatial iff L is spatial.

3.5. In a fuzzy topological space as defined in 3.1, the fuzzy points of the locale $\mathcal{T}$ are those fuzzy sets P with $P(x_0) = \lambda$, prime, and $P(x) = 1,\ x \neq x_0$, which are open sets of the fuzzy topology.

We may therefore consider the fuzzy points of a fuzzy topological space either to correspond to all prime elements of L as in Lemma 3.2 or to correspond only to those defined locale-wise by the fuzzy topology. We adopt the former alternative and proceed to describe non-topological set-theoretical fuzzy membership properties which are then carried over into pointwise contexts in fuzzy topology.

§4. Membership

From the localic definition of a point it follows that a point $p : L \to \{0,1\}$ is a member of the locale element a iff $p(a) = 1$, i.e. a $\notin p^{-1}(0)$. If the kernel $p^{-1}(0)$ is generated by the prime element b, then $p \in a$ iff $a \nleq b$. Translating this into fuzzy theory, a fuzzy point $x_{(\lambda)}$ is a member of the fuzzy set A iff $A(x) \nleq \lambda$.

In classical I-theory for the chain I the negation of $\leq$ is $>$, so that we may write $A(x) > \lambda$, $0 \leq \lambda < 1$. This accounts for the fact that strict inequality was found by the authors mentioned in the introduction to be necessary for fuzzy membership. Their theory is, however, incomplete unless the zero value is permitted. Pu and Liu's quasi-coincidence has no such gaps, since associating each point $x_{(\lambda)}$, $0 \leq \lambda < 1$, with its dual $x_{[\mu]}$, $\mu = 1 - \lambda$, $0 < \mu \leq 1$, quasi-coincidence is defined by $x_{[\mu]} q A$ iff $A(x) > 1 - \mu = \lambda$.

4.1. Returning to L-fuzzy points $x_{(\lambda)}$ for λ a prime element of the frame L, we have the following

LEMMA. For L-fuzzy sets A, B:

(i) if $A \subseteq B$, then $x_{(\lambda)} \in A \Rightarrow x_{(\lambda)} \in B$;

(ii) $x_{(\lambda)} \in A \cap B$ iff $x_{(\lambda)} \in A$ and $x_{(\lambda)} \in B$;

(iii) $x_{(\lambda)} \in \bigcup_{i \in I} A_i$ iff $x_{(\lambda)} \in A_j$ for some j in the indexing set I.

PROOF.

(i) $A(x) \leq B(x)$ and $A(x) \nleq \lambda$ together imply $B(x) \nleq \lambda$.

(ii) $A(x) \nleq \lambda$ and $B(x) \nleq \lambda$ imply $A(x) \wedge B(x) \nleq \lambda$ because λ is prime (section 2.2). The converse is trivial.

(iii) If $\bigvee_i A_i(x) \nleq \lambda$ and $A_j(x) \leq \lambda$ for all $j \in I$ then $\bigvee_i A_i(x) \leq \lambda$, giving a contradiction. Conversely, $A_j(x) \nleq \lambda$ and $\bigvee_i A_i(x) \leq \lambda$ imply $A_j(x) \leq \bigvee_i A_i(x) \leq \lambda$, another contradiction. □

4.2. The analogue of the set-theoretical statement that every set is the union of its singleton subsets should read that every element A of $\mathcal{F}(X)$ is the meet of prime elements $x_{(\lambda)}$ such that $x_{(\lambda)} \notin A$, i.e. $A(x) \leq \lambda$. But each element of a locale can be expressed as a meet of prime elements iff the locale is spatial [Johnstone 1982, p. 43]. Thus by Lemma 3.4 L is required to be spatial. Each element ℓ of L is then the meet of prime elements $\geq \ell$. So we have

LEMMA. Every fuzzy subset $A : X \to L$ is the meet of prime elements corresponding to its fuzzy points iff L is spatial.

§5. Fuzzy Topology

Fuzzy topological spaces have been extensively studied with respect to the lattice I, and as mentioned in the introduction, to a lesser extent with respect to completely distributive fuzzy lattices. Much of this theory remains valid for frame-fuzzy topology. In addition we show that point-based results in classical I-fuzzy theory generalise nicely to L-fuzzy points and membership. Assume henceforth that L is spatial.

5.1. For a fuzzy topological space $(X, \mathcal{T})$, a **basis** for $\mathcal{T}$ is a collection $\mathcal{B}$ of open sets of $\mathcal{T}$ such that for all $A \in \mathcal{T}, A = \bigcup_{i \in I} B_i$ for a subcollection $\{B_i\}_{i \in I}$ of $\mathcal{B}$.

LEMMA. A non-empty subfamily $\mathcal{B}$ of $\mathcal{T}$ is a basis for $\mathcal{T}$ iff for all $A \in \mathcal{T}$ and $x_{(\lambda)} \in A$ there is a $B_j \in \mathcal{B}$ such that $x(\lambda) \in B_j \subseteq A$.

PROOF. First, let $x_{(\lambda)} \in A \in \mathcal{T}$, and let $A = \bigcup_{i \in I} B_i$, $B_i \in \mathcal{B}$. Then $A(x) \leq \lambda$, and $A(x) = \bigvee_i B_i(x)$. So there exists $j \in I$ such that $B_j(x) \leq \lambda$ as in Lemma 4.1 (iii).

Conversely, assume that for all $A \in \mathcal{T}$ and $x_{(\lambda)} \in A$ there exists $B \in \mathcal{B}$ such that $x_{(\lambda)} \in B \subseteq A$. And assume that for all $B_j \subseteq A$, their union $C = \bigcup B_j \subset A$, i.e. C is a proper (fuzzy) subset of A. Then there exists $x \in X$ such $C(x) < A(x)$, so $A(x) \nleq C(x)$. By the spatiality of L there exists a prime element λ such that $A(x) \nleq \lambda$, $C(x) \leq \lambda$, i.e. $x_{(\lambda)} \in A$, $x_{(\lambda)} \notin C$, yielding a contradiction. □

5.2. A fuzzy topological space $(X, \mathcal{T})$ is, classically, second countable, C_2, iff the topology has a countable base. And it is first countable, C_1, iff for every fuzzy point $x_{(\lambda)}$ there exists a countable local base, i.e. a countable family of open sets containing $x_{(\lambda)}$ such that every neighbourhood of $x_{(\lambda)}$ contains a member of the family. Expressing this in L-membership terms, for every fuzzy set A with $A(x) \nleq \lambda$ there exists a member $B \in \mathcal{T}$ of the countable family such that $B(x) \nleq \lambda$ and $A(x) \geq B(x)$.

The proofs of the following lemmas then follow almost verbatim those given in [Mohannadi and Warner 1989].

LEMMA. The fuzzy topological space $(X, \mathcal{T})$ is first countable iff it has a nested countable local base.

LEMMA. A second countable fuzzy topological space is first countable.

Nothing is to be gained by repeating this exercise for other standard local properties such as convergence and sequentiality. For instance the proposition that L-first countability implies L-fuzzy sequentiality translates in the obvious way.

§6. Continuous Lattices

Taking L to be a continuous lattice with the Scott topology and (X, τ) to be a topological space, [Warner 1990] has proved that the set of continuous functions from X to L forms an L-fuzzy topology $\omega(\tau)$, (the analogue of that of lower semi-continuous functions with respect to I). The author then goes on to suggest that 'way-below' should provide the definition of fuzzy point membership, arguing that it is necessary to use the property that, in a continuous lattice, if $\ell \ll m$, then there exists ℓ' $(\neq \ell)$ such that $\ell \ll \ell' \ll m$. This suggestion is now withdrawn. In fact use can be made of the insertion of ℓ' whilst adhering to the correct version of fuzzy point membership.

This is hardly surprising since the relations $\ll$, $\nleq$ and the Scott topology are closely related. For instance the set $\{\ell \in L;\ \ell \nleq \lambda\}$ is Scott-open in L and sets of the form $\{\ell \in L; \delta \ll \ell\}$ form a basis for the Scott topology on L.

LEMMA [Warner 1990]. For a topological space (X, τ) and a continuous lattice L, the collection $\beta(\tau) = \{Z^{(\delta)} : Z \in \tau,\ \delta \in L\}$ (where $Z^{(\delta)}(x) = \delta,\ x \in Z,\ Z^{(\delta)}(x) = 0$ otherwise) is a base for $\omega(\tau)$.

PROOF. Let $x_{(\lambda)} \in A, A \in \omega(\tau)$. Then $A(x) \nleq \lambda$, (prime). By the continuity of L, there exists $\eta \in L$ such that $\eta \ll A(x)$, $\eta \nleq \lambda$. Take $\delta \in L$ such that $\eta \ll \delta \ll A(x)$. Then $A(x) \in \{\ell \in L : \delta \ll \ell\}$ which is Scott-open in L. Hence, by the continuity of A, there exists an open neighbourhood Z of x such $z \in Z \Rightarrow \delta \ll A(z)$. Then $\delta \leq A(z)$ for all $z \in Z$ and $\delta \nleq \lambda$ since $\delta \gg \eta \nleq \lambda$. So $x_{(\lambda)} \in Z^{(\delta)} \subseteq A$, and $\{Z^{(\delta)}\}$ forms a base (Lemma 5.1). □

COROLLARY. Every fuzzy topology on X which contains all constant sets and all crisp sets itself contains $\omega(\tau)$.

In other words, every Lowen fuzzy topology on X containing all crisp sets must contain $\omega(\tau)$.

§7. Separation Axioms

7.1. Regularity and normality are classically definable localically [e.g. Johnstone 1982, pp. 80, 128], and these definitions apply immediately to fuzzy topology. Normality does not involve points so need not concern us in this context.

DEFINITION. For a and b elements of a distributive lattice A we say a is **well inside** b (written $a \bar{<} b$) if there exists $c \in A$ with $c \wedge a = 0,\ c \vee b = 1$.

DEFINITION. A locale A is **regular** iff every element of A is the join of elements well inside itself, i.e. for all $a \in A$, $a = \{b \in A; b \bar{<} a\}$.

In classical topology this states that every open set U can be covered by open subsets whose closures are contained in U and is equivalent to the usual definition of regularity.

DEFINITION. A fuzzy topological space is **regular** iff every fuzzy open set V is the union of fuzzy open sets well inside V.

7.2. The definition of fuzzy membership in section 4 leads naturally to the following treatment of T_1 and T_2 spaces.

DEFINITION. An L-fuzzy topological space is T_1 iff all fuzzy sets of the form $A : X \to L$, $A(x) = \lambda$ (prime), $A(y) = 1,\ y \neq x$, are open [cf. Pu and Liu 1980a, p. 583].

DEFINITION. An L-fuzzy topological space is **Hausdorff** (T_2) iff each fuzzy set of the form $A : X \to L,\ A(x) = \lambda$ (prime), $A(y) = 1,\ y \neq x$, is the union of fuzzy open sets well inside A.

It is clear that these definitions are valid generalisations of those of classical topology and that the usual implications follow immediately, viz, $T_2 \Rightarrow T_1$, Regular $+\ T_1 \Rightarrow T_2$.

7.3. The use of 'well inside' has avoided an explicit definition of closure. Closed fuzzy sets and hence the closure of a fuzzy set should be defined in terms of limit points.

Pseudo-complementation is of no help in this context. The pseudo-complement, $\neg U$, of an open set U is another open set, corresponding classically to the interior of the complement of U. A follow-on to this paper will be devoted to closed sets, convergence, and the relation between the foregoing versions of fuzzy separation axioms and those which have already appeared in fuzzy literature.

Appendices

Round Table:

Model-Theoretic *vis-à-vis* Non-Model-Theoretic Fuzzy Set Theories

S. E. Rodabaugh et al. (eds.), Applications of Category Theory to Fuzzy Subsets, 337–345.

ROUNDTABLE:

MODEL–THEORETIC *VIS-À-VIS* NON–MODEL–THEORETIC FUZZY SET THEORIES

Transcribed and Edited from Audio Tapes and Written Sources by

S. E. Rodabaugh

PROF. U. HÖHLE. [I should like to explain an advantage of the] model-approach to fuzzy topology. This [approach] means categorically we pick up a "topos-like" category; this means a category in which we can classify extremal subobjects by fuzzy subsets. Then we have in this setting automatically the situation that we have fuzzy topologies on fuzzy subsets. And this, I think, solves one **major** problem in fuzzy topology. So I think that is an advantage of a topos-like approach to fuzzy topologies.

PROF. S. E. RODABAUGH. My comments are an illustrated response to Prof. Höhle's remarks. I agree with Prof. Höhle's position; indeed, I think **each** approach — model-theoretic and non-model-theoretic — has distinctive advantages.

Let me motivate what I am going to say from the standpoint of localic products. We had the question the other day: how to compare localic products [of a given collection of topologies] with topological products [of this same collection of topologies], and the question can be raised whether it be [a collection] of classical topologies or fuzzy topologies. So we have now three things to look at: localic products, classical product topologies, and fuzzy product topologies. ... We [can] take fuzzy product topologies in any of the usual non-model-theoretic settings [e.g. Goguen-Wong]; ... but with classical product topologies, we have a model-theoretic setting — we are in the topos **SET**; and finally we have localic products. So the question we were considering [in a previous session] was: **given (fuzzy) topological spaces**, which might be classical spaces, **when is the localic product of the topologies the same as the product (fuzzy) topology, where in the latter the lattice L of membership values be the same as that of the factor spaces**? The sequel uses terms and results from the abstract of my talk **Generalizations of Spatiality and Sobriety** [see Section 5 of my chapter in this volume], and is restricted, for convenience sake, to two factors, and the product (fuzzy) topology is that of Goguen-Wong.

Let (X_1, L, τ_1), (X_2, L, τ_2) be given, and let $\tau_1 \times_\ell \tau_2$, $\tau_1 \times \tau_2$ denote the localic product, product topology, respectively. The most "strategic" map for building a comparison between $\tau_1 \times_\ell \tau_2$ and $\tau_1 \times \tau_2$ is:

$$\Theta : Lpt(\tau_1 \times \tau_2) \longrightarrow Lpt(\tau_1) \times Lpt(\tau_2) \text{ by } [\Theta(p)]_k = p \circ (\pi_k, id)^{-1}$$

THEOREM (**general fuzzy case**). Θ is well-defined, and if surjective, then the bi-implication holds: $\tau_1 \times_\ell \tau_2 \approx \tau_1 \times \tau_2 \Leftrightarrow \tau_1 \times_\ell \tau_2$ is L-spatial.

COROLLARY (**classical case**). If $L = \{0, 1\}$, then the bi-implication holds: $\tau_1 \times_\ell \tau_2 \approx \tau_1 \times \tau_2 \Leftrightarrow \tau_1 \times \ell\tau_2$ is spatial.

The general fuzzy case has a **non-model-theoretic** proof, and the classical case has both this proof and a direct **model-theoretic** proof: the proof of the theorem cannot use subspaces as subobjects, and so relies on intensely lattice-theoretic methods; but the direct classical proof, due to Prof. P. T. Johnstone (personal communication), is subspace intensive and generally avoids lattice-theoretic methods. In order to further connect these matters with the topic of this roundtable, I need to give a short summary of each type of proof — the non-model-theoretic (lattice-theoretic) for the general fuzzy case, and the model-thoeretic (subspace-intensive) for the classical case. In each summary I shall only deal with the well-definedness and surjectivity of Θ.

SURJECTIVITY OF Θ: MODEL-THEORETIC APPROACH

INTRODUCTORY COMMENTS. In the crisp or classical case, $Pt(\tau)$ is in a bijection with the irreducible closed sets, a concept defined in terms of the subspace topology on a closed subset. The question of the surjectivity of Θ is equivalent to the following question: $\forall$ pair of irreducible closed subsets $(C_1 \subset (X_1, L, \tau_1), C_2 \subset (X_2, L, \tau_2))$, $\exists$ a unique irreducible closed set $C \subset (X_1 \times X_2, \tau_1 \times \tau_2)$?

OUTLINE OF ARGUMENT. Choose $C = C_1 \times C_2$. Apply the following facts:

(1) The subspace topology $\tau(C_1 \times C_2)$ is the product $\tau_1(C_1) \times \tau_2(C_2)$ of the subspace topologies.

(2) In any space, a subspace is irreducible closed iff any two of its nonempty open subsets have nonempty intersection.

DIFFICULTY. Fairly straightforward.

SURJECTIVITY OF Θ: NON-MODEL-THEORETIC APPROACH

INTRODUCTORY COMMENTS. $Lpt(\tau)$ is apparently unrelated to any notion of "irreducible fuzzy closed subset"; such would be linked to the notion of a fuzzy topology on a fuzzy subset, a notion not generally successful in the non-model-theoretic setting for non-singleton spaces. (For singleton spaces, relative fuzzy topologies on fuzzy subsets are indeed subobjects — see [Erceg 1980, Foster 1979]. We are forced to use lattice-theoretic means, namely construct a relationship between the L-points of a topology τ, $Lpt(\tau)$, and the prime principal ideals of τ.

OUTLINE OF ARGUMENT. It is difficult to give a brief outline, but Theorem 22 of my abstract **Generalizations of Spatiality and Sobriety** gives some flavor.

DIFFICULTY. Quite difficult.

WELL-DEFINEDNESS OF Θ: MODEL-THEORETIC APPROACH

INTRODUCTORY COMMENTS. The question of the well-definedness of Θ is the same as the following question: $\forall$ irreducible closed $C \subset (X_1 \times X_2, \tau_1 \times \tau_2)$, $\exists!$ pair of irreducible closed subsets $C_1 \subset (X_1, L, \tau_1), C_2 \subset (X_2, L, \tau_2))$?

OUTLINE OF ARGUMENT. Put $C_k = \pi_k(C)$. Then use the following facts:

(1) The subspace topology $\tau(C_1 \times C_2)$ is the product $\tau_1(C_1) \times \tau_2(C_2)$ of the subspace topologies.
(2) In any space, a subspace is irreducible closed iff any two of its nonempty open subsets have nonempty intersection.
(3) The continuous image of an irreducible closed set is "irreducible" w.r.t. images of closed sets, and the closure of such an irreducible set is still irreducible (and so irreducible closed).
(4) The following set-theoretic equation can be verified:

$$C = \pi_1[\mathrm{Cl}(c_1)] \times \pi_2[\mathrm{Cl}(c_2)]$$

DIFFICULTY. Rather hard.

WELL-DEFINEDNESS OF Θ: NON-MODEL-THEORETIC APPROACH

INTRODUCTORY COMMENTS. Being forced to use only lattice-theoretic means — e.g. that members of $Lpt(\tau)$ are maps preserving arbitrary $\bigvee$ and finite $\wedge$ — in this part of the proof is very advantageous.

OUTLINE OF ARGUMENT. Well-definedness is immediate from the definition of Θ.

DIFFICULTY. Trivial.

CONCLUSION

What emerges from this comparison of approaches to the question of comparing localic products with (fuzzy) topological products? Some things can best be done with the non-model-theoretic approach — especially issues that are inherently lattice-theoretic (e.g. the well-definedness of Θ) — and some things can best be done with the model-theoretic approach — especially issues involving subspaces of non-singleton spaces as subobjects (e.g. surjectivity of Θ). I conclude that each approach has its proper motivation and niche in fuzzy set theory.

RESPONSE OF PROF. L. N. STOUT. I want to say that we have a **trichotomy** which we ought to be considering [more on this below], not [just] a dichotomy [between model-theoretic and non-model-theoretic approaches]. You have pointed out nicely how the lattice-theoretic approach gives you some benefits. For the model-theoretic and topos-theoretic approach, when we start talking about change of base [= lattice of membership values] — little of that has been done yet, we are going to have to worry a lot about your lattice theory [approach to fuzzy sets and fuzzy topology]. ... In topos topology, things involving closed sets are very difficult, because closure doe not work nicely. I would expect that in the fuzzy [non-model-theoretic] case — because your order-reversing involution [as in Zadeh or Hutton] behaves much more nicely than the intuitionistic **NOT** — you might find that **that** difficulty in topos topology does not carry over and the closed sets [in fuzzy sets] work nicely; but they may have some of their own problems.

RESPONSE OF PROF. RODABAUGH. [In the setting of my talks and above comments] there is no order-reversing involution; I also avoid closed sets. One nice thing about this lattice-theoretic approach is that [an L-point on the (fuzzy) topology] is simply a [frame-like] mapping from the topology to the membership lattice [L]. You can rephrase [Prof. Johnstone's] arguments, still making use of subspaces, but without making use of closed [subsets]; [this can be done] by translating the outlines of his proofs into terms of sub-spaces based on open fuzzy subsets. In this case, it might be a little more awkward [than Prof. Johnstone's classical proof], but still easier than what goes on [in the argument for surjectivity of Θ in the fuzzy case].

RESPONSE OF PROF. STOUT. You probably have a proof [for the surjectivity of Θ] that uses a minimal set of hypotheses.

RESPONSE OF PROF. RODABAUGH. The order-reversing involution is stripped out, so this means I have thrown away closed sets ..., and I have also thrown away subobjects [in the non-model-theoretic approach].

RESPONSE OF PROF. STOUT. You have not got a whole lot left.

RESPONSE OF PROF. RODABAUGH. Simply the lattice-theoretic [framework].

PROF. L. N. STOUT. As I see it, we are not looking [only] at model-theoretic versus lattice-theoretic [non-model-theoretic], but [at] at least three aspects. It looks to me like we have the **lattice-theoretic approach**, which, as I see it, says: we understand sets very well, so let's keep using them; and let's talk about everything (sort of) within classical mathematics — we (sort of) understand sets and lattices in the classical sense ... [and] we don't have to worry about what reasoning we are using. We have the **model-theoretic approach** — the early parts of our categorical models sit in this model–theoretic approach, which is to say: how do we go [categorically] ... from scratch? There is also the **proof-**

theoretic approach. These three [approaches] are all tied together: if we have got the model-theoretic approaches right, then these [lattice-theoretic approaches] will be models of our reasoning; and if we have got the proof-theoretic approaches right, the reasoning you can do in here [in the model-theoretic approaches] will probably be described by a proof-theory; and if we have a completeness result, the proof-theory will give us a corresponding lattice-theoretic model that should look nice.

Prof. Stout's Mnemonic for Approaches to Fuzzy Set Theory

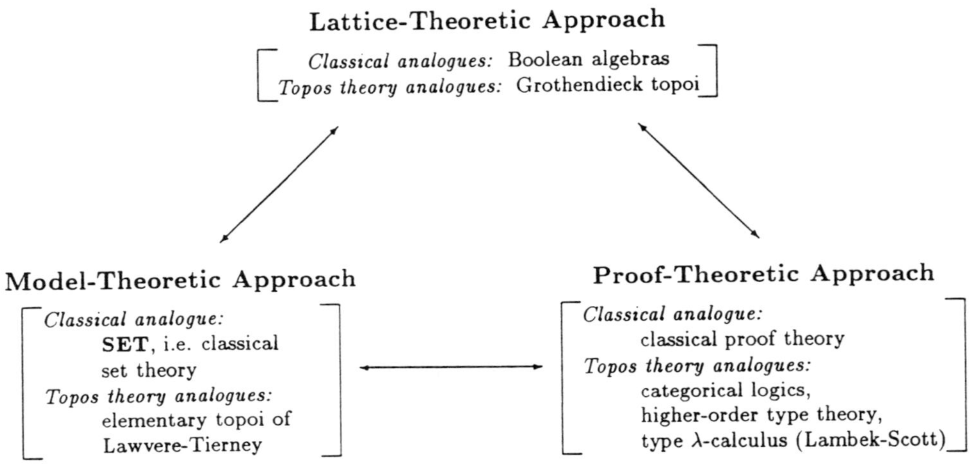

QUESTION OF PROF. RODABAUGH. Would an example [of the linkage between the lattice-theoretic and proof-theoretic approaches] be the fact that the Gödel-Heykin completeness theorem [for propositional logics] is the same as the theorem that coherent locales are spatial [Johnstone 1982, Notes to Chapter II, p. 78]?

ANSWER AND CONTINUATION OF PROF. STOUT. Yes. I think within fuzzy set theory, you have people working [primarily] in these two [lattice-theoretic and proof-theoretic]; the approximate reasoning folks are working [in] proof-theoretic in a very small fragment; [and] Prof. Höhle and I [are] alone over here [in model-theoretic].

The same kind of split happened within topos theory [see mnemonic above]. Up here [in lattice-theoretic] you have Grothendieck — ... everything in terms of representability in functors, so everything is sets based. The model-theoretic approach is what came next with elementary topoi — this is Lawvere-Tierney. [These] ... topoi are [constructed] without reference to sets and, in the process, gain a certain amount of generality; so that the (equivalent of the) lattice-theoretic approach ..., formulated within classical logic / classical mathematics, gave a fragment [of elementary topos theory] which is a very useful collection of examples which gives us a good feel for what is going on in topos theory. (Similarly, I expect that the lattice-theoretic approach in fuzzy sets may be generalized when you go to the model-theoretic approach ... There are two or three really useful elementary topoi which are not Grothendieck topoi.) Here [in the proof-theoretic approach] we have within elementary topoi the development of categorical logic, where you get (essentially) the development of higher-order type [theory], the Lambek-Scott approach. And you start getting a topos as a theory; from the theory you get models of that theory, and that frequently goes through Grothedieck topoi.

So I think we have three sides to the coin, not two, and that this proof-theoretic approach is of some intense interest; please see Section 5 of my paper [this is essentially Prof. Stout's chapter in this volume]. And we have the [historical] model within topos theory to show that, historically, this one [lattice-theoretic] came significantly before the other [two] developments.

Once you understand the classical situation, the model-theoretic situation is one of "how do I write that [example from the lattice-theoretic approach] down nicely and efficiently, capture the essential features here, getting perhaps some generalization as I go"; and then the proof-theory [motivation] is "Gee, it was hard to reason over there — maybe I ought to formalize that reasoning and get it straight."

The three turn out to be very closely related, and related to **different** applications. This [model-theoretic] will help us figure out what definitions might be the most general [leading to] ... clarification; this [proof-theoretic] is going to help us figure out how to do the reasoning within fuzzy sets without jumping off a cliff; and this [lattice-theoretic] is going to be where the examples we have got to take into account [are] and where our intuition is, and the completeness results in the proof theory tend to say that the proof theory has a lattice-theoretic model.

COMMENT BY THE EDITOR. The presentation of Prof. Stout stimulated much valuable discussion. Significant parts of this discussion are not, however, discernible from the audio recording of the roundtable; some voices are not sufficiently above ambient noise and cannot be extracted by means of available signal-processing equipment. The gist of this discussion can sometimes be discerned from the tape or inferred from Prof. Stout's comments: intuitionistic logic (Prof. Blanchard), Łukasiewicz logic (Prof. Höhle), mathematical semantics and Prolog (Prof. Eklund), computability, finiteness, and Grothendieck topoi (Prof. Stout), etc. There was **unanimous** agreement at the time that Prof. Stout's trinality was preferable to the duality implicit in the title of the Roundtable. The remainder of the Roundtable is devoted mainly to the question of change of lattice of membership values in fuzzy sets and topology [e.g. as in the my chapter in this volume] and the question of convergence: these discussions are abstracted below.

COMMENT OF PROF. STOUT. ... [Concerning] ... lattice-independence versus lattice-dependence, your [the editor's] foundations involve letting the lattice change at the same time you let everything else change.

RESPONSE OF PROF. RODABAUGH. I would **like** to work with [this simultaneous change]; my framework allows me that. But I find it frustrating.

CONTINUATION OF PROF. STOUT. It feels to me ..., with the intuition coming from the model-theoretic approach, generally I ought to be working with one lattice, and then worry about how you change the lattice separately from everything else; in other words, factor the problem into two pieces.

COMMENT OF PROF. RODABAUGH. This ... [reminds me of] a paper of Prof. Eklund [Eklund ****; also Chapter III of Eklund 1986b] in which he showed that each morphism in $\mathbb{T}$ [assuming that each L has an order-reversing involution] factored into a Chang morphism [with fixed lattice] followed by a Hutton morphism [with a fixed set]; that basically it [the morphism] takes place in two steps — changing the sets, and then changing the lattice. From what you are saying it is not surprising that we get these different Stone [Representation] Theories — within each theory we cannot change the lattice — and then in $\mathbb{T}$ we can shift from one subcategory fixed by one lattice to another one fixed by another lattice.

COMMENT OF PROF. RODABAUGH. I was wondering if Prof. Lowen wanted to ask the topoi people what happens to convergence.

COMMENT OF PROF. LOWEN. I already asked that [the other day]. [Laughter.]

COMMENT OF PROF. STOUT. I did something with it [convergence] in the last chapter of my dissertation, which I have not looked at in fifteen years, and neither has any else. ...

COMMENT OF PROF. RODABAUGH. In the [point-set] lattice-theoretic approach to convergence, [things] seem to work nicely only within Prof. Lowen's setting [see Prof. Lowen's chapter in this volume]. As you know, Prof. Lowen's spaces [each topology contains all the constant maps] are the externalization of the [topological space] objects of the topos **L-SET** [Höhle 1986], and I was thinking that might indicate there is a convergence going on down here [in the model-theoretic approach].

COMMENT OF PROF. STOUT. I am going to send [Prof. Lowen] that chapter [of Prof. Stout's dissertation] and we are going to write back and forth trying to figure out what it means in the fuzzy case [see Prof. Lowen's chapter in this volume].

COMMENT OF PROF. HÖHLE. I make one point that in the Boolean case, ... the internal convergence, from the model-theoretic side, is precisely for **IBUS** almost-everywhere convergence; so [we have] a notion of convergence for which there does not exist **any** classical topology, yet this convergence is "topological". My perspective is that from the model-theoretic side, we have some application in abstract probability theory; and if we take the Łukasiewicz logic, then the real numbers are precisely, as you also [have] pointed out, the probability distributions — and then I think maybe we get a new notion of convergence [for statistical applications].

COMMENT OF PROF. STOUT. Which would be a nice result of having the model-theoretic approach feed back to [the] classical. The parallel I would see there is the work showing that if you want to do several complex variables, you should do one complex variable in a topos.

Round Table:

Open Questions in Logic and Topology

S. E. Rodabaugh et al. (eds.), *Applications of Category Theory to Fuzzy Subsets*, 347–351.

ROUNDTABLE:

OPEN QUESTIONS IN LOGIC AND TOPOLOGY

Transcribed by S. E. Rodabaugh

PRELIMINARY COMMENT OF THE EDITOR. The open questions given below were either presented in a session, bearing the same name as this appendix, of the *Eleventh International Seminar on Fuzzy Set Theory* described in the Preface, or were specifically designated by a given author for this roundtable. The reader should consult the chapters of Parts I, II, and III for additional open questions related to the subject matter of this book.

OPEN QUESTION OF PROF. H. HERRLICH. Determine for the construct **FTOP**:

(1) its cartesian closed topological hull (cf. [Min 1989],
(2) its extensional topological hull,
(3) its topological universe hull.

FIRST OPEN QUESTION OF PROF T. KUBIAK. Is fuzzy complete normality a good extension of complete normality?

DISCUSSION BY PROF. KUBIAK. [Wang–Hu 1985] have proves (with $[0,1]$ as the range space for the fuzzy sets) that the fuzzy topological axioms T_0, T_1, T_2, regular, and normal (as introduced by [Hutton-Reilly 1980] are good extensions of their topological counterparts. Wang and Hu did not mention at all the other axioms introduced by Hutton and Reilly: complete regularity (CR), complete normality (CN), and perfect normality (CN). (That normality is a good extension is just another formulation of the Katětov-Tong insertion insertion theorem. This was observed earlier by [Martin 1981]. This author is able to supply the proofs in the case of CR (rather nontrivial) and PN (trivial provided one is aware of Tong's characterization of crisp PN), but unable to do this in case of CN.

LEMMA. Fuzzy CN is a good extension of crisp CN $\Leftrightarrow$ [$\forall$ topological space X, (1) $\Leftrightarrow$ (2)], where (1), (2) are given by:

(1) X is CN.
(2) If A, B, G, and H are functions from X into $[0,1]$ such that A and B are arbitrary, G is lower semicontinuous, H is upper semicontinuous, $A \leq G \leq B$, and $A \leq H \leq B$, then there exist a lower semicontinuous function U and an upper semicontinuous function U and an upper semicontinuous function K from X into $[0,1]$ such that $A \leq U \leq K \leq B$.

In fact, it is easy to see that condition (2) just says that the fuzzy topological space topologically generated from X is CN. Also, (2) $\Rightarrow$ (1) is obvious. Thus the open question is as follows: prove or disprove that (1) $\Rightarrow$ (2).

SECOND OPEN QUESTION OF PROF T. KUBIAK. The idea of this horrible question is borrowed from [Lang 1965, p. 105]. Take any work on normal topological spaces (e.g. [Alò and Shapiro 1974]), and restate and reprove as many as possible of the theorems for the case of extremally disconnected spaces. A more hopeless task: take any work on extremally disconnected spaces (e.g. Chapter 6 of [Porter and Woods 1988]) and act the other way.

DISCUSSION BY PROF. KUBIAK. No one (including a respectable extremally disconnected topologist I have spoken to) seems to have noticed a very remarkable duality between normality (N) and extremal disconnectedness (ED). For instance, the following statements about a topological space X are equivalent (to mention only a few):

(1) X is N [ED].

(2) (Urysohn's type lemmas) Every two disjoint closed [open] subsets of X are completely separated.

(3) (Tietze's type lemmas) Each closed [open] subsets of X is C*-embedded.

(4) The image of X under any continuous closed [open] map is N [ED].

What is the source of those dualities? One equivalent condition for ED is that disjoint open sets have disjoint closures. This is obviously equivalent to saying that each two disjoint open sets are contained in disjoint closed sets. Just replace "closed" for "open" and vice versa in the axiom of N. It is also interesting to observe that — unlike the case of N — all the above statements about ED are more or less trivial. [For instance, the following proof of (3) [ED] is to be compared with those of [Gillman and Jerison 1976] and [Porter and Woods 1988]: if U and V are disjoint and open, then $\chi_U \leq \chi_{\overline{U}} \leq \chi_{X\setminus V}$, and $\chi_{\overline{U}}$ is continuous and separates U and V.]

Both N and ED have been investigated in the context of L-fuzzy topologies (for (2), (3), and (4) [N] in fuzzy situation, see [Hutton 1975], [Kubiak 1986 and 1987], and [Rodabaugh 1983b]; (4) [ED] can easily be supplied) and frames (see [Johnstone 1982]). The definition of ED for frames involves pseudo-complementation. By the above discussion it is not surprising that, in fact, a frame L is ED iff L^{op} is a normal lattice (which incidentally shows how to define ED for bounded, not necessarily pseudo-complemented, distributive lattices). Thus a portion of results about N in a frame or lattice setting of [Dowker and Papert Strass 1975] and [Johnstone 1982] easily translates to ED.

For more "dualities" of that kind, see [Kubiak 1989b and 1990c], also Remark 7.15 of my chapter in this volume. A fuller account (including complete and perfect ED) is hoped to appear hereafter (clearly, without being a full answer to the question under discussion).

EXERCISE. State and prove an N-version of the following: each dense subset of an ED space is C*-embedded.

HINT. For a change, replace "closed" for "open".

THIRD OPEN QUESTION OF PROF T. KUBIAK. Give an internal characterization of fuzzy complete regularity.

DISUCSSION BY PROF. KUBIAK. The internal description of crisp complete regularity (CR) due to [Steiner 1966] is the best one, since it covers all CR spaces (i.e. T_1 separation not assumed). These spaces are characterized by possessing a **normal** and **separating** family for closed sets. A point-free form of these concepts gives rise to consider them in a fuzzy situation. For X an L-fuzzy topological space (with L a complete lattice possessing an order-reversing involution), $\mathcal{F} \subset L^X$ is **normal** iff, whenever $A, B \in \mathcal{F}$ and $A \leq B'$, there exist $C, D \in \mathcal{F}$ such that $A \leq C' \leq D \leq B'$. It is **separating** iff for each open V there exist $\{K_\gamma\}, \{F_\gamma\} \subset \mathcal{F}$ such that $V = V_\gamma K_\gamma = V_\gamma F'_\gamma$. Consider the following statements:

(1) X is CR (in the sense of [Hutton 1977],
(2) X has a normal and separating family of closed L-sets.

The proof that (2) implies (1) is easy (after [Hutton 1975]), so that the proper question is in proving (1) implies (2). Note that in the crisp case the required family is the one of all zero-sets. And the proof that it is normal depends upon the ring structure of $C(X)$. Thought it is quite clear what an L-zero-set is supposed to be, the fuzzy $C(X)$ (= all L-fuzzy real valued continuous functions on X) is merely a topological lattice in general, and its topological-algebraic structure (when L is linearly ordered) in the sense of [Rodabaugh 1985] is unsufficient.

FOURTH OPEN QUESTION OF PROF T. KUBIAK. Let **F-compact** and **F-Hausdorff** be two arbitrary L-topological invariants with the following properties:

(1) F-compactness (resp., F-hausdorffness) is a good extension of ordinary compactness (resp., hausdorffness),
(2) F-compactness (resp., F-hausdorffness) is preserved under weakening (resp., strengthening) of an L-topology,
(3) each L-space which is both F-compact and F-Hausdorff is maximal F-compact and minimal F-Hausdorff.

Prove or disprove the following statement: Each F-compact and F-Hausdorff L-space is topologically generated.

DISCUSSION BY PROF. KUBIAK. By 3.17(2) of my chapter in this volume, for a positive answer it suffices that some of (1) – (3) imply that F-compactness is equivalent to ultracompactness.

FIRST OPEN QUESTION OF PROF. S. E. RODABAUGH. Is there a localic approach to paths which will permit the composition of paths, and which can be generalized to give a lattice-theoretic approach to paths allowing the composition of paths?

DISCUSSION BY PROF. RODABAUGH. Prof. Klawonn in his talk [see Prof. Klawonn's chapter in this volume] has shown that if the codomain of the paths be the L-fuzzy unit interval with the Hutton topology stratified in the sense of Lowen, then the composition of such fuzzy paths is not meaningful in the sense of being a pushout square; but there is a model-theoretic solution in the category **IBUS**. So the issue is whether there is a lattice-theoretic solution to this question, and whether the pushout criterion is possible or desirable with a lattice-theoretic approach to fuzzy sets. We might see if there is a localic version of the classical situation, and then apply the $\mathbf{L}\Omega \dashv \mathbf{LPT}$ adjunction [see Section 5 of my chapter in this volume] to carry the localic version over to the fuzzy case.

SECOND OPEN QUESTION OF PROF. S. E. RODABAUGH. It was established in my talk on sobriety that a fuzzy topological space (X, L, τ) which is L-sober cannot be stratified in the sense of Lowen if L satisfies $|\mathbf{CSLF}(L, L)| > 1$. Characterize all such $L \in |\mathbf{CSLF}|$; i.e. characterize all complete lattices satisfying this condition. See [Rodabaugh 1986–7] and Section 6 of my chapter in this volume.

OPEN QUESTION OF PROF. L. N. STOUT. How do we incorporate intensifiers, hedges, and modalitites into the formalisms of the foundations for fuzzy logic?

DISCUSSION BY PROF. STOUT. One possibility is to use unary operators, but this has flaws. For a relevant critique of the current approach (*à la* Zadeh), see [McCawly, *What Linguists have always wanted to know about Logic, but were ashamed to ask*, University of Chicago Press].

Bibliography

BIBLIOGRAPHY

The notation **** indicates that the year in which the entry is to appear was not yet known when this book went to press.

J. Adámek

1983. *Theory of Mathematical Structures*, D. Reidel Publishing Co., Dordrecht.

1986a. **Classification of concrete categories**, *Houston J. Math.* **12**, 305–26.

1986b. **On a problem of Herrlich concerning topological categories**, *Top. Appl.* **22**, 175–82.

J. Adámek and H. Herrlich

1986. **Cartesian closed categories, quasitopoi, and topological universes**, *Comment. Math. Univ. Carolinae* **27**, 235–57.

J. Adámek, H. Herrlich, and G. E. Strecker

1979a. **Least and largest completions I**, *Comment. Math. Univ. Carolinae* **20**, 43–58.

1979b. **The structure of intial completions**, *Cahiers Top. Geom. Diff.* **20**, 333–52.

1990. *Abstract and Concrete Categories*, John Wiley & Sons, New York.

J. Adámek and J. Reiterman

1986. **Topological categories presented by small sets of axioms**, *J. Pure Appl. Alg.* **42**, 1–14.

1987. **The quasitopos hull of the category of uniform spaces**, *Top. Appl.* **27**, 97–104.

J. Adámek, J. Reiterman, and G. E. Strecker

1985. **Realizations of cartesian closed topological hulls**, *Manu. Math.* **53**, 1–33.

I. W. Alderton

1985. **Cartesian closedness and the MacNeille completion of an intially structured category**, *Quaest. Math.* 8, 63–78.

1989a. **Function spaces in fuzzy topology**, *Fuzzy Sets and Systems* **32**, 115–24.

1989b. **How many initial completions does a monotopological category have?**, *Quaest. Math.* **12**, 289–305.

R. A. Alò and H. L. Shapiro

1974. *Normal Topological Spaces*, Cambridge University Press, Cambridge.

P. Antoine

1966. **Etude élementaire des catégories dénsembles structures I/II**, *Bull. Soc. Math. Belg.* **18**, 142–64/387–414.

G. Artico and R. Moresco

1988. **a-Compactness in the fuzzy unit interval**, *Fuzzy Sets and Systems* **25**, 243–9.

R. Balbes and P. Dwinger

1974. *Distributive Lattices*, University of Missouri Press, Columbia.

B. Banaschewski and G. Bruns

1988. **The fundamental duality of partially ordered sets**, *Order* **5**, 61–74.

H. P. Barendegt

1984. **The lambda calculus**, *Studies in Logic* **103**, North Holland, Amsterdam.

M. Barr

1986. **Fuzzy set theory and topos theory**, *Bull. Canadian Math. Soc.* **29**, 501–8.

J. T. Bell

1988. *Toposes and Local Set Theories*, Oxford University Press, Oxford.

L. P. Belluce

1986. **Semi-simple algebras of infinite valued logic and bold fuzzy set theory**, *Canad. J. Math.* **38**, 1356–79.

1989. **Semi-simple and complete MV-algebras**, Preprint, Vancouver.

H. L. Bentley, H. Herrlich, and E. Lowen-Colebunders

1987. **The category of Cauchy spaces is cartesian closed**, *Top. Appl.* **27**, 105–12.

M. P. Berri, J. R. Porter, and R. M. Stephenson Jr.

1971. **A survey of minimal topological spaces**, *General Topology and its Relations to Modern Analysis and Algebra*, Proceedings of the Kanpur Topological Conference, 93–114, Academia, Prague.

G. Birkoff

1967. *Lattice Theory*, Amer. Math. Soc. Colloq. Publ. **XXXV**, Third Edition.

N. Blasco and R. Lowen

1988. **The Wallman compactification in FNS**, *Communications of the IFSA Mathematics Chapter* **2**, 1–3.

C. R. Borges
1985. **Helly space is not an ANR(normal)**, *Math. Japonica* **30**, 47–8.

N. Bourbaki
1965. *Topologie Générale*, Hermann, Paris.

G. Bourdaud
1976. **Some cartesian closed topological categories of convergence spaces**, LNM **540**, 93–108, Springer Verlag, Berlin.

G. C. L. Brümmer and R.-E. Hoffmann
1976. **An external characterization of topological functors**, LNM **540**, 136–51, Springer Verlag, Berlin.

B. Buckles, F. Petry, and H. Sachar
1989. **A domain calculus for fuzzy relational databases**, *Fuzzy Sets and Systems* **29**, 327–40.

P. Burmeister
1986. *A Model Theoretic Approach to Partial Algebra*, Akademie Verlag, Berlin.

M. H. Burton and J. J. Chadwick
1990. **Precompactness and completeness in fuzzy uniform spaces**, Preprint, Rhodes University, Grahamstown (South Africa)

H.-P. Butzmann and G. Kneis
1986. **Cech-Stone compactifications of pseudotopological spaces**, *Math. Nachr.* **128**, 259–64.

U. Cerruti
1981. **The Stone-Cech compactification in the category of fuzzy topological spaces**, *Fuzzy Sets and Systems* **6**, 197–204.

U. Cerruti and U. Höhle
1986. **An approach to uncertainty using algebras over a monoidal closed category**, *Rend. Circolo Mat. Series II* **12**, 47–63.

J. J. Chadwick
1988. **A generalized form of compactness in fuzzy topological spaces**, Preprint, Rhodes University, Grahamstown (South Africa), to appear in *J. Math. Anal. Appl.*

1989. **Relative compactness and compactness of general subsets in I-topological spaces**, Preprint, Rhodes University, Grahamstown (South Africa), to appear in *Quaest. Math.*

C. C. Chang

1958. **Algebraic analysis of many-valued logics**, *Trans. Amer. Math. Soc.* **88**, 467–90.

C. L. Chang

1968. **Fuzzy topological spaces**, *J. Math. Anal. Appl.* **24**, 182–90.

G. Choquet

1948. **Convergences**, *Ann. Univ. Grenoble* **23**, 57–112.

J. Coulon and J.-L. Coulon

1988. **Morphisms in the category JTF^{00} — weak classifier of monomorphisms**, *Portugaliae Mathematica* **45**, 29–47.

1989. **About some categories of Ω-valued sets**, Preprint, Lyon, to appear in *J. Math. Anal. Appl.*

A. Day

1975. **Filter monads, continuous lattices, and closure systems**, *Canad. J. Math.* **27**, 50–9.

A. De Luca and S. Termini

1972. **A definition of a non-probabilistic entropy in the setting of fuzzy sets theory**, *Inform. Control* **20**, 301–12.

C. H. Dowker and D. Papert

1967. **On Urysohn's Lemma**, *General Topolgy and its Relations to Modern Analysis and Algebra II*, 111–4, Academia, Prague.

1975. **Para compact frames and closed maps**, *Symposia Math.* **16**, 93–116.

E. J. Dubuc

1979. **Concrete quasitopoi**, 239–54, LNM **753**, Springer Verlag, Berlin.

E. J. Dubuc and H. Porta

1971. **Convenient categories of topological algebra and their duality theory**, *J. Pure Appl. Alg.* **1**, 281–316.

P. Eklund

1984a. **Category theoretic properties of fuzzy topological spaces**, *Fuzzy Sets and Systems* **13**, 303–10.

1984b. **A class of completely regular spaces**, *Internat. J. Math. & Math. Sci.* **7**, 197–200.

1986a. **A comparison of lattice-theoretic approaches to fuzzy topology**, *Fuzzy Sets and Systems* **19**, 81–7.

1986b. **Categorical fuzzy topology**, Ph.D. Thesis (1986), Åbo Akademi.

****. **Factorization of morphisms in fuzzy topology**, *J. Math. Anal. Appl.*, preprint version in 1986b.

P. Eklund and W. Gähler

1988. **Basic notions for topology I/II**, *Fuzzy Sets and Systems* **26/27**, 333–56/171–95.

****. **Generalized Cauchy spaces**, *Math. Nachr.*

M. A. Erceg

1980. **Functions, equivalence relations, quotient spaces, and subsets in fuzzy set theory**, *Fuzzy Sets and Systems* **3**, 75–92.

M. P. Fourman and D. S. Scott

1979. **Sheaves and logic**, *Applications of Sheaves*, Lecture Notes in Mathematics **753**, 302–401, Springer Verlag, Berlin.

P. J. Freyd

1972. **Aspects of topoi**, *Bull. Austral. Math. Soc.* **7**, 1–76 and 467–80.

R. Fric and D. C. Kent

1979. **Completion functors for Cauchy spaces**, *Internat. J. Math. & Math. Sci.* **2**, 589–604.

O. Frink

1938. **New algebras of logic**, *Amer. Math. Assoc. Monthly* **45**, 210–19.

J. L. Frith

1987. *Structured Frames*, Ph.D. Thesis, University of Cape Town (South Africa).

A. Frölicher

1980. **Catégories cartésienne fermées engendrées par des monoids**, *Cahiers Top. Geom. Diff.* **21**, 367–75.

N. Funayama

1959. **Imbedding infinitely distributive lattices completely isomorphically into Boolean algebras**, *Nagoya Math. J.* **15**, 71–81.

S. Gähler and W. Gähler

1989. **Discrete Cauchy spaces**, *Proc. Colloquium on Topology*, Pécs (Hungary).

W. Gähler

1977. *Grundstrukturen der Analysis I/II*, Akademie Verlag (Berlin).
1978. Birkhäuser Verlag (Basel, Stuttgart).
1981. **A topological appraoch to structure theory**, *Math. Nachr.* **100**, 93–144.
1984. **A general theory of convergence structures**, Proc. *Convergence Structures and Applications II* (Schwerin), Abh. Akad. Wiss. DDR N2, 57–64.
1985. **Axioms of structures and functor power series**, *Proc. Conference on Convergence Structures*, Bechyne 1984, 137–52, Akademie Verlag, Berlin.
1988. **Monads and convergences**, *Proc. Conference Generalized Functions, Convergence Structures, and their Applications*, Dubrovnik 1987 (Yugoslavia), 29–46, Plenum Press, New York.
****. **Generalized convergence groups**, *Math. Nachr.*

T. E. Gantner, R. C. Steinlage, and R. H. Warren

1978. **Compactness in fuzzy topological spaces**, *J. Math. Anal. Appl.* **62**, 547–62.

G. Gerla

1986. **Generalized fuzzy points**, *J. Math. Anal. Appl.* **120**, 761–8.
1989. **Pointless metric spaces**, *J. Symbolic Logic* **54**, 407–419.
1990. **Pointless geometries**, *Handbook of Incidence Geometry*, to appear, F. Buekenhout and W. Kantor, editors, North-Holland, Amsterdam.
1990. **Distances, diameters, and verisimilitude of theories**, Preprint.

G. Gerla and R. Volpe

1985. **Geometry without points**, *Amer. Math. Monthly* **92**, 707–11.
1986. **The definition of distance and diameter in fuzzy set theory**, *Studia Univ. Balbes-Bolyai Math.* **3**, 21–6.

G. Gierz, K. H. Hofmann, K. Keimel, J. D. Lawson, M. Mislove, and D. S. Scott

1980. *A Compendium of Continuous Lattices*, Springer Verlag, Berlin.

L. Gillman and M. Jerison

1976. *Rings of Continuous Functions*, Springer Verlag, New York.

J. A. Goguen

1967. **L-fuzzy sets**, *J. Math. Anal. Appl.* **18**, 145–74.

1973. **The fuzzy Tychonoff theorem**, *J. Math. Anal. Appl.* **43**, 734–42.

1974. **Concept representations in natural and artificial languages: axioms, extensions, and applications for fuzzy sets**, *Int. J. Man-Machine Studies* **6**, 513–61.

R. Goldblatt

1984. **Topoi: the Categorical Analysis of Logic**, revised edition, North-Holland Publishing Co., Amsterdam.

S. Gottwald

1981. **Fuzzy points and local properties in fuzzy topological spaces**, *Fuzzy Sets and Systems* **5**, 199–201.

1986. **Fuzzy set theory with t-norm and 0-operators**, *The Mathematics of Fuzzy Systems*, A. DiNola and A. G. S. Ventre, editors, TUV Rheinland, Köln.

G. Grätzer

1978. *General Lattice Theory*, Birkhäuser Verlag, Basel.

L. S. Hay

1963. **Axiomatization of the infinite-valued calculus**, *J. Symbolic Logic* **28**, 77–86.

Z. Hedrlin, A. Pultr, and V. Trnkova

1967. **Concerning a categorical approach to topological and algebraic theories**, *General Topology and its Relations to Modern Analysis and Algebra II*, 176–81, Academia, Prague.

H. Herrlich

1974. **Cartsian closed topological categories**, *Math. Colloq. Univ. Cape Town* **9**, 1–16.

1976. **Initial completions**, *Math. Zeitschr.* **150**, 101–10.

1983a. **Are there convenient subcategories of TOP**, *Top. Appl.* **15**, 263–71.

1983b. **Categorical topology 1971–81**, *General Topogy and its Relation to Anaylsis and Algebra IV*, 279–383, Heldermann Verlag, Berlin.

1984. **Universal topology**, *Categorical Topogy*, Sigma Ser. Pure Math. **5** 223–81, Heldermann Verlag, Berlin.

1988. **Hereditary topological constructs**, *General Topology and its Relations to Analysis and Algebra* **VI**, 249–62, Heldermann Verlag, Berlin.

H. Herrlich and L. D. Nel

1977. **Cartesian closed topological hulls**, *Proc. Amer. Math. Soc.* **62**, 215–22.

H. Herrlich and G. E. Strecker

1979. **Category Theory**, Sigma Ser. Pure Math. **1**, Heldermann Verlag, Berlin.

1986. **Cartesian closed topological hulls as injective hulls**, *Quaest. Math.* **9**, 262–80.

D. Higgs

1973. **A category approach to Boolean-valued set theory**, Lecture Notes, University of Waterloo.

U. Höhle

1978. **Probabilistische topologien**, *Manuscripta Math.* **26**, 223–45.

1979. **Probabilistische kompacte L-unschatfe Mengen**, *Manuscripta Math.* **26**, 331–47.

1980. **Example of a locally convex L-fuzzy topology on the set of almost everywhere defined random variables**, 129–33 in [Klement 1980].

1981. **Representation theorems for fuzzy quantities**, *Fuzzy Sets and Systems* **5**, 83–109.

1982. **Probabilistic metrization of fuzzy uniformities**, *Fuzzy Sets and Systems* **8**, 63–9.

1984. **Compact $\mathcal{G}$-fuzzy topological spaces**, *Fuzzy Sets and Systems* **13**, 39–63.

1985. **The category of IB-fuzzy topological spaces**, *Analysis of Fuzzy Information* **1**, 185–94.

1986. **Fuzzy topologies and the topological space objects in a topos**, *Fuzzy Sets and Systems* **19**, 299–304.

1987. **The category CM-SET – an algebraic approach to uncertainty**, Preprint, Bergische Universität (Wuppertal).

1988a. **Quotients with respect to similarity relations**, *Fuzzy Sets and Systems* **27**, 31–44.

1988b. **Microgeometric spaces**, Preprint, Wuppertal.

1989. **The Poincaré paradox and the cluster problem**, *Trees and Hierarchial Clustering*, A. Dress, Editor, Lecture Notes in Mathematical Biology, Springer Verlag, Berlin.

1990a. **Monoidal closed categories, weak topoi, and generalized logics**, *Fuzzy Sets and Systems*, to appear.

1990b. **Generalized logics – a common framework for intuitionistic logic and Łukasiewicz logic**, Preprint, Wuppertal.

U. Höhle and E. P. Klement

1984. **Plausibility measures – a general framework for possibility and fuzzy probability measures**, *Aspects of Vagueness*, 31–50, Reidel, Dordrecht.

U. Höhle and L. N. Stout

1991. **Foundations of fuzzy sets**, *Fuzzy Sets and Systems* **40**, 257–96.

J. M. Hu

1985. **Fuzzy topological spaces**, *J. Math. Anal. Appl.* **110**, 141–78.

B. Hutton

1975. **Normality in fuzzy topologicals spaces**, *J. Math. Anal. Appl.* **50**, 74–9.
1977. **Uniformities in fuzzy topological spaces**, *J. Math. Anal. Appl.* **58**, 559–71.
1980. **Products of fuzzy topological spaces**, *Top. Appl.* **11**, 59–67.

B. Hutton and I. Reilly

1980. **Separation axioms in fuzzy topological spaces**, *Fuzzy Sets and Systems* **3**, 93–104.

J. M. E. Hyland

1982. **The effective topos**, *The L. E. J. Brouwer Centenary Symposium*, Studies in Logic **110**, 165–216, North-Holland, Amsterdam.

J. M. E. Hyland, P. T. Johnstone, and A. M. Pitts

1980. **Topos theory**, *Math. Proc. Cambridge Phil. Soc.* **88**, 205–32.

J. M. E. Hyland, E. P. Robinson, and G. Rosolini

1990. **The discrete objects in the effective topos**, *Proc. Lond. Math. Soc.* (3)**60**, 1–36.

J. R. Isbell

1972. **Atomless parts of spaces**, *Math. Scand.* **31**, 5–32.

P. T. Johnstone

1977. *Topos Theory*, L. M. S. Mathematical Monographs Academic Press.
1979. **On a topological topos** *Proc. Lond. Math. Soc.* (3)**38**, 237–71.
1980. **Open maps of toposes**, *Manuscripta Math.* **31**, 217–47.
1981a. **Injective toposes**, *Continuous Lattices*, LNM **871**, 284–97, Springer, Berlin.
1981b. **Tychonoff's theorem without the axiom of choice**, *Fund. Math.* **113**, 21–35.
1982. *Stone Spaces*, Cambridge University Press, Cambridge.
1983. **The point of pointless topology**, *Bull. (New Ser.) Amer. Math. Soc.* **8**, 41–53.
1985. **When is a variety a topos?** *Algebra Universalis* **21**, 198–212.
1988. **An open letter to Ian Graham**, *Fuzzy Sets and Systems* **27**, 244–50.
1989. **Collapsed toposes as bitopological spaces**, *Categorical Topology* (World Scientific), 19–35.
1990. **Collapsed toposes and cartesian closed varieties**, *J. Algebra* **129**, 446–80.

A. K. Katsaras

1985. **Linear fuzzy neighbourhood spaces**, *Fuzzy Sets and Systems* **16**, 25–40.

1986. **Linear fuzzy neighbourhood spaces II**, *J. Math. Anal. Appl.* **115**, 560–72.

1988. **Totally bounded fuzzy syntopogeneous structures**, *Fuzzy Sets and Systems* **28**, 91–103.

1989. **Fuzzy neighborhood structures and fuzzy quasi-uniformities**, *Fuzzy Sets and Systems* **29**, 187–99.

H. H. Keller

1968. **Die Limesuniformisierbarkeit der Limesräume**, *Mat. Ann.* **176**, 334–41.

D. C. Kent and G. D. Richardson

1988. **Completions of Cauchy vector spaces**, *Proc. of the Conference on Generalized Functions, Convergence Structures, and their Applications*, Dubrovnik 1987 (Yugoslavia), 409–14, Plenum Press, New York.

E. E. Kerre and P. L. Ottoy

1987. **On the different notions of neighbourhood in Chang-Goguen fuzzy topological spaces**, *Simon Stevin* **61**, 131–46.

1989. **Lattice properties of neighbourhood systems in Chang fuzzy topological spaces**, *Fuzzy Sets and Systems* **30**, 205-213.

F. Klawonn

1989. **On the product of fuzzy paths**, *Proc. Third IFSA Congress*, 269–271, J. Bezdek, editor, Seattle.

A. J. Klein

1981. **α-Closure in fuzzy topology**, *Rocky Mount. J. Math.* **11**, 553–60.

1983. **Generating fuzzy topologies with semi-closure operators**, *Fuzzy Sets and Systems* **9**, 267–74.

1984. **Generalizing the L-fuzzy unit interval**, *Fuzzy Sets and Systems* **12**, 271–79.

E. P. Klement

1980. Editor, *Proceedings of the International Seminar on Fuzzy Set Theory*, January, Johannes Kepler Universität (Linz).

1981a. Editor, *Proceedings of the Second International Seminar on Fuzzy Set Theory*, September, Johannes Kepler Universität (Linz).

1981b. **Some remarks on T-norms, fuzzy s-algebras, and fuzzy measures**, 125–42 of [Klement 1981a].

1982a. **Construction of fuzzy σ-algebras using triangular norms**, *J. Math. Anal. Appl.* **85**, 543–65.

1982b. **Characterization of fuzzy measures by meansure of triangular norms**, *J. Math. Anal. Appl.* **86**, 345–58.

E. P. Klement, M. L. Puri, and D. Ralescu
1986. **Limit theorems for fuzzy random variables**, *Proc. Royal Soc. London A* **407**, 171–82.

A. Kock and G. C. Wraith
1971. **Elementary Toposes**, *Lecture Notes Series No.* **30**, Aarhus Universitet, Matematisk Institut.

W. Kotzé
1984. **Functions of class one and well-behaved spaces**, *Rend. Math. (Rome)* **4**(1), 139–56.
1985. **The fuzzy Hausdorff property**, *Glasnik Matematicki* (Zagreb) **20**(40), 409–17.
1986. **Quasi-coincidence and quasi fuzzy Hausdorff**, *J. Math. Anal. Appl.* **116**(2), 465–72.
1987. **Proximities in lattices**, *Communications of the IFSA Mathematics Chapter* **1**, 12–18.

G. Kneis
1986. **Completion functors for categories of convergence spaces I: acceptability functors**, *Math. Nachr.* **129**, 283–312.
1988. **Completion functor for categories of convergence spaces II: embedding of separated acceptable spaces into their completions**, *Math. Nachr.* **135**, 181–211.

H.-J. Kowalsky
1954. **Limesräume und Komplettierung**, *Math. Nachr.* **12**, 301–40.

T. Kubiak
1986. **Extending continuous L-real functions**, *Math. Japonica* **31**, 875–87.
1987. **L-fuzzy normal spaces and Tietze Extrnsion Theorem**, *J. Math. Anal. Appl.* **125**, 141–53.
1988a. **The fuzzy unit interval and the Helly space**, *Math. Japonica* **33**, 253–9.
1988b. **Compactness and connectedness of the L-fuzzy interval**, Preprint, Christian Albrecht University (Kiel).
1989a **Notes on the topological molification of the L-fuzzy unit interval**, *Communications of the Mathematics Chapter of IFSA* **4**, 26–30.
1989b. **On extremally disconnected subspaces**, *Fasc. Math.* **19**, 143–5.
1990a. **A quotient of the Helly space which is AR(normal)**, Preprint, Rhodes University (Grahamstown, South Africa), to appear in *Math. Japonica*.

1990b. **A unified treatment of various kinds of compactness in fuzzy topological spaces**, Preprint, Rhodes University (Grahamstown, South Africa), tc appear in *Math. Japonica.*

1990c. **Closed coverings of extremally disconnected spaces and partitions of unity**, *Notices South African Math. Soc.* **22 (1)**, 41.

L. Kucera and A. Pultr

1972. **On a mechanism of defining morphisms in concrete categories**,*Cahiers.*

1973. *Top. Géom. Diff.* **13**, 397–410.

K. Kuratowski

1970. **Some problems concerning semi-continuous set-valued mappings**, *Set-Valued Mappings, Selections, and Topological Properties of* 2^X, 45–8, LNM **171**, Springer Verlag, Berlin.

J. Lambek and P. J. Scott

1986. *Introduction to Higher Order Categorical Logic*, Cambridge Studies in Advanced Mathematics **7**, Cambridge University Press, Cambridge.

S. Lang

1965. *Algebra*, Addison-Wesley, Reading (Massachusetts).

F. W. Lawvere

1970a. **Equality in hyperdoctrines**, *Applications of Categorical Algebra*, Proc. Symp. Pure Math. **17**, AMS **1–14**, Providence.

1970b. **Quantifiers and sheaves**, *Actes du Congrès International des Maths. II*, 329–34, Nice.

1971. **Justifiers and sheaves**, *Actes du Congrès Intern. des Math., Nice 1970*, **tome I** (Gauthier-Villars), 329–34.

1973. **Metric spaces, generalized logic, and closed categories**, *Rend. Sem. Mat. Fis. Milano* **43**, 135–66.

Y.-M. Liu

1983. **Pointwise characterization of complete regularity and imbedding theorem in fuzzy topological spaces**, *Scientia Sinica (Series A)* **26**, 138–47.

Y.-M. Liu and M.-K Luo

1987. **Induced spaces and fuzzy Stone-Cech compactification**, *Scientia Sinica (Series A)* **30**, 1034–44.

1988. **On N-compactnesss in L-fuzzy unit interval**, *Science Bullentin* **33**. 1–4.

1989. **Fuzzy Stone-Cech compactification and the largest Tychonoff compactification**, *Chinese Ann. Math.* **10B**, 74–84.

E. Lowen-Colebunders

****. **On the regularity of the Kowalsky completion**, *Canad. J. Math.*

E. Lowen and R. Lowen

1985. **A characterization of convergence in fuzzy topological spaces**, *Internat. J. Math. Math. Sci.* **8**, 497–511.

1988. **The role of the diagonal operator in fuzzy topology**, *Fuzzy Sets and Systems* **27**, 53–61.

1989. **Topological quasitopos hulls of categories containing topological and metric objects**, *Cahiers Top. Geom. Diff.* **30**, 213–28.

R. Lowen

1974. **Theory of fuzzy topologies**, Ph.D. thesis, Free University of Brussels.

1976. **Fuzzy topological spaces and fuzzy compactness**, *J. Math. Anal. Appl.* **56**, 621–33.

1977. **Initial and final fuzzy topologies and the fuzzy Tychonoff Theorem**, *J. Math. Anal. Appl.* **58**, 11–21.

1978. **A comparison of different compactness notions in fuzzy topological spaces**, *J. Math. Anal. Appl.* **64**, 446–54.

1981a. **Convergence in fuzzy topological spaces**, *Gen. Top. Appl.* **10**, 147–60.

1981b. **Connectedness in fuzzy topological spaces**, *Rocky Mount. J. Math.* **11**, 427–33.

1981c. **Compact Hausdorff fuzzy topological spaces are topological**, *Top. Appl.* **12**, 65–74.

1982. **Fuzzy neighborhood spaces**, *Fuzzy Sets and Systems* **7**, 165–89.

1984. **Compactness properties in the fuzzy real line**, *Fuzzy Sets and Systems* **13**, 193–200.

1985. *On the Existence of Natural Non-Topogical Spaces*, Heldemann Verlag, Berlin.

1986. **A complete characterization of the level spaces of IR(L) and I(L)**, *Rocky Mount. J. Math.* **16**, 669–84.

R. Lowen and P. Wuyts

1982. **Completeness, compactness, and precompactness in fuzzy uniform spaces: Part I**, *J. Math. Anal. Appl.* **90**, 563–83.

1988. **Concerning the constants in fuzzy topology**, *J. Math. Anal. Appl.* **129**, 256–68.

1989. **Approach spaces, a common supercategory of TOP and MET**, *Math. Nachr.* **141**, 183–226.

R. Lowen, P. Wuyts, and E. Lowen

1989. **On the reflectiveness and coreflectiveness of subcategories in FTS**, *Math. Nachr.* **141**, 55–65.

J. Łukasiewicz and A. Tarski

1930. **Untersuchungen über den Aussagenkalkül**, *Comptes Rendus des Séances de la Société, des Sciences et des Lettres de Varsovie (III)* **23**, 30–50.

S. Mac Lane

1971. *Categories for the Working Mathematician*, Springer Verlag, Berlin.

W. A. MacCaull

1989. **Positive definite functions over regular f-rings and representations as sums of squares**, *Annals Pure Appl. Logic* **14**, 243–57.

H. MacNeille

1937. **Partially ordered sets**, *Trans. Amer. Math. Soc.* **42**, 416–60.

M. Makkai and G. Reyes

1977. *First Order Categorical Logic* Lecture Notes in Mathematics **611**, Springer Verlag, Berlin.

E. G. Manes

1976. *Algebraic Theories*, Graduate Texts in Mathematics **26**, Springer Verlag,

1982. **A class of fuzzy theories**, *J. Math. Anal. Appl.* **85**, 498–51.

H. W. Martin

1980a. **Weakly induced fuzzy topological spaces**, *J. Math. Anal. Appl.* **78**, 634–9.

1980b. **A Stone-Cech ultrafuzzy compactification**, *J. Math. Anal. Appl.* **73**, 453–6.

1981. **Normality and S-covers of fuzzy spaces**, 21–37 of [Klement 1981a].

K. Menger

1966. **Geometry and positivism — a probabilistic microgeometry**, *Selected Papers in Logic, Foundations, Didactics, Economics*, K. Menger, Reidel, Dordrecht, 1979.

G. H. J. Meßner

****. Ph.D. Thesis, Johannes Kepler Universität (Linz).

G. H. J. Meßner and S. E. Rodabaugh

****. **Stratification and generalized sobrieties — lattice independent constructions**, see [Rodabaugh ****].

K. C. Min

1989. **Fuzzy limit spaces**, *Fuzzy Sets and Systems* **32**, 343–57.

Paalman-de Miranda

1964. *Topological Semigroups*, Math. Centrum, Amsterdam.

F. K. Mohannadi and M. W. Warner

1989. **Countability and sequentiality in fuzzy topological spaces**, *Fuzzy Sets and Systems* **29**, 371–9.

G. P. Monro

1986a. **Quasitopoi, logic, and Heyting-valued models**, *J. Pure Appl. Alg.* **42**, 141–64.

1986b. **A category theoretic approach to Boolean valued models of set theory**, *J. Pure Appl. Alg.* **42** 245–74.

A. Muir

1986. **Relations between probability theory and fuzzy theory**, *The Mathematics of Fuzzy Systems*, A. DiNola and A. G. S. Ventre, editors, TUV Rheinland, Köln.

C. J. Mulvey

1974. **Intuitionistic algebra and representations of rings**, *Recent Advances in the Representation Theory of Rings and C*-Algebras by Continuous Sections*, Memoirs of the AMS **148**, Providence.

C. V. Negoita and D. A. Ralescu

1975. **Application of Fuzzy Sets to Systems Analysis**, John Wiley & Sons, New York.

L. D. Nel

1976. **Cartesian closed topological categories**, LNM **540**, 439–51, Springer-Verlag, Berlin.

1984. **Topological universes and smooth Gelfand-Naimark duality**, *Contemporary Math.* **30**, 244–76.

D. Papert and S. Papert

1958. **Sur les treillis des ouverts et les paratopologies**, *Séminaire Ehresmann* (topologie et géométrie différentielle), 1re année (1957–8), expose 1.

J. Penon

1973. **Quasitopos**, *C. R. Acad. Sci. Paris* **276**, A237–40.

1977. **Sur les quasitopos**, *Cahiers Top. Géom. Diff.* **18**, 181–218.

W. K. S. Phoa

1990. *The Synthetic Theory of Effective Domains*, J. T. Knight Prize essay, University of Cambridge.

Z. Piotrowski

1986. Editor, *Proceedings of the Second Youngstown State University Conference on Topology and Fuzzy Sets*, (16-18 May 1985), Youngstown State University Central Printing Office, Youngstown (Ohio).

A. M. Pitts

1982. **Fuzzy sets do not form a topos**, *Fuzzy Sets and Systems* **8**, 101–4.

H. Poincaré

1902. *La Science et l'Hypothése*, Flammarion, Paris.

1904. *La Valeur de la Science*, Flammarion, Paris.

D. Ponasse

1988. **Categorical studies of fuzzy sets**, *Fuzzy Sets and Systems* **28**, 235–44.

J. R. Porter and R. G. Wood

1988. *Extensions and Absolutes of Hausdorff Spaces*, Springer Verlag, New York.

P-M. Pu and Y.-M. Liu

1980a. **Fuzzy topology I: neighborhood structure of a point and More-Smith convergence**, *J. Math. Anal. Appl.* **76**, 571-99.

1980b. **Fuzzy topology II: product and quotient theorems**, *J. Math. Anal. Appl.* **77**, 20–37.

A. Pultr

1976a. **Fuzzy sets and fuzzy mappings**, *Comm. Math. Univ. Carolinae* **17–3**.

1976b. **Closed categories and L-fuzzy sets**, *Vortrage zurAutomaten und Algorithmentheorie*, TU Dresden.

K. V. S. V. N. Raju and A. K. Majumdar

1988. **Fuzzy functional dependencies and lossless join decomposition of fuzzy relational databases** *ACM Transactionson Database Systems* **13**, 129–66.

H. Rasiowa and R. Sikorski

1970. *The Mathematics of Metamathematics*, Third Edition, Polish Scientific Publishers, Warszawa.

G. Richter

1979. **Katorielle Algebra**, Akademie Verlag, Berlin.

S. E. Rodabaugh

1980. **The Hausdorff separation axiom for fuzzy topological spaces**, *Gen. Top. Appl.* **11**, 319–34.

1982a. **The L-fuzzy real line and its subspaces**, *Fuzzy Sets and Possibility Theory* 402–18, R. R. Yager, Editor, Pergemon Press, Oxford.

1982b. **Connectivity and the L-fuzzy unit interval**, *Rocky Mount. J. Math.* **12**, 113–21.

1982c. **Fuzzy addition in the L-fuzzy real line**, *Fuzzy Sets and Systems* **8** 39–52.

1983a. **The categorical accommodation of various notions to fuzzy topology**, *Fuzzy Sets and Systems* **9**, 241–65.

1983b. **Separation axioms and the fuzzy real lines**, *Fuzzy Sets and Systems* **11**, 163–83.

1985. **Complete fuzzy topological hyperfields and fuzzy multiplication in the fuzzy real lines**, *Fuzzy Sets and Systems* **15**, 285–310.

1986. Extended abstract of [Rodabaugh 1986-7] in 79–134 of [Piotrowski 1986].

1986-7. **A point-set lattice-theoretic framework $\mathbb{T}$ for topology which contains LOC as a subcategory of singleton spaces and in which there are general classes of Stone Representations and Compactification theorems**, Youngstown State University Central Printing Office (first edition February 1986, second edition April 1987), Youngstown (Ohio).

1987. **Realizations of events, points of a locale, and the Stone Representation Theorem**, *Communications of the IFSA Mathematics Chapter* **1**, 28–33.

1988a. **A theory of fuzzy uniformities with applications to the fuzzy real lines**, *J. Math. Anal. Appl.* **129**, 37–70.

1988b. **Dynamic topologies and their applications to crisp topologies, fuzzifications of crisp topologies, and fuzzy topologies on the crisp real line**, *J. Math. Anal. Appl.* **131**, 25–66.

1988c. **Lowen, para-Lowen, and a-level functors and fuzzy topologies on the crisp real line**, *J. Math. Anal. Appl.* **131**, 157–69.

1989. **Generalizations of spatiality and sobriety / Generalizations of coherence**, *Communications of the IFSA Mathematics Chapter* **4**, 45–50/41–4.

1991. **Point-Set lattice-theoretic topology**, *Fuzzy Sets and Systems* **40**, 297–345.

****. **Generalizations of the Stone Representations and Compactifications**, *J. Math. Anal. Appl.*, a four-part series. **Part I: Lattice dependent spatality and sobriety. Part II: Stratification and generalized sobrieties — lattice independent constructions**, see [Meßner-Rodabaugh ****]. **Part III: Comparison of localic and fuzzy topological products. Part IV: Separation and compactness axioms, and Stone representations and compactifications.**

A. Rose and J. B. Rosser

1958. **Fragments of many-valued statement calculi**, *Trans. Amer. Math. Soc.* **87**, 1–53.

A. Rosenfield

1979. **Digital topology**, *Amer. Math. Monthly* **86**, 621–30.

A. Rosenfield and R. Melter

1989. **Digital geometry**, *The Mathematical Intelligencer 11* **#3**, 69–72.

J. B. Rosser

1960. **Axiomatization of infinite-valued logics**, *Logique et Analyse L. N.* **3**, 137–53.

G. Salicrup

1988. **Epireflectivity and connectedness in topological categories**, *Aportaciones Matématicas*, Notas de Investigación **2**, 239–97.

G. Salicrup and R. Vasquez

1972. **Categorías de connexión**, *Ann. Inst. Nat. Univ. Nac. Autónoma México* **12**, 47–87.

M. Sarkar

1981. **On fuzzy topological spaces**, *J. Math. Anal. Appl.* **79**, 384–94.

H. Sato

1989. **E-ccc: between ccc and topos** *Thoret. Comput. Sci.* **64**, 55–66.

H. Schubert

1972. *Categories*, Springer Verlag, Berlin.

B. Schweizer and A. Sklar

1983. *Probabilistic Metric Spaces*, North-Holland, Amsterdam.

F. Schwarz

1986. **Hereditary topological categories and topological universes**, *Quaest. Math.* **10**, 197–216.

1989. **Description of the topological universe hull**, *Lect. Notes Comp. Sci.* **393**, 325–32.

D. Scott

1967. **A proof of the indepedence of the continuum hypothesis**, *Math. Syst. Theory* **1**, 89–111.

1979. **Identity and existence in intuitionistic logic**, *Applications of Sheaves*, LNM **753**, 660–96, Springer Verlag, Berlin.

R. Sikorski

1964. *Boolean Algebras*, Ergebnisse der Mathematik und ihrer Grenzgebiete, Neue Folge **25**, Springer Verlag, Berlin.

A. P. Šostak

1983. **A fuzzy modification of a linearly ordered space**, *Colloquia Math. Soc. Janos Bolyai* **41**: Topology and Applications, 581–604.

1985a. **A fuzzy modification of the category of linearly ordered spaces**, *Comment. Math. Univ. Carolinae* **26.3**, 421–22.

1985b. **On a fuzzy topological structure**, *Suppl. Rend. Circolo Matematico di Palermo (Serie II)* **11**, 89–103.

1988. **On compactness and connectedness degrees of fuzzy sets in fuzzy topological spaces**, *General Topology and its Relations to Modern Analysis and Algebra VI (Proc. Sixth Prague Top. Symp.* 1986, Z. Frolik, editor), 519–32, Heldermann Verlag, Berlin.

1989. **On the concept of E-regularity for fuzzy topology**, *Matematicki Vesnik* **41**, 199–209.

1990. **On completeness and precompactness spectra of fuzzy sets in fuzzy uniform spaces**, *Comment. Math. Univ. Carolinae* **31**, 149–57.

T. P. Speed

1973. **On rings of sets II: zero sets**, *J. Austral. Math. Soc. Ser. A* **16**(2), 185–99.

L. A. Steen and J. A. Seebach

1978. *Counterexamples in Topology*, Second Edition, Springer Verlag, Berlin.

E. F. Steiner

1966. **Normal families and completely regular spaces**, *Duke J. Math.* **33**, 743–5.

M. H. Stone

1936. **The theory of representations for Boolean algebras**, *Trans. Amer. Math. Soc.* **40**, 37–111.

1937a. **Applications of the theory of Boolean rings to general topology**, *Trans. Amer. Math. Soc.* **41**, 375–481.

1937b. **Topological characterization of distributive lattices and Brouwerian logics**, *Casopis Pest. Mat. Fys.* **67**, 1–25.

L. N. Stout

1984. **Topoi and categories of fuzzy sets**, *Fuzzy Sets and Systems* **12**, 169–84.

1987. **Fuzzy sets and topos theory**, *Proceedings of the Second IFSA Congress*, Tokyo.

1988. **Two discrete forms of the Jordan Curve Theorem**, *Amer. Math. Monthly* **95** (No. 4), 332–6.

1990. **A survey of fuzzy set and topos theory**, *Fuzzy Sets and Systems*, to appear.

G. E. Strecker

1984. **On cartesian closed topological hulls**, *Categorical Topology*, 523–39, Heldermann Verlag, Berlin.

V. Trnková,

1971. **On descriptive classification of set functors I**, *Comment. Math. Univ. Carolinae* **12**(1), 143–74.

J. D. Ullman

1982. *Principles of Database Systems*, second edition, Computer Science Press, Rockville (Maryland, USA).

G. Wang

1983. **A new fuzzy compactness defined by fuzzy nets**, *J. Math. Anal. Appl.* **94**, 1–23.

1984. **Topological molecular lattices**, *Monthly J. Sci.* **1**, 19–23 (Beijing).

G. Wang and L. Hu

1985. **On induced fuzzy topological spaces**, *J. Math. Anal. Appl.* **108**, 495–506.

M. W. Warner

1989. **A note on the relation between G- and Q-theories for first countability and sequentiality**, *Fuzzy Sets and Systems* **33**, 321–327.

1990. **Fuzzy topology with respect to continuous lattices**, *Fuzzy Sets and Systems* **35**, 85–91.

M. W. Warner and A. Muir

1981. **Homology theories and tolerance automata**, *Discrete Mathemathics* **33**, 209–21.

R. H. Warren

1978. **Neighborhoods, bases, and continuity in fuzzy topological spaces**, *Rocky Mountain J. Math.* **8**, 459–70.

K. Weihrauch and U. Schreiber

1981. **Embedding metric spaces into CPO's**, *Theoret. Computer Sci.* **16**, 5–24.

S. Williard

1970. *General Topology*, Addison Wesley, Reading (Massachusetts).

M. Wischnewsky

1972. **Initialkategorien**, dissertation, Ludwig-Maximilians Universität, München.

C. K. Wong

1974. **Fuzzy topology: product and quotient theorems**, *J. Math. Anal. Appl.* **45**, 512–21.

G. C. Wraith

1974. **Artin glueing**, *J. Pure Appl. Alg.* **4**, 345–8.

P. Wuyts and R. Lowen

1983. **On separation axioms in fuzzy topological spaces, fuzzy neighborhood spaces, and fuzzy uniform spaces**, *J. Math. Anal. Appl.* **93**, 27–41.

P. Wuyts, R. Lowen, and E. Lowen

1988. **Reflectors and coreflectors in the category of fuzzy topological spaces**, *Comput. Math. Appl.* **16**, 823–36.

O. Wyler

1973. **Filter space monads, regularity, completions**, *Proc. Second Pittsburgh Internat. Conf. on General Topology*.

1976. **Are there topoi in topology?** *Categorical Topology*, Lecture Notes in Mathematics **540**, 699–719, Springer Verlag, Berlin.

1979. **Function spaces in topological categories**, LNM **917**, 411–8, Springer-Verlag, Berlin.

1990. *Lecture Notes on Topoi and Quasitopoi*, Preprint, Carnegie Mellon University (Pittsburgh).

L. A. Zadeh

1965. **Fuzzy sets**, *Inform. Control* **8**, 338–53.

E. C. Zeeman and O. P. Buneman

1968. **Tolerance spaces and the brain**, *Towards a Theoretical Biology 1: Prolegomena*, C. H. Waddington, editor, Aldine Publising Co., Chicago.

D. Zhao

1987. **The N-compactness in L-fuzzy topological spaces**, *J. Math. Anal. Appl.* **128**, 64–79.

Index of Categories

INDEX OF CATEGORIES

Subject Index

SUBJECT INDEX

Addenda et Corrigenda

ADDENDA ET CORRIGENDA

The bibliographic data for [Höhle 1990a] and [Stout 1990] have been amended, respectively, as follows:

U. Höhle, **Monoidal closed categories, weak topoi, and generalized logics**, *Fuzzy Sets and Systems*, **42** (1991) 15-35.

L. N. Stout, **A survey of fuzzy set and topos theory**, *Fuzzy Sets and Systems*, **42** (1991) 3-14.

THEORY AND DECISION LIBRARY

SERIES B: MATHEMATICAL AND STATISTICAL METHODS

Editor: H. J. Skala, *University of Paderborn, Germany*

1. D. Rasch and M.L. Tiku (eds.): *Robustness of Statistical Methods and Nonparametric Statistics*. 1984 ISBN 90-277-2076-2
2. J.K. Sengupta: *Stochastic Optimization and Economic Models*. 1986 ISBN 90-277-2301-X
3. J. Aczél: *A Short Course on Functional Equations*. Based upon Recent Applications to the Social Behavioral Sciences. 1987 ISBN Hb 90-277-2376-1; Pb 90-277-2377-X
4. J. Kacprzyk and S.A. Orlovski (eds.): *Optimization Models Using Fuzzy Sets and Possibility Theory*. 1987 ISBN 90-277-2492-X
5. A.K. Gupta (ed.): *Advances in Multivariate Statistical Analysis*. Pillai Memorial Volume. 1987 ISBN 90-277-2531-4
6. R. Kruse and K.D. Meyer: *Statistics with Vague Data*. 1987 ISBN 90-277-2562-4
7. J.K. Sengupta: *Applied Mathematics for Economics*. 1987 ISBN 90-277-2588-8
8. H. Bozdogan and A.K. Gupta (eds.): *Multivariate Statistical Modeling and Data Analysis*. 1987 ISBN 90-277-2592-6
9. B.R. Munier (ed.): *Risk, Decision and Rationality*. 1988 ISBN 90-277-2624-8
10. F. Seo and M. Sakawa: *Multiple Criteria Decision Analysis in Regional Planning*. Concepts, Methods and Applications. 1988 ISBN 90-277-2641-8
11. I. Vajda: *Theory of Statistical Inference and Information*. 1989 ISBN 90-277-2781-3
12. J.K. Sengupta: *Efficiency Analysis by Production Frontiers*. The Nonparametric Approach. 1989 ISBN 0-7923-0028-9
13. A. Chikán (ed.): *Progress in Decision, Utility and Risk Theory*. 1991 ISBN 0-7923-1211-2
14. S.E. Rodabaugh, E.P. Klement and U. Höhle (eds.): *Applications of Category Theory to Fuzzy Subsets*. 1992 ISBN 0-7923-1511-1
15. A. Rapoport: *Decision Theory and Decision Behaviour*. Normative and Descriptive Approaches. 1989 ISBN 0-7923-0297-4
16. A. Chikán (ed.): *Inventory Models*. 1990 ISBN 0-7923-0494-2
17. T. Bromek and E. Pleszczyńska (eds.): *Statistical Inference*. Theory and Practice. 1991 ISBN 0-7923-0718-6
18. J. Kacprzyk and M. Fedrizzi (eds.): *Multiperson Decision Making Models Using Fuzzy Sets and Possibility Theory*. 1990 ISBN 0-7923-0884-0

KLUWER ACADEMIC PUBLISHERS – DORDRECHT / BOSTON / LONDON